Advances in Weather Radar

Related titles on radar:

Advances in Bistatic Radar Willis and Griffiths
Airborne Early Warning System Concepts, 3rd Edition Long
Bistatic Radar, 2nd Edition Willis
Design of Multi-Frequency CW Radars Jankiraman
Digital Techniques for Wideband Receivers, 2nd Edition Tsui
Electronic Warfare Pocket Guide Adamy
Foliage Penetration Radar: Detection and characterisation of objects under trees Davis
Fundamentals of Ground Radar for ATC Engineers and Technicians Bouwman
Fundamentals of Systems Engineering and Defense Systems Applications Jeffrey
Introduction to Electronic Warfare Modeling and Simulation Adamy
Introduction to Electronic Defense Systems Neri
Introduction to Sensors for Ranging and Imaging Brooker
Microwave Passive Direction Finding Lipsky
Microwave Receivers with Electronic Warfare Applications Tsui
Phased-Array Radar Design: Application of radar fundamentals Jeffrey
Pocket Radar Guide: Key facts, equations, and data Curry
Principles of Modern Radar, Volume 1: Basic principles Richards, Scheer and Holm
Principles of Modern Radar, Volume 2: Advanced techniques Melvin and Scheer
Principles of Modern Radar, Volume 3: Applications Scheer and Melvin
Principles of Waveform Diversity and Design Wicks *et al.*
Pulse Doppler Radar Alabaster
Radar Cross Section Measurements Knott
Radar Cross Section, 2nd Edition Knott *et al.*
Radar Design Principles: Signal processing and the environment, 2nd Edition Nathanson *et al.*
Radar Detection DiFranco and Rubin
Radar Essentials: A concise handbook for radar design and performance Curry
Radar Foundations for Imaging and Advanced Concepts Sullivan
Radar Principles for the Non-Specialist, 3rd Edition Toomay and Hannan
Test and Evaluation of Aircraft Avionics and Weapons Systems McShea
Understanding Radar Systems Kingsley and Quegan
Understanding Synthetic Aperture Radar Images Oliver and Quegan
Radar and Electronic Warfare Principles for the Non-Specialist, 4th Edition Hannen
Inverse Synthetic Aperture Radar Imaging: Principles, algorithms and applications Chen and Martorella
Stimson's Introduction to Airborne Radar, 3rd Edition Baker, Griffiths and Adamy
Test and Evaluation of Avionics and Weapon Systems, 2nd Edition McShea
Angle-of-Arrival Estimation Using Radar Interferometry: Methods and applications Holder
Biologically-Inspired Radar and Sonar: Lessons from nature Balleri, Griffiths and Baker
The Impact of Cognition on Radar Technology Farina, De Maio and Haykin
Novel Radar Techniques and Applications, Volume 1: Real aperture array radar, imaging radar, and passive and multistatic radar Klemm, Nickel, Gierull, Lombardo, Griffiths and Koch
Novel Radar Techniques and Applications, Volume 2: Waveform diversity and cognitive radar, and target tracking and data fusion Klemm, Nickel, Gierull, Lombardo, Griffiths and Koch
Radar and Communication Spectrum Sharing Blunt and Perrins
Systems Engineering for Ethical Autonomous Systems Gillespie
Shadowing Function from Randomly Rough Surfaces: Derivation and applications Bourlier and Li
Photo for Radar Networks and Electronic Warfare Systems Bogoni, Laghezza and Ghelfi

Multidimensional Radar Imaging Martorella
Radar Waveform Design Based on Optimization Theory Cui, De Maio, Farina and Li
Micro-Doppler Radar and Its Applications Fioranelli, Griffiths, Ritchie and Balleri
Maritime Surveillance with Synthetic Aperture Radar Di Martino and Antonio Iodice
Electronic Scanned Array Design Williams
Advanced Sparsity-Driven Models and Methods for Radar Applications Li
Deep Neural Network Design for Radar Applications Gurbuz
New Methodologies for Understanding Radar Data Mishra and Brüggenwirth
Radar Countermeasures for Unmanned Aerial Vehicles Clemente, Fioranelli, Colone and Li
Holographic Staring Radar Oswald and Baker
Polarimetric Radar Signal Processing Aubry, De Maio and Farina
Fundamentals of Inertial Navigation Systems and Aiding Braasch
Theory and Methods for Distributed Data Fusion Applications Govaers
Target Detection: A detection theory framework Zaimbashi and Nayebi

Advances in Weather Radar

Volume 1: Precipitation sensing platforms

Edited by
V.N. Bringi, Kumar Vijay Mishra and Merhala Thurai

The Institution of Engineering and Technology

Published by SciTech Publishing, an imprint of The Institution of Engineering and Technology, London, United Kingdom

The Institution of Engineering and Technology is registered as a Charity in England & Wales (no. 211014) and Scotland (no. SC038698).

First published 2023

The Institution of Engineering and Technology
Futures Place
Kings Way, Stevenage
Hertfordshire, SG1 2UA, United Kingdom

www.theiet.org

British Library Cataloguing in Publication Data
A catalogue record for this product is available from the British Library

ISBN Vol 1: 978-1-83953-622-9 (hardback)
ISBN Vol 2: 978-1-83953-624-3 (hardback)
ISBN Vol 3: 978-1-83953-626-7 (hardback)

ISBN 3 Vol Set: 978-1-83953-628-1 (hardback)

ISBN Vol 1: 978-1-83953-623-6 (PDF)
ISBN Vol 2: 978-1-83953-625-0 (PDF)
ISBN Vol 3: 978-1-83953-627-4 (PDF)

Typeset in India by MPS Limited

Cover Image credit: The Colorado State University-Illinois State Water Survey (CSU-CHILL) S-band weather radar with 9-m Gregorian design antenna. Photo credit: Kumar Vijay Mishra, with permission from CSU-CHILL Co-Principal Investigator V.N. Bringi.

This work is humbly placed at the Lotus Feet of Goddess Saraswati,
Patron of Knowledge and the Arts.
– V.N.B.

Dedicated, with the highest reverence and humility, to
Mahadeva – the Supreme Being.
To my Mom Shraddha Mishra.
– K.V.M.

Dedicated to my beloved parents.
– M.T.

Contents

About the editors xvii
Preface xix
Acknowledgments xxiii
List of editors xxv
List of contributors xxvii
List of reviewers xxix
Introduction to volume 1 xxxi

1 The decade of renaissance in weather radar research 1
Kumar Vijay Mishra, Merhala Thurai and V.N. Bringi
1.1 Introduction 1
1.1.1 Renaissance of weather radar technology 2
1.1.2 Shipborne, airborne and satellite radars 4
1.1.3 Advances in signal processing 8
1.1.4 Disdrometers supporting radar observations 19
1.1.5 Scattering from hydrometeors 22
1.1.6 Radiowave propagation applications 23
1.1.7 Concluding remarks 25
References 27

2 Doppler polarimetric radars for weather observations from 1995 to 2022: a historical perspective 37
Alexander V. Ryzhkov, Merhala Thurai and Dušan S. Zrnić
2.1 Introduction 37
2.2 Physical meaning of Doppler polarimetric radar variables 39
2.3 Polarimetric radars in research and operations since 1995 43
2.3.1 S-band research radars 43
2.3.2 C-band research radars 46
2.3.3 X-band radars 48
2.3.4 Short-wavelength polarimetric radars 49
2.3.5 Most recent developments 50
2.3.6 Polarimetric upgrade of operational weather radars 51
2.4 Weather applications of the Doppler polarimetric radars 52
2.4.1 Quantitative precipitation estimation 53
2.4.2 Classification of radar echo and hydrometeors 59

2.5 Radar polarimetry and cloud microphysics 64
2.5.1 Microphysical retrievals 66
2.5.2 Thermodynamic retrievals 70
2.6 Concluding remarks 70
References 73

3 Developments in solid-state weather radar 85
Stephen J. Frasier and Luca Facheris
3.1 Introduction 85
3.2 Enabling technologies 86
3.2.1 Solid-state sources 86
3.2.2 Digital transceivers and software radio technology 88
3.3 Survey of solid-state weather radar 90
3.4 Peak and average power considerations 92
3.5 Pulse compression and the matched filter 93
3.6 Range sidelobes and their reduction 96
3.6.1 Windowing 97
3.6.2 Nonlinear frequency modulation 100
3.6.3 Predistortion 103
3.6.4 Mismatched filters 106
3.6.5 Doppler tolerance 109
3.7 Blind range and its mitigation 110
3.8 Other techniques to increase average power 110
3.8.1 Frequency-modulated continuous-wave 111
3.8.2 Quadratic phase coding 113
3.9 Conclusion 116
References 116

4 Quality of polarimetric data in the WSR-88D system 121
Valery M. Melnikov and Dušan S. Zrnić
4.1 Introduction 121
4.2 Required accuracy for WSR-88D measurements 123
4.3 Quality of the WSR-88D dual-polarization measurements 125
4.3.1 Isolation of the dual-polarization channels 126
4.3.2 WSR-88D's dynamic range 128
4.3.3 Reflectivity calibration 130
4.3.4 Radar detectability 131
4.3.5 Z_{DR} calibration 134
4.3.6 Differential phases upon transmission and reception 136
4.3.7 Monitoring radar hardware 137
4.4 Estimating Z_{DRsys} using weather observations 139
4.4.1 Measurements in light rain 139
4.4.2 Measurements in dry snow aggregates 140
4.4.3 Measurements in regions of Bragg scatter 141
4.4.4 Z_{DRsys} across the network 142

4.5 Current calibration issues and some possible improvements 142
4.5.1 Current WSR-88D calibration issues 143
4.5.2 Other data quality enhancements 144
4.6 Concluding remarks 145
References 147

5 Improvement of GPM dual-frequency precipitation radar algorithms for Version 07 153
Shinta Seto
5.1 Introduction 153
5.1.1 Overview of the DPR 153
5.1.2 Overview of DPR algorithms 154
5.2 Basic theory 158
5.2.1 Drop size distribution 158
5.2.2 Retrieval of DSD parameters from unattenuated observations 160
5.2.3 Retrieval of DSD parameters from attenuated observations 162
5.3 History of the algorithm development 165
5.3.1 PR algorithm 165
5.3.2 KuPR and KaPR algorithms 167
5.3.3 Dual-frequency algorithm 168
5.4 Improvement in Version 07 170
5.4.1 R–D_m relation 170
5.4.2 Correction of SRT considering the soil moisture effect 172
5.4.3 Estimation of the vertical profile of ε 177
5.4.4 Correction of scattering tables for PR 181
5.4.5 Effects of the change in the precipitation detection method 182
5.5 Improvement plan for the next version 183
5.5.1 Extrapolation of the vertical profile in the main-lobe clutter region 183
5.5.2 Assumptions of precipitation particles 184
5.5.3 Differences between PR algorithm and KuPR algorithm 184
5.6 Summary 185
References 186

6 The NASA Polarimetric (NPOL) weather radar facility and some applications 189
David B. Wolff, David A. Marks, Charanjit S. Pabla, Jason L. Pippitt, Ali Tokay, Jianxin Wang and Michael Watson
6.1 The NPOL system 189
6.1.1 Transmitter 190
6.1.2 Receiver 190
6.1.3 Antenna 192
6.2 Radar calibration and quality control 192
6.2.1 Transmitter, receiver, and antenna calibration 192

6.2.2 Calibration techniques using observed data 193
6.2.3 Quality control 199
6.3 Tools for validation 202
6.3.1 Polarimetric radar retrievals 202
6.3.2 System for integrating multiplatform data to build the atmospheric column 202
6.4 In situ validation 203
6.4.1 The Wallops GPM Precipitation Research Facility 204
6.4.2 GPM remote field campaigns 207
6.5 Validating GPM precipitation retrievals 212
6.5.1 Using SIMBA to validate that GPM Level I requirements are met 212
6.5.2 Validation of IMERG precipitation using NPOL 216
6.5.3 Validation studies using NPOL at Wallops 221
6.6 Conclusion 222
References 224

7 NASA high altitude airborne weather radars 231
Gerald M. Heymsfield, Lihua Li, Matthew L. Walker McLinden, Liang Liao, Charles N. Helms and Stephen Guimond
7.1 Introduction 231
7.1.1 Motivation for high altitude radars 231
7.1.2 ER-2 Doppler radar (EDOP) 233
7.1.3 Cloud radar system (CRS) 235
7.1.4 High-altitude wind and rain airborne profiler (HIWRAP) 238
7.1.5 ER-2 X-band radar (EXRAD) 241
7.2 Airborne radar challenges 242
7.2.1 Pulse compression and surface backscatter 242
7.2.2 Doppler algorithms 243
7.3 Calibration 247
7.3.1 Internal calibration 247
7.3.2 Ocean calibration 249
7.4 Applications: precipitation physics 251
7.4.1 Differential measurements 251
7.4.2 Microphysical retrieval method 253
7.4.3 Attenuation correction 257
7.4.4 Four frequency nadir measurements 261
7.5 Applications: horizontal wind retrievals 261
7.5.1 Dual-Doppler technique 261
7.5.2 Coplane technique 264
7.5.3 Velocity-azimuth display (VAD) technique 265
7.5.4 Three-dimensional variational technique 269
7.6 Future developments 273
References 275

8 Ocean-going weather and profiling radar for clouds and precipitation **283**
P.T. May, B. Dolan, M. Katsumata, P.A. Kucera, V. Louf, A. Protat and C.R. Williams
8.1 Introduction 283
8.2 Some historical perspectives 284
8.2.1 Shipborne radars by the Japan Meteorological Agency (JMA) 284
8.2.2 GATE (Global Atmospheric Research Program (GARP) Atlantic Tropical Experiment – 1974) 284
8.2.3 TOGA-COARE: Doppler radars and profilers 287
8.2.4 Doppler weather radar applications post-TOGA COARE 290
8.2.5 The coming of shipborne dual-polarization radars 293
8.3 Three current shipborne C-band dual-polarization radars 294
8.4 Technical challenges and approaches for shipborne dual-polarization weather radar 297
8.4.1 Blockage 298
8.4.2 Antenna stabilization 298
8.4.3 Sea clutter 299
8.4.4 Quality control: calibration 300
8.4.5 Quality control: dual-polarization weather radar post-processing 304
8.4.6 Dual-polarization weather radar performance 307
8.4.7 Some examples of dual-polarization data and science applications 309
8.5 Weather and cloud profiling radars on ships 313
8.5.1 History of cloud radar deployments and outcomes 313
8.5.2 Technical descriptions 315
8.5.3 Capabilities and science applications 318
8.6 Current and future capabilities and opportunities 320
Acknowledgments 322
References 325

9 A versatile stratosphere–troposphere radar at 205 MHz in the tropics **337**
K. Mohanakumar, Titu K. Samson, P. Mohanan, K. Vasudevan, K.R. Santosh, V.K. Anandan, G. Viswanathan and B.M. Reddy
9.1 Stratosphere–troposphere (ST) radar 337
9.1.1 Concept of 200 MHz ST radar 338
9.1.2 Advantages of 200 MHz radar 340
9.1.3 ST radar at Cochin University of Science and Technology 340
9.1.4 Geographical relevance 341
9.1.5 Refractive index structure constant (C_n^2) over Cochin 343
9.2 Technical specifications of CUSAT ST radar 344
9.2.1 Radar-processing computer 345

9.2.2 Coherent signal generator 345
9.2.3 Signal processing 348
9.2.4 Mode of operation 348
9.2.5 Doppler beam swinging 348
9.2.6 Yagi–Uda antenna for ST radar 349
9.2.7 Total transmitted power 350
9.2.8 TRM 351
9.2.9 Radiation pattern measurement 352
9.2.10 Noise level survey between the radar site and a TV broadcasting station 354
9.3 Research to enhance the acceptability of ST radar data 356
9.3.1 Validation of radar wind observations with GPS radiosonde 356
9.3.2 Evaluation of ST radar wind observations with reanalysis and model outputs 357
9.3.3 Radar observations under rainy conditions 358
9.3.4 New technique for the identification of clear air and rain echoes 359
9.3.5 Validation of Aeolus satellite wind observations with the CUSAT ST radar 360
9.3.6 Higher height coverage of CUSAT ST radar 362
9.4 Atmospheric studies using the 205 MHz ST radar 362
9.4.1 Vertical structure and evolution of Indian summer monsoon 363
9.4.2 Diurnal variations during the active and break phases of monsoon 364
9.4.3 Thunderstorms observations from ST radar 366
9.4.4 Determination of tropical tropopause using ST radar observations 367
9.4.5 Tropopause variability and associated dynamics during monsoon 368
9.4.6 Features of atmospheric circulation observed during sudden stratospheric warming 370
9.4.7 Detection of inertia gravity waves 371
9.5 Potential of 205 MHz ST radar for ionospheric observations 372
9.5.1 Configuration of 205 MHz ST radar for ionospheric observations 373
9.5.2 Significant ionospheric observations using the 205 MHz ST radar 373
9.5.3 Detection of ionospheric disturbances with volcanic eruption and Tsunami 375
9.6 Other observations by 205 MHz ST radar 377
9.6.1 Detection of electric charges in the upper atmosphere 377
9.6.2 Flight tracking using ST radar 378
9.6.3 Meteor observation from ST radar 379

9.7 Curious observational evidence from ST radar 380
9.7.1 Rotational Doppler at melting layer 380
9.7.2 Single-sided Doppler at 90 km 380
9.7.3 Anomalous precipitation-like pattern observed from the tropopause level 380
9.8 Application of 205 MHz radar for radio astronomical studies 383
9.9 Future perspective 383
9.9.1 Imaging of the atmosphere 383
Acknowledgments 384
References 388

10 An integrated future US weather radar architecture for aviation 393
Mark E. Weber, John Y.N. Cho, Henry G. Thomas and James M. Kurdzo
10.1 Introduction 393
10.2 Airport meteorological radar development 396
10.2.1 Terminal Doppler Weather Radar 396
10.2.2 Airport surveillance radar 398
10.3 Multi-agency weather and aircraft surveillance radar concept development 403
10.3.1 Multifunction phased array radar 403
10.3.2 Benefits for public weather warning and water management 410
10.4 The future national weather radar system 416
10.4.1 Opportunities 416
10.4.2 Architecture framework 419
10.4.3 Roadmap for transition to the future architecture 421
Appendix 423
References 428

11 The mitigation of ground clutter 433
J.C. Hubbert, S. Ellis and G. Meymaris
11.1 Historical perspective and background 433
11.1.1 Ground-clutter statistics 434
11.1.2 Ground-clutter identification 434
11.1.3 Ground-clutter filtering 436
11.1.4 Wind turbine clutter 438
11.1.5 Phased array technology 439
11.2 A comparison of spectral and regression clutter filtering 439
11.2.1 Spectral clutter filter and Fourier basis functions 439
11.2.2 Regression filtering and orthogonal polynomials 441
11.2.3 The frequency response of the regression filter 443
11.3 Operational aspects of regression clutter filtering 445
11.3.1 Automated selection of the polynomial order 446
11.3.2 Interpolation across the zero-velocity gap 448

11.4 Identifying low CSR clutter 452
11.4.1 RHOHV-test example 454
11.5 Regression and spectral clutter filter comparison using experimental data 457
11.6 Summary 463
Acknowledgment 464
References 465

12 Polarimetric planar phased array radar – challenges for observing weather 473
Dušan S. Zrnić, Igor I. Ivić, Dordje Mirković, Lesya Borowska and Guifu Zhang
12.1 Introduction 473
12.2 Radiating elements 474
12.2.1 Electric dipole 475
12.2.2 Magnetic dipole 476
12.2.3 Collinear electric and magnetic dipoles 476
12.2.4 Patch radiators 476
12.2.5 Array antenna 477
12.3 Polarimetric modes and effects of pattern on bias in the polarimetric variables 480
12.3.1 Operational polarimetric modes 480
12.3.2 Cross-polar pattern effects on bias in polarimetric variables 482
12.3.3 Copolar pattern effects 484
12.3.4 Phase coding 486
12.3.5 Alternate transmission of H and V field (AHV) 487
12.4 Vertically oriented array 487
12.4.1 Relations 487
12.4.2 Simultaneous transmission with phase coding 489
12.4.3 Polarimetric variables 490
12.4.4 Simultaneous transmission without phase coding 494
12.4.5 Alternate transmission AHV 495
12.4.6 Principal planes 497
12.5 Tilted antenna array 497
12.6 Examples 498
12.7 Conclusions 500
Appendix 502
References 504

About the editors

V.N. Bringi is an emeritus professor at Colorado State University, USA. He received his PhD from The Ohio State University, then joined the faculty at Colorado University where he spent his entire career. He is a fellow of the American Meteorological Society (AMS) and recipient of the AMS Remote Sensing Prize in 2013.

Kumar Vijay Mishra is a senior fellow at the United States DEVCOM Army Research Laboratory. He received his PhD in electrical engineering and MS in mathematics from The University of Iowa, Iowa City. He is the recipient of many best paper awards and fellowships including the IET Premium Award for Best Paper (2021), US National Academies ARL Harry Diamond Distinguished Fellowship (2018), and Andrew and Erna Finci Viterbi Postdoctoral Fellowship (2015–2017).

Merhala Thurai is a research scientist at Colorado State University, USA. She received her BSc in Physics from Imperial College, London, and her PhD from King's College, London. She worked at Rutherford Appleton Laboratory, Oxon, UK, for 14 years, and has been with Colorado State University since 2004.

Preface

This edited three-volume set of books published by the Institution of Engineering and Technology (IET) on "Advances in Weather Radar" falls under the prestigious SciTech/IET book series on radar, sonar, and navigation. We are delighted to edit this new book consisting of 32 original chapters reviewed by one of the editors and one external reviewer. They are written by exceptionally well-qualified experts from academia, research laboratories, and national weather services covering a breathtaking range of topics. The focus of this edited book is on dual-polarization radar because of its important application in rainfall measurement, in elucidating details of cloud physical processes, classification of meteorological and non-meteorological echo types, the validation and evaluation of bulk microphysical processes that predict number density and mixing ratio and radar hydrology among other more recent applications. The chapters were selected by the editors who have a combined experience of 80 years in prestigious research laboratories while the lead editor's entire career in academia gives strong credibility to our reputation.

There are only two other edited volumes available: one on "Radar in Meteorology" published by the American Meteorological Society in 1990 and edited by the late David Atlas consisting of 44 chapters. This volume gives detailed accounts of the history of radar in the meteorology post World War II, research and operational topics using Doppler radar, and one chapter on dual-polarization radar. The other edited volume on "Radar and Atmospheric Sciences" was a collection of essays published by the AMS in 2003 in honor of the late David Atlas edited by Roger Wakimoto and Ramesh Srivastava but contained only one devoted to dual-polarization. It is, therefore, very appropriate that the next edited volumes cover the scientific progress made since 2003–2023 a period of two decades when Doppler dual-polarization radar became *de rigueur* in any national weather service agency.

Several books have been published since 1983 which give an intermediate to advanced level exposition of polarimetric Doppler radar theory with only small sub-sections in Doviak and Zrnic (1984) which was expanded to about a chapter in the 2nd edition published in 1993. The material for the 2nd edition came from a course taught by the authors at the University of Oklahoma to a senior and the first-year graduate students in meteorology and electrical engineering. In 2001, the first book exclusively devoted to polarimetric Doppler radar principles and applications was published (Bringi and Chandrasekar) which soon became the standard reference text. The book was an outcome of a graduate-level course in Radar and Electromagnetics taught by the authors for many years at Colorado State University. For the next 15 years, it was the textbook of choice for academics and researchers due to its development of the theory from first principles in a self-contained format. It was also the time when polarimetric radar research was accelerating at a scorching pace.

Two more books appeared recently (Zhang 2017) and Ryzhkov and Zrnic (2019), the former based on sound theory rooted in mathematics and based on a class taught by Zhang at the University of Oklahoma. The book by Ryzhkov and Zrnic is a monograph written at an intermediate level for meteorologists and researchers enabling them to gain a quick understanding of polarimetry without going into the mathematical theory which can be found in Bringi and Chandrasekar (2001) or Zhang (2017). It is our belief which is perhaps obvious that books written and edited by authors themselves are a different breed altogether compared to edited volumes where each chapter is written by experts giving a level of detail that cannot be found in textbooks or monographs. With rapid advances in the application of polarimetric radar to topics that cannot be covered by conventional books, the editors of the current 3-volume set of books felt, there was a vacuum in documenting recent advances in dual-polarized radar resulting in the current edited volumes.

It is well known that the electromagnetic wave is defined by its amplitude, phase, and polarization state. The interaction of the wave and the precipitation particles changes the amplitude, phase, and polarization state. The magnitude of the square of the amplitudes corresponds to the power scattered by precipitation, the phase change due to changes in the electrical path length of moving precipitation along the radial direction which is related to the Doppler frequency, and the scattering matrix due to interaction with precipitation such as raindrops which are non-spherical and oriented in space. These changes in the EM waves when quantified with particle microphysical states are one of the main drivers of dual-polarized Doppler weather radar.

Recognizing the importance of dual-polarization the US National Weather Service (NWS) in a major effort installed the Next Generation (NEXRAD) WSR-88 Doppler network of 160 radars beginning around 2008. The main NWS goals were two-fold: (a) detection of tornadoes by the mesoscale circulations aloft in supercell storms prior to damaging winds at the ground with lead times of around 20 min, and (b) measuring hourly rain amounts in pixels 4 × 4 km out to ranges of 230 km for hydrological applications (flash floods and landslides) and land-falling hurricanes which are the most prolific rain producers causing coastal ecological damage and loss of life.

Most of the chapters in this book describe the physical properties of the interactions also known as the "inverse" problem which is poorly constrained. Measuring the change of polarization state due to interactions with precipitation is considered as arguably the most important advance since Doppler technology was introduced on a NEXRAD prototype. The few dozen pioneers (one of those pioneers is the lead editor of this three-volume set) who were by and large electrical engineers were astonished by the meteorological and non-meteorological applications that they could not have possibly imagined at that time (1982). This edited three-volume set has many chapters that describe these innovative applications. We hope that these edited volumes will be useful to graduate students, to radar systems designers, to high-level managers of National Meteorological Services, and other research scientists who need to delve deeper into specific topics which cannot be found elsewhere.

This volume on "Precipitation Sensing Platforms" focuses on major advances in designing, operating, and deploying weather radars across the globe. The second volume "Precipitation Science, Scattering, and Processing Algorithms" considers theoretical milestones achieved in microphysics, electromagnetics, and signal processing of radar meteorology. The final volume "Emerging Applications" includes applications of weather radars in novel as well as non-meteorological applications.

We thank all contributing authors for submitting their high-quality contributions. We sincerely acknowledge the support and help from all the reviewers for their timely and comprehensive evaluations of the manuscripts that improved the quality of this book.

Finally, we are grateful to the IET Press Editorial Board and the staff members Nicki Dennis, Olivia Wilkins, and Sarah Lynch for their support, feedback, and guidance.

V.N. Bringi
Fort Collins, CO, USA

Kumar Vijay Mishra
Adelphi, MD, USA

Merhala Thurai
Fort Collins, CO, USA

Acknowledgments

V.N.B. acknowledges continuous support from the Atmospheric Sciences Program of the National Science Foundation to Colorado State University, the Army Research Office, and NASA.

K.V.M. acknowledges support from the National Academies of Sciences, Engineering, and Medicine via the Army Research Laboratory Harry Diamond Distinguished Fellowship.

M.T. acknowledges support from the National Science Foundation, grant number AGS-1901585 to Colorado State University.

List of editors

V.N. Bringi, Colorado State University, Fort Collins, CO, USA
Kumar Vijay Mishra, United States DEVCOM Army Research Laboratory, Adelphi, MD, USA
Merhala Thurai, Colorado State University, Fort Collins, CO, USA

List of contributors

V.K. Anandan, ISRO Telemetry, Tracking and Command Network
Lesya Borowska, NOAA National Severe Storms Laboratory
V.N. Bringi, Colorado State University
John Y.N. Cho, MIT Lincoln Laboratory
Brenda Dolan, Colorado State University
Scott M. Ellis, National Center for Atmospheric Research, Boulder
Luca Facheris, University of Florence
Stephen J. Frasier, University of Massachusetts, Amherst
Stephen Guimond, NASA Goddard Space Flight Center & University of Maryland Baltimore County
Charles N. Helms, NASA Goddard Space Flight Center & University of Maryland, College Park
Gerald M. Heymsfield, NASA Goddard Space Flight Center
John C. Hubbert, National Center for Atmospheric Research, Boulder
Igor I. Ivić, NOAA National Severe Storms Laboratory
Masaki Katsumata, Japan Agency for Marine-Earth Science and Technology
Paul A. Kucera, National Center for Atmospheric Research, Boulder
James M. Kurdzo, MIT Lincoln Laboratory
Lihua Li, NASA Goddard Space Flight Center
Liang Liao, NASA Goddard Space Flight Center & Morgan State University
Valentin Louf, Bureau of Meteorology, Australia
David A. Marks, NASA Wallops Flight Facility & Science Systems & Applications, Inc.
Peter T. May, Monash University
Matthew L. Walker McLinden, NASA Goddard Space Flight Center
Dordje Mirković, NOAA National Severe Storms Laboratory
Valery Melnikov, NOAA National Severe Storms Laboratory & The University of Oklahoma
Greg Meymaris, National Center for Atmospheric Research, Boulder
Kumar Vijay Mishra, United States DEVCOM Army Research Laboratory
K. Mohanakumar, Cochin University of Science and Technology
P. Mohanan, Cochin University of Science and Technology
Charanjit S. Pabla, NASA Wallops Flight Facility & Science Systems & Applications, Inc.
Jason L. Pippitt, NASA Goddard Space Flight Center & Science Systems & Applications, Inc.

Alain Protat, Bureau of Meteorology, Australia
B.M. Reddy, National Geophysical Research Institute, India
Alexander Ryzhkov, NOAA National Severe Storms Laboratory & The University of Oklahoma
Titu K. Samson, Arizona State University
K.R. Santosh, Cochin University of Science and Technology
Shinta Seto, Nagasaki University
Henry G. Thomas, MIT Lincoln Laboratory
Merhala Thurai, Colorado State University
Ali Tokay, NASA Goddard Space Flight Center & University of Maryland Baltimore County
K. Vasudevan, Cochin University of Science and Technology
G. Viswanathan ISRO Telemetry, Tracking and Command Network
Jianxin Wang, NASA Goddard Space Flight Center & Science Systems & Applications, Inc.
Michael Watson, NASA Wallops Flight Facility & Peraton Inc.
Mark E. Weber, MIT Lincoln Laboratory
Christoper R. Williams, University of Colorado, Boulder
David B. Wolff, NASA Wallops Flight Facility
Guifu Zhang, The University of Oklahoma
Dušan S. Zrnić, NOAA National Severe Storms Laboratory & The University of Oklahoma

List of reviewers

V.N. Bringi, Colorado State University
Stephen L. Durden, Jet Propulsion Laboratory
Christopher W. Fairall, NOAA Physical Sciences Laboratory
Mircea Grecu, Morgan State University & NASA Goddard Space Flight Center
Gerald M. Heymsfield, NASA Goddard Space Flight Center
John C. Hubbert, National Center for Atmospheric Research, Boulder
R. Jeffrey Keeler, Dr Radar Consulting
Lihua Li, NASA Goddard Space Flight Center
William P. Mahoney III, National Center for Atmospheric Research, Boulder
Kumar Vijay Mishra, United States DEVCOM Army Research Laboratory
Nobuhiro Takahashi, Nagoya University
Elizabeth Thompson, NOAA Physical Sciences Laboratory
Merhala Thurai, Colorado State University
Ali Tokay, NASA Goddard Space Flight Center & University of Maryland Baltimore County
Tomoo Ushio, Osaka University
David A. Warde, NOAA National Severe Storms Laboratory & The University of Oklahoma
Christoper R. Williams, University of Colorado, Boulder
David B. Wolff, NASA Wallops Flight Facility

Introduction to volume 1

After nearly 50 years of sustained research and 30 years of operational deployment in the developed regions of the world, research in weather radars has witnessed tremendous growth over the past decade and is now spilling over to novel applications and geographies. The earliest weather radars in the era of the 1950s were mainly used to measure the backscatter power at a single polarization state. During the 1960s–1970s, the Doppler shift principle was used to measure the radial velocity of hydrometeors in addition to the backscattered power although still using only a single polarization state. By the mid-1970s, the Doppler radar's importance for operational applications was widely recognized. In the late 1980s, the deployment of Doppler radars for operational weather forecasting by the US National Weather Service (NWS) accelerated and has continued to the present time.

Before the early 1970s, there was limited research in dual-polarized radars mainly for radio wave propagation studies and much less for weather applications. The measurements were limited by the poor cross-polar performance of the radar hardware, particularly the antenna and its feed. The modern era of polarimetric radar using precision microwave hardware began at the National Research Council in Ottawa with the Canadian radars starting in the late 1960s–early 1970s using dual-circular polarization techniques for propagation research (Ka and X-bands) and hail detection (at S-band). This research clearly showed that oblate raindrops formed a highly oriented medium and that the differential propagation phase and differential attenuation between the horizontal and vertical polarizations through such a highly oriented medium could be measured at Ka and S-bands using dual-receiver circular polarization technology without requiring any Doppler processing. The use of dual-linear polarizations (horizontal and vertical) was proposed in the late 1970s primarily because at low elevation angles, these states are aligned along the principal axes of the highly oriented oblate rain medium.

During the decades of 1990s and 2000s, weather radar engineering and scientific understanding of meteorological phenomena matured to such an extent that it led to the installation of national weather radar networks in several countries. However, it was during the past decade (2010s) that this technology realistically entered a renaissance and witnessed rapid growth in the weather radar industry and research, including deployment on satellites, aircraft, autonomous vehicles, and ships. This decade also experienced successful cross-fertilization of ideas, such as phased array antennas and solid-state transmitters, from hard target radars to meteorological radar systems not to mention multiple breakthroughs in signal processing, machine learning, and electromagnetic theory of weather radar.

The first volume of this book brings together the important weather radar engineering milestones during the last decade.

Chapter 1 "The decade of Renaissance in weather radar research" features editors' picks on the key technical and scientific ideas that have propelled the field forward at such a rapid rate and will likely be the mainstay of its research community in the coming years.

Chapter 2 "Doppler polarimetric radars for weather observations from 1995 to 2020: a historical perspective" provides a historical overview of the period 1995–2010 that marked the innovation and intense research in weather radars. This chapter summarizes these developments in the post-NEXRAD deployment era. The first satellite-borne radar was launched aboard NASA TRMM satellite. Multiple X-band weather radar testbeds were developed and deployed in the United States and Japan. The standards set during this period continue to form the backbone of weather radar operations today.

Chapter 3 "Developments in solid-state weather radar" covers one of the foremost engineering developments of employing solid-state transceivers in weather radars. This has enabled pulse-compression technology leading to better monitoring of the atmospheric boundary layer. While pulse-compression was initially proposed for lidars and later made its way to military surveillance radars, it took almost three decades for weather radars to employ this technique because of the challenges arising from the volumetric nature of the target.

Chapter 4 "Quality of polarimetric data in the WSR-88D system" takes the focus toward the NWS WSR-88D network of weather radars deployed in the United States since the 1990s. It captures the current calibration techniques used by the NWS WSR-88D system and the achieved accuracies in reflectivity and dual-polarization variables. Equipped with dual-polarization, the lifetime of the WSR-88Ds has been extended beyond the year 2030. With additional features such as accurate measurements of the rain and snow rates, measurements of convection parameters in pre-storm environment, and observations of the boundary layer, the WSR-88D calibration assumes a higher significance than before.

Chapter 5 "Improvement of GPM dual-frequency precipitation radar algorithms for version 07" is concerned with important developments in space-borne weather radar. It describes the retrieval algorithms for vertical precipitation rate profiles from the measurements of the dual-frequency precipitation radar (DPR) on the core satellite of the global precipitation measurement (GPM) mission launched in 2014. The DPR is the successor of the first space-borne precipitation radar (PR) on the Tropical Rainfall Measurement Mission (TRMM) satellite that operated from December 1997 to April 2015.

Chapter 6 "The NASA Polarimetric (NPOL) weather radar facility and some applications" is also related to the GPM mission. NPOL is an S-band, dual-polarimetric Doppler weather radar that provides well-calibrated high-quality data for GPM ground validation. The chapter includes NPOL's design, applications, and validation studies.

Chapter 7 "NASA high-altitude airborne weather radars" describes NASA's GPM-related weather radars for airborne platforms such as the X-band (9.6 GHz)

ER-2 Doppler Radar (EDOP), the W-band (94 GHz) Cloud Radar System (CRS), the Ku (13 GHz)/Ka-band (35 GHz) High-altitude Wind and Rain Airborne Profiler (HIWRAP), and the X-band (9.6 GHz) ER-2 X-band Radar (EXRAD).

Chapter 8 "Ocean-going weather and profiling radar for clouds and precipitation" switches from airborne to seaborne deployment. During the past 20 years, the deployment of weather radar on ships has been increasing. The chapter provides a historical and contemporary perspective and studies in this context.

Chapter 9 "A versatile stratosphere–troposphere radar at 205 MHz in the tropics" is concerned with the atmospheric radar or wind profiler radar (WPR), which monitors horizontal and vertical winds in the troposphere and stratosphere in all weather conditions. These systems operate at the lower end of the spectrum in the very high-frequency (VHF) and ultra-high frequency (UHF) bands. The chapter focuses on the design and applications of the first VHF stratosphere–troposphere (ST) radar in the 200 MHz range that has been deployed in Kochi, India.

Chapter 10 "An integrated future US weather radar architecture for aviation" reflects on integrating various national weather radar network in the United States that are operated by the Federal Aviation Administration (FAA), the National Oceanic and Atmospheric Administration (NOAA), and the Department of Defense (DoD). The chapter also includes the latest enhancements in FAA weather radar systems such as Terminal Doppler Weather Radar (TDWR), Airport Surveillance Radar (ASR), and Air Route Surveillance Radar (ARSR).

Chapter 11 "The mitigation of ground clutter" examines the research on ground clutter filtering algorithms using data from the S-band Polarimetric radar (S-Pol) operated by the National Center of Atmospheric Research (NCAR). The chapter covers regression filtering in detail and compares the results with popular spectral-domain filters such as Gaussian Model Adaptive Processing (GMAP) used by NEXRAD.

Finally, Chapter 12 "Polarimetric planar phased array radar – challenges for observing weather" captures the ongoing frenetic developments in using an array of antennas for weather observations. Although the phased array weather radar is yet to reach full deployment worthiness, the chapter examines its advantages in electronic beam-steering, adaptive scanning, time multiplexing, clutter suppression, and faster volume updates via bespoke developed by the University of Oklahoma.

Chapter 1
The decade of renaissance in weather radar research

Kumar Vijay Mishra[1], *Merhala Thurai*[2] *and V.N. Bringi*[2]

1.1 Introduction

Research in weather radar systems has become very popular in the 21st century. Although mainly used by national and international meteorological bodies, several other applications of weather radars have become possible. Since the deployment of the first operational weather radar network during the early 1990s by the US National Weather Service (NWS), the meteorological radar world has changed considerably. While 20 years ago, radar meteorology was still considered an experimental and futuristic field because many technologies were still in an early stage and confined to academia, it is nowadays state of the art that is also intertwined with other radar disciplines. Today, radar meteorology is a flourishing and vibrant area supported by a significantly expanded community comprising academia, government, industry, and standards communities.

After over four decades of niche applications, weather radar technology during the last decade realistically entered a renaissance after crossing several milestones. This timeframe witnessed a dramatic progress in weather modeling, antennas, and digital technologies, which is leading to fast scanning weather radars equipped with powerful signal processors. Moreover, a lot of theoretical research has been devoted to understanding of precipitation science, extreme weather phenomena, performance guarantees, and adoption of novel algorithms. Weather radar has been successfully deployed across platforms such as aircraft, satellites, ships, and autonomous vehicles. In processing algorithms, new mathematical tools such as optimization theory, sparse reconstruction, robust beamforming, and machine learning (ML) have gained salience in weather radar research.

Computational electromagnetics, once the domain of academia, progressed from research to commercially available codes, thereby allowing accurate manufacture of specialized microwave components, for example, symmetric orthomode transducers and switchable circulators. These codes even encouraged the nonexpert

[1]United States DEVCOM Army Research Laboratory, Adelphi, USA
[2]Colorado State University, Fort Collins, USA

to compute scattering from complex-shaped crystals and snowflakes. Other applications such as automotive radar, wireless communications, satellite-based exploration, hydrology, entomology, ornithology, and volcanology have greatly benefitted from these advances. The commercialization of radar meteorology has accelerated in recent years with several weather radar enterprises launched around the world. Standardization efforts have also picked up a good pace in various standards groups supported by the IEEE, American Meteorological Society (AMS), and International Telecommunication Union (ITU).

This chapter introduces the reader to important breakthroughs in weather radars during the last ten years that we term as the decade of renaissance. We summarize these developments in the context of broader radar theory, engineering, and applications. The chapter is an assessment of the editors about the major breakthroughs during the last ten years that will continue to positively influence our field in the coming years. The chapter breaks down along three distinct topical lines: engineering, signal processing, and precipitation science. While, undoubtedly, the current era is very exciting for the radar meteorology community, it is not feasible to exhaustively list all innovations in this chapter alone. We refer the reader to the remaining chapters for a detailed exposition of other aspects.

1.1.1 Renaissance of weather radar technology

The Next Generation Weather Radar (NEXRAD) agencies that funded the nationwide WSR-88D network are the National Oceanic and Atmospheric Administration (NOAA), the Federal Aviation Administration (FAA), and Department of Defense (DoD). A Technical Advisory Committee was formed to provide guidance for the upgrade of the WSR-88D over the expected lifetime to 2035 [1].

The renaissance of weather radar technology and applications referred to here is the period of intense research which laid the foundation for dual-polarization technology leading to the installation of radar networks for the operational services for many countries, the largest being the US NWS with 160 radars. A historical perspective of radar systems operated by Universities and Research Laboratories (post-1995) was described including system specifications and evaluation of product algorithms [2]. A historical pre-1995 account can be found in [3], which emphasizes the personnel some of whom were pioneers of dual-polarization radar.

By 1995, the race was on to determine the optimal method of measuring the dual-pol signals and their error characteristics. Chapter 4 in Volume 1 of this book gives the detailed algorithms used in the WSR-88D and how the engineers at the National Severe Storms Laboratory [NSSL] improved data quality. Radar manufacturers began to offer dual-polarization Doppler systems with modern receivers, digital signal processor, and data systems. Many application-specific products were developed specifically for forecasters with little experience. Since Volume 1 contains two chapters (Chapter 2 and Chapter 4 of Volume 1) that are devoted to this Renaissance period, our introductory description will cover some of the advances specifically related to the WSR-88D.

1.1.1.1 The WSR-88D

The renaissance in weather radar is defined here as a period of intense research to weather service operations beginning with the deployment of single-polarization highly coherent systems nationwide by the NWS which was completed in 1997. This was the single most complex system replacement in the history of the NWS. In our Renaissance period, we focused on the upgrade of the single-polarization WSR-88D to dual polarization and beyond. The late Dr. David Atlas and Professor Thomas Seliga persuaded the NWS to add a clause in the procurement contract that required vendors to demonstrate how they would upgrade the system to dual polarization. Specifically, the vendor built a dual-polarization feed and an orthomode coupler and obtained antenna patterns at horizontal and vertical polarizations for demonstration purposes. This was a remarkable foresight as the subsequent company which was selected for the upgrade to dual polarization (2009–2013) had to demonstrate this in the field during installation. The design had to be compatible with all of the WSR-88D functions. NSSL suggested that the upgrade be a system that transmits simultaneously both horizontally and vertically polarized components so that the transmitted polarization be in general an unknown elliptical type [4]. But the requirement was that the magnitude of both components on transmission and reception be calibrated to a relative accuracy of 0.1 dB. In the specifications for the upgrade, the detectability (sensitivity) of the system was reduced from the original stipulation that a −7.5 dBZ at 50 km range results in an SNR of 0 dB per pulse. Nonetheless, neither NSSL nor the NWS realized that the actual detectability of the system was much better, viz., about −10 to −13 dBZ per pulse.

The vendor's choice design was the antenna-mounted receiver with extensive temperature control. Although this design increases the detectability, the error in the specification led the vendor to choose front-end low noise amplifiers that just about met the perceived specifications. Tests in Norman, Oklahoma, revealed significantly lower detectability compared to the collocated WSR-88D (code name KCRI) used by the Radar Operation Center (ROC) to support the network. This misunderstanding between the NWS and the vendor was quickly resolved so that the current detectability of the upgraded fleet is on the average −11 dBZ at 50 km. There is, however, a few dB variation about this average mainly due to the different frequencies on the network and the different heights of the radar towers.

1.1.1.2 Weather radar antennas

The renaissance of weather radar antennas with high performance was based on earth-satellite antenna design (about a decade ago) which required low sidelobe envelopes and low on-axis cross-polar levels. Feed design [5] was an area of intense research with a profiled corrugated horn [6] being the optimal choice of driving single-offset and dual-offset antennas that have no feed blockage. These antennas were designed with sidelobe envelopes that fall off with $\log_{10}(\Phi)$ dB for close-in sidelobes and −50 dB for $10° < \Phi < 180°$ in any plane. Prime-focus fed parabolic antennas that are the most common weather radar antennas cannot satisfy the log envelope in any plane due to blockage by the feed/orthomode transducer (OMT) and the feed support struts. A high degree of beam symmetry of the main lobe is desirable in any pattern as the antenna is rotated about its polarization axis independent of the port excited (H- or V-port), which is a characteristic of the profiled corrugated horn primary pattern. (Note: the earth-satellite antennas must also have very low cross-polar level on boresight, as low as < -45 dB.) Weather radar antennas are generally prime focus parabolic reflectors that are inherently symmetric, and a well-designed feed can give very low on-axis cross-pol but blockages and struts deteriorate this figure. Nevertheless, a few newer weather radar antennas were single-offset design and one dual-offset Gregorian design (see front jacket of Volume 1) mainly for research [7]. The latter, named (CSU-CHILL) radar, is operated by Colorado State University.

A renaissance in the methodology of antenna measurements is the evaluation of antenna performance on a test range with a more stringent set of specifications for dual-polarization radar. Previously, the antenna specifications were limited to principal plane cut patterns at both polarizations, gain and differential gain between H and V, close-in sidelobe envelopes, and 3-dB beamwidth in E- and H-planes. Symmetry of the main lobe as the antenna is rotated about its polarization axis is not a usual specification as the antenna under test (AUT) cannot be rotated about its polarization axis. The AUT should have all the measurements with radome in place which is not done in a test range. The most comprehensive antenna testing with and without a radome is in [8].

The other set of comprehensive testing was conducted for the above-mentioned 9-m dual-offset Gregorian antenna (Figure 1.1) which was carried out at the manufacturer's calibrated range. The AUT was illuminated by the transmit antenna which had the same feed as the AUT. The cross-pol at boresight was very low, < -50 dB. The four cross-pol lobes maximum were less than -38 dB. Measurements in drizzle showed that the system linear depolarization ratio (LDR) was close to the upper bound of -39 dB.

In recent years, phased array antennas (see Chapter 12 of Volume 1 and Chapter 1 of Volume 2) are being tested for performance in rapidly evolving storms. While single-polarization phased arrays may be comparable to the reflector antenna performance, there is no dual-pol version that can compare with reflector antennas in terms of accuracy needed for microphysical retrievals. It appears that a fully dual-polarized phased array antenna does not appear to be viable in the foreseeable future. The NWS predicts the life expectancy of the WSR-88D to the year 2035 and beyond.

1.1.2 Shipborne, airborne and satellite radars

1.1.2.1 Shipborne radars

The past ten years has also seen the development of shipborne polarimetric weather radars. Amongst them is the C-band Australian OceanPOL research radar which was commissioned in 2016. It was mounted on "Research Vessel Investigator," which was being designed in 2012 for the CSIRO Marine National Facility. With dual-polarization capability, it was the first of its kind on an ocean-going platform. The radar is operated on all RV *Investigator* cruises regardless of the voyage's primary goals. As such, it samples a wide area of the oceans in the Australian

Figure 1.1 *The Colorado State University-Illinois State Water Survey (CSU-CHILL) S-band weather radar with a 9-m Gregorian design antenna shown here with the pedestal and radome. Photo credit: Kumar Vijay Mishra, with permission from CSU-CHILL Co-Principal Investigator V. N. Bringi.*

region from the tropics to the ice edge near Antarctica. The radar is a C-band 1.3° beamwidth system with a 250 kW peak transmitter power located at the highest point to allow continuous operation and full 360° volumes to be collected. Furthermore, the data from the radar are open access via www.openradar.io with quality control as described in [9]. The radar is providing data for studies of precipitation system over the remote Southern Ocean as well as crucial validation data for reanalysis and satellite precipitation products over this remote data sparse area [10,11] (Figure 1.2).

Other recently developed shipborne polarimetric radars include the C-band radar mounted on the Japanese research vessel "Mirai" and another C-band radar, SEA-POL, operated by Colorado State University, USA. The former has been used for studying the internal structure of typhoons in the western pacific (equivalent to Hurricanes over the Atlantic) [12]. Radar reflectivity, copolar correlation

(a)

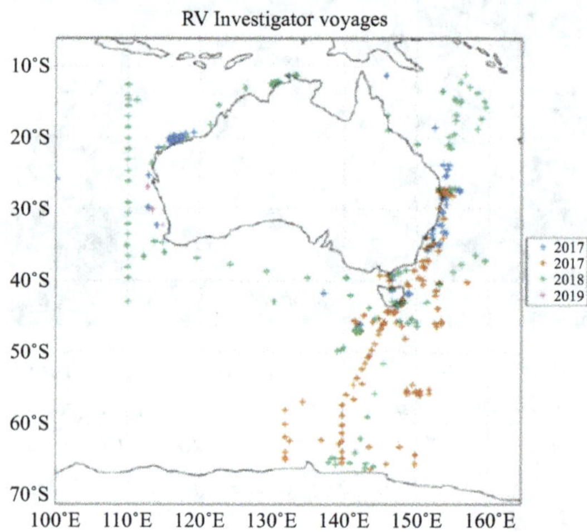

(b)

Figure 1.2 *(a) The RV Investigator at sea. The OceanPOL radar antenna is in the large dome at the top of the ship. Other instrumentation includes routine flux measurements as well as aerosol and greenhouse gas measurements. Image courtesy of CSIRO. (b) Map showing RV Investigator cruises between 2017 and 2019; OceanPOL radar data are currently available at openradar.io. Data from more recent cruises are being added to the archive. Image courtesy of CSIRO.*

coefficient, and differential phase were used. The latter has utilized the additional (and important) parameter, Z_{DR}, the differential reflectivity, to "assess microphysical processes within oceanic convection" [13]. The impact of ship movement on Z_{DR} measurements can be significant but correction methods can be applied to reduce/mitigate these effects [14].

Dual-polarization capability on research vessels is growing with more countries other than US, Australia, and Japan developing their own systems (such as a C-band shipborne polarimetric radar which is installed on Research Vessel (R/V) New Ocean Researcher 1 or NOR1 from Taiwan), as well as Australia's new ice-breaker (RSV Nuyina).

1.1.2.2 Spaceborne radars

There has been several spaceborne radars developed for precipitation and cloud sensing in the last 25 years. The first instrument to observe 3D storm structures from space was the Ku-band precipitation radar (PR) onboard the Tropical Rainfall Measurement Mission (TRMM) satellite [15]. This research satellite was launched in November 1997 and had operated for 17 years. The PR had a spatial resolution of 4.3 km from 350 km altitude (low earth orbit). The measurements were confined to 35° North and 35° South latitudes. The PR delivered a wealth of information on tropical rainfall; see, e.g., [16]. Because of attenuation effects at Ku-band, the initial focus of research was to develop suitable attenuation correction procedures, not just in the rain region but also in the melting layer at heights around the zero-degree isotherm [17]. One of the PR products—as a result of the implementation of attenuation correction procedures—was the melting layer heights and the thicknesses; these have been used for radiowave propagation-related studies [18] for earth-space links.

The follow-on mission to TRMM was the Global Precipitation Measurement Mission (GPM). It was launched in February 2014. It carries dual-wavelength precipitation radars (DPR) operating at Ku- and Ka-bands. The configuration is such that both radar footprints (5 km) coincide at earth's surface. The measurements were extended to 35° North and 35° South latitudes. Differential attenuation between the two frequency bands enables better rain-snow discrimination. DPR is also more sensitive to light rain and snow [19,20].

There have also been satellite radars developed to observe cloud cover and the vertical structure from space, CloudSat being one of them [21]. It was launched in April 2006. CloudSat's cloud profiling radar operates at 94 GHz so that it can provide information on cloud optical depth, cloud type, ice content at cloud top, cloud base height, and cloud cover. The data have also been used for snow retrievals and compared with DPR-based retrievals for coincident/overlap observations [22]. EarthCARE is another (planned) mission which has another cloud profiling radar at 94 GH, but it has additionally Doppler capability for cloud measurements [23]. The satellite is due to be launched in 2024.

An overview of the recent cloud and precipitation radars from space can be seen in [24].

1.1.2.3 Airborne radars

Airborne radars have also made considerable contributions to precipitation remote sensing. Some were developed to support spaceborne radars. The Airborne Rain Mapping Radar (ARMAR) is one such example [25]. It operated at 13.8 GHz and had supported the TRMM program. There were several other radars, including dual-wavelength radars at 10 GHz and 35 GHz which were used for experimental observations to support algorithm development for the spaceborne systems [26]. More recently, the Airborne Second Generation Precipitation Radar (APR-2) operating at Ku- and Ka-bands was developed to support GPM-related activities [27]. It has scanning capability, Doppler capability, and can also detect cross-polar back scatter. The system has been used for several campaigns [28,29] as well as ground-validation campaigns such as the aforementioned GCPEx field campaign [30].

There are also airborne radars operating at frequencies as high as W-band focused more on cloud physics as well as light precipitation and snow studies. The Wyoming Cloud Radar is one of the more well-known systems. It provides high-resolution measurements of reflectivity, differential reflectivity, LDR, and radial velocity. Examples of campaigns can be found in [31] and [29].

Engineering aspects of high-altitude airborne weather radars include specialized hardware that operates at very low temperatures apart from pressurization of transmitter and front-end to survive low-pressure environment. This necessitates designing pressure vessels and conduction cooling for high-power transmitters. Chapter 7 describes the engineering challenges of several airborne weather radars developed by the NASA Goddard Space Flight Center for high-altitude (>18 km) operations during the last 15 years. These systems range in operating frequencies from X- to W-band; beams from fixed nadir to scanning that entails various innovations in Doppler algorithms [32]; transmitters from Traveling Wave Tube (TWT) to solid-state; and antenna structures from a reflector to reflect-arrays. The systems include X-band (9.6 GHz) ER-2 Doppler Radar (EDOP), W-band (94 GHz) Cloud Radar System (CRS), Ku (13 GHz)/Ka-band (35 GHz) High-altitude Wind and Rain Airborne Profiler (HIWRAP), and the X-band (9.6 GHz) ER-2 X-band Radar (EXRAD). These have been mounted on various aircraft such as the ER-2, WB-57, and Global Hawk Unmanned Aerial System (UAS). In particular, HIWRAP has also participated in select GPM campaigns [33].

1.1.3 Advances in signal processing

Traditionally, weather radar signal processing has been characterized by the random and volumetric nature of meteorological targets. Further, the dual-polarimetric received echoes from weather are characterized by a narrowband, bivariate, complex Gaussian process. These characteristics had a major bearing on various aspects of transmit waveform and receiver design of weather radar processors such as spectral moment estimation, Doppler algorithms, clutter filtering, and pulse compression [34–36]. There have been several improvements in each one of these steps over the past few decades. Chapter 3 “Developments in solid-state weather radar” provides an overview of advances in pulse compression algorithms. Chapter 7 “NASA high altitude airborne weather radars” mentions the challenges and new techniques for Doppler estimation. Chapter 11 “The mitigation of ground clutter” summarizes the whitening and clutter mitigation techniques employed in weather radars. In the following, we highlight some major leaps in recent times that have brought a new perspective to weather radar signal processing.

1.1.3.1 Waveform design: from convex to nonconvex optimization

Conventional pulse compression technique is used to achieve high range resolution while also maintaining the maximum unambiguous range and reducing the peak transmit power. Here, the transmitter sends a long-duration modulated waveform while achieving the range resolution of an unmodulated short pulse. Then, in the digital domain, the sampled received signal is correlated with a replica of the transmit signal such that it optimizes a certain performance metric such as the signal-to-noise ratio (SNR) or peak-to-sidelobe ratio of the output. First proposed in the 1950s, pulse compression has been widely used in hard-target radars [37] and lidars [38]. However, its application to weather radars gained salience with the advent of solid-state transmitters in mid-2000s [39]. Since then, waveform design based on various pulse compression has been developed in many different forms.

Initial pulse compression-based waveform design employed known codes. However, this approach does not allow for flexibility to choose other performance metrics. This paves the way for the use of mathematical optimization to obtain customized codes. In the late 1990s and early 2000s, significant progress was made toward waveform design by exploiting *convex optimization* [40], wherein difficult-to-design waveforms are recast into computationally tractable problems to obtain either exact or approximate solutions through techniques such as interior-point methods. However, as more hardware- or application-specific constraints are included in such problems, often the resulting optimization becomes *nonconvex*.

Assume that the transmit fast-time code with N coded sub-pulses is $\mathrm{x} = [x_1, x_2, \dots, x_N]^T \in \mathbb{C}^N$. Then, the discrete-time received signal $\mathrm{y} \in \mathbb{C}^N$ after sampling from any range bin is [41]

$$\mathrm{y} = \alpha_0 \mathrm{x} + \underbrace{\sum_{\substack{k=-N+1 \\ k\neq 0}}^{N-1} \alpha_k \mathrm{J}_k \mathrm{x}}_{\text{interference due to radar waveform}} + \nu,$$

where α_0 is the complex-valued scalar that is proportional to the reflectivity of the scatterer of interest, the complex-valued coefficients α_k are proportional to the reflectivity of the adjacent range bins illuminated by the radar waveform, J_k be $N \times N$ shift matrix, and ν is the additive white Gaussian noise (AWGN) vector. The second (interference) term is not present in conventional weather radars that do not employ pulse compression. Our goal is to design a linear finite impulse response receive filter $\mathrm{w} \in \mathbb{C}^N$ to estimate α_0. The output of the filter becomes $z = \mathrm{w}^H \mathrm{y}$, where $(\cdot)^H$ denotes conjugate transpose.

Setting $\mathrm{w} = \mathrm{x}$ results in a classical matched filter such that

$$\mathrm{z} = \mathrm{x}^H \mathrm{y} = \alpha_0 \mathrm{x}^H \mathrm{x} + \sum_{\substack{k=-N+1 \\ k\neq 0}}^{N-1} \alpha_k \mathrm{x}^H \mathrm{J}_k \mathrm{x} + \mathrm{x}^H \nu.$$

By setting $\|\mathrm{x}\|^2 = 1$, we readily obtain the estimate of α_0 as $\widehat{\alpha}_0 = \mathrm{x}^H\mathrm{y}$. This estimation requires the absence of sidelobes and the performance of the estimator is measured by the received signal-to-interference-plus-noise ratio (SINR). Alternatively, one may consider minimizing the sidelobes by solving the following optimization problem for the transmit waveform design:

$$\begin{cases} \underset{\mathrm{x}}{\text{minimize}} & \sum_{k=1}^{N-1} \left|\mathrm{x}^H \mathrm{J}_k \mathrm{x}\right|^2 \\ \text{subject to} & \mathrm{x} \in \Omega \end{cases}$$

where application- or transceiver-specific constraints are encapsulated in the set Ω [42]. The problem above is the conventional integrated sidelobe (ISL) minimization that is more prevalent in weather radar waveform design.

In *mismatched filtering*, larger filter lengths are used to spread the autocorrelation sidelobe power across larger lag coefficients. Here, assume the augmented waveform and receive filter as $\tilde{\mathrm{x}} = \left[0_M^T, \mathrm{x}^T, 0_M^T\right]^T$ and $\tilde{\mathrm{w}} \in \mathbb{C}^{\tilde{N}}$, with filter length $\tilde{N} = 2M + N$, $M \in \mathbb{Z}_+$., longer than that of the original waveform x. The receiver filter output is

$$\tilde{z} = \alpha_0 \tilde{\mathrm{w}}^H \tilde{\mathrm{x}} + \sum_{\substack{k = -\tilde{N}+1 \\ k \neq 0}}^{\tilde{N}-1} \alpha_k \tilde{\mathrm{w}}^H \tilde{\mathrm{J}}_k \tilde{\mathrm{x}} + \tilde{\mathrm{w}}^H \tilde{v},$$

where $\tilde{\mathrm{J}}_k$ is the analogous $\tilde{N} \times \tilde{N}$ shift matrix and is the AWGN vector. In this case, the reflectivity estimate becomes $\widehat{\alpha}_0 = \frac{\tilde{z}}{\tilde{\mathrm{w}}^H} \tilde{\mathrm{x}}$. The performance metric may also be chosen as mean-square error because minimizing the MSE is equivalent to maximizing the SINR [43]. The optimum filter is obtained by solving the optimization problem,

$$\underset{\tilde{\mathrm{w}}}{\text{minimize}} \frac{\tilde{\mathbf{w}}^H \mathbf{R} \tilde{\mathbf{w}}}{|\mathbf{w}^H \tilde{\mathbf{x}}|^2},$$

whose solution has a closed-form known as *minimum variance distortionless response* (MVDR) : $\tilde{\mathrm{w}}^\star = \mathrm{R}^{-1}\tilde{\mathrm{x}}$.

However, the MSE-based filter requires a prior knowledge of noise variance. This is avoided through the design of mismatch ISL or PSL filters that usually lead to better estimates. In the case of the former, define the mismatch ISL $= \sum_{\substack{k = -\tilde{N}+1 \\ k \neq 0}}^{\tilde{N}-1} \frac{\left|\tilde{\mathrm{w}}^H \tilde{\mathrm{J}}_k \tilde{\mathrm{x}}\right|^2}{\left|\tilde{\mathrm{w}}^H \tilde{\mathrm{x}}\right|^2}$ [41]. The resulting optimization usually jointly design the transmit waveform and filters as

$$\begin{cases} \underset{\tilde{\mathrm{x}}, \tilde{\mathrm{w}}}{\text{minimize}} & \dfrac{\sum_{\substack{k = -\tilde{N}+1 \\ k \neq 0}}^{\tilde{N}-1} \left|\tilde{\mathrm{w}}^H \tilde{\mathrm{J}}_k \tilde{\mathrm{x}}\right|^2}{\left|\tilde{\mathrm{w}}^H \tilde{\mathrm{x}}\right|^2} \\ \text{subject to} & \mathrm{x} \in \Omega \end{cases}$$

On the other hand, the mismatch PSL optimization takes the form [41,44]:

$$\begin{cases} \underset{\tilde{\mathrm{x}},\tilde{\mathrm{w}}}{\text{minimize}} & \max_{k\neq 0} \dfrac{\left|\tilde{\mathrm{w}}^H \tilde{\mathrm{J}}_k \tilde{\mathrm{x}}\right|^2}{\left|\tilde{\mathrm{w}}^H \tilde{\mathrm{x}}\right|^2} . \\ \text{subject to} & \mathrm{x} \in \Omega \end{cases}$$

This is a nonconvex min-max optimization, which is tackled through various relaxation methods such as recently proposed methods such as block coordinate descent (BCD) and second-order cone programming (SOCP).

1.1.3.2 Beamforming algorithms for phased array weather radar

Recently, phased array weather radars (PAWRs) have been introduced to bring agility and efficiency in the scanning operations of weather radars that usually relied on huge parabolic dish antennas and mechanical scanning. A phased array radar comprises several antenna elements that form a highly directional beam without requiring any mechanical motion. The array achieves beam-steering electronically by adjusting the relative phase of excitation in its constituent elements. Chapters 12 of Volume 1 “Polarimetric planar phased array radar – challenges for observing weather” and Chapter 1 of Volume 2 “Phased array weather radar developed in Japan” describe the system-wide PAWR developments.

Consider a uniformly spaced phased array antenna that has N elements. Then, for the lth snapshot (or pulse repetition interval), the complex-valued discrete-time received signal $N \times 1$ column vector $\mathbf{x}_l = \left[x_{l,0}, \cdots, x_{l,N-1}\right]^T$ are

$$\mathbf{x}_l = \mathbf{A}\mathbf{s}_l + \mathbf{n}_l,$$

where $\mathbf{s}_l = \left[s_{l,0}, \cdots, s_{l,M-1}\right]^T$ is the source signal vector of M samples from a given range bin, $\mathbf{n}_l$ is additive spatio-temporal white Gaussian noise, $\mathbf{A} = [\mathbf{a}(\theta_0), \cdots, \mathbf{a}(\theta_{M-1})]$ is the $N \times M$ receive steering matrix with the ith column

$$\mathbf{a}(\theta_i) = \begin{bmatrix} 1 \\ e^{-j\frac{2\pi}{\lambda}\sin\theta_i} \\ \vdots \\ e^{-j\frac{2\pi}{\lambda}(N-1)\sin\theta_i} \end{bmatrix},$$

where d is the channel inter-element spacing, λ is the operating wavelength of the radar, and $(\cdot)^T$ denotes transpose operation. The received signals $\mathbf{x}_l$ are combined to form complex estimates $\{y_{l,m}\}_{m=1}^{M-1}$ of the precipitation profile from the direction θ_m as

$$y_{l,m} = \mathbf{w}^H(\theta_m)\mathbf{x}_l,$$

where $(\cdot)^H$ is the conjugate transpose and $\mathbf{w}(\theta_m), m = 1, \cdots, M$ are carefully selected weight vectors. The goal of the beamforming algorithm is to determine $\mathbf{w}(\theta_m)$.

Once the estimates $\{y_{l,m}\}_{m=1}^{M-1}$ are available for L snapshots, standard moments such as the reflectivity (or power), mean Doppler velocity, and spectrum width are computed from the autocovariance of $\{y_{l,m}\}_{m=1}^{M-1}$ through the pulse-pair processing method.

The radar estimates the direction-of-arrival (DoA) along the precipitation profiles by adopting spatial filters called *beamformers*. A common beamforming method to determine the weight vector is the Fourier Beamformer. This is equivalent to spatial matched filtering. So, the weight vector is simply a copy of the corresponding steering vector, i.e., $\mathbf{w_{FR}}(\theta_m) = \mathbf{a}(\theta_m)/N$. Here, a uniform phase shift is applied to steer the beam in a specified direction. This computationally simple method is the most optimal in the presence of white noise. However, its precipitation estimates could be considerably biased because of high sidelobe levels. Recently, adaptive beamforming techniques have been employed in phased-array-based atmospheric and weather radars to mitigate the sidelobes of conventional beamformers [45,46]. These methods have been previously investigated for point-target surveillance radars in detail; their application to meteorological targets is very recent.

A few studies [45,47,48] discuss improvement in sidelobe levels by applying Capon and MVDR beamformers that are capable of rejecting interfering signals such as clutter from directions other than the direction-of-interest. This method requires only the knowledge of the desired signal direction of arrival and determines the weight vector by solving the optimization

$$\begin{aligned} &\underset{\mathbf{w}(\theta_m)}{\text{minimize}}\ \mathbf{w}^H(\theta_m)\mathbf{R}_{xx}(0)\mathbf{w}(\theta_m) \\ &\text{subject to } \mathbf{w}^H(\theta_m)\mathbf{a}(\theta_m) = 1, \end{aligned}$$

where $\mathbf{R}_{xx}(0) = \frac{1}{l}\sum_{l=1}^{L}\mathbf{x}_l\mathbf{x}_l^H$ is the lag-0 autocorrelation of the received signal. The solution of this problem is

$$\mathbf{w_{CP}}(\theta_\mathbf{m}) = \frac{\mathbf{R}_{xx}^{-1}(0)\mathbf{a}(\theta_m)}{\mathbf{a}^H(\theta_m)\mathbf{R}_{xx}^{-1}(0)\mathbf{a}(\theta_m)}$$

This approach requires sufficiently high number of samples to accurately estimate the sample covariance matrix and is prone to steering vector mismatch. In [46], a minimum-mean-squared-error (MMSE) beamformer is proposed that employs prior information to reduce the required number of pulses in the covariance estimation process.

The performance of most adaptive beamformers degrades when there is an imprecise knowledge of the steering vector, sample size is small, or prior information is unavailable. To mitigate these disadvantages, robust adaptive beamforming (RAB) such as diagonal loading [49] may also be used. Since the beamformers involve inversion of a covariance matrix, the numerical stability of the weight vectors is affected while inverting a matrix with small eigen values. Diagonal loading helps with the stability and reduces the spread in weight amplitudes. In this method, a quadratic inequality constraint of bounding the weight coefficients is added as follows:

$$\begin{aligned} \underset{\mathbf{w}(\theta_m)}{\text{minimize}} \quad & \mathbf{w}^H(\theta_m)\mathbf{R}_{xx}(0)\mathbf{w}(\theta_m) \\ \text{subject to} \quad & \mathbf{w}^H(\theta_m)\mathbf{a}(\theta_m) = 1 \\ & \mathbf{w}^H(\theta_m)\mathbf{w}(\theta_m) \le T, \end{aligned}$$

where T is the constant norm constraint. Solving this problem yields

$$\mathbf{w}_{DL}(\theta_m) = \frac{(\mathbf{R}_{xx} + \alpha\mathbf{I})^{-1}\mathbf{a}(\theta_m)}{\mathbf{a}^H(\theta_m)(\mathbf{R}_{xx} + \alpha\mathbf{I})^{-1}\mathbf{a}(\theta_m)},$$

where α is the diagonal-loading parameter and $\mathbf{I}$ is an identity matrix. The value of $\alpha = 0$ corresponds to Capon/MVDR beamformer while $\alpha = \infty$ denotes non-adaptive beamforming. The intermediate values of α that minimize the total output power are determined based on a set of constraints such that the SNR and SINR degradation is minimum.

Diagonal loading has been employed for other volumetric target sensors such as wind profilers. For example, the norm-constrained directionally-constrained minimization-of-power (NC-DCMP) beamformer used in [48,50,51] for middle and upper troposphere measurements is similar to the classical diagonal loading. It is dependent on the norm constraints whose values are empirically determined in advance. An optimized version of the NC-DCMP was proposed in [47] to increase SINR for enhanced probability of detection. This algorithm evaluates residual clutter and noise powers and is especially useful for weak received signals.

1.1.3.3 Sparse reconstruction methods

While matched filtering or pulse compression reliably estimates target parameters, their resolution is inversely proportional to the support of the ambiguity function of the transmit pulse, thereby restricting ability to super-resolve targets that are closely spaced. Further, matched filtering requires the signal to be sampled at or above the Nyquist sampling rate that guarantees perfect reconstruction of a bandlimited analog signal. As the bandwidth increases, the Nyquist sampling rate being twice the baseband bandwidth also increases, thereby requiring the radar receiver to employ expensive, high-rate analog-to-digital converters (ADCs). The sampled signal is then also processed at high rates, resulting in significant power, cost, storage, and computational overhead. Analogous trade-offs are present in dwell-time and Doppler resolution; array elements and DoA resolution; and sweep rate and resolution [52].

Despite the initial massive time series data acquisition by the weather radar, the end-products are often reasonably downsized in volume leading to a pertinent question: Could lesser samples have been acquired in the first place? During the past 15 years, this bottleneck has led to the development of sparse reconstruction algorithms that exploit the theory of *compressed sensing* (CS) [53]. CS is a novel signal processing technique that unites sampling and digital data compression in a single step by relying on the inherent sparsity of the data in some dictionaries. While conventional signal processing methods sample at Nyquist-Shannon rate and then compress the data for minimal storage, CS allows sampling of only useful information at lower sampling rates. CS-based radar signal processing is shown to reduce receiver hardware complexity by eliminating pulse compression [52], decrease data sampling cost [54], and enhance target detection resolution [55].

While the initial goal of sparse processing in weather radars was to allow for smaller dwell times without significant loss of target information [56], other applications include three-dimensional imaging [57], frequency-modulated continuous-wave weather radar [58], and refractivity retrieval [59]. The CS techniques developed for the hard target radars cannot be directly applied to weather radars because the precipitation echoes may not necessarily have sparsity in either range-time (see Figure 1.3) or even Doppler domain. In [56] and [57], this issue is resolved by modeling the remote sensing of weather as a low-rank matrix completion problem because weather radar returns exhibit high spatial and temporal correlations. Low-rank matrices are the multidimensional equivalents of one-dimensional sparse vectors. The sparsely sampling of weather scenario would lead to the missing entries that are recovered using matrix completion methods. The spectral moments are then estimated through classic radar signal processing.

Given a matrix $M \in \mathbb{R}^{m \times n}$, its singular value decomposition (SVD) is given by $M = USV^T, U \in \mathbb{R}^{m \times \imath}, V \in \mathbb{R}^{\imath \times} n$, and $S = \text{diag}(\sigma_1, \cdots, \sigma_\imath)$, where $\sigma_1 \geq \cdots \geq \sigma_\imath > 0$ is the unique singular values and $\imath \leq \min(m, n)$ is the rank of the matrix. For a low-rank matrix, most of the diagonal elements of S are zero such that $\imath \ll \min(m, n)$. The best $\imath'$-rank approximation $\tilde{M}$ of the matrix M is given by zeroing out the $\imath - \imath'$ smallest singular values so that $\tilde{M} = U\tilde{S}V^T$ and $\tilde{S} = \text{diag}\left(\sigma_1, \cdots, \sigma_{\imath'}, 0, \cdots, 0\right)$.

As an illustration, in Figure 1.4 we plot the singular values of the spatially non-sparse 1930 (range gates) by 413 (azimuthal rays) horizontal reflectivity Z_h data matrix corresponding to the actual Iowa XPOL-2 radar data of Figure 1.3(d). We observe that most of the singular values are very small or close to zero (due to highly correlated spatial samples of weather backscatter). This makes it possible to apply matrix completion to fully recover a low-rank matrix from a randomly observed sample of its entries.

Let Ω denote the set of the random locations of the partially observed entries of the original low-rank matrix M. Then, recovering M corresponds to the rank minimization problem:

$$(P_0) \quad \begin{array}{c} \text{minimize } rank(\boldsymbol{X}) \\ \text{subject to } X_{ij} = M_{ij}, (i,j) \in \Omega \end{array} \tag{1.1}$$

However, like ℓ_0 minimization, rank minimization is also intractable. The approach to low-rank matrix completion is therefore to minimize the matrix equivalent of ℓ_1, i.e., the nuclear-norm $\|\text{X}\|_* = \sum_{\text{k}} \sigma_{\text{k}}$,

$$(P_1) \quad \begin{array}{c} \text{minimize } \|\boldsymbol{X}_*\| \\ \text{subject to } X_{ij} = M_{ij}, (i,j) \in \Omega \end{array} \tag{1.2}$$

In Figure 1.5, we reconstruct the sparsely sampled meteorological reflectivity using singular value thresholding (SVT) for nuclear norm minimization. Although

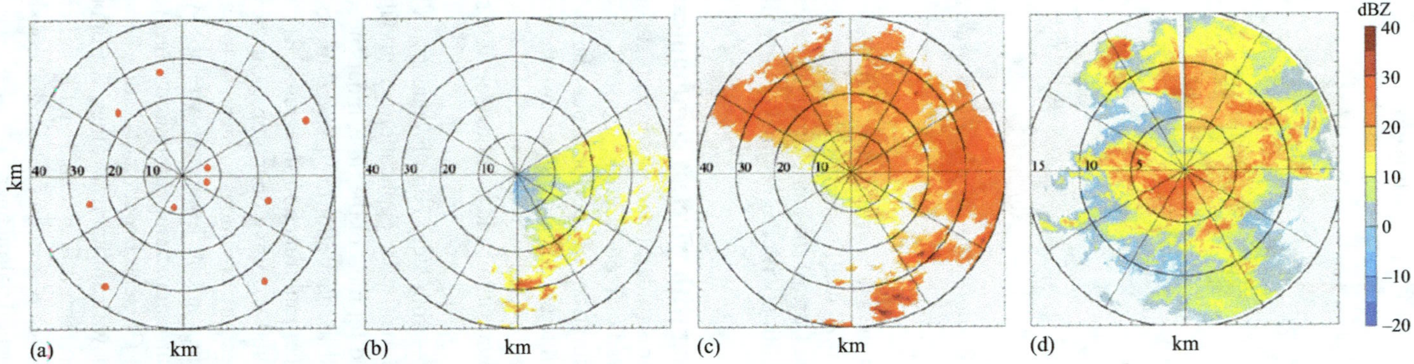

Figure 1.3 *(a) Point targets on a radar display. (b) Spatially sparse precipitation (Iowa XPOL-2 radar data observed on 0139 UTC, June 13, 2013). (c) Precipitation echoes are sparse along only a few range profiles (0150 UTC, June 13, 2013). (d) Precipitation returns do not exhibit spatial sparsity (2308 UTC, June 12, 2013).*

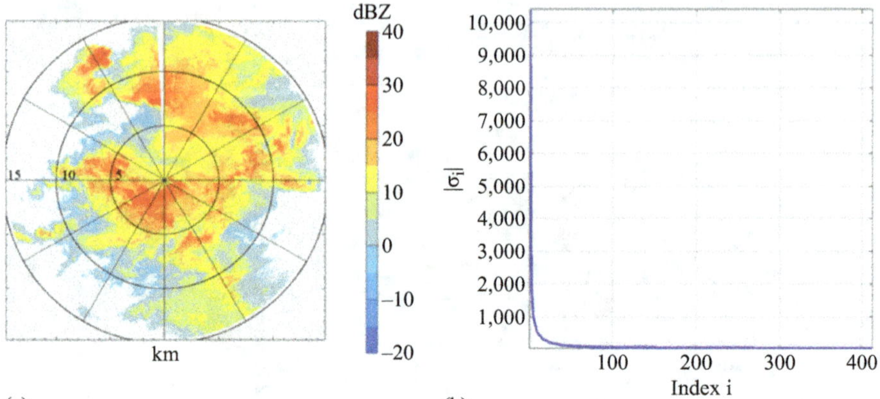

(a) (b)

Figure 1.4 (a) Original Z_H data and (b) plot of its ordered singular values.

the illustrated reconstruction uses a low-rank approximation $\tilde{Z}_h$ as matrix M, the results are not very different if Z_h itself is used with some modifications to problem P_1 (the relative errors ε_1 and ε_2 are of the same order). The very close similarity of the reconstructed data distribution with the original clearly illustrates the potential of CS for weather radars.

1.1.3.4 Spectrum sharing with wireless communications

In recent years, sensing systems (radar, lidar, or sonar) that share the spectrum with wireless communications (radio-frequency/RF, optical, or acoustical) and still operate without any deterioration of performance have captured significant research interest [60]. The interest in such spectrum sharing systems is largely because the spectrum offered by the wireless media is a scarce resource, while performance of both communication and remote sensing systems improves by exploiting wider spectrum.

Several frequency bands—from Very High Frequency (VHF) to terahertz—are allocated exclusively for different radar applications. Although a large fraction of these bands remains underused for most of the time, radars need to maintain constant access to these bands as well as gain additional spectrum to accomplish missions such as secondary surveillance, multifunction integrated RF, and communications-enabled autonomous driving and cognitive operations. On the other hand, the wireless industry's demand for spectrum continues to increase toward serving more users with data rates higher than the current. Emerging wireless systems such as Long-Term Evolution (LTE) cellular commercial communications technology, 5G, WiFi, Internet-of-Things (IoT), and Citizens Broadband Radio Services already face the problem of sharing spectrum with military, weather, astronomy, and aircraft surveillance radars (ASRs). Therefore, it is essential and beneficial for both radar and communications systems to develop strategies to simultaneously operate in the same spectral range in a mutually beneficial manner.

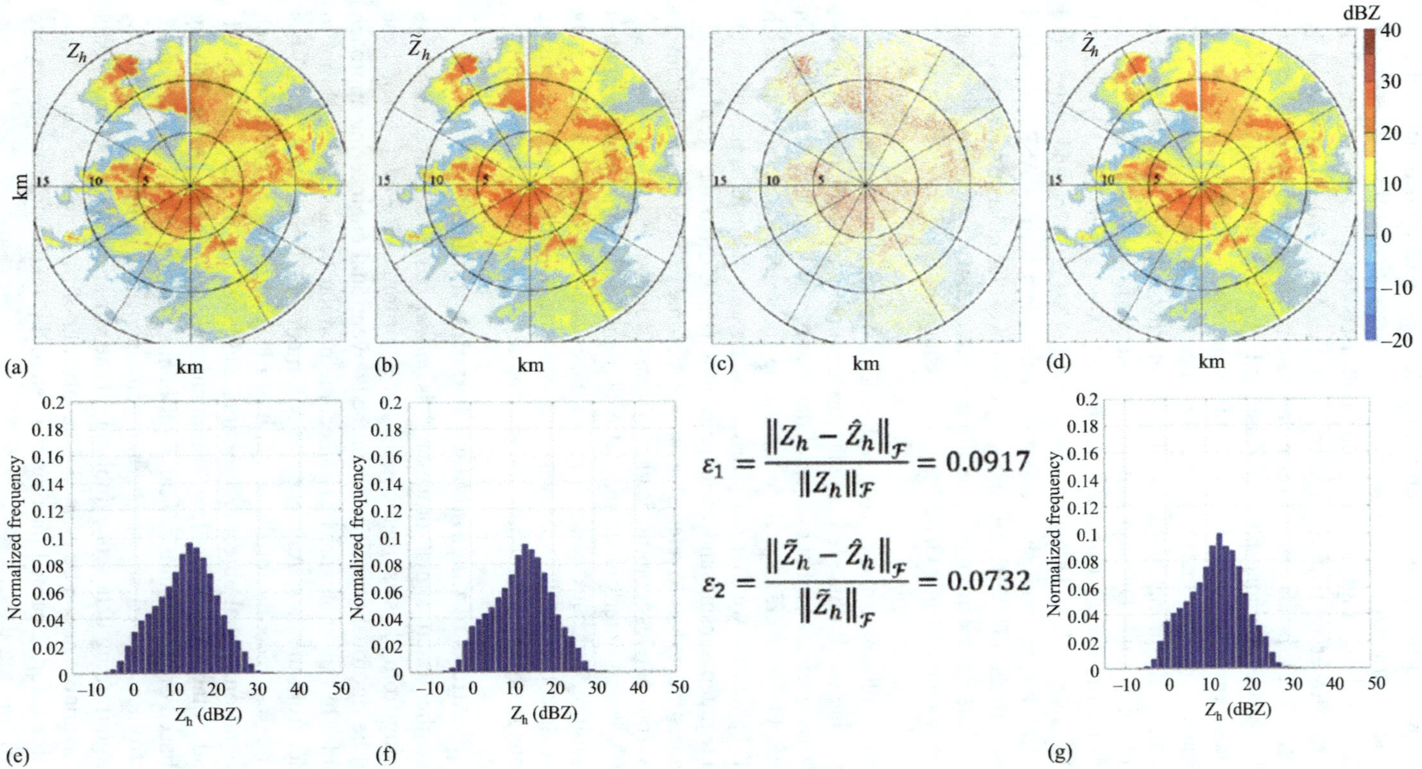

Figure 1.5 (a) Original Z_h, (b) Low-rank approximation $\tilde{Z}_h$ of original Z_h obtained by retaining 25% of the most significant singular values, (c) Randomly sampled one-third entries of $\tilde{Z}_h$, (d) Reconstructed reflectivity $\widehat{Z}_h$ using SVT. Histograms of data corresponding to (e) Z_h (f) $\tilde{Z}_h$, and (g) $\widehat{Z}_h$. Two relative error metrics ε_1 and ε_2 are computed. $\|\cdot\|_{\mathcal{F}}$ denotes the Frobenius norm.

Significant synergistic efforts are currently underway by major national entities for efficient radio spectrum utilization. The National Science Foundation has sponsored several projects that bring together many different users for a flexible access to the spectrum. At the lower end of the spectrum in the VHF (30–300 MHz), UHF (300–1000 MHz), and L-bands (1–2 GHz), radar systems such as FOliage PENetration (FOPEN) radar, astronomy/ionosphere radars, and Air Route Surveillance Radar (ARSR) have been encountering and managing interference from the broadcast and TV stations for decades now. The spectral congestion in centimeter wave bands (S-, C-, X-, Ku-, and K-) arose later, primarily due to LTE waveforms, e.g. 802.11b/g/n (2.4 GHz) WCDMA (Wide-band Code Division Multiplexing Access), WiMAX LTE, LTE GSM (Global System for Mobile comm), EDGE (Enhanced Data rates for GSM Evolution), 802.11a/ac VHT (Very High Throughput) wireless LAN (WLAN), and commercial flight communications that now share spectrum with radars such as ASRs, Terminal Weather Doppler Radar (TDWR) network, and other weather radars. Nowadays, it is the higher end of the RF spectrum or the millimeter wave (mmWave), formally defined with the frequency range 30–300 GHz, that is witnessing concerted efforts for spectrum management [60].

In [61], a comprehensive analysis of interference from wireless communications to weather radars is included. While few cellular services do not currently majorly impact weather radar frequencies, a partial overlap of the spectrum still leads to degradation in the performance. Ongoing efforts include sensitizing spectrum regulatory bodies such as FCC and ITU about the weather radar operations; a more futuristic approach would be to adopt integrated sensing and communications (ISAC) into the weather radar design. We suggest the reader to refer Chapter 6 of Volume 3 “Spectral interference in weather radars from wireless communication systems” for details.

However, wireless interference may also be beneficial. Some current studies show that exploiting the backscatter from wireless communications signals may enhance target detection [62]. Moreover, opportunistic sensing of weather using communications is now a considerably large community (see also Section 1.6).

1.1.3.5 Applications of machine/deep learning models

In general, weather radar signal processing is *model-based* because it follows models stemming from physics, mathematics, or algorithms. However, when these models cannot be rigorously specified, then the conventional engineering approach is not applicable. In this context, ML has become popular because of its ability to learn the model from the data. As a class of ML methods, deep learning (DL) has gained much interest recently for the solution of many challenging problems such as speech recognition, visual object recognition, and language processing [63]. DL has several advantages such as low computational complexity when solving optimization-based or combinatorial search problems and the ability to extrapolate new features from a limited set of features contained in a training set. Usually, sufficiently large training data sets exhibiting all the variation in the observed data sets are either available or can be created. The labeling of training data can be done with a reasonable effort. Learned phenomenon should remain stationary to acquire large amount of training data.

Typical meteorological radar applications of DL include rainfall estimation [64,65], nowcasting [66], object classification [67], and opportunistic sensing [68]. Note that most scenarios employ DL in a narrow sense rather than replacing the entire processing with a learning network. Tasks that do not require explicit reasoning based on broader background knowledge, a rigorous quantitative performance guarantees, or explicit explanations for how the result was found are appropriate for DL applications. Chapter 12 in Volume 2: "Deep-learning-aided rainfall estimation from communications satellite links".

1.1.3.6 THz-band signal processing

Lately, there is a greater interest in terahertz (THz)-band weather observations. The atmospheric radar community has intensified research on G-band radars (110–170 GHz) [69,70]. However, interest in high-resolution low-THz (100–300 GHz) automotive radar is also catalyzing research on THz-band weather sensing (see Chapter 5 of Volume 3 "The effect of weather on the performance of mm-wave and sub-THz automotive radar"). Further, the availability of very wide contiguous bandwidths in this band is also of interest to ISAC community [71].

The unique nature of THz propagation, scattering, attenuation, and bandwidth necessitates tailoring the signal processing for automotive radars to address and exploit two key features of this band: attenuation and wide bandwidth. The first characteristic mandates short-range deployment of the radar. Further, high propagation losses and power limitations at THz are compensated by the beamforming gains obtained through the deployment of extremely dense antenna arrays [71]. Another THz characteristic, i.e., wide modulation bandwidth of up to 10 GHz introduces significant linear amplitude and phase impairments from both the radar hardware and the channel causing range-profile blurring and/or synchronization issues. Further, wideband Doppler processing, distance-dependent bandwidth, unusual nature of THz-band noise, and wideband beamforming are additional features of THz radar signal processing [72]. At THz band, *specular* scattering becomes less dominant and there is an increasing sensitivity to surface texture/roughness [73].

1.1.4 Disdrometers supporting radar observations

Disdrometers have played an important role in radar meteorology. It is well known that the weather radar measures the backscatter from dielectric scatterers within a given pulse volume. In the case of rain, this backscatter reflectivity depends on the drop size distribution (DSD). Some of the oldest and the most quoted studies about rain DSDs were conducted by Marshall and Palmer in 1948 [74], and even earlier by Laws and Parsons in 1943 [75].

DSDs govern the relationship between radar reflectivity (Z) and rainfall rate (R). The problem is that there are countless numbers of Z-R relationships that have been quoted in the literature [76]. Many of these relationships were based on ground-based measurements with instruments such as disdrometers. There are different types of disdrometers. One of the earliest instruments developed for DSD measurements is the (impact-type) [77]. This was extensively used for many different field campaigns for more than 30 years. In the last 25 years however, several other disdrometers, with different techniques, have been developed. These include Parsivel disdrometer [78], 2D video disdrometer [79], often abbreviated to 2DVD, and optical disdrometer (ODM) [80], and the meteorological particle spectrometer (MPS) [81]. There are other instruments such as micro rain radar (MRR) [82] and Precipitation Occurrence Sensor System (POSS) [83] which can provide useful information on DSDs. Whilst MRR is a vertically pointing radar (with Doppler capability), POSS is a bistatic radar designed to derive DSDs as well as other hydrometeor types within a "common volume" [83–85].

Of these, the 2DVD gives the most comprehensive information about size, shape, and fall velocity for each drop falling through its sensor area [79,86]. It has been used for DSD studies in many different countries [87,88] and for investigating drop shapes and fall velocities [89,90], especially during turbulent conditions. Another study [91] which examined the prevalence of large drops from 2DVD measurements of more than 240 million drops from 18 climatically diverse geographical locations found a giant drop with equivalent drop diameter of 9.7 mm. Figure 1.6 shows the shape of this drop reconstructed from the contours recorded by the two orthogonal cameras of a 2DVD installed in northern Oklahoma. This was verified as a fully melted hydrometer and "occurred beneath a hail-producing cloud" that moved across the disdrometer site on April 29, 2012. The shape is very close to the theoretically expected equilibrium shape [92] and has a fall velocity to the Gunn-Kinzer [93] value for the specified equivalent drop diameter, D_{eq}. Drop shapes are of course important for estimating rainfall rates from polarimetric radar measurements [34].

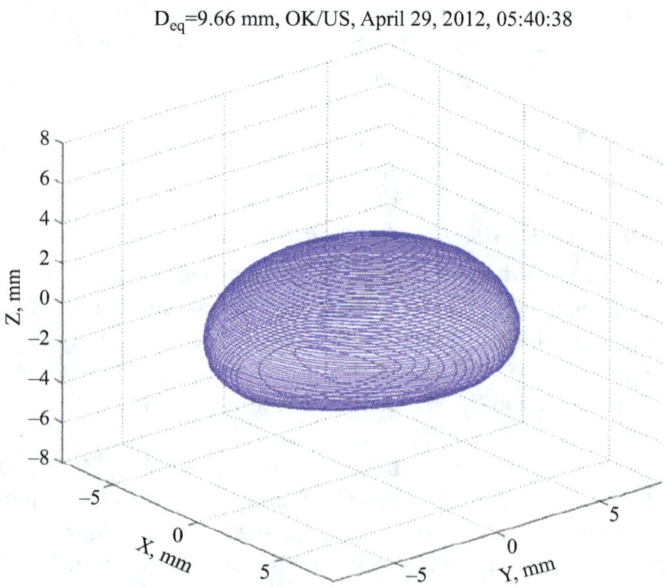

Figure 1.6 A 3D reconstruction of the 9.7 mm raindrop that was recorded by the 2DVD at a site in Oklahoma at 0540 UTC, April 29, 2012 [89].

Disdrometers have been used not only for deriving DSDs in rain but also for external radar calibration [94,95]. However, spatial and temporal correlations of DSDs need to be considered since radar pulse-volume is much larger than the sensor areas of disdrometers and since radar measures instantaneous "snapshot" of reflectivity whereas disdrometers require integration times of a few minutes, viz., 1–3 minutes.

Many field experiments have made use of combined radar observations and ground-based disdrometer measurements. For example, there have been several ground validation (GV) campaigns to support satellite radar measurements such as the GPM-DPR. The GV campaigns were conducted in several climatically different locations within the US, e.g. (a) the Mid-latitude Continental Convective Clouds Experiment (MC3E) in central Oklahoma, and (b) Iowa Flood Studies (IFloodS) covering parts of north-eastern Iowa, as well as (c) outside the US such as the GPM Cold Season Precipitation Experiment (GCPEX) conducted in Ontario, Canada.

MC3E was aimed at improving the representation of convective lifecycle in atmospheric models via 3D characterization of convective clouds. This took place in the late spring of 2011. It also assessed the reliability of satellite-based retrievals of precipitation for such systems. Polarimetric radars at S-, C-, X-, Ku-, Ka-, and W-bands, as well as dense networks of rain gauges and disdrometers were deployed. More information can be found in [96].

IFloodS was focused more on hydrology-related applications and to assess strengths and limitations of satellite products for flood forecasting and other related applications. This took place in the spring of 2013. Polarimeteric radars at S-, Ku-, and Ka-bands as well as several University of Iowa X-band radars were deployed along with a large network of 2DVDs, Parsivel disdrometers, rain gauges, and devices for soil-moisture measurements. Further details can be found in [97] and [98].

GCPEx was conducted during the winter season of 2011–2012 in Ontario, primarily (a) to address the limitations of GPM snowfall retrieval algorithms and (b) to assess the ability of active and passive sensors to detect and estimate falling snow. Instrumentations included C-band, Ku-band, and Ka-band, radars along with 2DVDs, Parsivel disdrometers, POSS, MRR, and snowflake video imager. The campaign also included aircraft measurements using Ku-band and Ka-band radars as well as many instruments for in-situ measurements. More details can be seen in [30].

The above-mentioned snowflake video imager [99] records greyscale images of frozen particles falling through its sensor area. Another instrument that can measure some of the characteristics of falling snow is MASC [100]. It has three cameras which get activated when a snow particle falls through the "common" view area. Several radar observation winter campaigns have included MASC [101,102]. The images have also been used to examine the variation in snow crystal riming and observed Z_{DR} at X-band [103]. Image processing and ML techniques were used for deriving the "degree of riming" from MASC digital images of snow particles during a snow event in Greeley, Colorado, and correlated with the CSU-CHILL X-band radar Z_{DR} over and surrounding the MASC location, extracted from low-elevation PPI scans. Panel (a) in Figure 1.7 shows a scatterplot of the CSU-CHILL X-band Z_{DR} over the MASC location versus the riming index determined from MASC images for two different time periods. During the first time period, more heavily rimed particles were inferred from the MASC image

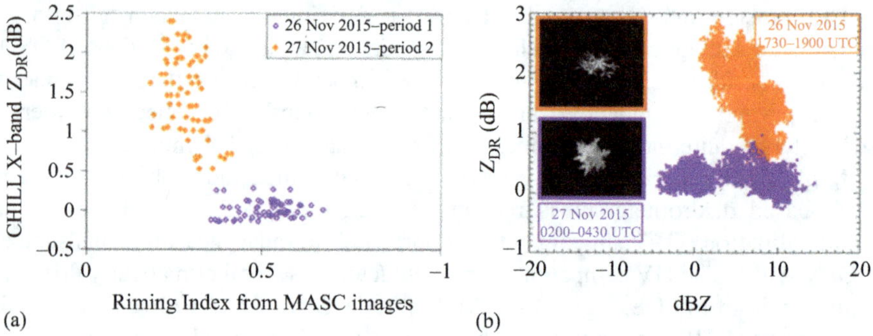

Figure 1.7 (a) Scatterplot of CSU–CHILL X-band Z_{DR} over the MASC location versus the riming index derived from the MASC particle images [103]. The Z_{DR} data were extracted from the 0.9° elevation PPI scans' azimuths that immediately flanked the MASC location. The event occurred on November 26–27, 2015. (b) Z_H–Z_{DR} variation for the same time periods from the CSU-CHILL radar. The top and bottom images inside the plot show an example from MASC for the second period (very little riming) and for the first period (significantly more riming).

analyses and are seen to be associated with near-zero Z_{DR} whereas during the second time period, particles with much less degree of riming were inferred from MASC images and are seen to be associated with distinctly positive Z_{DR}, of around +2 to 3 dB. Panel (b) shows the X-band Z_H–Z_{DR} variation for the same time periods, with the same color-code. Two MASC images are shown as examples for the two cases.

The above examples (and the references provided) have demonstrated how radar observations combined with ground instrumentations can be synergistically used to yield better understanding of precipitation microphysical properties and processes. Disdrometer data have also been used to develop algorithms for retrieving DSD moments, particularly the lower-order moments (which are important for understanding the dominance of the various microphysical processes) from polarimetric radar measurements [104,105]. The retrieval methods entail the estimation of two optimal reference moments, and a representative function for the underlying shape of the DSDs based on the generalized gamma (G-G) model [106,107]. For the reference moments, the sixth and the third moments were found to be the best pair for retrievals from X-band polarimetric radars, and their estimations were found to be best derived from the radar reflectivity for the former and a combination of differential reflectivity and specific differential propagation phase or specific attenuation for the latter. Long-term measurements of full DSD spectra from disdrometers have been used as input to G-G model fitting to determine the optimal characterization [108].

1.1.5 Scattering from hydrometeors

The last 20 years have also seen the improvement of computational techniques for calculating scattering cross-sections for hydrometeors. Earlier work made use of techniques such as point-matching algorithms [109], Fredholm integral method [110], and transition or T-matrix formulation [111]. The first two were extensively used for spheroidal-shaped raindrops, whereas the last one was additionally used for two-layer (e.g. ice core surrounded by water layer for melting ice particles) models [112]. A comprehensive survey of the methods was done by Oguchi in 1981 [113].

For non-melted hydrometeors, which tend to have more complicated 3D shapes, methods such as discrete dipole approximation (DDA) have been utilized [114]. This method partitions the dielectric scatterer into "cells," each representing a dipole. The method can be applied to inhomogeneous particles and those with sharp edges, but its main disadvantage is that it can be computationally time-consuming if high accuracy is required. Nevertheless, it is a more commonly used method for scattering calculations for most hydrometeors other than rain, including hail particles [115].

More recently, a higher-order method of moments (MoM) in the surface integral equation (SIE) formulation has been proposed for deriving single-particle (complex) scattering amplitudes for snow and ice hydrometeors [116,117]. The method has been shown to be more efficient than DDA and more versatile than T-matrix method. The method has been successfully applied to 3D shapes of snow particles reconstructed from a multiangle snowflake camera [118].

For raindrops, the T-matrix method has been utilized very often. It can be applied to drops even with finite (i.e. non-zero) orientation angles with respect to zenith. However, when raindrops undergo large amplitude oscillations including the transverse and horizontal mode oscillations [119], their shape may become "rotationally asymmetric," i.e. with shapes which don't have axis of rotational symmetry. For such cases, the above-mentioned MoM-SIE method has been applied, again successfully, but this time using reconstructed 3D shapes from 2DVD data. The technique was demonstrated using 2DVD measurements during a widespread event with an embedded line convection [120].

Scattering calculations for such asymmetric raindrops have also been performed using commercially available 3D electromagnetic (EM) analysis software packages which can provide EM solvers for wide-ranging applications. In one case study, drop shapes derived from 2DVD measurements during the passage of outer rainbands of a category-1 Hurricane were used to compute the S-band radar cross-sections (RCS) for horizontal and vertical polarizations [121]. Figure 1.8 shows the single-particle differential reflectivity for all drops > 2 mm equi-volume drop diameter, D_{eq}, shown as green dots. The $\pm$ one standard deviation (σ) are represented by red lines. The "+" points in blue represent the Z_{DR} calculations using the most probable shapes from [86]. Although they are seen to be lying within the $\pm\sigma$ lines, they seem closer to the upper red line, indicating that the reconstructed shapes result in Z_{DR} values which tend to be somewhat closer to those for more "spherical-like" shapes. Also worth noting is the $\pm\sigma$ increase with increasing drop diameter implying that the drop oscillation amplitudes increase with D_{eq}. This increase has also been noted with wind-tunnel observations [122].

1.1.6 Radiowave propagation applications

Polarimetric radar data have also been utilized successfully for evaluating radiowave propagation effects along terrestrial line-of-sight links as well as earth-space

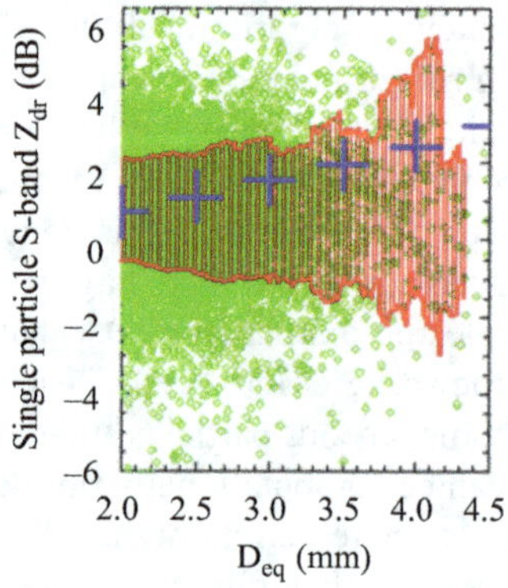

Figure 1.8 Single-particle Z_{DR} variation with D_{eq} calculated using reconstructed drop shapes (green dots), their +/− one standard deviation (shaded red), and the expected Z_{DR} from the most probable shapes (blue + points) [121].

links. Much of this work was carried out in the 1990s but work has also continued beyond 2000. A recent review article outlining the various uses of radar data for propagation research can be found in [123].

For terrestrial links, radar data have been used for evaluating not only rain attenuation effects but also rain-induced depolarizations. Examples can be found in [124] and [125]. Observations from the Chilbolton radar located in southern UK were used for both cases.

Chilbolton radar has also been used to support several other propagation programs, including the Olympus propagation experiments [126,127] conducted by the European Space Agency [128], and the Italsat program [129] conducted by the Italian Space Agency [130]. The radar was used as a "diagnostic tool" for measurements of earth-space attenuation from the satellite beacon receivers installed near the Chilbolton radar. Another topic of research where this radar was extensively used is rain-scatter interference (between terrestrial and earth-space links using the same frequency bands) [131]. In this example, Chilbolton radar data were used for some case studies to provide rainfall information for 131 km transhorizon experiment with a "bistatic" common volume (rain-filled or otherwise) between the transmit and receive antenna beams.

The CSU-CHILL S-band polarimetric radar located in Greeley, Colorado is another example where the data were used for earth-space propagation research [132]. This was conducted as part of NASA's Advanced Communications Technology Satellite (ACTS) program [133]. Convective rain cases and "bright-band" rain cases were used as case studies for evaluating earth-space propagation effects at Ka-band.

Weather radar images have also been used to determine spatial and temporal characteristics of rain cells for developing model-based approach and for predicting radiowave propagation effects [134]. Based on the analyses, a global model for generating realistic synthetic rain fields had been developed [135]. The data used for this work were obtained from the research radar based in Spino d'Adda, Italy.

The spaceborne TRMM-PR data have also provided much-needed information for earth-space link budgets. One of the input parameters required for the prediction method given in ITU-R Recommendation P.618-13 ("Propagation Data and Prediction Method Required for the Design of Earth-Space") is the freezing-height data given in ITU-R Recommendation P.839-3 ("Rain Height Model for Prediction Methods"). The latter provides a model for predicting the "rain-height" for a given location based on its latitude and longitude. This model has been tested using the melting layer heights from the TRMM database for tropical and subtropical locations [136]. It has also been tested for specific locations [137].

Microwave and millimeter wave links have also provided data which have been used to estimate rainfall rates on a "path-averaged" sense. In [138], a demonstration is provided to show how signal attenuation data from a transmitter-receiver network "can be used to retrieve the space–time dynamics of rainfall for an entire country (The Netherlands, $\sim$35,500 km^2)". Further details can be seen in Chapter 2 of Volume 3 of this book.

1.1.7 Concluding remarks

We foresee that the most anticipated development of the field is connected to a proliferation of weather radar technologies across other scientific domains. The chapter also highlights some specific issues emerging from past and current studies. Progress in understanding rain and ice microphysics is continuing with radar observations being combined with ground-based instruments. The quest for highly accurate rainfall estimates is also expected to continue and yet there is no combination of polarimetric measurements that can clearly outperform any of the others for a given spatial and temporal resolution. The other issue is the question of representativeness or point-to-area variance that depends on the spatial correlation function which is not well characterized. Finally, the algorithm and measurement errors must be estimated to determine which component is dominant.

The chapter demonstrates that weather radars continue to benefit from new theories of signal processing and learning techniques. We anticipate that this process will also go the other way around: that is, novel signal processing methods and applications derived from the more widespread use of weather radars will also contribute to other research topics. For example, nonlinear, non-Gaussian, model-free, spectrum-sharing, and distributed signal processing techniques are likely to yield new algorithms in the context of weather sensing.

There is also much room to improve global-scale modeling of rainfall through the use of current and future satellite-borne weather radar missions. The relationships of observables and rainfall are also expected to be different for sea and land; tropics, mid-latitudes, and polar regions; and across all parts of microwave spectrum.

List of acronyms/abbreviations

2DVD	2D video disdrometer
ACTS	Advanced Communications Technology Satellite
APR-2	Airborne Second Generation Precipitation Radar

ARMAR	Airborne Rain Mapping Radar
AUT	antenna under test
CSIRO	The Commonwealth Scientific and Industrial Research Organisation
CSU-CHILL	Colorado State University CHILL radar
DDA	Discrete dipole approximation
DoD	Department of Defense
DPR	Dual-wavelength Precipitation Radar
DSD	Drop size distribution.
EarthCARE	Earth Cloud Aerosol and Radiation Explorer
EM	electromagnetic
FAA	Federal Aviation Administration
GCPEx	GPM Cold Season Precipitation Experiment
GPM	Global Precipitation Measurement Mission
GV	Ground validation
IFloodS	Iowa Flood Studies
ITU-R	ITU Radiocommunication Sector
MASC	multiangle snowflake camera
MC3E	Mid-latitude Continental Convective Clouds Experiment
MoM	method of moments
MPS	Meteorological Particle Spectrometer
MRR	Micro Rain Radar
NASA	The National Aeronautics and Space Administration
NEXRAD	Next Generation Weather Radar
NOAA	National Oceanic and Atmospheric Administration
NSSL	National Severe Storms Laboratory
NWS	National Weather Service
ODM	optical disdrometer
OMT	Orthomode Transducer
POSS	Precipitation Occurrence Sensor System
PR	Precipitation Radar
ROC	Radar Operation Center
RCS	radar cross-section
SIE	Surface integral equation
T-matrix method	Transition-matrix method
TAC	Technical Advisory Committee
TRMM	Tropical Rainfall Measurement Mission
WSR-88D	Weather Surveillance Radar 88 Doppler

References

[1] NEXRAD Technical Advisory Committee, "*In support of the next generation weather radar (WSR-88D)*," 2002. [Online]. Available: https://www.roc.noaa.gov/WSR88D/Applications/TAC_charter.aspx.

[2] A. V. Ryzhkov and D. S. Zrnic, *Radar Polarimetry for Weather Observations*. Cham: Springer International Publishing, 2019.

[3] V. Bringi and D. Zrnic, "Polarization weather radar development from 1970–1995: Personal reflections," *Atmosphere*, vol. 10, no. 11, p. 714, 2019.

[4] R. J. Doviak, V. Bringi, A. Ryzhkov, A. Zahrai, and D. Zrnić, "Considerations for polarimetric upgrades to operational WSR-88D radars," *Journal of Atmospheric and Oceanic Technology*, vol. 17, no. 3, pp. 257–278, 2000.

[5] M. I. Skolnik, *Radar Handbook*, 3rd ed. McGraw Hill, 2008. ISBN: 9780071485470.

[6] P. J. B. Clarricoats and A. D. Olver, *Corrugated Horns for Microwave Antennas*, vol. 18, 1984. London: Peter. Peregrinus Press.

[7] V. N. Bringi, R. Hoferer, D. A. Brunkow, *et al.*, "Design and performance characteristics of the new 8.5-m dual-offset Gregorian antenna for the CSU-CHILL radar," *Journal of Atmospheric and Oceanic Technology*, vol. 28, no. 7, pp. 907–920, 2011.

[8] M. Frech, B. Lange, T. Mammen, J. Seltmann, C. Morehead, and J. Rowan, "Influence of a radome on antenna performance," *Journal of Atmospheric and Oceanic Technology*, vol. 30, no. 2, pp. 313–324, 2013.

[9] V. Louf, A. Protat, R. A. Warren, *et al.*, "An integrated approach to weather radar calibration and monitoring using ground clutter and satellite comparisons," *Journal of Atmospheric and Oceanic Technology*, vol. 36, no. 1, pp. 2285–2301, 2019.

[10] G. G. Mace, A. Protat, S. Benson, and P. McGlynn, "Inferring the properties of snow in southern ocean shallow convection and frontal systems using dual-polarization C-band radar," *Journal of Applied Meteorology and Climatology*, vol. 62, no. 4, pp. 467–487, 2023.

[11] E. Montoya-Duque, Y. Huang, P. T. May, and S. T. Siems, "An evaluation of IMERG and ERA5 quantitative precipitation estimates over the Southern Ocean using shipborne observations," *Journal of Applied Meteorology and Climatology*," vol. 62, no. 11, pp. 1479–1495, 2023.

[12] M. Katsumata, S. Mori, B. Geng, and J. Inoue, "Internal structure of ex-Typhoon Phanfone (2014) under an extratropical transition as observed by the research vessel Mirai," *Geophysical Research Letters*, vol. 43, no. 17, pp. 9333–9341, 2016.

[13] S. A. Rutledge, V. Chandrasekar, B. Fuchs, *et al.*, "Sea-pol goes to sea," *Bulletin of the American Meteorological Society*, vol. 100, no. 11, pp. 2285–2301, 2019.

[14] M. Thurai, P. T. May, and A. Protat, "Shipborne polarimetric weather radar: Impact of ship movement on polarimetric variables at C band," *Journal of Atmospheric and Oceanic Technology*, vol. 31, no. 7, pp. 1557–1563, 2014.

[15] C. Kummerow, W. Barnes, T. Kozu, J. Shiue, and J. Simpson, "The tropical rainfall measuring mission (TRMM) sensor package," *Journal of Atmospheric and Oceanic Technology*, vol. 15, no. 3, pp. 809–817, 1998.

[16] N. Li, Z. Wang, X. Chen, and G. Austin, "Studies of general precipitation features with TRMM PR data: An extensive overview," *Remote Sensing*, vol. 11, no. 1, p. 80, 2019.

[17] T. Iguchi, T. Kozu, R. Meneghini, J. Awaka, and K. I. Okamoto, "Rain-profiling algorithm for the TRMM precipitation radar," *Journal of Applied Meteorology*, vol. 39, no. 12 PART 1, pp. 2038–2052, 2000.

[18] M. Thurai, E. Deguchi, T. Iguchi, and K. Okamoto, "Freezing height distribution in the tropics," *International Journal of Satellite Communications and Networking*, vol. 21, no. 6, pp. 533–545, 2003.

[19] A. Hamada and Y. N. Takayabu, "Improvements in detection of light precipitation with the global precipitation measurement dual-frequency precipitation radar (GPM DPR)," *Journal of Atmospheric and Oceanic Technology*, vol. 33, no. 4, pp. 653–667, 2016.

[20] D. Casella, G. Panegrossi, P. Sanò, *et al.*, "Evaluation of the GPM-DPR snowfall detection capability: Comparison with CloudSat-CPR," *Atmospheric Research*, vol. 197, pp. 64–75, 2017.

[21] S. Tanelli, S. L. Durden, E. Im, *et al.*, "CloudSat's cloud profiling radar after two years in orbit: Performance, calibration, and processing," *IEEE Transactions on Geoscience and Remote Sensing*, vol. 46, no. 11, pp. 3560–3573, 2008.

[22] R. J. Chase, S. W. Nesbitt, G. M. McFarquhar, N. B. Wood, and G. M. Heymsfield, "Direct comparisons between GPM-DPR and CloudSat snowfall retrievals," *Journal of Applied Meteorology and Climatology*, vol. 61, no. 9, pp. 1257–1271, 2022.

[23] T. Wehr, T. Kubota, G. Tzeremes, *et al.*, "The EarthCARE mission – science and system overview," *Atmospheric Measurement Techniques*, vol. 16, no. 15, pp. 3581–3608, 2023.

[24] A. Battaglia, P. Kollias, R. Dhillon, *et al.*, "Spaceborne cloud and precipitation radars: Status, challenges, and ways forward," *Reviews of Geophysics*, vol. 58, no. 3, 2020. https://doi.org/10.1029/2019RG000686.

[25] S. L. Durden, E. Im, F. K. Li, W. Ricketts, A. Tanner, and W. Wilson, "ARMAR: an airborne rain-mapping radar," *Journal of Atmospheric & Oceanic Technology*, vol. 11, no. 3, pp. 727–737, 1994.

[26] T. Kozu, K. Nakamura, R. Meneghini, and W. C. Boncyk, "Dual-parameter radar rainfall measurement from space: A test result from an aircraft experiment," *IEEE Transactions on Geoscience and Remote Sensing*, vol. 29, no. 5, pp. 690–703, 1991.

[27] S. L. Durden, S. Tanelli, and E. Im, "Recent observations of clouds and precipitation by the airborne precipitation radar 2nd generation in support of the GPM and ACE missions," in *Proceedings of SPIE*, 2012, vol. 8523, p. 85230M.

[28] F. Joseph Turk, S. Hristova-Veleva, S. L. Durden, *et al.*, "Joint analysis of convective structure from the APR-2 precipitation radar and the DAWN

Doppler wind lidar during the 2017 convective processes experiment (CPEX)," *Atmospheric Measurement Techniques*, vol. 13, no. 8, 2020.

[29] R. J. Chase, S. W. Nesbitt, and G. M. McFarquhar, "A dual-frequency radar retrieval of two parameters of the snowfall particle size distribution using a neural network," *Journal of Applied Meteorology and Climatology*, vol. 60, no. 3, pp. 341–359, 2021.

[30] G. Skofronick-Jackson, D. Hudak, W. Petersen, *et al.*, "Global precipitation measurement cold season precipitation experiment (GCPEX): For measurement's sake, let it snow," *Bulletin of the American Meteorological Society*, vol. 96, no. 10, pp. 1719–1741, 2015.

[31] J. Wang and B. Geerts, "Identifying drizzle within marine stratus with W-band radar reflectivity," *Atmospheric Research*, vol. 69, no. 1–2, pp. 1–27, 2003.

[32] L. Tian, G. M. Heymsfield, A. C. Didlake, S. Guimond, and L. Li, "Velocity-Azimuth display analysis of doppler velocity for HIWRAP," *Journal of Applied Meteorology and Climatology*, vol. 54, no. 8, pp. 1792–1808, 2015.

[33] L. Li, G. Heymsfield, J. Carswell, D. Schaubert, J. Creticos, and M. Vega, "High-altitude imaging wind and rain airborne radar (HIWRAP)," *IGARSS 2008 – 2008 IEEE International Geoscience and Remote Sensing Symposium* , Boston, MA, USA, 2008, pp. III-354–III-357, doi: 10.1109/IGARSS.2008.4779356.

[34] V. N. Bringi and V. Chandrasekar, *Polarimetric Doppler Weather Radar: Principles and Applications*, Cambridge University Press, 2001, Online ISBN: 9780511541094, DOI: https://doi.org/10.1017/CBO9780511541094

[35] R. J. Keeler and R. J. Serafin, "Meteorological radar," in *Radar Handbook*, M. I. Skolnik (eds.), McGraw Hill, 2008, pp. 19.1–19.49. ISBN 978-0-07-148547-0.

[36] R. J. Keeler and R. E. Pasarrelli, "Signal processing for atmospheric radars," in *Radar in Meteorology*, D. Atlas (ed.). Boston, MA: American Meteorological Society, 1990, pp. 199–229.

[37] W. W. S. B. L. Lewis, and F. F. Kretschmer, Jr., *Aspects of Radar Signal Processing*. Artech House, 1986. ISBN: 9781580531290.

[38] C. J. Oliver, "Pulse compression in optical radar," *IEEE Transactions on Aerospace and Electronic Systems*, vol. AES-15, no. 3, pp. 306–324, 1979.

[39] K. V. Mishra, *Frequency Diversity Wideband Digital Receiver and Signal Processor for Solid-State Dual-Polarimetric Weather Radars*. Fort Collins, CO: Colorado State University, 2012.

[40] L. Vandenberghe and S. P. Boyd, *Convex Optimization*. Cambridge: Cambridge University Press, 2004.

[41] P. Stoica, J. Li, and M. Xue, "Transmit codes and receive filters for radar," *IEEE Signal Processing Magazine*, vol. 25, no. 6, pp. 94–109, 2009.

[42] S. J. Frasier, F. Argenti, and L. Facheris, "Predistortion for very low pulse-compression sidelobes in solid-state meteorological radar," *IEEE Geoscience and Remote Sensing Letters*, vol. 20, 2023. doi: 10.1109/LGRS.2023.3263678.

[43] P. Stoica, H. He, and J. Li, "Optimization of the receive filter and transmit sequence for active sensing," *IEEE Transactions on Signal Processing*, vol. 60, no. 4, pp. 1730–1740, 2012.

[44] A. De Maio, Y. Huang, M. Piezzo, S. Zhang, and A. Farina, "Design of radar receive filters optimized according to Lp-norm based criteria," *IEEE Transactions on Signal Processing*, vol. 59, no. 8, pp. 4023–4029, 2011.

[45] F. Nai, S. M. Torres, and R. D. Palmer, "Adaptive beamspace processing for phased-array weather radars," *IEEE Transactions on Geoscience and Remote Sensing*, vol. 54, no. 10, pp. 5688–5698, 2016.

[46] E. Yoshikawa, T. Ushio, Z. Kawasaki, *et al.*, "MMSE beam forming on fast-scanning phased array weather radar," *IEEE Transactions on Geoscience and Remote Sensing*, vol. 51, no. 5, pp. 3077–3088, 2013.

[47] T. Hashimoto, K. Nishimura, M. Tsutsumi, K. Sato, and T. Sato, "A user parameter-free diagonal-loading scheme for clutter rejection on radar wind profilers," *Journal of Atmospheric and Oceanic Technology*, vol. 34, no. 5, pp. 1139–1153, 2017.

[48] C. D. Curtis, M. Yeary, and J. L. Lake, "Adaptive nullforming to mitigate ground clutter on the national weather radar testbed phased array radar," *IEEE Transactions on Geoscience and Remote Sensing*, vol. 54, no. 3, pp. 1282–1291, 2016.

[49] S. A. Vorobyov, "Principles of minimum variance robust adaptive beamforming design," *Signal Processing*, vol. 93, no. 12, pp. 3264–3277, 2013.

[50] K. Kamio, K. Nishimura, and T. Sato, "Adaptive sidelobe control for clutter rejection of atmospheric radars," *Annales Geophysicae*, vol. 22, no. 11, pp. 4005–4012, 2004.

[51] K. Nishimura, T. Nakamura, T. Sato, and K. Sato, "Adaptive beamforming technique for accurate vertical wind measurements with multichannel MST radar," *Journal of Atmospheric and Oceanic Technology*, vol. 29, no. 12, pp. 1769–1775, 2012.

[52] K. V. Mishra and Y. C. Eldar, "Sub-nyquist radar: Principles and prototypes," in *Compressed Sensing in Radar Signal Processing*. Cambridge: Cambridge University Press, pp. 4023–4029, 2019.

[53] R. G. Baraniuk, "Compressive sensing [Lecture Notes]," *IEEE Signal Processing Magazine*, vol. 24, no. 4, pp. 118–121, 2007.

[54] J. H. G. Ender, "On compressive sensing applied to radar," *Signal Processing*, vol. 90, no. 5, pp. 1402–1414, 2010.

[55] K. V. Mishra, M. Cho, A. Kruger, and W. Xu, "Spectral super-resolution with prior knowledge," *IEEE Transactions on Signal Processing*, vol. 63, no. 20, pp. 5342–5357, 2015.

[56] K. V. Mishra, A. Kruger, and W. F. Krajewski, "Compressed sensing applied to weather radar," in *International Geoscience and Remote Sensing Symposium (IGARSS)*, 2014.

[57] R. Kawami, A. Hirabayashi, T. Ijiri, S. Shimamura, H. Kikuchi, and T. Ushio, "3-dimensional compressive sensing and high-quality recovery for

phased array weather radar," in *2017 12th International Conference on Sampling Theory and Applications, SampTA 2017*, 2017.

[58] R. Purnamasari, A. B. Suksmono, I. Zakia, and I. J. M. Edward, "Compressive sampling of polarimetric doppler weather radar processing via inverse fast fourier transform," *IEEE Journal of Selected Topics in Applied Earth Observations and Remote Sensing*, vol. 14, pp. 5269–5284, 2021.

[59] T. Y. Yu, L. Ding, and S. Ozturk, "Application of compressive sensing to refractivity retrieval with a network of radars," in *IEEE National Radar Conference – Proceedings*, 2011.

[60] K. V. Mishra, M. R. Bhavani Shankar, V. Koivunen, B. Ottersten, and S. A. Vorobyov, "Toward millimeter-wave joint radar communications: A signal processing perspective," *IEEE Signal Processing Magazine*, vol. 36, no. 5. pp. 100–114, 2019.

[61] E. Saltikoff, J. Y. N. Cho, P. Tristant, *et al.*, "The threat to weather radars by wireless technology," *Bulletin of the American Meteorological Society*, vol. 97, no. 7, pp. 1159–1167, 2016.

[62] J. Liu, K. V. Mishra, and M. Saquib, "Co-designing statistical MIMO radar and in-band full-duplex multi-user MIMO communications," arXiv preprint arXiv:2006.14774, 2020.

[63] Y. LeCun, G. Hinton, and Y. Bengio, "Deep learning (2015), Y. LeCun, Y. Bengio and G. Hinton," *Nature*, vol. 521, 2015.

[64] G. Vulpiani, S. Giangrande, and F. S. Marzano, "Rainfall estimation from polarimetric S-band radar measurements: Validation of a neural network approach," *Journal of Applied Meteorology and Climatology*, vol. 48, no. 10, pp. 2022–2036, 2009.

[65] F. Zhang, C. Lai, and W. Chen, "Weather radar echo extrapolation method based on deep learning," *Atmosphere*, vol. 13, no. 5, p. 815, 2022.

[66] S. M. Bonnet, A. Evsukoff, and C. A. M. Rodriguez, "Precipitation nowcasting with weather radar images and deep learning in são paulo, brasil," *Atmosphere*, vol. 11, no. 11, p. 1157, 2020.

[67] K. Cui, C. Hu, R. Wang, Y. Sui, H. Mao, and H. Li, "Deep-learning-based extraction of the animal migration patterns from weather radar images," *Science China Information Sciences*, vol. 63, no. 4, p. 140304, 2020.

[68] A. Gharanjik, K. V. Mishra, B. Shankar, and B. Ottersten, "Learning-based rainfall estimation via communication satellite links," in *2018 IEEE Statistical Signal Processing Workshop, SSP 2018*, 2018.

[69] K. Lamer, M. Oue, A. Battaglia, *et al.*, "Multifrequency radar observations of clouds and precipitation including the G-band," *Atmospheric Measurement Techniques*, vol. 14, no. 5, pp. 36159–3629, 2021.

[70] B. M. Courtier, A. Battaglia, P. G. Huggard, *et al.*, "First observations of G-band radar doppler spectra," *Geophysical Research Letters*, vol. 49, no. 4, 2022. doi: 10.1029/2021GL096475

[71] A. M. Elbir, K. V. Mishra, and S. Chatzinotas, "Terahertz-band joint ultra-massive MIMO radar-communications: Model-based and model-free hybrid

beamforming," *IEEE Journal on Selected Topics in Signal Processing*, vol. 15, no. 6, pp. 1468–1483, 2021.

[72] S. Bhattacharjee, K. V. Mishra, R. Annavajjala, and C. R. Murthy, "Multi-carrier wideband OCDM-based THZ automotive radar," *IEEE International Conference on Acoustics, Speech and Signal Processing*, Rhodes Island, Greece, 2023, pp. 1–5.

[73] F. T. Ulaby, R. K. Moore, and A. K. Fung, "Microwave remote sensing: Active and passive. Volume I: Microwave remote sensing fundamentals and radiometry," *Microwave Remote Sensing: Active and Passive. Volume I: Microwave Remote Sensing Fundamentals and Radiometry*, 1981.

[74] J. S. Marshall and W. M. K. Palmer, "The distribution of raindrops with size," *Journal of Meteorology*, vol. 5, no. 4, pp. 165–166, 1948.

[75] J. O. Laws and D. A. Parsons, "The relation of raindrop-size to intensity," *Eos, Transactions American Geophysical Union*, vol. 24, no. 2, pp. 452–460, 1943.

[76] L. J. Battan, *Radar Observation of the Atmosphere*. Chicago, IL: University of Chicago Press, 1973.

[77] J. Joss and A. Waldvogel, "Ein Spektrograph für Niederschlagstropfen mit automatischer Auswertung," *Pure and Applied Geophysics PAGEOPH*, vol. 68, no. 1, pp. 240–246, 1967.

[78] A. Tokay, D. B. Wolff, and W. A. Petersen, "Evaluation of the new version of the laser-optical disdrometer, OTT parsivel," *Journal of Atmospheric and Oceanic Technology*, vol. 31, no. 6, pp. 1276–1288, 2014.

[79] M. Schönhuber, G. Lammer, and W. L. Randeu, "One decade of imaging precipitation measurement by 2D-Video-Distrometer," *Advances in Geosciences*, vol. 10, pp. 85–90, 2007.

[80] C. Klepp, S. Michel, A. Protat, *et al.*, "OceanRAIN, a new in-situ shipboard global ocean surface-reference dataset of all water cycle components," *Scientific Data*, vol. 5, 2018.

[81] R. N. D. Baumgardner, G. Kok, W. Dawson, and D. O'Connor, "A new ground-based precipitation spectrometer: The meteorological particle sensor (MPS)," in *AMS Conference on Cloud Physics*, 2002.

[82] A. Ferrone, A. C. Billault-Roux, and A. Berne, "ERUO: A spectral processing routine for the micro rain radar PRO (MRR-PRO)," *Atmospheric Measurement Techniques*, vol. 15, no. 11, pp. 3569–3592, 2022.

[83] B. E. Sheppard, "Measurement of raindrop size distributions using a small doppler radar," *Journal of Atmospheric and Oceanic Technology*, vol. 7, no. 2, pp. 255–268, 1990.

[84] B. E. Sheppard, M. Thurai, P. Rodriguez, P. C. Kennedy, and D. R. Hudak, "Improved precipitation typing using poss spectral modal analysis," *Journal of Atmospheric and Oceanic Technology*, vol. 38, no. 3, pp. 537–554, 2021.

[85] B. E. Sheppard and P. I. Joe, "Comparison of raindrop size distribution measurements by a Joss-Waldvogel disdrometer, a PMS 2DG spectrometer, and a POSS Doppler radar," *Journal of Atmospheric & Oceanic Technology*, vol. 11, no. 4 part 1, pp. 874–887, 1994.

[86] M. Thurai, G. J. Huang, V. N. Bringi, W. L. Randeu, and M. Schönhuber, "Drop shapes, model comparisons, and calculations of polarimetric radar parameters in rain," *Journal of Atmospheric and Oceanic Technology*, vol. 24, no. 6, pp. 1019–1032, 2007.

[87] E. J. Thompson, S. A. Rutledge, B. Dolan, and M. Thurai, "Drop size distributions and radar observations of convective and stratiform rain over the equatorial Indian and West Pacific Oceans," *Journal of the Atmospheric Sciences*, vol. 72, no. 11, pp. 4091–4125, 2015.

[88] W. Bang, S. Kwon, and G. Lee, "Characteristic of raindrop size distribution using two-dimensional video disdrometer data in Daegu, Korea," *Journal of the Korean Earth Science Society*, vol. 38, no. 7, pp. 511–521, 2017.

[89] M. Thurai, V. N. Bringi, A. B. Manic, N. J. Šekeljic, and B. M. Notaroš, "Investigating raindrop shapes, oscillation modes, and implications for radio wave propagation," *Radio Science*, vol. 49, no. 10, pp. 921–932, 2014.

[90] M. Thurai, V. Bringi, P. Gatlin, and M. Wingo, "Raindrop fall velocity in turbulent flow: An observational study," *Advances in Science and Research*, 2021, vol. 18, pp. 33–39.

[91] P. N. Gatlin, M. Thurai, V. N. Bringi, *et al.*, "Searching for large raindrops: A global summary of two-dimensional video disdrometer observations," *Journal of Applied Meteorology and Climatology*, vol. 54, no. 5, pp. 1069–1089, 2015.

[92] K. V. Beard and C. Chuang, "A new model for the equilibrium shape of raindrops," *Journal of the Atmospheric Sciences*, vol. 44, no. 11, pp. 1509–1524, 1987.

[93] R. Gunn and G. D. Kinzer, "The terminal velocity of fall for water droplets in stagnant air," *Journal of Meteorology*, vol. 6, no. 4, pp. 243–248, 1949.

[94] G. W. Lee and I. Zawadzki, "Radar calibration by gage, disdrometer, and polarimetry: Theoretical limit caused by the variability of drop size distribution and application to fast scanning operational radar data," *Journal of Hydrology*, vol. 328, no. 1–2, pp. 83–97, 2006.

[95] A. Adachi, "Radar calibration using polarimetric observations with rain attenuation correction," in *AMS Conference on Radar Meteorology*, 2013.

[96] M. P. Jensen, W. A. Petersen, A. Bansemer, *et al.*, "The midlatitude continental convective clouds experiment (MC3E)," *Bulletin of the American Meteorological Society*, vol. 97, no. 9, pp. 1667–1686, 2016.

[97] B. C. Seo, W. F. Krajewski, F. Quintero, *et al.*, "Comprehensive evaluation of the IFloodS radar rainfall products for hydrologic applications," *Journal of Hydrometeorology*, vol. 19, no. 11, pp. 1793–1813, 2018.

[98] K. V. Mishra, W. F. Krajewski, R. Goska, *et al.*, "Deployment and performance analyses of high-resolution Iowa XPOL radar system during the NASA IFloodS campaign," *Journal of Hydrometeorology*, vol. 17, no. 2, pp. 455–479, 2016.

[99] A. J. Newman, P. A. Kucera, and L. F. Bliven, "Presenting the snowflake video imager (SVI)," *Journal of Atmospheric and Oceanic Technology*, vol. 26, no. 2, pp. 167–179, 2009.

[100] T. J. Garrett, C. Fallgatter, K. Shkurko, and D. Howlett, "Fall speed measurement and high-resolution multi-angle photography of hydrometeors in free fall," *Atmospheric Measurement Techniques*, vol. 5, no. 11, pp. 2625–2633, 2012.

[101] M. Schaer, C. Praz, and A. Berne, "Identification of blowing snow particles in images from a multi-angle snowflake camera," *Cryosphere*, vol. 14, no. 1, pp. 367–384, 2020.

[102] B. Notaros, "Meteorological electromagnetics: Optical and radar measurements, modeling, and characterization of snowflakes and snow," *IEEE Antennas and Propagation Magazine*, vol. 63, no. 2, pp. 14–27, 2021.

[103] P. Kennedy, M. Thurai, C. Praz, V. N. Bringi, A. Berne, and B. M. Notaroš, "Variations in snow crystal riming and Z_{DR}: A case analysis," *Journal of Applied Meteorology and Climatology*, vol. 57, no. 3, pp. 695–707, 2018.

[104] T. H. Raupach and A. Berne, "Retrieval of the raindrop size distribution from polarimetric radar data using double-moment normalisation," *Atmospheric Measurement Techniques*, vol. 10, no. 7, pp. 2573–2594, 2017.

[105] V. Bringi, K. V. Mishra, M. Thurai, P. C. Kennedy, and T. H. Raupach, "Retrieval of lower-order moments of the drop size distribution using CSU-CHILL X-band polarimetric radar: A case study," *Atmospheric Measurement Techniques*, vol. 13, no. 9, pp. 4727–4750, 2020.

[106] A. N. Auf Der Maur, "Statistical tools for drop size distributions: Moments and generalized gamma," *Journal of the Atmospheric Sciences*, vol. 58, no. 4, pp. 407–418, 2001.

[107] G. W. Lee, I. Zawadzki, W. Szyrmer, D. Sempere-Torres, and R. Uijlenhoet, "A general approach to double-moment normalization of drop size distributions," *Journal of Applied Meteorology*, vol. 43, no. 2, pp. 264–281, 2004.

[108] M. Thurai, V. N. Bringi, E. Adirosi, F. Lombardo, and P. N. Gatlin, "Variability of raindrop size distribution as characterized by the generalized gamma formulation," in *Precipitation Science*, 1st ed., S. Michaelides (ed.), Elsevier, 2022, pp. 473–501. ISBN: 978-0-12-822973-6.

[109] D. Klugmann and O. Fiŝer, "Mie versus point matching algorithm for radar rain properties retrieval," in *International Geoscience and Remote Sensing Symposium (IGARSS)*, 2008, vol. 4, no. 1.

[110] B. G. Evans, N. K. Uzunoglu, and A. R. Holt, "Two new approaches to the calculation of rain induced attenuation and cross polarization," *Annales Des Télécommunications*, vol. 32, no. 11–12, pp. 398–403, 1977.

[111] M. I. Mishchenko, L. D. Travis, and D. W. Mackowski, "T-matrix computations of light scattering by nonspherical particles: A review," *Journal of Quantitative Spectroscopy and Radiative Transfer*, vol. 55, no. 5, pp. 535–575, 1996.

[112] V. N. Bringi and T. A. Seliga, "Scattering from non-spherical hydrometeors," *Annales Des Télécommunications*, vol. 32, no. 11–12, pp. 392–397, 1977.

[113] T. Oguchi, "Scattering from hydrometeors: A survey," *Radio Science*, vol. 16, no. 5, pp. 691–730, 1981.

[114] B. T. Draine and P. J. Flatau, "Discrete-dipole approximation for scattering calculations," *Journal of the Optical Society of America A*, vol. 11, no. 4, pp. 1491–1499, 1994.

[115] Z. Jiang, M. R. Kumjian, R. S. Schrom, *et al.*, "Comparisons of electromagnetic scattering properties of real hailstones and spheroids," *Journal of Applied Meteorology and Climatology*, vol. 58, no. 1, pp. 93–112, 2019.

[116] B. M. Notaroš, V. N. Bringi, C. Kleinkort, *et al.*, "MoM-SIE scattering models of snow and ice hydrometeors based on 3D shape reconstructions from MASC images," in *2017 International Applied Computational Electromagnetics Society Symposium – Italy, ACES 2017*, 2017.

[117] E. Chobanyan, N. J. Šekeljić, A. B. Manić, M. M. Ilić, V. N. Bringi, and B. M. Notaroš, "Efficient and accurate computational electromagnetics approach to precipitation particle scattering analysis based on higher-order method of moments integral equation modeling," *Journal of Atmospheric and Oceanic Technology*, vol. 32, no. 10, pp. 1745–1758, 2015.

[118] C. Kleinkort, G. J. Huang, V. N. Bringi, and B. M. Notaros, "Visual hull method for realistic 3D particle shape reconstruction based on high-resolution photographs of snowflakes in free fall from multiple views," *Journal of Atmospheric and Oceanic Technology*, vol. 34, no. 3, pp. 679–702, 2017.

[119] R. J. Kubesh and K. V. Beard, "Laboratory measurements of spontaneous oscillations for moderate-size raindrops," *Journal of the Atmospheric Sciences*, vol. 50, no. 8, pp. 1089–1098, 1993.

[120] S. B. Manić, M. Thurai, V. N. Bringi, and B. M. Notaroš, "Scattering calculations for asymmetric raindrops during a line convection event: Comparison with radar measurements," *Journal of Atmospheric and Oceanic Technology*, vol. 35, no. 6, pp. 1169–1180, 2018.

[121] M. Thurai, S. Steger, F. Teschl, M. Schonhuber, and D. B. Wolff, "Rain drop shapes and scattering calculations: A case study using 2D video disdrometer measurements and polarimetric radar observations at S-band during hurricane Dorian rain-bands," in *15th European Conference on Antennas and Propagation, EuCAP 2021*, 2021.

[122] M. Thurai, M. Szakáll, V. N. Bringi, K. V. Beard, S. K. Mitra, and S. Borrmann, "Drop shapes and axis ratio distributions: Comparison between 2D video disdrometer and wind-tunnel measurements," *Journal of Atmospheric and Oceanic Technology*, vol. 26, no. 7. pp. 1427–1432, 2009.

[123] C. Capsoni, L. Castanet, N. Jeannin, *et al.*, "Review of radar measurements of precipitation for the characterization of propagation effects on terrestrial and slant path radio links," in *Proceedings of 6th European Conference on Antennas and Propagation, EuCAP 2012,* 2012.

[124] J. W. F. Goddard and M. Thurai, "Radar-derived path reduction factors for terrestrial systems," in *IEE Conference Publication*, 1997, no. 436, pt 2.

[125] J. Tan and M. Thurai, "Calculations of rain-induced cross-polarisation on line-of-sight systems at 38 GHz," *IEE Colloquium (Digest)*, no. 239, 1996.

[126] B. Arbesser-Rastburg, "Olympus propagation experiments," in *18th NASA Propagation Experimenters Meeting (NAPEX 18)*, 1994, pp. 7–10.

[127] P. J. Hardaker, A. R. Holt, and J. W. F. Goddard, "Comparing model and measured rainfall rates obtained from a combination of remotely sensed and in situ observations," *Radio Science*, vol. 32, no. 5, pp. 1785–1796, 1997.

[128] B. R. Arresser-Rastburg and G. Brussaard, "Propagation research in Europe using the OLYMPUS satellite," *Proceedings of the IEEE*, vol. 81, no. 6, pp. 865–875, 1993.

[129] S. Ventouras, C. L. Wrench, and P. G. Davies, "Propagation measurements using ITALSAT beacons together with associated measurements of radar reflectivity from rain and clouds at 3 GHz and 35 GHz," in *IEE Conference Publication*, 1997, no. 436, pt 2.

[130] R. Polonio and C. Riva, "Italsat propagation experiment at 18.7, 39.6, and 49.5 ghz at spino d'adda: Three years of cpa statistics," *IEEE Transactions on Antennas and Propagation*, vol. 46, no. 5, pp. 631–635, 1998.

[131] M. Thural and J. W. F. Goddard, "Precipitation scatter measurements from a transhorizon experiment at 11.2 GHz," *IEE Proceedings H: Microwaves, Antennas and Propagation*, vol. 139, no. 1, pp. 53–58, 1992.

[132] J. D. Beaver and V. N. Bringi, "The application of s-band polarimetric radar measurements to ka-band attenuation prediction," *Proceedings of the IEEE*, vol. 85, no. 6, pp. 893–909, 1997.

[133] J. Beaver, J. Turk, and V. N. Bringi, "Ka-band propagation measurements from the advanced communications technology satellite (ACTS)," in *International Geoscience and Remote Sensing Symposium (IGARSS)*, 1994, vol. 1.

[134] C. Capsoni and L. Luini, "Use of radar images for the development of a propagation oriented space-time rain model," *in International Geoscience and Remote Sensing Symposium (IGARSS)*, 2009, vol. 4.

[135] L. Luini and C. Capsoni, "MultiEXCELL: A new rainfall model for the analysis of the millimetre wave propagation through the atmosphere," in *European Conference on Antennas and Propagation, EuCAP 2009, Proceedings*, 2009.

[136] M. Thurai and T. Iguchi, "Rain height information from TRMM precipitation radar," *Electronics Letters*, vol. 36, no. 12, pp. 1059–1061, 2000.

[137] M. A. N. Azlan, J. Din, and H. Y. Lam, "Rain height information from TRMM precipitation radar for satellite communication in Malaysia," in *2011 IEEE International Conference on Space Science and Communication: "Towards Exploring the Equatorial Phenomena," IconSpace 2011 – Proceedings*, 2011.

[138] A. Overeem, H. Leijnse, and R. Uijlenhoet, "Country-wide rainfall maps from cellular communication networks," *Proceedings of the National Academy of Sciences of the United States of America*, vol. 110, no. 8, pp. 2741–2745, 2013.

Chapter 2

Doppler polarimetric radars for weather observations from 1995 to 2022: a historical perspective

Alexander V. Ryzhkov[1,3,4], Merhala Thurai[2] and Dušan S. Zrnić[3,4,5]

2.1 Introduction

Doppler polarimetric radars have become a standard for modern operational weather radars around the world. Compared to single-polarization radars, dual-polarization radars substantially improve data quality, precipitation estimation, and severe weather warnings. It took more than 40 years from the inception of the idea of using polarimetry for weather radar observations until its current massive operational use. Bringi and Zrnić [1] provided a historical overview of the polarization weather radar research during the early period from 1970 to 1995. In this chapter, we review developments in the Doppler polarimetric weather radar technology and science for the last 25 years from the mid-1990s until 2022, thus complementing a review in [1]. However, we differ from [1] in the exposition of the material. In [1], the authors devote separate sections to describe polarimetric radar research by various organizations, whereas we chose to chronicle various subject themes. Since 1995, the number of organizations engaged in dual-polarization research has been steadily increasing to the point that it is almost impossible to track.

During this period, three important developments have happened in the evolution of dual-polarization technology. First, this technology has reached maturity and wide credibility exemplified by the published number of books on weather radar polarimetry. The first book fully devoted to the subject is by Bringi and Chandrasekar [2] followed by the monographs of Zhang [3] and Ryzhkov and Zrnić [4]. The first two of these grew out of lectures at universities. Second, operational

[1]Cooperative Institute for Severe and High-Impact Weather Research and Operations, Norman, OK, USA
[2]Colorado State University, Fort Collins, CO, USA
[3]NOAA National Severe Storms Laboratory, Norman, OK, USA
[4]School of Meteorology, University of Oklahoma, Norman, OK, USA
[5]School of Electrical and Computer Engineering, University of Oklahoma, Norman, OK, USA

WSR-88Ds (Weather Surveillance Radars 1988 Doppler) in the US have been upgraded to dual polarization followed by a similar upgrade by many national weather services (NWS). Hence, a tremendous variety of polarimetric radar observables from various weather events is continually captured and made available to the public. Finally, the massive availability of polarimetric radar data enabled fundamental discoveries in understanding microphysical processes of precipitation formation resulting in significant improvement of the operational methodologies for quantitative precipitation estimation, hydrometeor classification, and timely warnings of severe weather associated with tornadoes, large hail, flash floods, freezing rain, icing, etc.

We continue with a brief review of textbooks devoted to weather radar polarimetry. The book by Doviak and Zrnić [5] remained influential and was reprinted in an affordable paperback edition. Its chapter "Precipitation Measurements" contains information sufficient for understanding the basic principles of radar polarimetry and learning about early successes of this emerging technology. The first book devoted primarily to polarimetry [2] quickly became *the reference* for the subject. It devotes a fair amount of text to the electromagnetic (EM) aspects of polarimetry and highlights its principles and applications. Both are backed by vigorous theoretical arguments. The book by Zhang [3] draws from class lectures on weather radars at the University of Oklahoma. It provides a reference for the polarimetric radar data user including fundamentals and tools. The Zhang monograph covers a wide range of topics including advanced polarimetric signal processing, retrievals of raindrop size distributions, and even contains a full chapter on phased array radar polarimetry.

The latest in the series [4] is primarily focused on the utilization of radar polarimetry for weather observations including advanced methods for polarimetric estimation of rain and snow, hydrometeor classification, latest techniques for operational detection of tornadoes, hail of different sizes, hazardous winter precipitation, and nonmeteorological echoes such as ground/sea clutter, biota, and others. The authors discuss measurement errors of polarimetric radar variables and various artifacts affecting the quality of radar measurements such as attenuation, partial beam blockage, nonuniform beam filling, and depolarization on propagation among others. The monograph also emphasizes a link between the cloud microphysics and polarimetric observations by introducing a concept of "polarimetric fingerprints" of various microphysical processes and suggesting novel methods for ice microphysical retrievals.

Succinct and informative overviews of the basics and practical applications of radar polarimetry can be found in [6–11] and in recently published general textbooks on radar meteorology [12–14]. The fact that polarimetry turned into a standard subject of weather radar courses at universities proves it is becoming a routine requiring a cadre of trained personnel to use it effectively. Wide acceptance timing is evident in the number of papers dealing with dual-polarization issues at the Radar Meteorology Conferences (Figure 2.1). The largest increase from a few to about 45 per year occurs in the 1990–2000 decade. Beyond about 2000 (not plotted), polarimetry-devoted sections disappeared but became integral part of combined weather analysis with other techniques.

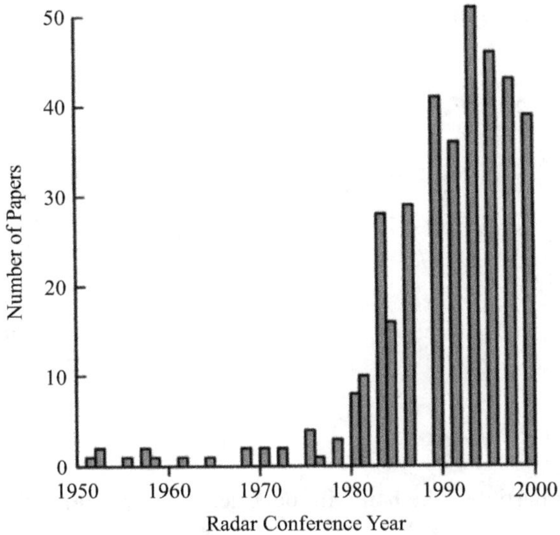

Figure 2.1 Number of papers devoted to dual-polarization issues between 1950 and 2000 at the American Meteorological Society's Conferences on Radar Meteorology.

2.2 Physical meaning of Doppler polarimetric radar variables

Single-polarization Doppler radars measure three variables: radar reflectivity factor Z, mean Doppler velocity v, and Doppler velocity spectrum width σ_v. In addition, dual-polarization radars can directly measure differential reflectivity Z_{DR}, differential phase Φ_{DP}, linear or circular depolarization ratios (*LDR* or *CDR*), and cross-correlation coefficient ρ_{hv}. Two very important variables, specific differential phase K_{DP} and specific attenuation A, can be derived from the radial profiles of Φ_{DP} and Z. For definitions and detailed analysis of Doppler polarimetric variables, a reader can consult basic monographs [2–5]. Herein, we provide a brief description of their physical meaning.

Radar reflectivity is defined as

$$Z = \frac{\lambda^4}{\pi^5 |K_w|^2} \int \sigma(D) N(D)\, dD \tag{2.1}$$

where $\sigma(D)$ is the radar backscattering cross-section of a hydrometeor with equivolume diameter D, $N(D)$ is the size distribution (SD) of hydrometeors, λ is the radar wavelength, and $|K_w|^2 = |(\varepsilon_w - 1)/(\varepsilon_w + 2)|^2$ where ε_w is the dielectric constant of *water*. For spherical particles with dielectric constant ε and sizes much smaller than the radar wavelength, a simple Rayleigh approximation of the cross-section σ is:

$$\sigma = \frac{\pi^5 D^6}{\lambda^4} |K|^2 \tag{2.2}$$

where $|K|^2 = |(\varepsilon - 1)/(\varepsilon + 2)|^2$. Therefore,

$$Z = \frac{|K|^2}{|K_w|^2} \int D^6 N(D)\, dD. \tag{2.3}$$

For liquid particles such as raindrops, $K = K_w$ and Z is equal to the sixth moment of SD:

$$Z_w = \int D^6 N(D)\, dD. \tag{2.4}$$

For solid ice particles like hail with bulk density of ρ_i and dielectric constant ε_i

$$Z_i = \frac{|K_i|^2}{|K_w|^2} \int D^6 N(D) dD. \tag{2.5}$$

For snow that is characterized by the inverse dependence of its bulk density ρ_s on the diameter

$$\frac{\varepsilon_s - 1}{\varepsilon_s + 2} = \frac{\rho_s}{\rho_i} \frac{\varepsilon_i - 1}{\varepsilon_i + 2}, \tag{2.6}$$

and the following expression can be written for its reflectivity Z_s:

$$Z_s = \frac{|K_i|^2}{|K_w|^2 \rho_i^2} \int \rho_s^2(D) D^6 N(D)\, dD, \tag{2.7}$$

and radar reflectivity is proportional to the fourth moment of SD. This explains why Z of snow (Z_s) is much lower than Z of rain (Z_w) for a given $N(D)$.

Differential reflectivity Z_{DR} is defined as the ratio of radar reflectivity factors Z_h and Z_v at horizontal and vertical polarizations, respectively, if they are expressed in a linear scale or their difference in a logarithmic scale

$$Z_{DR}(dB) = Z_h(dBZ) - Z_v(dBZ). \tag{2.8}$$

Z_{DR} depends on the particle size, shape, orientation, density, and phase composition but not on the concentration of particles. $Z_{DR} = 0$ dB if the hydrometeors are spherical or randomly oriented in the polarization plane. Generally, Z_{DR} increases with increasing oblateness, density, or water fraction. Because raindrops become more oblate with increasing size, and their aspect ratio is a well-determined function of diameter, Z_{DR} is a good measure of the mean or median raindrop diameter. It is important that differential reflectivity is significantly higher for raindrops than for snow particles with the same sizes, shapes, and orientations, which makes Z_{DR} an excellent discriminator between rain and snow. In snow, Z_{DR} is crucially dependent on its density with dry aggregated snow characterized by low Z_{DR} which is close to 0 dB as opposed to ice crystals that may have quite high Z_{DR} due to their high density and often very nonspherical shape. Differential reflectivity is a key parameter for discrimination between meteorological and nonmeteorological radar echo and for classification of hydrometeors of various habits. Ground clutter and atmospheric biota (insects and birds) have a wide distribution of Z_{DR} including extremely high positive (and occasionally negative) values which are not typical for meteorological scatterers. Z_{DR} is very efficient for distinguishing between hail and rain. Hail is generally characterized by higher Z and lower Z_{DR} than rain due to the difference in size, dielectric constant, and orientations of hailstones and raindrops.

The cross-correlation coefficient ρ_{hv} between the radar signals at horizontal and vertical polarizations is another polarimetric radar variable that is very instrumental for radar echo and hydrometeor classification. It is particularly useful for distinguishing between meteorological and nonmeteorological radar echoes. The latter have much lower value of ρ_{hv}, whereas the majority of weather scatterers is characterized by ρ_{hv} very close to 1. Mixed-phase hydrometeors like wet snow, melting hail, and hail growing in the wet growth regime are the only types of hydrometeors for which ρ_{hv} can significantly deviate from 1.

Linear and circular depolarization ratios in the linear and circular polarization bases (*LDR* and *CDR*, respectively) were historically the first polarimetric variables used for hydrometeor classification, specifically for hail detection. These variables, however, are not directly measured by the modern operational Doppler polarimetric radars because these utilize simultaneous transmission and reception of the waves with orthogonal polarizations. Nevertheless, a proxy of depolarization ratio can be roughly estimated from the combination of Z_{DR} and ρ_{hv} [15].

Differential phase Φ_{DP} is the difference between the phases of the radar returns at horizontal and vertical polarizations. It plays a very important role for quantification of precipitation and liquid and ice water contents (*LWC* and *IWC*). A radial profile of Φ_{DP} can be used to compute two radar parameters closely related to rain rate R and *LWC* / *IWC* – a specific differential phase K_{DP} and specific attenuation A. K_{DP} is estimated as a half of the radial derivative of Φ_{DP}. Its magnitude depends on the size distribution of hydrometeors as well as their shapes and orientation. In the Rayleigh approximation, K_{DP} in rain can be written as [4]

$$K_{DP} \sim \frac{F_o}{\lambda} \int F_s(D) D^3 N(D)\, dD, \tag{2.9}$$

where F_o and F_s are the factors describing dependencies of K_{DP} on raindrops' orientation (F_o) and shape (F_s). It can be shown that the shape factor F_s for raindrops is roughly proportional to D^2 so that K_{DP} is determined by the fifth moment of drop size distribution (DSD). In other words, K_{DP} is proportional to a lower order moment of DSD than Z and, therefore is better correlated with rain rate which is proportional to the 3.67th order moment of DSD.

In snow [4],

$$K_{DP} \sim \frac{F_o F_s}{\lambda} \int \rho_s^2(D) D^3 N(D)\, dD, \tag{2.10}$$

so that K_{DP} is proportional to the first moment of SD because ρ_s is inversely dependent on the snowflake diameter. This emphasizes the benefit of using K_{DP} for quantification of ice and snow because ice water content is determined by the lower end of the size distribution.

A radial profile of A can be retrieved from a radial profile of attenuated Z_a and two-way path integrated attenuation PIA along the propagation path (r_1, r_2) in rain as [2,4]

$$A(r) = \frac{[Z_a(r)]^b C(b, PIA)}{I(r_1, r_2) + C(b, PIA) I(r, r_2)}, \tag{2.11}$$

where

$$I(r_1, r_2) = 0.46b \int_{r_1}^{r_2} [Z_a(s)]^b ds, \tag{2.12}$$

$$I(r, r_2) = 0.46b \int_{r}^{r_2} [Z_a(s)]^b ds, \tag{2.13}$$

$$C(b, PIA) = \exp(0.23bPIA) - 1. \tag{2.14}$$

In (2.11)–(2.14), the factor b is a constant parameter (usually between 0.6 and 0.9 at microwave frequencies). The value of PIA (path integrated attenuation) cannot be estimated with a single-polarization radar but can be computed from the total span of differential phase $\Delta\Phi_{DP}$ measured by a dual-polarization radar:

$$PIA(r_1, r_2) = \alpha[\Phi_{DP}(r_2) - \Phi_{DP}(r_1)], \tag{2.15}$$

where α is equal to the average ratio of A and K_{DP} along the propagation path. As shown in Section 2.4, specific attenuation A is particularly beneficial for rainfall estimation.

Because Φ_{DP} is the phase measurement, its estimate and the estimates of its derivatives K_{DP} and A are immune to radar miscalibration, partial beam blockage, attenuation in rain, and the impact of a wet radome that cause biases in the measurements of Z, Z_{DR}, LDR, and CDR. Moreover, Φ_{DP} can be used for attenuation/differential attenuation correction of radar reflectivity and differential reflectivity using simple formulas

$$\Delta Z(r) = \alpha\, [\Phi_{DP}(r) - \Phi_{DP}(r_1)], \tag{2.16}$$

and

$$\Delta Z_{DR}(r) = \beta\, [\Phi_{DP}(r) - \Phi_{DP}(r_1)], \tag{2.17}$$

where $\Delta Z(r)$ and $\Delta Z_{DR}(r)$ are the biases of Z and Z_{DR} caused by attenuation along propagation path in rain between ranges r_1 and r.

2.3 Polarimetric radars in research and operations since 1995

Bringi and Zrnić [1] provide an overview of the pioneering polarimetric weather radar studies for the period ending in 1995. Next 25 years constitute the "golden age" of weather radar polarimetry marked by significant research achievements and operational implementation of the polarimetric technology on the national weather radar networks across the globe. Various research teams in different countries entered the "polarimetric race" that accelerated in mid-1990s. Investigations have been performed using surveillance weather radars operating at three primary microwave bands: S, C, and X. Toward the end of this quarter-century survey period, scanning polarimetric radars have been built and utilized for research at Ka and W bands traditionally applied for cloud research rather than weather surveillance. In this section, we overview various radars used for polarimetric research and operations since 1995.

2.3.1 S-band research radars

The first research polarimetric radars operated mainly at S and C bands to minimize the impact of attenuation in precipitation. Herein, we provide a brief overview of a few S-band radars that played very important role at early stages of the polarimetric research and some of which continue serving as a valuable resource for in-depth studies and important observational components of various field campaigns.

2.3.1.1 Cimarron and KOUN radars in Oklahoma

A bulk of polarimetric radar research in Oklahoma was done with the Cimarron and KOUN radars. The latter one transferred from the NWS for research at the National Severe Storms Laboratory (NSSL) became a prototype of the operational polarimetric WSR-88D. The Cimarron radar [16] used a traditional scheme with alternative transmission/reception of the orthogonally polarized waves with horizontal and vertical polarizations. A novel mode of operation (called the SHV mode) when the radar simultaneously transmits and receives H- and V-waves was first implemented on the KOUN radar [17] and later became a standard for operational polarimetric weather radars. There were several reasons favoring the SHV mode. These are:

(1) The SHV mode avoids the expense of a high-power microwave switch. We remind the readers that the one-time engineering development plus switch manufacturing is only a part of the expense. A very significant cost is in the aftermath expenses including keeping documentation and spare parts.
(2) The errors in the estimates of the polarimetric variables in the SHV mode [18] can be considerably smaller (especially at significant spectrum widths or surveillance scans) than the corresponding errors in the alternate mode.
(3) Computation of the cross-correlation coefficient is direct and avoids possible bias and increased errors of alternate mode.
(4) The unambiguous interval in computations of differential phase Φ_{DP} is the full 360° compared to 180° in the alternate mode.

(5) Computation of the mean Doppler velocity is not coupled to the computation of the differential phase. Therefore, velocity estimates do not alias if the differential phase is subject to aliasing.

One of the disadvantages of the SHV mode is its inability to measure the linear depolarization ratio (*LDR*). However, Ryzhkov *et al.* [15] suggested estimating a proxy of the depolarization ratio at circular polarization (*CDR*) from the combination of Z_{DR} and ρ_{hv} by the radars utilizing the SHV scheme.

Zrnić and Ryzhkov [19] summarized the results of the Oklahoma studies during the nineties. One of the notable achievements was the demonstrated utility of specific differential phase K_{DP} for more accurate estimation of rainfall compared to the conventional methods based on the use of the $R(Z)$ relations [20,21]. Further observational studies in Oklahoma with the KOUN radar were instrumental for developing the algorithms for detection of tornadoes and hail as well as determination of hail size [22–27].

2.3.1.2 Chilbolton radar (UK)

The magnetron-based polarimetric radar located in Chilbolton, UK has the narrowest beam (0.25°) among the S-band weather radars due to its fully steerable 25 m antenna. A purpose-built, fast mechanical switch enabled alternate radar pulses to be transmitted with vertical and horizontal polarizations. This allows the measurement of *LDR* which proved to be very efficient for the detection of the melting layer [28]. Polarimetric observations with Chilbolton radar were among the first used for demonstrating the concept of self-consistency between Z, Z_{DR}, and K_{DP} in rain, allowing absolute calibration of Z using polarimetric variables Z_{DR} and K_{DP} [29]. The self-consistency technique then was applied not just at S band but also at C and X bands. Hogan [30] used Chilbolton data to develop one of the first variational methods for separation of hail-dominant regions in convective storms and for rainfall rate retrievals using a combination of Z, Z_{DR}, and Φ_{DP}. More recently, a separate Ka-band radar as well as a W-band radar were added to the Chilbolton facility, all of which were used for cloud-related studies in projects such as CloudNet [31].

2.3.1.3 CSU-CHILL dual-wavelength polarimetric radar (Colorado)

The Colorado State University-Chicago ILLinois (CSU-CHILL) radar run by the CSU currently uses a unique dual-offset feed, Gregorian type antenna (Figure 2.2) that provides very low sidelobe levels and high polarization purity [32]. The construction of this antenna allows the use of two different single-wavelength feed horns (S-band or X-band/11 or 3 cm wavelength) as well as a dual wavelength (simultaneous S- and X-band) horn. All three of these horns are designed for dual-polarization (horizontal and vertical) operation [33]. The CSU-CHILL radar's evolving dual-polarization measurement capabilities have had a number of research applications ranging from in-depth hail polarimetric studies [34] to the investigation of polarimetric signatures within the dendritic growth layer in stratiform clouds

Figure 2.2 CSU-CHILL radar Gregorian antenna.

and winter storms. Kennedy and Rutledge [35] expanded the use of differential phase measurements by examining Φ_{DP} field structures observed by the CSU-CHILL radar in winter storms. In 2014 and 2015, the CSU-CHILL S- and X-band radars were used extensively for winter storm studies as part of the "MASCRAD" project [36]. A variety of ground instruments were installed at a site 13 km south of the radar.

In one case study [37], the CSU-CHILL X-band Z_{DR} over the instrumentation site was found to vary with the degree of riming of snow crystals determined from the images provided by the Multi-Angle Snow Camera (MASC). Riming index derived by processing the MASC digital images was well correlated with X-band Z_{DR} over the site. In another case study, the CSU-CHILL S-band radar showed negative Z_{DR} column which was associated with an unusual graupel-shower event [38] that passed over the ground instrument site. MASC images and 2DVD measurements had identified the graupel particles to be primarily of the lump type with larger particles having slightly prolate mean shapes.

2.3.1.4 The National Center for Atmospheric Research (NCAR) S-PolKa radar

Another unique polarimetric radar, S-PolKa, used for research belongs to the National Center for Atmospheric Research in Boulder, Colorado. The S-PolKa is a transportable, polarimetric, dual-wavelength Doppler weather radar. S-PolKa transmits and receives electromagnetic waves simultaneously at S band and Ka band (at 0.86 cm wavelength) with matched beamwidths and range resolutions, making it one of the two transportable S-band weather radars worldwide along with the NASA N-Pol radar that are available to the scientific community (see a separate chapter on the N-Pol radar in this book). The S-PolKa radar was designed to operate under high winds without a radome. The pedestal was stabilized by four 6 m containers making it relatively easy to assemble and operate since no concrete pad was needed. This mechanical design is unique and was completely fabricated by the NCAR machine shop [39]. The radar participated in 20 field campaigns during the period from 1996 to 2015 including overseas deployments as part of large international experiments in Europe (MAP), Taiwan (TIMREX), Caribbean (RICO), and Indian Ocean (DYNAMO).

The polarization state of the transmitted wave can be changed from H to V on a pulse-to-pulse basis (also known as H-V alternating) using an in-house designed and constructed fast mechanical switch with 49 dB of isolation on transmission and 40 dB on reception. With two receivers, the radar allows estimation of the elements of a full 2×2 coherency matrix.

The S-PolKa radar was effectively used in a wide range of studies including detailed analysis of the microphysical structures of deep convective storms [40], dual-wavelength retrievals of liquid water content [41], and restoration of the near-the-ground fields of refractive index and humidity [42], just to mention a few.

2.3.2 C-band research radars

The first polarimetric C-band radar for weather observations called POLDIRAD operated in Germany in the mid-1980s [43]. The radar provided very informative and reliable measurements of Z_{DR} and LDR. A decade later, several C-band polarimetric weather radars came into play with the design similar to the S-band polarimetric radars utilized in the US. Some of these C-band radars are described in the next subsection.

2.3.2.1 C-Pol (Australia)

One of the first polarimetric C-band radars, the Australian C-Pol radar was developed in the mid-1990s as a mobile radar in collaboration between the Australian Bureau of Meteorology, CSIRO Radio-physics, and NCAR. The radar is described in [44] and was the critical platform for field campaigns studying intense convection on tropical islands (MCTEX, [45]), monsoon convection in the South China Sea (SCSMEX, [46]) and for advanced nowcasting applications during the Sydney 2000 Olympics as part of the first WMO Forecast Demonstration Project [47] as well as its long-term deployment near Darwin where it operated until 2017. The radar also provided pivotal data for major international field programs exploring convection and its impact on the environment and troposphere-stratosphere exchange [48,49].

Early work with C-Pol focused on fundamental developments of dual-polarization rainfall estimation [50–52]. The C-Pol radar was also used in tandem with the Darwin 50 and 920 MHz wind profilers to study the retrieval of raindrop size distributions and to explore hail measurements with the C-band radar [53,54] as well as the development of some DSD climatologies and documenting systematic differences between convective and stratiform DSD [54,55]. The C-Pol data showed that hail was present in many tropical thunderstorms even though it rarely reached the ground [49,56] and demonstrated complex signatures associated with melting hail [56]. The profiler-C-Pol combination was also used to provide unique validation of microphysical retrievals [57].

2.3.2.2 King City radar (Canada)

The C-band King City radar located in King City 40 km north of Toronto was installed in 1984 and upgraded by adding dual polarization in 2004. The radar has a 0.65° beam making it one of the most sensitive C-band research radars. After polarimetric upgrade, the focus was on the adaptation of the S-band hydrometeor classification algorithm for C band in partnership with NSSL in the US. Additionally, the King radar served as the focus for two major field experiments to validate satellite-based cold season products. The first was the Canadian CloudSat CALIPSO Validation (C3VP) project in 2006. The second was the GPM Cold Season Precipitation Experiment (GCPEx) in 2012. For both campaigns, the King radar invoked unique scanning features coordinated with aircraft in-situ measurements and was the primary source of weather radar information for field study. The result of this work was the development of improved data quality practices, particle type identification and quantitative precipitation methods, particularly as applied to snow for both ground- and space-based remote sensing.

2.3.2.3 OU-PRIME radar (University of Oklahoma)

The OU-PRIME polarimetric radar was the most powerful research C-band radar with a half-degree antenna beam and 1-MW transmitted power. It was commissioned in Norman, Oklahoma, in 2009. The OU-PRIME observations were paired with the ones by the S-band KOUN radar which is located just 6.9 km from OU-PRIME. This created a unique opportunity to compare polarimetric measurements in severe storms with almost collocated C-band and S-band radars. Of the primary interest was comparison of the polarimetric S- and C-band signatures in tornadic and hail-bearing storms [58–60] and utilization of the weakly attenuated radar signals at S band to quantify much stronger attenuation at C band and develop attenuation/differential attenuation correction techniques at C band [61,62].

2.3.2.4 University of Alabama in Huntsville (UAH) Advanced Radar for Meteorological and Operational Research (ARMOR)

The C-band dual-polarization ARMOR radar was introduced for research at UAH in 2005. This radar was extensively used for the investigation of polarimetric signatures of hail [63] and tornadic debris at C band [64], and polarimetric radar and lightning observations in deep moist convection [65] among other research topics.

2.3.2.5 MRI solid-state polarimetric radar (Japan)

The polarimetric radar belonging to the Japanese Meteorological Research Institute (MRI) is the first C-band solid-state weather radar utilized for research. The transmitted peak power is only 3.5 kW requiring use of a long transmitted pulse (129 μs) with its subsequent compression to ensure fine radial resolution at the distances beyond 20 km from the radar. A short transmitted pulse of 1 μs is used to cover closer ranges. The focus of the research studies with the MRI radar was on rainfall estimation (possibly mixed with hail) and retrieval of raindrop size distributions at attenuating wavelength [66,67].

2.3.3 X-band radars

The first research X-band Doppler polarimetric radar was built in 1997 at the NOAA/Environmental Technology Laboratory in Boulder, Colorado (Figure 2.3). The name of the radar (HYDROX) implicates that its primary purpose was rainfall estimation and hydrological applications. HYDROX was a mobile radar mounted on a truck. Matrosov *et al.* [68,69] quickly demonstrated the advantages of rainfall estimation using X-band specific differential phase K_{DP}. With attenuation correction issues properly addressed, certain advantages of X-band radars over longer-wavelength radars operating at C or S band became evident. These advantages include higher mobility, smaller size and cost, lower power requirements, potentially higher spatial resolution, and stronger differential phase signals.

This initial success of the X-band radar weather observations encouraged academic laboratories and private companies to manufacture X-band radars for a wide range of weather applications including investigations of tornadic storms and aviation hazards in addition to quantitative precipitation estimation. During the first

Figure 2.3 First X-band Doppler polarimetric radar built at the NOAA/Environmental Technology Laboratory (circa 1998).

decade of the 21st century, multiple X-band polarimetric radars have been manufactured and tested in academic research centers. These include the X-POL and RaXPol radars built by the private company ProSensing based at the University of Massachusetts, the MP-X radar designed at the National Research Institute for Earth Science and Disaster Prevention (NIED) in Japan, the NOAA NSSL NOXP radar in Norman, Oklahoma, the University of Alabama in Huntsville Alabama X-band (MAX) radar, a family of Doppler on Wheels (DOW) mobile radars to mention a few. Traditional radar manufacturers also joined the race. The Enterprise Electronics Corporation (EEC) built a stationary X-band polarimetric radar (BoXPol) for the Meteorological Institute at the University of Bonn, Germany. The German company, Gematronik, started developing its own brand of mobile X-band radars of the METEOR series (e.g., Meteor 50DX). French company NOVIMET designed its HYDRIX X-band radar for rainfall estimates in the mountainous regions of Alps.

It is widely recognized that X-band polarimetric radars can be efficiently used as gap-fillers complementing larger surveillance S- and C-band radars in the areas of their poor coverage, often caused by the complex terrain [70]. There is also an apparent benefit of utilizing X-band radars for hydrological purposes to monitor the precipitation variability over small basins at scales smaller than typically observed with the longer-wavelength radars. Employing a network of several identical units can compensate for X-band radars' lack of spatial coverage. This strategy has proved economically and operationally feasible, as demonstrated by X-band networks like those used by the NSF-funded Center for Collaborative Adaptive Sensing of the Atmosphere (CASA) Integrated Project 1(IP1) [70], X-RAIN (X-band Polarimetric RAdar Information [71], and the Tropical Radar Network (TropiNet) [72]. The objective of these X-band weather radar networks was to retrieve meteorological echoes in the lower troposphere and adapt to rapidly changing severe weather.

As an example of effective utilization of a small network of X-band radars for hydrological studies, we highlight the University of Iowa XPOL system consisting of four mobile polarimetric radars manufactured by ProSensing Inc. [73]. In 2013, Iowa XPOLs participated in the field campaign Iowa Flood Studies (IFloodS) which was organized in central and northeastern Iowa in the Midwestern US by NASA in collaboration with the Iowa Flood Center. The IFloodS field campaign provided a unique opportunity to develop and assess rainfall estimators for mobile Iowa XPOL radars.

2.3.4 Short-wavelength polarimetric radars

Traditionally, millimeter-wave radars used for cloud research looked vertically and the only polarimetric variable they measured was *LDR*. In the last decade, several research scanning polarimetric radars at Ku, Ka, and W bands have been built to also measure Z_{DR}, Φ_{DP}, and ρ_{hv} [74]. As an example, the Dual-frequency Dual-polarized Doppler Radar (D3R) operating at Ku and Ka bands was manufactured to support NASA for ground validation of the Global Precipitation Measurement (GPM) mission satellite radar DPR operating at the same pair of wavelengths [75].

Ka-band scanning polarimetric radars prove to be very efficient for studying snow and ice where attenuation of millimeter waves is relatively small and the short wavelength ensures high radar sensitivity combined with good angular resolution. The examples are the MIRA-35 radar in Germany [76] and the KASPR radar in the US at the Stony Brook University Radar Observatory [77]. A dual-wavelength scanning polarimetric radar Ka-W SACR2 manufactured as part of the US Department of Energy (DOE) Atmospheric Radiation Measurements (ARM) program for the studies of polar clouds in Alaska operates at Ka-band and W-band frequencies [78].

2.3.5 Most recent developments

Most recent developments in weather radar technology include the use of the solid-state and phased array radars. During the last decade, the solid-state X-band Doppler polarimetric weather radars became available. The magnetron transmitters of the conventional radars need to be periodically replaced and the high power of the transmitted signal may cause premature wear and tear of the waveguides and microwave circuits. The solid-state radars operate with much lower voltages and peak powers (similar to a light bulb) which result in the improvement of their durability, extending their longevity, and lower maintenance.

Solid-state radars require transmitting very long pulses (up to 100 μs) to achieve the appropriate average power as opposed to the conventional magnetron or klystron radars with pulse lengths τ not exceeding 2–4 μs. Radial resolution of the weather radar is equal to $c\tau/2$, where c is the speed of light. Hence, the pulse length of 1 μs is equivalent to the resolution of 150 m. For pulse length of 100 μs, the radial resolution is 15 km which is not acceptable. Therefore, the received radar pulse has to be compressed 100 times by a special waveforming to ensure 150 m resolution and sufficiently high radar sensitivity. The problem is that the radar cannot receive a return signal while it is transmitting. Therefore, the range equal to $c\tau/2$ is not suitable for full compression of the long transmitted pulse and may be blind to the radar. One solution to this problem is transmission of a shorter pulse (or a sequence of shorter pulses) after a long pulse. This often causes discontinuity at range where processing of the short pulse is replaced with processing of the long pulse (i.e., 15 km if a 100 μs long pulse is transmitted) which is an issue for many solid-state weather radars. A variant solution is simultaneous transmission of the short pulse at an offset frequency [79]. A recent solution entailing progressive pulse compression eliminates the "blind" range without the fill-in pulses [80].

Toshiba (Japan) became the first company to manufacture solid-state weather radars for research and operational purposes. The operational network of X-band polarimetric radars owned by the Ministry of Land, Infrastructure, Transport, and Tourism (MLIT) of Japan is partially equipped with the solid-state radars. These radars have 200 W transmitter power and a pulse length up to 32 μs. With such pulse length and transmitted power, the radar detectability is quite modest but the blind area is relatively small. However, a high density of the MLIT radars avoids the blind range and modest detectability via significantly overlapping the coverage areas of the neighboring radars.

The current Service Life Extension Program (SLEP) will keep the US WSR-88D network operating through about 2035. It is very likely that SLEP will be further extended beyond 2035. The WSR-88Ds are very sophisticated but have some inherent limitations. The use of mechanically rotating dish antennas that have big inertia constrains the radar beam agility.

As a consequence of using "rigid" scanning strategies, the radar may needlessly scan huge volumes of the atmosphere where no weather phenomena of interest exist, whereas the areas of rapidly developing hazardous storms are sampled at a slow rate. Moreover, quick scans in vertical planes are not practical. Continuous rotation of massive antennas causes premature wear and tear of the mechanical parts. All these factors motivate an effort to evaluate the utility of the phased array technology with electronic scanning of the radar beam.

The radars with electronic beam steering have been used for decades in a military sector. However, phased array radars (PAR) have seldom been used for observing weather. The military PARs cannot be used for modern meteorological needs because technical requirements for polarimetric weather radars are more stringent. The greatest challenge is to enable polarimetric diversity in PARs that matches existing dual-polarization capability implemented on the WSR-88Ds.

The research on the possible utilization of the PAR technology for weather radars started at the NSSL in 2003 when the lab acquired an old SPY-1A PAR originally built for the US Navy and repurposed for weather observations. The SPY-1A radar did not have polarimetric capability and used a passive array of radiators and a WSR-88D transmitter not suitable for multibeam electronic steering.

During the last decade, NSSL in partnership with industry continued experimenting with different versions of polarimetric PAR like the 10-panel demonstrator and a larger-scale S-band 76-panel Advanced Technology Demonstrator (ATD) that replaced SPY-1A PAR in 2018. A more detailed description of the history of PAR development and capabilities of the modern phased array weather radars can be found in [81–83] and Chapter 1 of Volume 2 "Phased Array Weather Radar Developed in Japan".

2.3.6 Polarimetric upgrade of operational weather radars

The NEXRAD network of operational weather radars is a backbone of the US NWS observational infrastructure. It consists of 159 WSR-88Ds. The original network of single-polarization WSR-88Ds was designed in 1988 and completed in 1997. The initial version of the WSR-88D radar had Doppler capability and allowed measuring a radial component of the wind velocity as opposed to the earlier generations of weather radars capable of measuring only a power of the reflected radar signal.

After the benefits of radar polarimetry have been demonstrated during the 1990s in the research sector, the stage was set to demonstrate the value of dual-polarization data in an operational setting. The three agencies owning the WSR-88D network, NWS, Air Weather Service of the Air Force, and Federal Aviation Administration agreed to it. A Joint POLarization Experiment (JPOLE) was organized with the demonstration using the KOUN radar involving real-time transferring the Z, Z_{DR}, ρ_{hv}, and Φ_{DP} fields to the local NWS forecast office [84]. The success of the experiment satisfied the necessary conditions for approval of the upgrade by the US National Oceanic and Atmospheric Administration. The polarimetric upgrade of the WSR-88D radars started in 2010 and was completed in 2013.

NWS of developed countries around the world either have completed polarimetric upgrade of their operational weather radars or are in the process of doing so. In Europe, operational networks of polarimetric C-band weather radars have been deployed in a number of countries. The biggest ones are operated in France (29 radars), Turkey (19 radars), Germany (17 radars), and UK (15 radars). 17 C-band polarimetric radars in France are complemented by 7 S-band radars deployed in southern France and 5 X-band radars cover the Alpine region of France at its southeast. Another modern network of polarimetric weather radars is operated by the Korea Meteorological Administration (KMA). It consists of 11 S-band radars. China is quickly modernizing its network of operational weather radars by adding polarimetric capability to more than 100 of them. Japan is the only country in the world that runs an operational network of X-band polarimetric radars (in addition to the C-band network). Japanese X-band network numbers 35 radars and has been in operation since 2010. Some of the recently deployed radars use solid-state transmitters with lower peak transmitted power that implies the need for pulse compression. A primary function of the Japanese X-band network is rainfall estimation because Japan is prone to frequent flash floods caused by heavy rain. A regional composite of rain totals retrieved from the X-band radar measurements is updated every minute and such product proved to be very useful for a large number of customers. Longer wavelength radars have significantly higher power of a transmitted signal and larger range of operations. A limited range of X-band radars is dictated by two factors. One is the difference in the transmitted power and radar sensitivity. The second one, likely most important, is a stronger impact of attenuation in precipitation at shorter radar wavelengths. This is a primary reason why either S- or C-band radars are preferred for weather surveillance.

Although X-band radars become almost blind in the proximity of very intense bands of precipitation or hailstorms, such periods of "blindness" are relatively short and the corresponding loss of radar data is relatively rare. The situation when the X-band radar signal is attenuated but not lost completely is much more common. Under this scenario, polarimetric capability allows to effectively correct radar reflectivity by using differential phase Φ_{DP} (2.16). A similar correction for differential attenuation of differential reflectivity Z_{DR} can also be applied (2.17).

2.4 Weather applications of the Doppler polarimetric radars

Quantitative precipitation estimation (QPE) as well as classification of radar echo and hydrometeors were traditionally the primary tasks of weather radars. It is hard to overestimate a societal impact of accurate precipitation measurements in the context of global warming with rapidly increasing frequency of devastating flash flood events, particularly those associated with landfalling hurricanes and typhoons causing massive damage to property and infrastructure as well as human fatalities. While accurate precipitation measurements are critical for storm warnings and hydrology, they are also important for assimilation into the NWP models and optimization of microphysical parameterization.

Polarimetric radar is uniquely suited for discrimination between meteorological and nonmeteorological radar echoes and for classification of different types of atmospheric hydrometeors because they have very different polarimetric characteristics. Reliable identification of tornadic debris, large hail, and freezing rain is critical for timely severe weather warnings. In this section, we will provide a brief overview of the polarimetric QPE and classification methods.

2.4.1 Quantitative precipitation estimation

Traditional radar methods for rain and snow estimation were based on the use of a single radar variable, the radar reflectivity factor Z. Significant progress in radar QPE was achieved during last two to three decades with the introduction of polarimetric weather radars capable to complement measurements of Z with additional variables such as differential reflectivity Z_{DR}, specific differential phase K_{DP}, and most recently, specific attenuation A. The advantages of polarimetric radar measurements for rainfall estimation have been demonstrated in a number of research studies back in the 1990s and early 2000s [21,50,84–89]. It was shown that the use of multiple radar variables helps to reduce the QPE uncertainty caused by the DSD variability, whereas the specific differential phase K_{DP} addresses the issues of radar miscalibration, attenuation, and partial beam blockage.

In the 2000s, the polarimetric technology and methodology for rainfall estimation matured to the point of justifying their operational implementation on the nationwide weather radar networks. The US spearheaded this effort after the successful completion of the Joint Polarization Experiment (JPOLE) demonstration project [84] which culminated in the polarimetric upgrade of all WSR-88Ds in 2013. Another important step was the integration of the information from the WSR-88Ds using the Multi-Radar Multi-Sensor (MRMS) platform that combines the polarimetric radar data with atmospheric environment data, satellite data, and lightning and rain gauge observations to generate a suite of severe weather and QPE products [90].

2.4.1.1 Polarimetric rainfall relations

The variability of DSDs is a primary source of uncertainty in the radar rainfall estimation. Because radar reflectivity Z is a sixth moment of DSD and rain rate R is approximately proportional to its 3.67th moment [4], the magnitude of Z is primarily determined by the contribution of a few largest drops in the raindrop size spectrum, whereas a bulk of smaller drops dominate rain rate. This spurred the use of multiple $R(Z)$ relations for single-polarization radars. The choice of a particular relation depends on the type of rain (convective vs stratiform, tropical vs continental, etc.), season, and climate region.

Historically, the first polarimetric rainfall relations used a combination of Z and Z_{DR} [91,92]. Utilization of Z_{DR}, a polarimetric radar variable that depends on the median volume diameter of raindrops and is insensitive to their concentration, together with Z was expected to reduce the sensitivity of the rain rate estimate to the DSD variability. However, Z_{DR} is even more affected by a few largest raindrops than Z. Therefore, the benefit of the $R(Z, Z_{DR})$ estimator instead of the $R(Z)$ relation turned out to be relatively marginal. The radar variables K_{DP} and A provided better improvement because both are proportional to the DSD moments that are much closer to the 3.67th moment than Z and are immune to the radar calibration errors, attenuation in rain, and partial beam blockage.

A more quantitative assessment of relative performance of different rainfall relations can be obtained using Figure 2.4 where the dependencies of the fractional mean absolute error (fMAE) on rain rate for the $R(Z)$, $R(Z, Z_{DR})$, $R(K_{DP})$, and $R(A)$ relations optimized for the DSDs measured in Oklahoma at S, C, and X bands are shown. These relations are listed in Table 2.1. The advantages of all three polarimetric relations compared to the $R(Z)$ relation are obvious in a wide range of rain intensities at all three microwave bands. The $R(A)$ is an apparent winner at low-to-moderate rain rates. Its advantage is particularly impressive at S band where the fMAE caused by the variability of DSD is less than 10–15% in a wide range of rain intensities. For higher rain rates exceeding 10 mm/h, fMAE tends to increase at C and X bands for all relations except $R(K_{DP})$ that is least sensitive to the effect of the resonance scattering on large raindrops which is primarily responsible for such an increase. Therefore, the K_{DP}-based relation is an obvious choice for heavy rain at C and X bands. The K_{DP}-based algorithms for rainfall estimation were used in a large number of studies and operational applications. A long list of various $R(K_{DP})$ relations can be found in [4] (their Table 10.3) and in the recent review of the polarimetric QPE methods by Ryzhkov *et al.* [93].

A very important advantage of the K_{DP}- and A-based QPE algorithms is their immunity to the radar miscalibration, attenuation in rain, partial beam blockage, and impact of a wet radome. All discussed radar rainfall relations have advantages and disadvantages at different rain rates and radar wavelengths. Hence, it is beneficial to use composite algorithms by combining various relations. The operational polarimetric rainfall estimation algorithm currently implemented on the WSR-88D radar network uses a combination of the $R(Z)$, $R(A)$, and $R(K_{DP})$ relations [94–96] and is called the Q3DP algorithm [94]. The $R(K_{DP})$ relations are used in areas of high Z where rain may be mixed with hail and the $R(A)$ relation is utilized otherwise. The QPE algorithm resorts to various $R(Z)$ relations if the radar sample is within the melting layer or above or if the total span of differential phase $\Delta\Phi_{DP}$ in rain is too small or beam blockage is severe.

The Q3DP algorithm became operational from October 2020 and demonstrated very good performance, particularly in heavy rain associated with flash floods. Figure 2.5 illustrates its performance for the most significant flash flood events of the 2021 warm season in the US. One occurred in Tennessee on August 21 and caused 27 fatalities and another one was associated with the hurricane Ida during the period from August 30 till September 2 with a death toll of 67 people. The

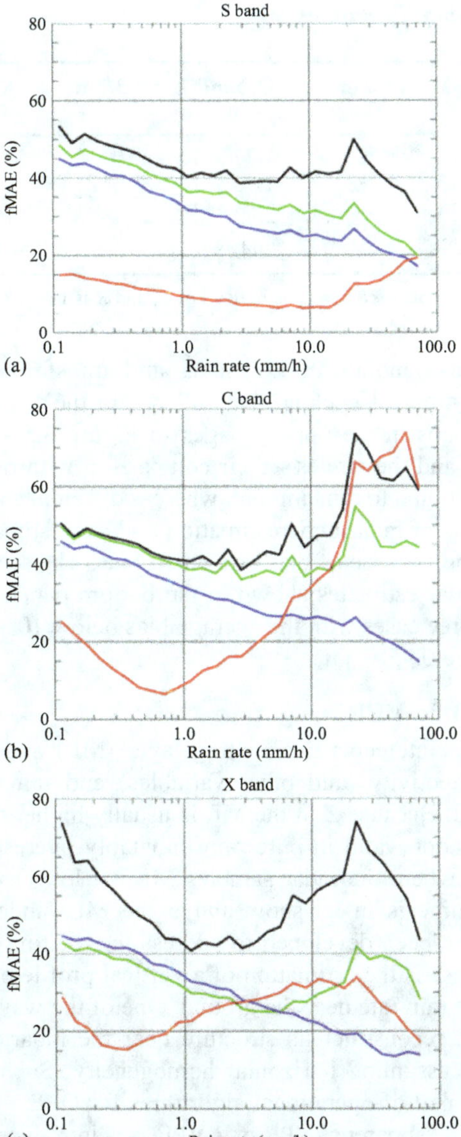

Figure 2.4 Dependencies of the fractional mean absolute error (fMAE) on rain rate for the R(Z) (black curves), R(Z, Z_{DR}) (green curves), R(K_{DP}) (blue curves), and R(A) (red curves) relations optimized for the Oklahoma DSDs at S, C, and X bands (see Table 2.1). From [93].

Table 2.1 Radar rainfall relations

Radar relation	S band λ = 11.0 cm	C band λ = 5.3 cm	X band λ = 3.2 cm
$R(Z)$	$1.94 \times 10^{-2} Z^{0.694}$	$1.77 \times 10^{-2} Z^{0.712}$	$3.06 \times 10^{-2} Z^{0.639}$
$R(Z, Z_{dr})$	$1.66 \times 10^{-2} Z^{0.738} Z_{dr}^{-0.979}$	$1.70 \times 10^{-2} Z^{0.731} Z_{dr}^{-0.459}$	$1.50 \times 10^{-2} Z^{0.790} Z_{dr}^{-2.17}$
$R(K_{DP})$	$47.1 K_{DP}^{0.790}$	$25.3 K_{DP}^{0.776}$	$16.9 K_{DP}^{0.801}$
$R(A)$	$4120 A^{1.03}$	$294 A^{0.89}$	$43.5 A^{0.79}$

Z is in mm^{6}m^{-3}, Z_{dr} is in a linear scale, K_{DP} is in deg km^{-1}, and A is in dB km^{-1}.

hurricane Ida had its landfall in Louisiana and transformed from a tropical depression to a post-tropical cyclone that rolled into the US Northeast two days later. Color-coded maps in the upper images in Figure 2.5 show 24-hour radar-estimated rain totals and the circles—their comparison with rain gauges. The size of the circle is proportional to rain amount, whereas different shades of red and blue indicate various levels of radar underestimation and overestimation. White or pale colors mean either the absence of bias or a small bias. The scatterplots of 24-hour totals versus their radar estimates shown in the bottom panels indicate very good performance in all three cases with the average bias below 10% and the correlation coefficient well above 0.9.

2.4.1.2 Polarimetric VPR

Once the radar beam intercepts the melting layer (ML) and snow/ice aloft, the relation between reflectivity (and other variables) and rain rate at the surface becomes complicated. Because Z in the ML is usually higher than in rain near the surface, the $R(Z)$ relation valid in pure rain inevitably overestimates surface rain rate at the distances where the radar samples ML. At longer distances where the radar resolution volume is in ice/snow above the ML, underestimation occurs. Various methods have been developed to address the melting layer problem [97–100]. Common to these is the estimation of a vertical profile of reflectivity (VPR) and its use to retrieve rain rate near the ground. One of the ways to obtain the VPR is to scan the storm, reconstruct its structure near the radar, and extrapolate its effect to long range assuming horizontal homogeneity. Such volume-scan VPR corrections are often part of operational algorithms for QPE.

The concept of an "apparent" VPR or AVPR was introduced in [101,102]. The VPR correction is performed in the "bright band affected area" (BBA) within the stratiform part of the storm that is identified using radar reflectivity Z and cross-correlation coefficient ρ_{hv}. The average radial profile of Z at the lowest antenna tilt is determined within BBA and subtracted from the radial profiles of Z at each particular azimuth. Then, the standard $R(Z)$ relation is applied to estimate rain rate. Therefore, it is expected that the radial dependency of Z is entirely determined by the vertical profile of Z through the ML and the vertical structure if a storm is horizontally uniform within BBA. This methodology assumes that the BBA is relatively homogeneous azimuthally and is called "tilt-VPR." Such an assumption

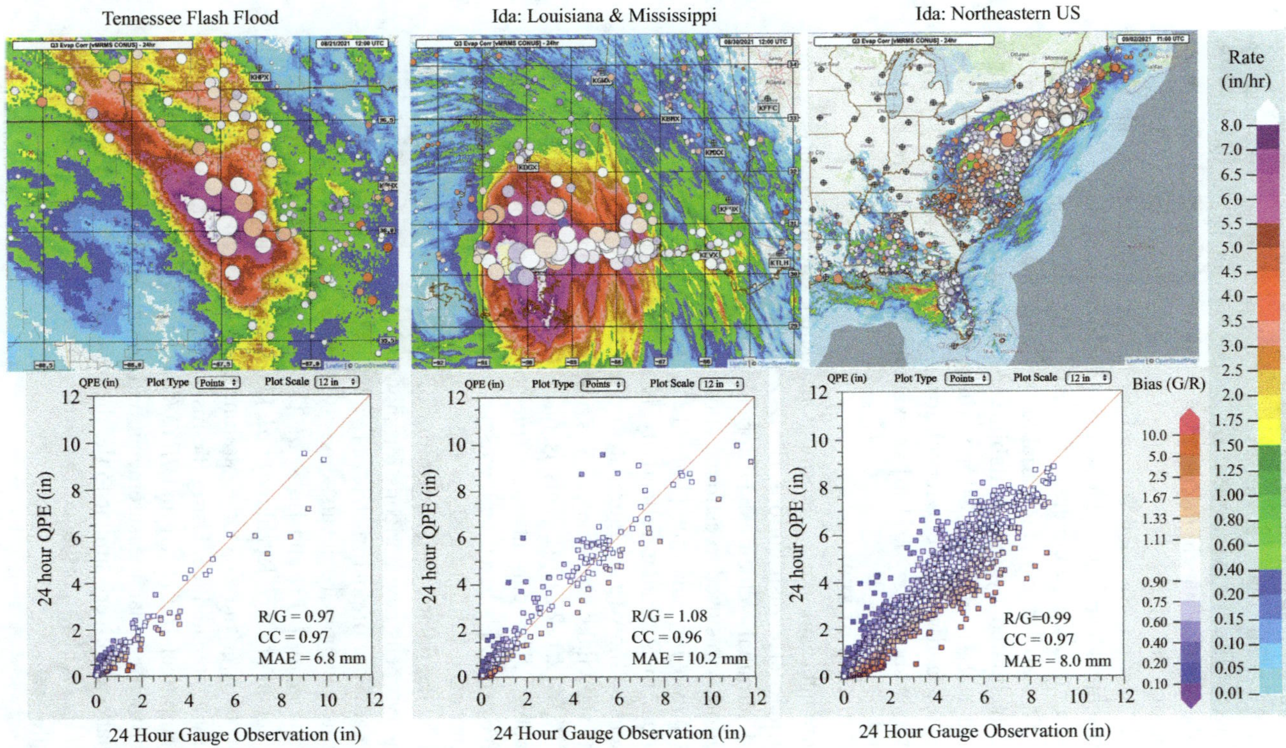

Figure 2.5 *Top panels: Maps of 24 h rain total estimated using Q3DP rainfall algorithm in Tennessee during the period ending at 1200 UTC on August 21, 2021, in Louisiana during the period ending at 1200 UTC on August 30, 2021, and US Northeast during the period ending at 1100 UTC on September 2, 2021. Colored dots represent CoCoRaHS gauge sites. The size of the dots represents gauge observed amounts and the color represents the gauge/QPE (G/R) bias ratios. Bottom panels: Scatterplots of radar 24-hour rain totals versus gauge estimates. The statistic scores in the scatterplots are the bias ratio (R/G), mean absolute error (MAE), and correlation coefficient (CC). From [93].*

can fail in the presence of frontal boundaries with large variations of the ML height which usually happens during cold season with lower ML. Hanft *et al.* [103] suggested the so-called dpVPR methodology to derive azimuthally-dependent VPRs, which results in the overall improvement of the VPR correction. Note that, in addition to reducing artificial enhancement of Z and $R(Z)$ in BBA, the dpVPR algorithm reduces the negative bias in $R(Z)$ at longer distances from the radar where the radar beam is above the ML.

Concluding this section, we need to mention recent studies where artificial intelligence (AI)/machine learning (ML) algorithms have been used for polarimetric QPE. This is a rapidly growing area of research stimulated by accumulation of large datasets of polarimetric radar variables combined with surface rainfall measurements and facilitated by the availability of abundant modern computer resources. A detailed analysis of the AI/ML polarimetric QPE algorithms is beyond the scope of this review and we just cite recent articles by Zhang *et al.* [104], Shin *et al.* [105], and Wolfensberger *et al.* [106] as examples for interested readers.

2.4.1.3 Measurements of snow

Radar snow measurements present an enormous challenge due to immense variability of snow particle size distributions (PSDs), density, shapes, orientations, crystal habits, and terminal velocities. Historically, radar reflectivity factor at horizontal polarization Z was used to estimate snow water equivalent rates S in the form of numerous power-law relations [107–112]. The majority of these assume that Z is proportional to S^2. Radar reflectivity is very close to the fourth PSD moment for low-density snow (e.g., aggregates) due to the inverse dependence of snow density on the snowflake size whereas S is proportional to significantly lower order moment of PSD (2.1–2.2) [4]. This is one of the primary problems with existing $S(Z)$ relations. The current radar methodology for snow estimation employed on the WSR-88Ds is based on the usage of a multitude of climatological $S(Z)$ relations optimized for different snow types and geographical areas and has not yet capitalized on the polarimetric capability of the WSR-88D radars.

Radar polarimetry opens new horizons for snow habit classification and quantification. Since the advance of polarimetry, only a handful of studies have explored polarimetric quantitative snow retrievals—mostly focused on ice water content (*IWC*). Vivekanandan *et al.* [113] and Lu *et al.* [114] utilized specific differential phase K_{DP} for estimation of *IWC*. Ryzhkov *et al.* [115] and Ryzhkov and Zrnić [4] used a combination of K_{DP} and Z_{DR} or K_{DP} alone to quantify *IWC* of pristine or lightly-to-moderately aggregated ice. Using the instrumented aircraft, Nguyen *et al.* [116] derived the empirical $IWC(K_{DP})$ and $IWC(K_{DP}, Z_{DR})$ relations from the airborne X-band polarimetric radar measurements and in situ microphysical probes. These empirical relations are very similar to the theoretical ones derived in [4].

Bukovcic *et al.* [117,118] introduced the first polarimetric relations for snow water equivalent rate S:

$$S(K_{DP}, Z) = \frac{27.9 \times 10^{-3}}{(F_o F_s)^{0.62}} \left(\frac{p_0}{p}\right)^{0.5} (K_{DP}\lambda)^{0.62} Z^{0.33} \tag{2.18}$$

and

$$S(K_{\mathrm{DP}}, Z_{\mathrm{dr}}) = 10.8 \times 10^{-3} \left(\frac{p_0}{p}\right)^{1/2} \frac{K_{\mathrm{DP}}\lambda}{1 - Z_{\mathrm{dr}}^{-1}} D_{\mathrm{m}}^{0.15}. \tag{2.19}$$

In (2.18) and (2.19), p is the atmospheric pressure, $p_0 = 1013$ mb, λ is the radar wavelength in mm, K_{DP} is in deg/km, Z is in $\mathrm{mm^6 m^{-3}}$, Z_{dr} is the differential reflectivity in a linear scale, and D_{m} is the mean volume diameter of snowflakes determined as [4]

$$D_{\mathrm{m}} = -0.1 + 2.0\left(\frac{Z(1 - Z_{\mathrm{dr}}^{-1})}{K_{\mathrm{DP}}\lambda}\right)^{1/2}. \tag{2.20}$$

In (2.18), F_{o} and F_{s} are the factors characterizing orientation and shapes of snowflakes. The factor F_{o} is determined by the width of the canting angle distribution and the factor F_{s} is a function of the aspect ratio of snowflakes.

Both relations in (2.18) and (2.19) are less sensitive to the variability of snow size distributions than the $S(Z)$ relations. The biggest downside of the $S(K_{\mathrm{DP}}, Z)$ snow estimator is its sensitivity to the shape and orientation factors F_{s} and F_{o}. This emphasizes the need for realistic assumptions about snow particles' shapes and orientations either from in situ microphysical measurements or from polarimetric radar retrievals. Great advantage of the $S(K_{\mathrm{DP}}, Z_{\mathrm{dr}})$ estimate is its immunity to the variability of the orientations and shapes of ice particles or snowflakes. However, it is quite sensitive to the degree of snow riming, whereas the $S(K_{\mathrm{DP}}, Z)$ estimate is not. It is also quite vulnerable to possible biases of the Z_{DR} measurements, especially if Z_{DR} is low. Hence, relation (2.19) is not recommended if $Z_{\mathrm{DR}} < 0.3$ dB.

The performance of polarimetric algorithm for snow estimation is illustrated for heavy snowfall event that occurred in the US Northeast on March 14–15, 2017. Temporal dependencies of instantaneous snowfall rates and accumulations estimated using the standard nonpolarimetric relation $S(Z) = 0.088Z^{0.5}$ and two polarimetric relations specified by (2.18) and (2.19) are shown in Figure 2.6. The $S(Z)$ relation moderately underestimates snowfall rates throughout the event, with peaks not greater than 3 mm $\mathrm{h^{-1}}$ (gauge maximum is 5.5 mm $\mathrm{h^{-1}}$). Polarimetric relations are in good agreement with the gauge, especially $S(K_{\mathrm{DP}}, Z)$, which more accurately reproduces the peaks in rates. Polarimetric-based accumulations are in good agreement with the gauge estimate, whereas the $S(Z)$ relation underestimates the SWE amounts by about 40%. Additional examples of polarimetric measurements of snow can be found in [118].

The results in Bukovcic *et al.* [117,118] are quite encouraging but some serious challenges still remain. The values of K_{DP} and Z_{DR} in heavily aggregated dry snow are close to zero, which limits the applicability of (2.18) and (2.19). Moreover, a single snow event may have both wet and dry snow segments with possible presence of graupel or plates during different time intervals, which makes polarimetric snow QPE quite difficult under such a scenario.

2.4.2 Classification of radar echo and hydrometeors

There are various classification methods that utilize statistical decision theory, machine learning/neural networks, fuzzy logic, etc. The fuzzy logic methodology is

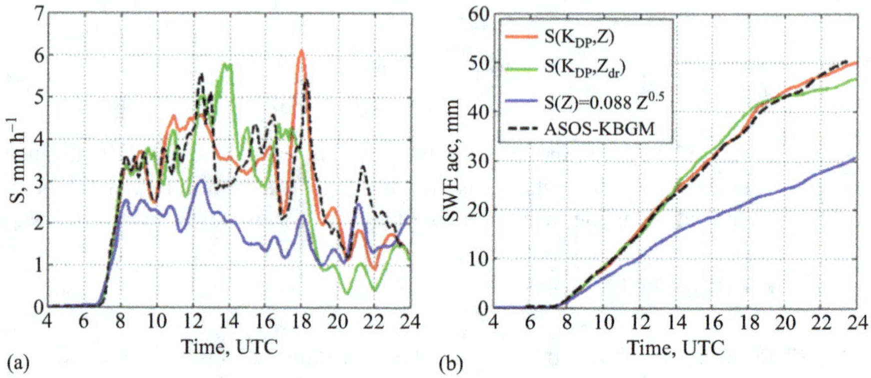

Figure 2.6 Instantaneous snow rates S (a) and snow water equivalent (SWE) accumulations (b) for the snow event observed by the KBGM WSR-88D radar on March 14, 2017. Black dashed curves indicate the estimates from the ASOS extinction coefficient and red, green, and blue curves—the radar estimates using the S(K_{DP},Z), S(K_{DP},Z_{DR}), and S(Z) relations, respectively. From [93].

currently the most popular classification technique implemented on the modern operational weather radar networks. It can be easily implemented and tailored for identification of the classes for which the domains in the space of polarimetric variables overlap and the variables may be contaminated by measurement noise. Furthermore, in the fuzzy logic approach, the confidence factors can be conveniently assigned to various types of measurements.

The first fuzzy logic classification schemes for polarimetric radars have been introduced in the 1990s [19,119–121] followed by the majority of the studies during subsequent two decades [122–131]. The first hydrometeor classification algorithm (HCA) operationally implemented on the US WSR-88D radar network utilizes the scheme described in [125]. The HCA discriminates between ten classes of radar returns: (1) ground clutter including the one due to anomalous propagation (GC/AP), (2) biological scatterers (BS), (3) dry aggregated snow (DS), (4) wet snow (WS), (5) crystals of various orientation (CR), (6) graupel (GR), (7) "big drops" (BD), (8) light and moderate rain (RA), (9) heavy rain (HR), and (10) a mixture of rain and hail (RH). The "big drops" category designates rain with a DSD skewed toward large sizes. This usually implies the presence of drops bigger than 3 mm and the deficit of smaller drops. "Big drop" signatures are generally caused by size sorting associated with convective updrafts often exemplified with Z_{DR} enhancement.

Six polarimetric variables are directly utilized for discrimination. These are (1) reflectivity factor at horizontal polarization Z, (2) differential reflectivity Z_{DR}, (3) cross-correlation coefficient ρ_{hv} between horizontally and vertically polarized radar returns, (4) specific differential phase K_{DP}, (5) texture parameter SD(Z) of the Z field, and (6) texture parameter SD (Φ_{DP}) of the field of differential phase Φ_{DP}.

In the fuzzy logic methodology, different classes of radar echo (or hydrometeors) are described by the sets of the membership functions $P^{(i)}(V_j)$ characterizing distributions of polarimetric variables V_j for the ith class. The parameters of the membership functions are determined from theoretical simulations and observations. Different radar variables may have different classification potential with respect to various classes, which is reflected in the matrix of weights W. Any degradation in the quality of radar measurements attributed to bias or noise can be addressed via the confidence vector Q.

The fuzzy logic HCA for WSR-88D prescribes computing the aggregation value A_i for each ith class of the radar echo via the weighted sum of the membership functions:

$$A_i = \frac{\sum_j W_{\text{ij}} Q_j P^{(i)}(V_j)}{\sum_j Q_j W_{\text{ij}}}, \tag{2.21}$$

where the factor W_{ij} characterizes classification efficiency of variable V_j with respect to ith class and Q_j is the confidence index quantifying possible degradation of the measurement of the variable V_j caused by (1) radar miscalibration, (2) attenuation, (3) nonuniform beam filling (NBF), (4) partial beam blockage (PBB), (5) the magnitude of ρ_{hv} (which determines the statistical errors of all polarimetric variables), and (6) receiver noise. The parameters of the membership functions $P^{(i)}$ (V_j) for the initial version of HCA are specified in [125] and in [4] (their Chapter 9).

Because polarimetric characteristics of rain below the melting layer and snow above it are very similar, effective performance of the hydrometeor classification algorithm is contingent on the accurate melting layer designation which is a crucial module of any classification scheme. The melting layer (ML) in stratiform clouds exhibits very pronounced polarimetric signatures that ensure its reliable detection. In addition to the classical Z enhancement known as the "bright band," it is characterized by the strong increase of Z_{DR} near the bottom of the ML and sharp depression of ρ_{hv} coupled with the *LDR* enhancement in the middle of the ML. The values of Z, Z_{DR}, and ρ_{hv} have been used to design the first ML detection algorithm implemented on WSR-88Ds [132]. There are alternative techniques for detecting the melting layer based on slightly different principles. The Brandes and Ikeda [133] method matches observed polarimetric radar measurement profiles with idealized model profiles of Z, *LDR*, and ρ_{hv} expected in the ML, whereas Tabary *et al.* [134] present a similar technique for operational ML identification at C band but use solely profiles of ρ_{hv} as well as Matrosov *et al.* [135] at X band. Wolfensberger *et al.* [136] utilize the product of Z (in linear scale) and $(1-\rho_{\text{hv}})$ for detection of the melting layer at X band.

The problem with all these methodologies for ML detection is that only the data collected in a close proximity to the radar are utilized and, as a result, designation of the ML is made at relatively close distances from the radar and does not account for spatial nonuniformity of the ML far away from the radar. To detect and quantify the ML at longer ranges, the data from lower antenna tilts (down to 0.5°) have to be used. Recently Ryzhkov and Krause [137] suggested a more sophisticated method, which allows to produce the maps of the heights of the ML top and bottom up to 150–200 km from the radar.

According to the NEXRAD HCA, classification is performed at each antenna elevation and a typical example of the vertical cross-sections of Z, Z_{DR}, ρ_{hv}, and classification results constructed from the series of conical scans is illustrated in Figure 2.7. The melting layer is identified as wet snow (light green) and separates dry snow and rain in the stratiform part of the storm. In the storm's convective part, graupel is shown in yellow and hail in pink. Hail is detected aloft (two small areas) and likely melts completely before reaching the ground. Biological scatterers are recognized in the boundary layer ahead of the storm at the distances beyond 95 km from the radar for this case.

In a modified version of the WSR-88D HCA, the class "rain mixed with hail" was split into three separate size categories of small, large, and giant hail [25–27]. The term "small hail" is used for size smaller than 2.5 cm, "large hail"—for maximal sizes between 2.5 and 5.0 cm, and hail in excess of 5.0 cm is considered "giant." The corresponding module in the WSR-88D HCA algorithm is called "Hail Size Discrimination Algorithm" (HSDA). The HSDA primarily capitalizes on the fact that the magnitudes and vertical profiles of Z, Z_{DR}, and ρ_{hv} in the rain/hail mixture below the freezing level depend on the dominant size of hail undergoing melting.

Polarimetric radar detects tornado efficiently because lofted tornadic debris typically has large size, irregular shape, nearly random orientation, and high dielectric constant that produce high Z, low Z_{DR}, and anomalously low ρ_{hv} that distinguish it from meteorological scatterers. The original WSR-88D HCA algorithm [125] was modified by Snyder and Ryzhkov [23] to add a tornado class or "tornadic debris signature" (TDS). Identification of tornado requires some estimate of vertical vorticity, which is accomplished by adding azimuthal wind shear as an input variable to the classification scheme.

In deep convective storms, their ability to produce large hail or heavy precipitation depends on the strength and size of a convective updraft. Fortunately, such updrafts manifest themselves as the so-called "Z_{DR} columns"—the appearance of an upward extension of enhanced Z_{DR} above the ambient 0°C level. The Z_{DR} column signature is attributed to partially frozen large raindrops lofted by the updraft and water-coated graupel or hail which grow there [138]. Higher Z_{DR} columns signify stronger updrafts and their detection is crucial to estimate potential severity of the storm. This motivated development of an automated Z_{DR} column detection algorithm for the operational WSR-88D [139].

The initial HCA algorithm implemented on the WSR-88D was developed for warm-season weather and, therefore, required modification to address classification issues related to transitional winter weather such as detection of freezing rain and discrimination between rain, ice pellets/sleet, and different types of snow. Ordinary warm rain and freezing rain have identical polarimetric characteristics and their discrimination requires knowledge of ambient temperature that motivates

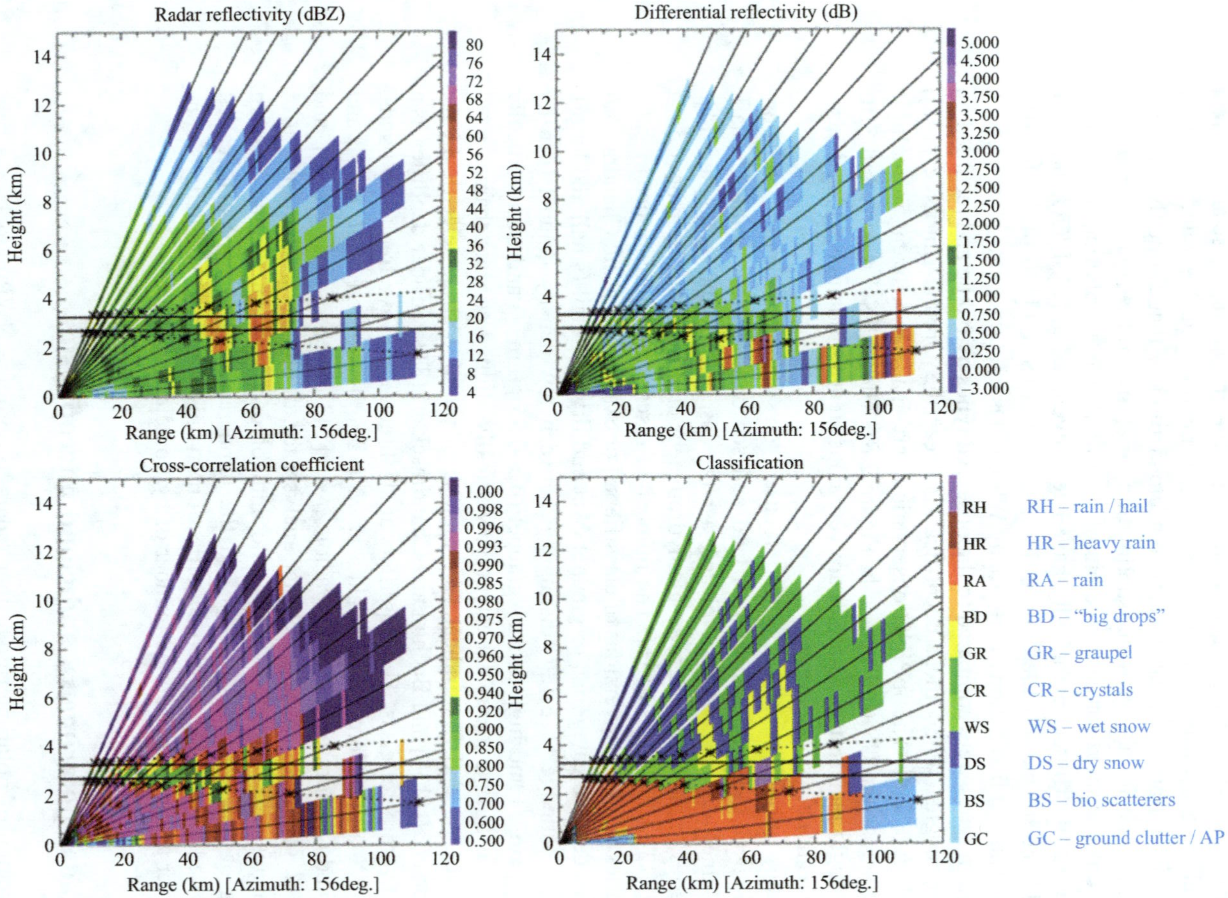

Figure 2.7 *Composite plot of Z, Z_{DR}, ρ_{hv}, and the results of classification in the RHI cross-section for the Oklahoma storm observed by the KOUN radar on May 13, 2005. Overlaid solid horizontal lines indicate the top and bottom of the "true" melting layer and dotted lines—the boundaries of the melting layer within the radar beam as functions of the distance from the radar. From [125]. © American Meteorological Society. Used with permission.*

combining polarimetric radar data with information about vertical profiles of temperature and humidity which can be retrieved from numerical weather prediction (NWP) models. Ice particles generated at higher levels in the clouds undergo various microphysical changes before they fall to the ground and these are different for particles with different initial sizes. Their microphysical evolution can be described in a framework of spectral bin cloud models that explicitly treat the impact of depositional growth, sublimation, melting, refreezing, riming, and aggregation separately for each size bin of the initial size distribution of ice aloft. Such models of different complexity have been developed at NSSL/OU and serve as a backbone for the so-called Spectral Bin Classifier (SBC) [140,141]. The SBC algorithm was applied nationally (to many WSR-88Ds) and the fields of designated precipitation types were compared to the observed types by public and recorded on the meteorological Phenomena Identification Near the Ground (mPING; [142]) system (Figure 2.8). The mPING observations are made by citizens in multiple locations across the US (left panels in Figure 2.8). The mPING reports and SBC analyses (right panels in Figure 2.8) are very consistent in the three selected cases of the transitional winter weather.

Various modifications of HCA focused on the identification of non-meteorological radar echoes such as land and sea clutter, biota, military chaff, smoke plumes, dust storms, and volcanic ash are described in the monograph [4].

A series of alternative fuzzy logic hydrometeor classifiers has been devised at the CSU [121,126,127,130,131]. The number of discriminated hydrometeor categories varies in different versions of classifiers. For example, Liu and Chandrasekar [121] consider two categories of graupel (wet and dry) and hail (small and large). Dolan and Rutledge [126,127] distinguish between low-density graupel (density $\rho < 0.55$ g cm^{-3}) and high-density graupel ($\rho > 0.55$ g cm^{-3}) and treat vertically aligned ice as a separate crystal category. Vertically aligned crystals characterized by negative Z_{DR} and K_{DP} indicate the regions of possible strong electric fields. Thompson *et al.* [130] and Bechini and Chandrasekar [131] consider dendrites as a separate hydrometer class. The classifiers at all three microwave frequency bands, S, C, and X are explored by Al-Sakka *et al.* [129], Thompson *et al.* [130], and Bechini and Chandrasekar [131]. Grazioli *et al.* [143], Wen *et al.* [144,145], and Besic *et al.* [146] suggest using cluster analysis for an objective determination of the distinguishable classes of radar echo and for derivation of the membership functions of polarimetric variables and the temperature. This approach, called unsupervised or semi-supervised classification, avoids scattering simulations conducted over an arbitrary defined number of hydrometeor classes.

2.5 Radar polarimetry and cloud microphysics

The utilization of polarimetric weather radars to optimize NWP models is a new frontier of research. It is widely understood that inadequate microphysical parameterization schemes in NWP models are a primary source of forecast uncertainties [147,148]. Due to its ability to distinguish between hydrometeors with different

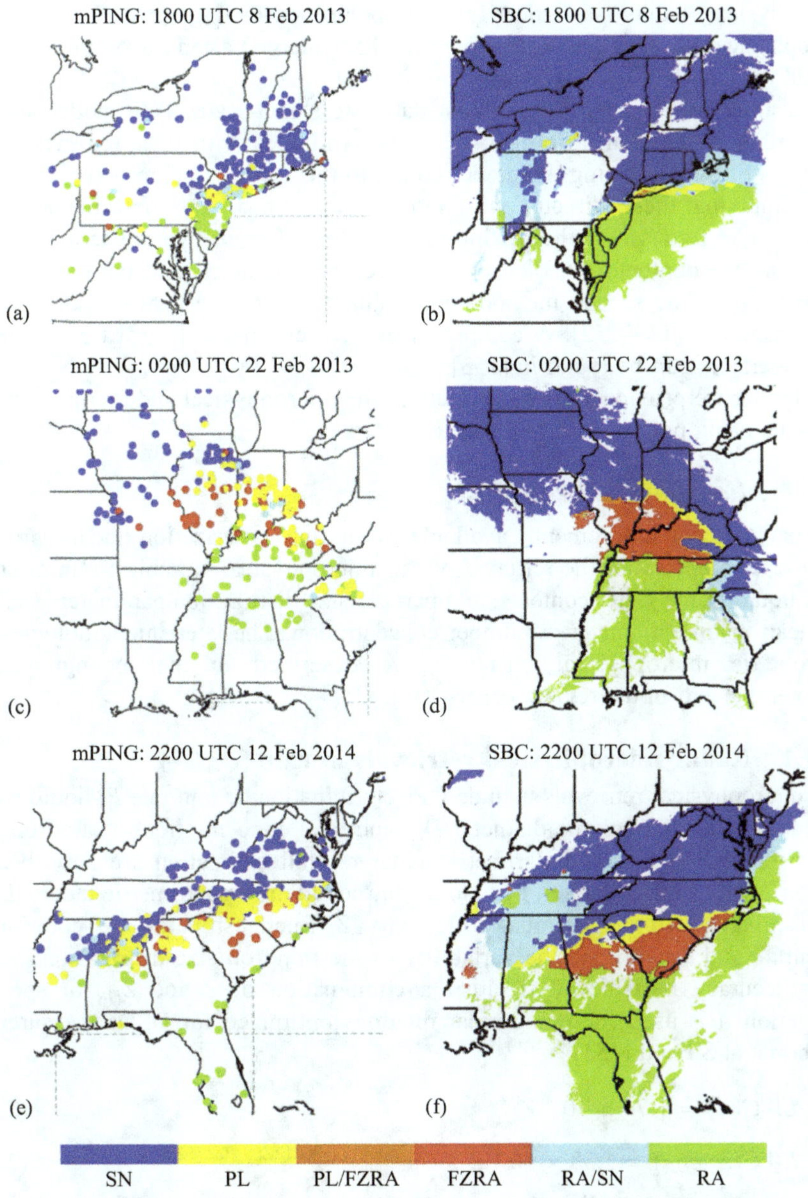

Figure 2.8 (a), (c), (e) mPING distributions of observed precipitation type for the 2-hour period surrounding the time indicated at the top of each panel, and (b), (d), (f) the corresponding SBC analyses. In (b), (d), and (f), only the areas with observed composite reflectivity larger than 0 dBZ are shaded. From [140]. © American Meteorological Society. Used with permission.

microphysical habits and identify the "polarimetric fingerprints" of various microphysical processes, polarimetric radar has emerged as an important source of needed information.

Assimilation of polarimetric radar data into storm-scale NWP models can be done either by comparing prognostic model variables with their retrieved radar estimates or by converting the model output to the fields of polarimetric variables and comparing these directly with radar observations. The first methodology implies radar retrieval of the mixing ratios q_i for different hydrometeor types (or water and ice contents), median or mean volume diameter, and particle number concentration. The second methodology requires use of the forward polarimetric radar operators [149–152]. A comprehensive review of the forward polarimetric radar operators can be found in [153].

Herein, we provide a brief overview of the microphysical and thermodynamic retrievals using polarimetric radar data.

2.5.1 Microphysical retrievals

Multiparameter measurements available from dual-polarization radars provide ample opportunities for the retrieval of key microphysical variables: rain or snow rates, liquid or ice water contents, and particle size distribution parameters such as the mean volume diameter or number concentration. The latest methodologies for polarimetric microphysical retrievals are described in the monograph [4] (Chapter 11) and in the review paper [153].

2.5.1.1 Radar microphysical retrievals in rain

Rain microphysical retrievals include the determination of rain rate R, liquid water content LWC, mean volume diameter D_m, and total concentration of raindrops N_t. Novel methodologies for polarimetric radar rainfall estimation are described in Section 2.4.1.1 of this chapter. Prior to the introduction of polarimetric radars, LWC was estimated using only radar reflectivity Z. Such estimates are notoriously inaccurate and sensitive to the variability of the raindrop size distribution. Much better accuracy is achieved if either a combination of Z and Z_{DR} or specific attenuation A is used following radar relations optimized for DSDs measured in Oklahoma at S band [153]

$$LWC(Z) = 1.74 \times 10^{-3} Z^{0.64} \tag{2.22}$$

$$LWC(Z, Z_{DR}) = 1.38 \times 10^{-3} Z \times 10^{(-2.43 Z_{DR} + 1.12 Z_{DR}^2 - 0.176 Z_{DR}^3)} \tag{2.23}$$

$$LWC(A) = 115 A^{0.92} \tag{2.24}$$

where LWC is expressed in g m^{-3}, Z is in mm^6 m^{-3}, and Z_{DR} is in dB. The advantage of using the $LWC(A)$ relation is that it is least sensitive to the DSD variability and is immune to the biases of Z and Z_{DR} measurements. Our analysis of the errors in the radar estimates of LWC using a large disdrometer dataset in Oklahoma shows that the fractional standard deviation (FSD) of the $LWC(Z, Z_{DR})$ estimate is about 35% for LWC varying from 0.1 to 1.0 g m^{-3}. The $LWC(A)$

relation yields FSD about two times lower—17%. A table listing radar relations for *LWC* at S, C, and X bands is in [4] (Chapter 11).

Owing to the monotonic relation between raindrop size and oblateness, the mean volume diameter D_m of raindrops, defined as the ratio of the fourth and third moments of the DSD, or median volume diameter D_0 (which is very close to D_m) is traditionally estimated from Z_{DR}. A summary of various proposed $D_0(Z_{DR})$ relations is in Chapter 11 of [4] for different radar wavelengths. Herein, we present two popular $D_0(Z_{DR})$ relations often used at S band:

$$D_0 = 0.171Z_{DR}^3 - 0.725Z_{DR}^2 + 1.48Z_{DR} + 0.717 \tag{2.25}$$

suggested by Brandes *et al.* [154] and

$$D_0 = 0.0436Z_{DR}^3 - 0.216Z_{DR}^2 + 1.08Z_{DR} + 0.659 \tag{2.26}$$

from Cao *et al.* [155]. In (2.25) and (2.26), D_0 is in mm and Z_{DR} is in dB. Relations (2.25) and (2.26) were obtained using disdrometer measurements in Florida and Oklahoma. These two estimates are relatively close and can be utilized interchangeably. However, all rain retrieval relations may need some minor modifications depending on a climate region. It would be a good practice to check the validity of the suggested relations in a concrete geographical region where sizeable DSD datasets exist and modify these if needed. The total number concentration of raindrops N_t (in units of number per *L*) can be estimated from the combination of *Z* and Z_{DR} as

$$\log(N_t) = -2.37 + 0.1Z - 2.89Z_{DR} + 1.28Z_{DR}^2 - 0.213Z_{DR}^3 \tag{2.27}$$

where *Z* is expressed in dBZ and Z_{DR} is in dB.

The standard deviation of the estimates $D_0(Z_{DR})$ in (2.25) and (2.26) increases with D_0 but the fractional standard deviation (FSD) is constant at about 10%. The standard deviation of the $\log(N_t)$ estimate is about 0.3 for the majority of DSDs and tends to be larger for very high [$\log(N_t) > 3$] and very low [$\log(N_t) < 1$] raindrop concentrations.

2.5.1.2 Radar microphysical retrievals in ice and snow

Recently, substantial progress has been made in microphysical retrievals of ice and snow from polarimetric radar data. Two sets of relations for estimation of ice water content (*IWC*) and mean volume diameter D_m have been suggested. The first set proposed by Ryzhkov and Zrnić [4] is:

$$IWC = 4.0 \times 10^{-3} \frac{K_{DP}\lambda}{1 - Z_{dr}^{-1}} \tag{2.28}$$

$$D_m = -0.1 + 2.0\left(\frac{Z(1 - Z_{dr}^{-1})}{K_{DP}\lambda}\right)^{1/2} \tag{2.29}$$

and the second described in Bukovcic *et al.* [117,118] is:

$$IWC(Z, K_{DP}) = 3.3 \times 10^{-2}(K_{DP}\lambda)^{0.67}Z^{0.33} \quad (2.30)$$

$$D_m = 0.67\left(\frac{Z}{K_{DP}\lambda}\right)^{1/3}. \quad (2.31)$$

The first set of equations benefits from being immune to variations of snowflake shape and orientation but is sensitive to the degree of riming and is prone to possible Z_{DR} miscalibration biases. The second one does not use Z_{DR} but is sensitive to the variability of particles' shapes and orientations. The total concentration if ice N_t can be found as

$$\log(N_t) = 3.69 + 2\log(IWC) - 0.1Z(\text{dBZ}), \quad (2.32)$$

where N_t is in L^{-1} and IWC is in g m^{-3}.

It is convenient to display the results of radar microphysical retrievals in the Quasi-Vertical Profiles (QVP) format. The QVP profiles are obtained via azimuthal averaging of the polarimetric variables and derived microphysical attributes at high antenna elevation angles to significantly reduce statistical errors of estimates [156,157]. The QVPs from individual volume scans stacked in a height versus time format represent the vertical microphysical structure of the storm and its temporal evolution clearly. While the QVP is a radar-centric product, the Columnar Vertical Profiles (CVP) product represents the evolution of the vertical profiles of the polarimetric radar variables and retrieved microphysical parameters LWC/IWC, D_m, and N_t within the column centered at an arbitrary distance from the radar [158]. The examples of the CVP microphysical products from two landfalling hurricanes, Harvey and Florence, are displayed in Figure 2.9 [159].

Hu and Ryzhkov [160] built a climatology of the vertical profiles of *LWC* / *IWC*, D_m, and N_t for three different types of weather systems: hurricanes, continental, and marine mesoscale convective systems (MCS) using polarimetric data collected by a multitude of WSR-88Ds from dozens of events in each category. Such climatology reveals significant microphysical differences between hurricanes/marine MCS and continental MCSs with tropical systems characterized by higher concentrations and smaller ice particles compared to the continental storms. This information can serve as an observational reference for cloud modelers. The retrieval relations (2.28)–(2.32) will likely be refined in the course of further research via comparison with in situ microphysical probes' measurements at the ground and onboard research aircrafts [158,161].

2.5.1.3 Dual-frequency microphysical retrievals

The dual-frequency or triple-frequency microphysical retrievals capitalize on the difference between reflectivity factors at different radar frequencies or wavelengths. They are based on the fact that the hydrometeors scatter in the Rayleigh regime at longer wavelengths and the resonance scattering takes place at shorter wavelengths. As a result, reflectivity Z is usually higher at longer wavelength and the difference between reflectivity factors at longer and shorter wavelengths is a

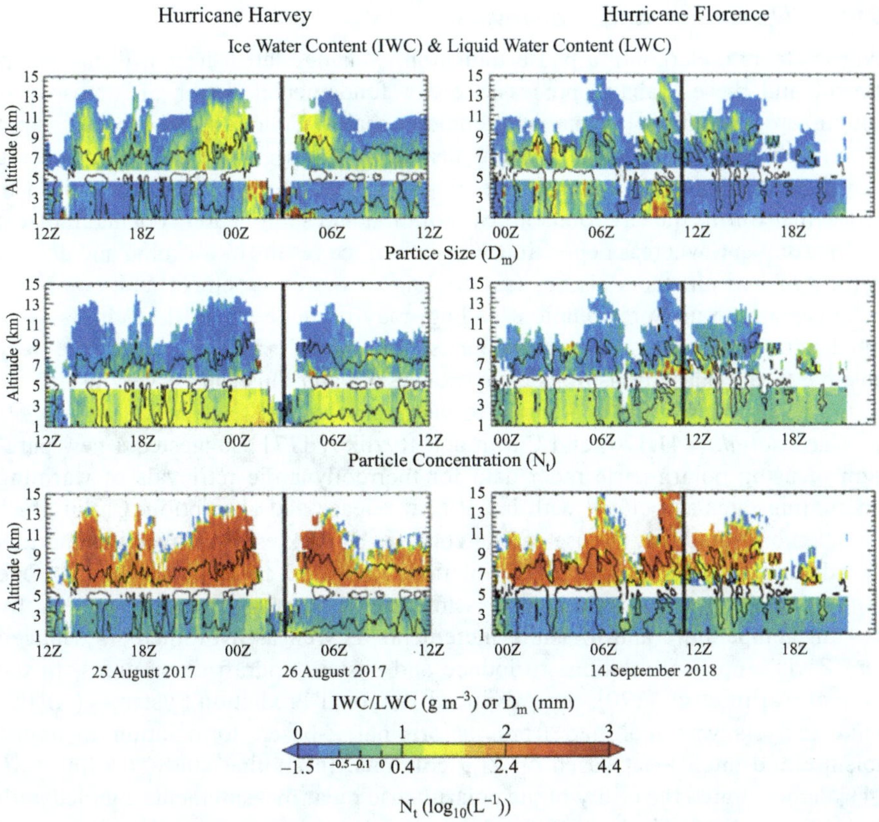

Figure 2.9 Columnar Vertical Profiles (CVP) of retrieved LWC/IWC, D_m, and N_t for the hurricanes Harvey and Florence. From [159]. © American Meteorological Society. Used with permission.

good measure of the characteristic size of hydrometeors. Needless to say that the dual-wavelength ratio (DWR) or dual-frequency ratio (DFR) should be used for estimation of the mean particles size after the difference in attenuation at the two radar wavelengths is taken into account.

Historically, the dual-wavelength ratio between S-band and X-band reflectivities was first utilized for the detection of hail and estimation of its size [162]. Later on, the primary focus of multifrequency radar studies shifted toward estimation of the key microphysical parameters of snow and ice such as snow rate and characteristic size of snowflakes [163–168].

A dual-frequency precipitation radar (DPR) operating at Ku and Ka bands is deployed onboard the Global Precipitation Measurement (GPM) Core Observatory satellite. Its primary goal is to infer precipitation rate and raindrop/particle size distributions. A more detailed information about DPR algorithms and observations can be found in [169].

2.5.2 Thermodynamic retrievals

Hydrometeors undergoing a phase transition exchange latent heat with the environment and these diabatic processes are a fundamental driver of atmospheric motion across a range of spatial and temporal scales. Condensation of water vapor in convective updrafts releases latent heat and drives convection by increasing buoyancy. Riming/accretion and refreezing represent another source of latent heat release that warms the environment. Evaporation and sublimation cool and moisten the environment, whereas depositional growth of ice results in warming and drying. Melting of hydrometeors also causes cooling of the environment and, combined with evaporation and precipitation loading, may produce strong downdrafts associated with "cold pools" and microbursts. Polarimetric radar variables are very sensitive to the phase transitions of hydrometeors and thus can be very useful for thermodynamic retrievals in addition to the microphysical retrievals.

Carlin *et al.* [141,170] and Carlin and Ryzhkov [171] suggested a new paradigm of using polarimetric radar data for thermodynamic retrievals of warming and cooling rates associated with latent heat release and absorption. Carlin *et al.* [170] demonstrated that the use of Z_{DR} columns as proxies for convective updrafts has advantages for radar data assimilation compared to utilization of Z. One common technique of reflectivity assimilation is through cloud analysis by inserting temperature and moisture increments as well as hydrometeors deduced from Z via empirical relations to induce and sustain updraft circulations. In the study of Carlin *et al.* [170], the Advanced Regional Prediction System's (ARPS) cloud analysis was modified from its original Z-based formulation to adjust moisture and latent heat based on Z_{DR} columns. In another study, Carlin *et al.* [141] demonstrated the utility of the polarimetric radar measurements coupled with the 1D bin microphysical model for predicting the onset of snowfall at the ground with a lead time up to six hours. Such a forecast was possible due to the accurate prediction of the moistening of the atmospheric surface layer caused by sublimation of falling snow.

2.6 Concluding remarks

The basic principles for the utilization of Doppler polarimetric radars in meteorology were laid down during the period from 1970 to 1995 [1]. Next 25 years were marked by a rapid acceleration of the polarimetric radar studies and transition of research to operations. This was a result of collective efforts of several scientific teams around the world that used sophisticated research polarimetric radars some of which were uniquely designed to measure the full polarimetric covariance matrix. At the turn of the century, the scheme for simultaneous transmission/reception of horizontally and vertically polarized waves was introduced and accepted as the most practical for operational applications.

The first research polarimetric radars operated mainly at S and C bands to minimize the impact of attenuation. However, smaller-size and less expensive shorter-wavelength polarimetric radars operating at X band (and even at Ka and W bands) were massively introduced and utilized for weather observations during the first and second decades of the century. This was possible (at X band) owing to effective polarimetric techniques for attenuation correction based on the measurements of differential phase.

Numerous demonstrated advantages of polarimetric weather radars like QPE improvement, ability to discriminate between different types of precipitation, tornado, and hail detection convinced decision-makers at national weather services to start massive polarimetric upgrade of the operational weather radars around the globe. This transition was spearheaded by the US NWS via modernization of its network of the WSR-88Ds.

Although weather radar polarimetry has already matured as a discipline in the research and operational domains, its further refinements and novel applications are underway. On a technological side, these include the low-power solid-state radars to complement large surveillance weather radar networks and development of phased array polarimetric weather radars. On a basic research side, a more active use of polarimetric radar data is anticipated via assimilation into NWP models and in better microphysical parameterization of cloud models. This will be accomplished via polarimetric microphysical and thermodynamic retrievals.

List of acronyms/abbreviations

A	Specific attenuation (dB/km)
ASOS	Automated Surface Observing Systems
c	Speed of light
CDR	Circular depolarization ratio (dB)
CVP	Columnar vertical profiles
D	Equivolume diameter
$\mathbf{D_0}$	Median volume diameter
$\mathbf{D_m}$	Mean volume diameter
DFR	Dual-frequency ratio (dB)
DSD	Drop size distribution
DWR	Dual-wavelength ratio (dB)
DYNAMO	DYNAMics of the Madden-Julian Oscillation
$\mathbf{F_s}$	Shape factor
$\mathbf{F_o}$	Orientation factor
fMAE	Fractional mean absolute error
HCA	Hydrometeor classification algorithm
HSDA	Hail size discrimination algorithm
IWC	Ice water content
JPOLE	Joint POLarization Experiment
$\mathbf{K_{DP}}$	Specific differential phase (deg/km)

LDR	Linear depolarization ratio (dB)
LWC	Liquid water content (g/m^3)
MAP	Mesoscale Alpine Program
MCS	Mesoscale convective system
MCTEX	Maritime Continent Thunderstorm Experiment
MRMS	Multi-radar multi-sensor platform
N(D)	Particle size distribution
$\mathbf{N_t}$	Total number concentration (1/L)
p	Atmospheric pressure
PIA	Two-way path-integrated attenuation (dB)
PSD	Particle size distribution
QPE	Quantitative precipitation estimation
QVP	Quasi-vertical profiles
R	Rain rate (mm/h)
RICO	Rain In Cumulus over Ocean
S	Snow water equivalent (mm/h)
SCSMEX	South China Sea Monsoon EXperiment
TDS	Tornadic debris signature
TIMREX	Terrain-Influenced Monsoon Rainfall EXperiment
v	Doppler velocity (m/s)
VPR	Vertical profile of reflectivity
WSR-88D	Weather Surveillance Radars 1988 Doppler
Z	Radar reflectivity (dBZ)
$\mathbf{Z_{DR}}$	Differential reflectivity (dB)
$\mathbf{Z_{dr}}$	Differential reflectivity in linear units
$\mathbf{Z_i}$	Radar reflectivity of ice
$\mathbf{Z_s}$	Radar reflectivity of snow
$\mathbf{Z_w}$	Radar reflectivity of water
α	Ratio A/K_{DP}
β	Ratio A_{DP}/K_{DP}
ε	Dielectric constant
ε_i	Dielectric constant of solid ice
ε_s	Dielectric constant of snow
ε_w	Dielectric constant of water
ρ	Particle density
ρ_{hv}	Cross-correlation coefficient
ρ_i	Density of solid ice
ρ_s	Density of snow

σ	Radar cross-section
σ_v	Doppler spectrum width
λ	Radar wavelength
τ	Radar pulse length
Φ_{DP}	Total differential phase (deg)

References

[1] Bringi, V. and Zrnić, D. Polarization weather radar development from 1970–1995: Personal reflections. *Atmosphere*. 2019; 10:714.

[2] Bringi, V.N. and Chandrasekar, V. *Polarimetric Doppler Weather Radar. Principles and Applications*. Cambridge, UK: Cambridge University Press; 2001, p. 636.

[3] Zhang, G. *Weather Radar Polarimetry*. Boca Raton, FL: CRC Press; 2016, p. 304.

[4] Ryzhkov, A. and Zrnić, D. *Radar Polarimetry for Weather Observations*. Cham, Switzerland: Springer Atmospheric Sciences; 2019, p. 486.

[5] Doviak, R. and Zrnić, D. *Doppler Radar and Weather Observations*. New York: Dover Publications Inc.; 2006, p. 562.

[6] Illingworth, A. 'Improved precipitation rates and data quality by using polarimetric measurements' in Meischner P. (ed.). *Weather Radar Principles and Applications*. Berlin: Springer; 2004, pp. 130–66.

[7] Meischner, P. 'Understanding severe weather system using Doppler and polarization radar' in Meischner P. (ed.). *Weather Radar Principles and Applications*. Berlin: Springer; 2004, pp. 167–98.

[8] Kumjian, M. R. Principles and applications of dual-polarization weather radar. Part I: Description of the polarimetric radar variables. *J. Oper. Meteorol.*, 2013; 1(19), 226–42.

[9] Kumjian, M. R. Principles and applications of dual-polarization weather radar. Part II: Warm- and cold-season applications. *J. Oper. Meteorol.*, 2013;1(20), 243–64.

[10] Kumjian, M. R. Principles and applications of dual-polarization weather radar. Part III: Artifacts. *J. Oper. Meteorol.*, 2013;1(21), 265–74.

[11] Kumjian, M. R. Weather radars. In C. Andronache (ed.), *Remote Sensing of Clouds and Precipitation*. New York: Springer; 2018, pp. 15–63.

[12] Fabry, F. *Radar Meteorology. Principles and Practice.* Cambridge, UK: Cambridge University Press. 2015, p. 256.

[13] Rauber, R. and Nesbitt, S. *Radar Meteorology. A First Course (Advancing Weather and Climate Science)*. Hoboken, NJ: Wiley-Blackwell; 2018, p. 488.

[14] Fukao, S. and Hamazu, K. *Radar for Meteorological and Atmospheric Observations.* Tokyo: Springer; 2014; p. 537.

[15] Ryzhkov, A., Matrosov, S., Melnikov, V., *et al.* Estimation of depolarization ratio using radars with simultaneous transmission/reception. *J. Appl. Meteorol. Climatol.*, 2017; 56: 1797–816.

[16] Zahrai, A. and Zrnić D. S. The 10-cm-wavelength polarimetric weather radar at NOAA's National Severe Storms Laboratory. *J. Atmos. Oceanic Technol.*, 1993; 10: 649–62.

[17] Doviak, R., Bringi, V., Ryzhkov, A., Zahrai, A., and Zrnić, D. Considerations for polarimetric upgrades of operational WSR-88D radars. *J. Atmos. Oceanic Technol.*, 2000; 17: 257–78.

[18] Melnikov, V. and Zrnić D. 2004: Simultaneous transmission mode for the polarimetric WSR-88D: Statistical biases and standard deviations of polarimetric variables. NOAA/NSSL Report, 84 pp. [Available online at https://www.nssl.noaa.gov/publications/wsr88d_reports/SHV_statistics.pdf]

[19] Zrnić, D. and Ryzhkov, A. Polarimetry for weather surveillance radars. *Bull. Amer. Meteorol. Soc.* 1999; 80: 389–406.

[20] Zrnić, D. and Ryzhkov, A. Advantages of rain measurements using specific differential phase. *J. Atmos. Oceanic Technol.* 1996; 13: 454–64.

[21] Ryzhkov, A. and Zrnić, D. Assessment of rainfall measurement that uses specific differential phase. *J. Appl. Meteorol.* 1996; 35: 2080–90.

[22] Ryzhkov A., Schuur, T., Burgess, D. and Zrnić, D. Polarimetric tornado detection. *J. Appl. Meteorol.*, 2005; 44: 557–70.

[23] Snyder, J. and Ryzhkov, A. Automated detection of polarimetric tornado debris signatures. *J. Appl. Meteorol. Climatol.* 2015; 54: 1861–70.

[24] Heinselman, P. and Ryzhkov, A. Validation of polarimetric hail detection. *Weather Forecast.* 2006; 21: 839–50.

[25] Ryzhkov, A., Kumjian, M., Ganson, S. and Khain, A. Polarimetric radar characteristics of melting hail. Pt I: Theoretical simulations using spectral microphysical modeling. *J. Appl. Meteorol. Climatol.*, 2013; 52: 2849–70.

[26] Ryzhkov, A., Kumjian, M. Ganson, S. and Zhang, P. Polarimetric radar characteristics of melting hail. Pt II: Practical implications. *J. Appl. Meteorol. Climatol.*, 2013; 52: 2871–86.

[27] Ortega, K., Krause, J. and Ryzhkov, A. Polarimetric radar characteristics of melting hail. Part III: Validation of the algorithm for hail size discrimination. *J. Appl. Meteorol. Climatol.*, 2016; 55: 829–48.

[28] Illingworth, A. and Thompson, R. Radar bright band correction using the linear depolarization ratio. In *International Symposium on Weather Radar and Hydrology*, 2012; Exeter, UK, pp. 64–68.

[29] Goddard, J., Tan, J. and Thurai, M. Technique for calibration of meteorological radars using differential phase. *Electron. Lett.* 1994; 30: 166–67.

[30] Hogan, R. A variational scheme for retrieving rainfall rate and hail reflectivity fraction from polarization radar. *J. Appl. Meteorol. Climatol.* 2007; 46: 1544–64.

[31] Illingworth, A., Hogan, R., O'Connor, E., *et al.* Cloudnet. Continuous evaluation of cloud profiles in seven operational models using ground-based observations. *Bull. Amer. Meteorol. Soc.* 2007; 88: 883–98.

[32] Bringi, V., Hoferer R., Brunkov D., *et al.* Antenna design and performance characteristics of the new 8.5-m dual-offset Gregorian antenna for the CSU–CHILL radar. *J. Atmos. Oceanic Technol.* 2011; 28: 907–20.

[33] Junyent, F., Chandrasekar, V., Bringi, V., *et al.* Transformation of the CSU-CHILL radar facility to a dual-frequency, dual-polarization Doppler system. *Bull. Amer. Meteorol. Soc.* 2015; 96: 975–96.

[34] Hubbert, J., Bringi, V., Carey, L., and Bolen, S. CSU-CHILL polarimetric radar measurements from a severe hail storm in eastern Colorado. *J. Appl. Meteorol.*, 1998; 37: 749–75.

[35] Kennedy, P. and Rutledge, S. S-band dual-polarization radar observations of winter storms. *J. Appl. Meteorol. Climatol.* 2011; 50: 844–58.

[36] Notaros, B., Bringi, V., Kleinkort, C., *et al.* Accurate characterization of winter precipitation using Multi-Angle Snowflake Camera, Visual Hull, Advanced scattering methods and polarimetric radar (Special Issue Advances in Clouds and Precipitation). *Atmosphere* 2016; 7(6): 81–111.

[37] Kennedy, P., Thurai, M., Praz, C., Bringi, V., Berne, A. and Notaros, B. Variations in snow crystal riming and ZDR: A case analysis. *J. Appl. Meteorol. Climatol.* 57; 2018: 695–707.

[38] Bringi, V., Kennedy, P., Huang, G.-J., Kleinkort, C., Thurai, M., and Notaros, B. Dual-polarized radar and surface observations of a winter graupel shower with negative Zdr column. *J. Appl. Meteorol. Climatol.* 2017; 56: 455–70.

[39] Lutz, J., Johnson, P., Lewis, B., Loew, E., Randall, M., and Van Andel, J. NCAR SPol: Portable polarimetric S-band radar. in: *Preprints, Ninth Symposium on Meteorological Observations and Instrumentation*, Charlotte, NC, American Meteorological Society, 1995, pp. 408–10.

[40] Hubbert, J., Wilson, J., Weckwerth, T., Ellis, S., Dixon, M., and Loew, E. S-Pol's polarimetric data reveal detailed storm features (and insect behavior). *Bull. Amer. Meteor. Soc.* 2018; 2045–60.

[41] Ellis, S. and Vivekanadan, J. Liquid water content estimates using simultaneous S and Ka band radar measurements. *Radio Sci.* 2011; 46(2): RS2021.

[42] Weckwerth, T., Pettet, C., Fabry, F., Park, S., LeMone, M., and Wilson J. Radar refractivity retrieval: validation and application to short-term forecasting. *J. Appl. Meteorol.* 2005; 44: 285–300.

[43] Schroth, A., Chandra, M., and Meischner, P. A C-band coherent polarimetric radar for propagation and cloud physics research. *J. Atmos. Oceanic Technol.* 1988; 5: 803–22.

[44] Keenan, T., Glasson, K., Cummings, F., Bird, T., Keeler, J., and Lutz, J. The BMRC/NCAR C-band polarimetric (C-pol) radar system. *J. Atmos. Oceanic Technol.* 1998; 15: 871–86.

[45] Keenan, T., Rutledge, S., Carbone, R., *et al.* The Maritime Continent Thunderstorm Experiment (MCTEX): overview and some results. *Bull. Amer. Meteor. Soc.* 2000; 81: 2433–55.

[46] Lau, K. Ding, Y., Wang, J-T., *et al.* A report on the field operations and early results of the South China Sea Monsoon Experiment (SCSMEX). *Bull. Amer. Meteorol. Soc.* 1999; 81: 1261–70.

[47] Keenan, T., Joe, P., Wilson, J., *et al.* The Sydney 2000 World Weather Research Program Forecast Demonstration Project: overview and current status. *Bull. Amer. Meteorol. Soc.* 2003; 84(8): 1041–54.

[48] Vaughan, G., Schiller, C., MacKenzie, A., *et al.* SCOUT-03/ACTIVE: High-altitude aircraft measurements around deep tropical convection. *Bull. Amer. Meteorol. Soc.* 2008; 89: 647–62.

[49] May, P., Mather, J., Vaughan, G., *et al.* The Tropical Warm Pool International Cloud Experiment. *Bull. Amer. Meteorol. Soc.* 2008; 89: 629–46.

[50] May, P., Keenan, T., Zrnić, D., Carey, L., and Rutledge, S. Polarimetric radar measurements of tropical rain at 5-cm wavelength. *J. Appl. Meteorol.* 1999; 38: 750–65.

[51] Keenan, T., Carey, L., Zrnić, D., and May, P. Sensitivity of 5-cm wavelength polarimetric radar variables to raindrop axial ratio and drop size distribution. *J. Appl. Meteorol.* 2001; 40: 526–45.

[52] Zrnić, D., Keenan, T., Carey, L., and May, P. Sensitivity analysis of polarimetric variables at a 5-cm wavelength in rain. *J. Appl. Meteorol.* 2000; 39: 1514–26.

[53] Bringi, V., Williams, C., Thurai, M., and May, P. Using dual-polarized radar and dual-frequency profiler for DSD characterization: a case study from Darwin, Australia. *J. Atmos. Oceanic Technol.* 2009; 26: 2107–22.

[54] Thurai, M., Bringi, V., and May, P. CPOL radar-derived drop size distribution statistics of stratiform and convective rain for two regimes in Darwin, Australia. *J. Atmos. Oceanic Technol.* 2010; 27: 932–42.

[55] Penide, G., Protat, A., Kumar, V., and May, P. Comparison of two convective/stratiform precipitation classification techniques: Radar reflectivity texture versus drop size distribution-based approach. *J. Atmos. Oceanic Technol.* 2013; 30: 2788–97.

[56] May, P., Jameson, A., Keenan, T., and Johnston, P. A comparison between polarimetric radar and wind profiler observations of precipitation in tropical showers. *J. Appl. Meteorol. Climatol.* 2001; 40: 1702–17.

[57] May, P. and Keenan, T. Evaluation of microphysical retrievals from polarimetric radar with wind profiler data. *J. Appl. Meteorol. Climatol.* 2005; 44: 827–38.

[58] Palmer, R., Bodine, D., Kumjian, M., *et al.* Observations of the 10 May 2010 Tornado outbreak using OU-PRIME: potential for new science with high-resolution polarimetric radar. *Bull. Amer. Meteorol. Soc.*, 2011; 92: 871–91.

[59] Picca, J. and Ryzhkov, A. A dual-wavelength polarimetric analysis of the 16 May 2010 Oklahoma City extreme hailstorm. *Mon. Weather Rev.*, 2012; 140: 1385–403.

[60] Kaltenboeck, R. and Ryzhkov, A. Comparison of polarimetric signatures of hail at S and C bands for different hail sizes. *Atm. Res.*, 2013; 123: 323–36.

[61] Borowska, L., Zrnić, D., Ryzhkov, A., Zhang, P., and Simmer, C. Polarimetric estimates of a 1-month accumulation of light rain with a 3-cm wavelength radar. *J. Hydrometeorol.*, 2011; 12: 1024–39.

[62] Gu, J.-Y., Ryzhkov, A., Zhang, P., *et al.* 2011 Polarimetric attenuation correction in heavy rain at C band. *J. Appl. Meteorol.*, 2011; 50: 39–58.

[63] Anderson, M., Carey, L., Petersen, W., and Knupp, K. C-band dual-polarimetric signatures of hail. *Electron. J. Oper. Meteor.* 2011; 12: 1–30.

[64] Crowe, C., Schultz, C., Kumjian, M., Carey, L., and Petersen, W. Use of dual-polarization signatures in diagnosing tornadic potential. *Electron. J. Oper. Meteorol.* 2012; 13: 57–78.

[65] Mecikalski, R., Bain, A., and Carey, L., Radar and lightning observations of deep moist convection across Northern Alabama during DC3: 21 May 2012. *Mon. Weather Rev.* 2015; 143: 2774–94.

[66] Adachi, A., Kobayashi, T., Yamauchi, H., and Onogi, S., Detection of potentially hazardous convective clouds with a dual-polarized C-band radar. *Atmos. Meas. Technol.* 2013; 6: 2741–60.

[67] Adachi, A., Kobayashi, T., and Yamauchi, H. Estimation of raindrop size distribution and rainfall rate from polarimetric radar measurements at attenuating frequency based on the self-consistency principle. *J. Meteor. Soc. Japan*, 2015; 93: 359–88.

[68] Matrosov, S., Kropfli, R., Reinking, R., and Martner, B. Prospects for measuring rainfall using propagation differential phase in X- and Ka-radar bands. *J. Appl. Meteor.* 1999; 38: 766–76.

[69] Matrosov, S., Clark, K., Martner, B., and Tokay, A. 2002: X-band polarimetric radar measurements of rainfall. *J. Appl. Meteor.*, 2002; 41: 941–52.

[70] McLaughlin, D., Pepyne, D., Chandrasekar, V., *et al.* Short-wavelength technology and the potential for distributed networks of small radar systems. *Bull. Amer. Meteor. Soc.* 2009; 90: 1797–817.

[71] Maki, M., Maesaka, T., Kato, A., Kim, D.-S., and Iwanami, K. Developing a composite rainfall map based on observations from an X-band polarimetric radar network and conventional C-band radar. *Indian J. Radio Space Phys.* 2012; 41: 461–70.

[72] Galvez, M., Colom, J., Chandrasekar, V., Junyent, F., Cruz-Pol, S., and Rodriguez Solis, R. Salient features of the radar nodes in the Puerto Rico Tropical Weather Testbed. *IEEE International Geoscience and Remote Sensing Symposium (IGARSS)*, Cape Town, South Africa; 2009, pp. 841–44.

[73] Mishra, K., Krajewski, W., Goska, R., *et al.* Deployment and performance analyses of high-resolution Iowa XPOL radar system during the NASA IFloodS campaign. *J. Hydrmeteorol.* 2016; 17: 455–79.

[74] Kollias, P., Bharadwaj, N., Clothiaux, E., *et al.* The ARM radar network. At the leading edge of cloud and precipitation observations. *Bull. Amer. Meteorol. Soc.* 2020; 101: E588–E60.

[75] Chandrasekar, V., Schwaller, M., Vega, M., *et al.* Scientific and engineering overview of the NASA Dual-Frequency Dual-Polarized Doppler Radar (D3R) system for GPM Ground Validation, in: *2010 IEEE International Geoscience and Remote Sensing Symposium*, pp. 1308–1311, https://doi.org/10.1109/IGARSS.2010.5649440, 2010.

[76] Myagkov, A., Seifert, P., Bauer-Pfundstein, M., and Wandinger, U. Cloud radar with hybrid mode towards estimation of shape and orientation of ice crystals, *Atmos. Meas. Technol.* 2016; 9: 469–89.

[77] Oue, M., Kollias, P., Matrosov, S., Battaglia, A., and Ryzhkov, A. 2021: Analysis of the microphysical properties of snowfall using scanning polarimetric and vertically pointing multi-frequency Doppler radars. *Atm. Meas. Technol.* 2021; 14: 4893–913.

[78] Oue, M., Kollias, P., Ryzhkov, A. and Luke, E. 2018: Towards exploring the synergy between cloud radar polarimetry and Doppler spectral analysis in deep cold precipitating systems in the Arctic. *J. Geophys. Res. Atmos.* 2018; 123: 2797–815.

[79] Bharadwaj, N. and Chandrasekar, V. Waveband waveform design principles for solid-state weather radars. *J. Atmos. Oceanic Technol.* 2012; 29: 14–31.

[80] Salazar Aquino, C., Cheong, B., and Palmer, R. Progressive pulse compression: a novel technique for blind range recovery for solid-state radars. *J. Atmos. Oceanic Technol.*, 2021; 38: 1599–611.

[81] Zrnić, D., Kimpel, J., Forsyth, D., *et al.* Agile-beam phased array radar for weather observations. *Bull. Amer. Meteorol. Soc.* 2007; 88: 1753–66.

[82] Weber, M., Cho, J., Herd, J., Flavin, J., Benner, W., and Torok, G. The next generation multimission U.S. surveillance radar network. *Bull. Amer. Meteorol. Soc.* 2007; 88: 1739–52.

[83] Weber, M., Hondl, K., Yussouf, N., *et al.* Towards the next generation operational meteorological radar. *Bull. Amer. Meteorol. Soc.* 2021; E1357–E383.

[84] Ryzhkov, A., Schuur, T., Burgess, D., Giangrande, S., and Zrnić, D. The Joint Polarization Experiment: polarimetric rainfall measurements and hydrometeor classification. *Bull. Amer. Meteorol. Soc.* 2005; 86: 809–24.

[85] Ryzhkov, A. and Zrnić, D. Comparison of dual-polarization radar estimators of rain. *J. Atmos. Oceanic Technol.* 1995; 12: 249–56.

[86] Testud, J., Le Bouar, E., Obligis, E., and Ali-Mehenni, M. The rain profiling algorithm applied to polarimetric weather radar. *J. Atmos. Oceanic Technol.* 2000; 17: 332–56.

[87] Brandes, E., Zhang, G., and Vivekanandan, J. Experiments in rainfall estimation with a polarimetric radar in a subtropical environment. *J. Appl. Meteorol.* 2002; 41: 674–85.

[88] Ryzhkov, A., Giangrande, S., and Schuur, T. Rainfall estimation with a polarimetric prototype of the WSR-88D radar. *J. Appl. Meteorol.* 2005; 44: 502–15.

[89] Matrosov, S., Kingsmill, D., and Ralph, F. The utility of X-band polarimetric radar for quantitative estimates of rainfall parameters. *J. Hydrometeorol.* 2005; 6: 248–62.

[90] Zhang, J. Howard, K., Langston, C., *et al.* Multi-Radar Multi-Sensor (MRMS) quantitative precipitation estimation: initial operating capabilities. *Bull. Amer. Meteorol. Soc.* 2016; 97: 621–38.

[91] Seliga, T., Bringi, V., and Al-Khatib, H. Differential reflectivity measurements in rain: first experiments. *IEEE Trans. Geosci. Electron.* 1979; GE-17: 240–4.

[92] Seliga, T., Bringi, V., and Al-Khatib, H. A preliminary study of comparative measurement of rainfall rate using the differential reflectivity radar technique and a raingauge network. *J. Appl. Meteor.* 1981; 20: 1362–8.

[93] Ryzhkov, A., Zhang, P., Bukovcic, P., Zhang, J., and Cocks, S. Polarimetric radar quantitative precipitation estimation. *Remote Sens.* 2022; 14: 1695.

[94] Zhang, J., Tang, L., Cocks, S., *et al.* A dual-polarization radar synthetic QPE for operations. *J. Hydrometeorol.* 2020; 21: 2507–21.

[95] Wang, Y., Cocks, S., Tang, L., *et al.* A prototype quantitative precipitation estimation algorithm for operational S-band polarimetric radar utilizing specific attenuation and specific differential phase: Part I—Algorithm description and initial results. *J. Hydrometeorol.* 2019; 20: 985–97.

[96] Cocks, S., Wang, Y., Tang, L., *et al.* A prototype quantitative precipitation estimation algorithm for operational S-band polarimetric radar utilizing specific attenuation and specific differential phase: Part II—Case study analysis and performance verification. *J. Hydrometeorol.* 2019; 20: 999–1014.

[97] Kitchen, M., Brown, R., and Davies, A. Real-time correction of weather radar data for the effects of bright band, range and orographic growth in widespread precipitation. *Quart. J. Roy. Meteorol. Soc.* 1994; 120: 1231–54.

[98] Seo, D.-J., Breidenbach, J., Fulton, R., Miller, D., and O'Bannon, T. Real-time adjustment of range-dependent biases in WSR-88D rainfall estimates due to nonuniform vertical profile of reflectivity. *J. Hydrometeorol.* 2000; 1: 222–40.

[99] Germann, U. and Joss, J. Mesobeta profiles to extrapolate radar precipitation measurements above the Alps to the ground level. *J. Appl. Meteorol.* 2002; 41: 542–57.

[100] Vignal, B. and Krajewski, W. Large-sample evaluation of two methods to correct range-dependent error for WSR-88D rainfall estimates. *J. Hydrometeorol.* 2001; 2: 490–504.

[101] Zhang, J. and Qi, Y. A real-time algorithm for the correction of brightband effects in radar-derived QPE. *J. Hydrometeorol.* 2010; 11: 1157–71.

[102] Qi, Y., Zhang, J., Zhang, P., and Cao, Q. VPR correction of bright band effects in radar QPEs using polarimetric radar observations. *J. Geophys. Res. Atmos.* 2013; 118: 3627–33.

[103] Hanft, W., Zhang, J., and Simpson, M. Dual-pol VPR corrections for improved operational radar QPE in MRMS. *J. Hydrometeorol.* 2023; 24: 353–71.

[104] Zhang, Y., Bi, S., Liu, L., *et al.* Deep learning for polarimetric radar quantitative precipitation estimation during landfalling typhoons in South China. *Remote Sens.* 2021; 13: 3157.

[105] Shin, K., Song, J., Bang, W., and Lee, G. Quantitative Precipitation Estimates using machine learning approaches with operational dual-polarization radar data. *Remote Sens.* 2021; 13: 694.

[106] Wolfensberger, D., Gabella, M., Boscacci, M., Germann, U., and Berne, A. RainForest: A random forest algorithm for quantitative precipitation estimation over Switzerland. *Atmos. Meas. Technol.* 2021; 14: 3169–93.

[107] Gunn, K. and Marshall, J. The distribution with size of aggregate snowflakes. *J. Meteorol.* 1958; 15: 452–61.
[108] Sekhon, R., and Srivastava, R. Snow size spectra and radar reflectivity. *J. Atmos. Sci.* 1970; 27: 299–307.
[109] Fujiyoshi, Y., Endoh, T., Yamada, T., Tsuboki, K., Tachibana, Y., and Wakahama, G. Determination of a Z–R relationship for snowfall using a radar and sensitive snow gauges. *J. Appl. Meteor.* 1990; 29: 147–52.
[110] Matrosov, S., Campbell, C., Kingsmill, D., and Sukovich, E. Assessing snowfall rates from X-band radar reflectivity measurements. *J. Atmos. Oceanic Technol.* 2009; 26: 2324–39.
[111] Huang, G.-J., Bringi, V., Cifelli, R., Hudak, D., and Petersen, W. A methodology to derive radar reflectivity-liquid equivalent snow rate relations using C-band radar and a 2D video disdrometer. *J. Atmos. Oceanic Technol.* 2010; 27: 637–51.
[112] Von Lerber, A., Moisseev, D., Bliven, L., Petersen, W., Harri, A., and Chandrasekar, V. Microphysical properties of snow and their link to Ze–S relations during BAECC 2014. *J. Appl. Meteorol. Climatol.* 2017; 56: 1561–82.
[113] Vivekanandan, J., Bringi, V., Hagen, M., and Meischner, P. Polarimetric radar studies of atmospheric ice particles. *IEEE Trans. Geosci. Remote Sens.* 1994; 32: 1–10.
[114] Lu, Y., Aydin, K., Cothiaux, E., and Verlinde, J. Retrieving cloud ice water content using millimeter- and centimeter wavelength radar polarimetric observables. *J. Appl. Meteorol. Climatol.* 2015; 54: 596–604.
[115] Ryzhkov, A., Zrnić, D., and Gordon, B. Polarimetric method for ice water content determination. *J. Appl. Meteorol.* 1998; 37: 125–34.
[116] Nguyen, C., Wolde, M., and Korolev, A. Determination of ice water content (*IWC*) in tropical convective clouds from X-band dual-polarization airborne radar. *Atmos. Meas. Technol.* 2019; 12: 5897–911.
[117] Bukovčić, P., Ryzhkov, A., Zrnić, D., and Zhang, G. Polarimetric radar relations for quantification of snow based on disdrometer data. *J. Appl. Meteorol. Climatol.* 2018; 57: 103–20.
[118] Bukovčić, P., Ryzhkov, A., and Zrnić, D. Polarimetric relations for snow estimation—radar verification. *J. Appl. Meteorol. Climatol.* 2020; 59: 991–1009.
[119] Straka, J. and Zrnić, D. An algorithm to deduce hydrometeor types and contents from multiparameter radar data, in *Preprints, 26th International Conference on Radar Meteorology*, Norman, OK, American Meteorological Society, 1993; 513–6.
[120] Vivekanandan, J., Zrnić, D., Ellis, S., Oye, D., Ryzhkov, A., and Straka, J. Cloud microphysics retrieval using S-band dual-polarization radar measurements. *Bull. Amer. Meteorol. Soc.* 1999; 80: 381–8.
[121] Liu, H. and Chandrasekar, V. Classification of hydrometeors based on polarimetric radar measurements: development of fuzzy logic and neuro-fuzzy systems, and in situ verification. *J. Atmos. Oceanic Technol.* 2000; 17: 140–67.
[122] Zrnić, D., Ryzhkov, A., Straka, J., Liu, Y., and Vivekanandan, J. Testing a procedure for the automatic classification of hydrometeor types. *J. Atmos. Oceanic Technol.*, 2001; 18: 892–913.

[123] Keenan, T. Hydrometeor classification with a C-band polarimetric radar. *Aust. Meteorol. Mag.* 2003; 52: 23–31.

[124] Marzano, F., Scaranari, D., and Vulpiani, G. Supervised fuzzy-logic classification of hydrometeors using C-band weather radars. *IEEE Trans. Geosci. Remote Sens.* 2007; 45: 3784–99.

[125] Park, H.-S., Ryzhkov, A., Zrnić, D., and Kim, K.-E. The hydrometeor classification algorithm for the polarimetric WSR-88D. Description and application to an MCS. *Weather Forecast.*, 2009; 24: 730–48.

[126] Dolan, B., and Rutledge, S. A theory-based hydrometeor identification algorithm for X-band polarimetric radars. *J. Atmos. Oceanic Technol.* 2009; 26: 2071–88.

[127] Dolan, B., Rutledge, S., Lim, S., Chandrasekar, V., and Thurai, M. A robust C-band hydrometeor identification algorithm and application to a long-term polarimetric radar dataset. *J. Appl. Meteorol. Climatol.* 2013; 52: 2162–86.

[128] Chandrasekar, V., Keranen, R., Lim, S., and Moisseev, D. 2013: Recent advances in classification of observations from dual-polarization weather radars. *Atmos. Res.* 2013; 119: 97–111.

[129] Al-Sakka, H., Boumahmoud, A., Fradon, B., Frasier, S., and Tabary, P. A new fuzzy logic hydrometeor classification scheme applied to the French X-, C-, and S-band polarimetric radars. *J. Appl. Meteorol. Climatol.* 2013; 52: 2328–44.

[130] Thompson, E., Rutledge, S., Dolan, B., Chandrasekar, V., and Cheong, B.-L. A dual-polarization radar hydrometeor classification algorithm for winter precipitation. *J. Atmos. Oceanic Technol.* 2014; 31: 1457–81.

[131] Bechini, R. and Chandrasekar, V. A semisupervised robust hydrometeor classification method for dual-polarization radar applications. *J. Atmos. Oceanic Technol.* 2015; 32: 22–47.

[132] Giangrande, S., Krause, J. and Ryzhkov, A. Automatic designation of the melting layer with a polarimetric prototype of the WSR-88D radar. *J. Appl. Meteorol. Climatol.* 2008; 47: 1354–64.

[133] Brandes, E. and Ikeda, K. Freezing-level estimation with polarimetric radar. *J. Appl. Meteorol.*, 2004; 43: 1541–53.

[134] Tabary, P., Le Henaff, A. Vulpiani, G. Parent-du-Chatelet, J. and Gourley, J.J. 2006. Melting layer characterization and identification with a C-band dual-polarization radar: A long-term analysis, in: *Preprints Fourth European Conference on Radar in Meteorology and Hydrology (ERAD 2006)*, Barcelona, Spain, 17–20.

[135] Matrosov, S., Clark, K., and Kingsmill, D. A polarimetric radar approach to identify rain, melting layer, and snow regions for applying corrections to vertical profiles of reflectivity. *J. Appl. Meteorol. Climatol.* 2007; 46: 154–66.

[136] Wolfensberger, D., Scipion, D., and Berne, A. Detection and characterization of the melting layer based on polarimetric radar scans. *Quart. J. Roy. Meteorol. Soc.* 2016; 142: 108–24.

[137] Ryzhkov, A. and Krause, J. New polarimetric radar algorithm for melting layer detection and determination of its height. *J. Atmos. Oceanic Technol.*, 2022; 39: 529–43.

[138] Kumjian, M., Khain, A., Benmoshe, N., Ilotoviz, E., Ryzhkov, A., and Phillips, V. The anatomy and physics of Z_{DR} columns: investigating a polarimetric radar signature with a spectral bin microphysical model. *J. Appl. Meteorol. Climatol.* 2014; 53: 1820–43.

[139] Snyder, J., Ryzhkov, A., Kumjian, M., Picca, J., and Khain, A. 2015: Developing a Z_{DR} column detection algorithm to examine convective storm updrafts. *Weather Forecast.* 2015; 30: 1819–44.

[140] Reeves, H., Ryzhkov, A. and Krause, J. Discrimination between winter precipitation types based on spectral-bin microphysical modeling. *J. Appl. Meteorol. Climatol.* 2016; 55: 1747–61.

[141] Carlin, J., Reeves, H., and Ryzhkov, A. Polarimetric observations and simulations of sublimating snow: implications for nowcasting. *J. Appl. Meteorol. Climatol.* 2021; 60: 1035–54.

[142] Elmore, K., Flamig, Z., Lakshmanan, V., *et al.* mPING: Crowd-sourcing weather reports for research. *Bull. Amer. Meteorol. Soc.*, 2014; 95: 1335–42.

[143] Grazioli, J., Turia, D., and Berne, A. Hydrometeor classification from polarimetric radar measurements: a clustering approach. *Atmos. Meas. Technol.* 2015; 8: 149–70.

[144] Wen, G., Protat, A., May, P., Wang, X., and Moran, W. Cluster-based method for hydrometeor classification using polarimetric variables. Part I: Interpretation and analysis, *J. Atmos. Oceanic Technol.* 2015; 32: 1320–30.

[145] Wen, G., Protat, A., May, P., Moran, W., and Dixon, M. Cluster-based method for hydrometeor classification using polarimetric variables. Part II: Classification, *J. Atmos. Oceanic Technol.* 2016; 33: 45–59.

[146] Besic, N., Figueras i Ventura, J., Grazioli, J., Gabella, M., Germann, U., and Berne, A. Hydrometeor classification through statistical clustering of polarimetric radar measurements: a semi-supervised approach. *Atmos. Meas. Technol.* 2016; 9: 4425–45.

[147] Morrison, H., Milbrandt, J., Bryan, G., Ikeda, S., Tessendorf, S., and Thompson, G. Parameterization of cloud microphysics based on the prediction of bulk ice particle properties. Part II: Case study comparisons with observations and other schemes. *J. Atmos. Sci.* 2015; 72: 312–39.

[148] Fan, J., Han, B., Varble, A., *et al.* Cloud-resolving model intercomparisons of an MC3E squall line case: Part I—Convective updrafts. *J. Geophys. Res. Atmos.* 2017; 122: 9351–78.

[149] Pfeifer, M., Craig, G., Hagen, M., and Keil, C. A polarimetric radar forward operator for model evaluation. *J. Appl. Meteorol. Climatol.* 2008; 47: 3203–20.

[150] Jung, Y., Zhang, G., and Xue, M. Assimilation of simulated polarimetric radar data for a convective storm using the Ensemble Kalman Filter. Part I: Observation operators for reflectivity and polarimetric variables. *Mon. Weather Rev.* 2008; 136: 2228–45.

[151] Ryzhkov, A., Pinsky, M., Pokrovsky, A., and Khain, A. Polarimetric radar observation operator for a cloud model with spectral microphysics. *J. Appl. Meteorol. Climatol.* 2011; 50: 873–94.

[152] Wolfensberger, D. and Berne, A. From model to radar variables: a new forward polarimetric radar operator for COSMO. *Atmos. Meas. Technol.* 2018; 11: 3883–916.

[153] Ryzhkov, A., Snyder, J., Carlin, J., Khain, A., and Pinsky, M. What polarimetric weather radars offer to cloud modelers: Forward radar operators and microphysical/thermodynamic retrievals. *Atmosphere.* 2020; 11: 362.

[154] Brandes, E., Zhang, G., and Vivekanandan, J. Drop size distribution retrieval with polarimetric radar: model and application. *J. Appl. Meteorol.* 2004; 43: 461–75.

[155] Cao, Q., Zhang, G., Brandes, E., and Schuur, T. Polarimetric radar rain estimation through retrieval of drop size distribution using a Bayesian approach. *J. Appl. Meteorol. Climatol.* 2010; 49: 973–90.

[156] Ryzhkov, A., Zhang, P., Reeves, H., *et al.* Quasi-vertical profiles – a new way to look at polarimetric radar data. *J. Atmos. Oceanic Technol.* 2016; 33: 551–62.

[157] Tobin, D. and Kumjian, M. Polarimetric radar and surface-based precipitation-type observations of ice pellet to freezing rain transitions. *Wea. Forecasting.* 2017; 32: 2065–82.

[158] Murphy, A., Ryzhkov, A, and Zhang, P. Columnar Vertical Profiles (CVP) methodology for validating polarimetric radar retrievals in ice using in situ aircraft measurements. *J. Atmos. Oceanic Tech.*, 2020; 37: 1623–42.

[159] Homeyer, C., Fierro, A., Schenkel, B., *et al.* Polarimetric signatures in landfalling tropical cyclones. *Mon. Weather Rev.*, 2021; 149: 131–54.

[160] Hu, J. and Ryzhkov, A. Climatology of the vertical profiles of polarimetric radar variables and retrieved microphysical parameters in continental/tropical MCSs and landfalling hurricanes. *J. Geophys. Res. Atmos.*, 2022; 127: e2021JD035498.

[161] Dunnavan, E., Carlin, J., Hu, J., *et al.* Radar retrieval evaluation and investigation of dendritic growth layer polarimetric signatures in a winter storm. *J. Appl. Meteorol. Climatol.*, 2022; 61: 1679–705.

[162] Eccles, P. and Atlas, D. A dual-wavelength radar hail indicator. *J. Appl. Meteorol.* 1973; 12: 847–54.

[163] Matrosov, S. Possibilities of cirrus particle sizing from dual-frequency radar measurements, *J. Geophys. Res.*, 1993; 98, D11, 20675–83.

[164] Matrosov, S. A dual-wavelength radar method to measure snowfall rate. *J. Appl. Meteorol.* 1998; 37: 1510–21.

[165] Liao, L., Meneghini, R., Iguchi, T., and Detwiler, A. Use of dual-wavelength radar for snow parameter estimates. *J. Atmos. Ocean. Technol.*, 2005; 22, 1494–506.

[166] Kneifel, S., Von Lerber, A., Tiira, J., and Moisseev, D. Observed relations between snowfall microphysics and triple-frequency radar measurements. *J. Geophys. Res. Atmos.* 2015; 120: 6034–55.

[167] Leinonen, J., Tanelli, S., Sy, O., *et al.* Retrieval of snowflake microphysical properties from multifrequency radar observations. *Atmos. Meas. Technol.* 2018; 11: 5471–88.

[168] Matrosov, S., Korolev, A., Wolde, M., and Nguyen, C. Sizing ice hydrometeor populations using dual-wavelength radar ratio. *Atmos. Meas. Technol.* 2022; 15: 6373–86.

[169] Liao, L. and Meneghini, R. GPM DPR retrievals: algorithm, evaluation, and validation. *Remote Sens.* 2022; 14: 843.

[170] Carlin, J., Gao, J., Snyder, J., and Ryzhkov, A. Assimilation of Z_{DR} columns for improving the spin-up and forecast of convective storms in storm-scale models: proof-of-concept experiments. *Mon. Weather Rev.*, 2017; 145: 5033–57.

[171] Carlin, J. and Ryzhkov, A. Estimation of melting layer cooling rate from dual-polarization radar: spectral bin model simulations. *J. Appl. Meteorol. Climatol.*, 2019; 58: 1485–508.

Chapter 3
Developments in solid-state weather radar

Stephen J. Frasier[1] and Luca Facheris[2]

3.1 Introduction

Since the initial development of microwave radar, enabled by the invention of the magnetron, most radar systems have relied upon vacuum tube technologies for their high power sources. Even today, devices such as the magnetron, klystron, and traveling wave tube (TWT) serve in the vast majority of operational radar systems for surveillance and navigation. All tube technologies are characterized by high operating voltages, typically a few to several kilovolts, which are necessary to support an electron beam flowing from cathode to anode. Interactions of this electron beam with resonant structures in the tube and, in the case of the klystron and TWT, with an input radio frequency (RF) signal, yield the high power output. Tubes are noteworthy for relatively high reliability, but they have a finite lifetime dictated by the depletion of the cathode as a source of electrons.

Solid-state sources, on the other hand, are characterized by low to moderate operating voltages, typically tens of volts, and comparatively high operating currents (or current densities within the semiconductor material). Solid-state sources do not have finite lifetimes as do the tube technologies, at least in principle, and the lower operating voltages imply a higher degree of safety and reliability.

The rise of solid-state weather radar coincides with research into networks of short-range radar systems as a means to better monitor the atmospheric boundary layer [1]. In such networks, maximum range is limited to around 50 km or less so as to avoid earth curvature and terrain blockage issues. The limited range permits the deployment of smaller, less costly radar systems often operating at X-band. Demonstration networks such as those of the CASA project deployed in Oklahoma [2] and in Dallas-Fort Worth, Texas [3] have shown the benefit of such regional networks. Similar X-band networks have been deployed in mountainous regions of France [4] and in coastal regions of Italy [5]. These recent networks have still relied primarily on magnetron-based transmitters (the Dallas-Fort Worth network employs a mix of magnetron and solid-state radars). Nonetheless, these networks

[1]Microwave Remote Sensing Laboratory, Department of Electrical and Computer Engineering, University of Massachusetts, Amherst, MA, USA
[2]Department of Information Engineering, University of Florence, Florence, Italy

have spurred industry interest to develop both solid-state weather radar systems [6–8] and subscription-based services based on ground-based and spaceborne solid-state radar technology [9,10].

Solid-state technology is also key to the development of phased-array radars which are now being developed and evaluated for meteorological applications [11–14]. These radars realize their transmit power through the combination of hundreds to thousands of individual radiating elements. Each element is driven by a transmit/receive (TR) module containing a small power amplifier (PA) for transmission, a low noise amplifier (LNA) for reception, and a circuit for amplitude and phase control.

In this chapter, we describe the salient properties of solid-state weather radars. We begin with a summary of the state of the art in solid-state power sources and related enabling technologies. We then survey known solid-state radar systems and describe the methods necessary to realize systems with sufficient sensitivity for weather applications.

Tube-based technologies are capable of high peak power, which is desirable for radar. Typical peak power levels range from tens to hundreds of kiloWatts, whereas for solid-state sources, typical power levels range from tens to hundreds of Watts. We will show that radar sensitivity is fundamentally dictated by pulse energy. Because the peak power of a solid-state transmitter is substantially lower than that of a tube, the pulse energy must come from the pulse-duration rather than from its amplitude. Normally, a longer pulse implies coarser range resolution. In order to retain range resolution, the pulse must be modulated to occupy a bandwidth concomitant with the desired resolution, and a technique known as pulse compression must be employed to realize the desired resolution. We describe several aspects of pulse compression that must be dealt with in weather radar applications. Finally, we also consider alternatives to the conventional pulse-compression radar, namely the frequency-modulated continuous-wave (FMCW) radar and the use of quadratic phase coding to realize high duty cycle.

3.2 Enabling technologies

3.2.1 Solid-state sources

Because solid-state sources employ lower operating voltages, they operate with much higher current densities within the semiconductor material. Any losses within the material generate heat, and dissipating or removing the heat is the principle difficulty in realizing high power levels. Indeed, most of the weight and volume of a high-power solid-state amplifier is taken up by cooling through heat sinks, fans, and possibly liquid circulation. Figure 3.1, reproduced from [15], summarizes the power generation capabilities versus frequency for several semiconductor technologies.

The most common and least expensive semiconductor technology, silicon-based complementary metal-oxide-semiconductor (CMOS), is generally too lossy at microwave frequencies to be effective for high-power operation. Laterally diffused metal-oxide-semiconductor (LDMOS) is a planar silicon technology widely used for

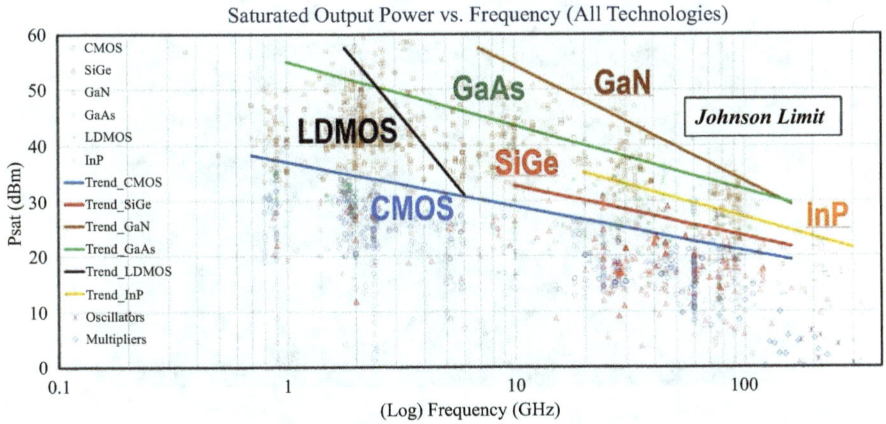

Figure 3.1 Summary of saturated output powers for various solid-state technologies as of July 2019 (from [15], © 2020 reproduced by permission of H. Wang).

mobile network base stations at the lower end of the microwave frequency range [16]. Lower loss semiconductors at microwave frequencies include III-V compounds such as gallium arsenide (GaAs) and indium phosphide (InP) for millimeter-wave applications. These compounds are comparatively expensive to silicon.

In applications such as phased-array radars, it is desirable to integrate both digital and analog functions into a single monolithic microwave integrated circuit (MMIC) chip that can be deployed near the individual antenna elements. Silicon-Germanium (SiGe) has been proposed for applications not requiring high power as it permits slightly higher power than CMOS and the possibility of integration with digital circuits controlling components such as phase shifters and variable attenuators [17]. SiGe transmit-receive MMICs have been incorporated into phased arrays designed for manufacturabilty [18] such as the X-band Skyler phased array developed by Raytheon [19–21]. CMOS control chips are proposed for use in a C-band phased-array weather radar [13]. In these applications, it is possible to use these lower-cost, lower-performance technologies because the transmit power is distributed among thousands of low-power TR-modules. Figure 3.2 illustrates two examples of TR-modules. The first is an X-band circuit card employing discrete components including a 2W GaAs PA, LNA, switches, and a common leg circuit (CLC) containing the amplitude and phase control. The signal is routed through the CLC on both transmission and reception. The second is the block diagram for the first generation of a SiGe MMIC chip incorporating all the same functions. The PA for this TR chip produces 100 mW.

For more conventional weather radar architectures employing reflector antennas, affordable high-power microwave amplifiers are a relatively recent development enabled principally by advances in gallium nitride (GaN) technology. Amplifiers based on high electron mobility transistor (HEMT) in GaN producing

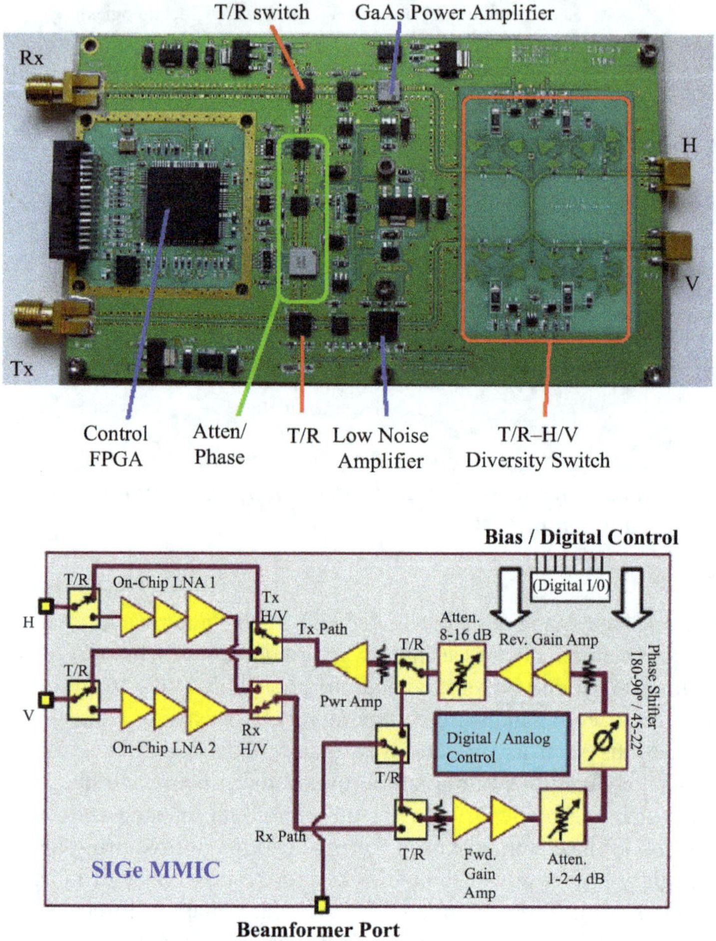

Figure 3.2 Annotated photo (top) of a TR-module for the UMass Phase-Tilt Weather Radar, an X-band phased array with 64 active elements. Block diagram (bottom) of a mixed-signal SiGe MMIC chip for the X-band Skyler radar developed by Raytheon (from [22] © 2014 Raytheon Company, used with permission of Raytheon Intelligence and Space).

hundreds to thousands of Watts are now being offered as replacements for traveling-wave-tube amplifiers. In the mm-wave range, GaAs amplifiers produce tens of Watts [23].

3.2.2 Digital transceivers and software radio technology

Another enabling technology is related to the advancement of digital signal generation and signal processing. The ability to generate precisely controlled waveforms did not really materialize until the 1990s with the advent of commercially available direct digital synthesizers (DDSs) and arbitrary waveform generators (AWGs). Prior to this time, the generation of pulse-compression waveforms, particularly frequency-modulated waveforms, required either agile voltage-controlled oscillators or surface acoustic wave (SAW) or bulk acoustic wave (BAW) devices. It is now commonplace to specify a transmit waveform simply as a sequence of digital samples which are played through a high-speed digital-to-analog converter (DAC). On the receiver side, advances in analog to digital conversion and in programmable logic devices such as field programmable gate arrays (FPGAs)) have made the digitization and subsequent processing of digital signals at the intermediate frequency of the receiver more common. The current state of the art is the so-called "Radio Frequency Systems on a Chip" (RFSoCs) that incorporate analog functions such as amplification and frequency conversion tightly coupled with the FPGA [24]. These currently offer essentially direct sampling of RF signals up to C-band frequencies.

Most recently, the proliferation of software radio platforms and open-source standards have enabled the easy generation and processing of almost any desired waveform [25]. When connected to a standard personal computer or server via ethernet, software radio devices provide the necessary real-time interface for capturing radar signals without the need for a real-time operating system on the computer. Often, both transmit and receive functions are incorporated in a digital receiver/exciter (DREX), i.e., in a single unit operating at the radar's intermediate frequency (IF). Finally, advances in CPU technology have ensured the real-time (or near real-time) post-processing of radar data routine.

A typical solid-state weather radar block diagram is depicted in Figure 3.3. It consists of a DREX that produces the transmit waveform at the radar's intermediate frequency. Frequency conversion from baseband (I and Q) waveforms to the intermediate frequency is achieved through digital upconversion with a numerically controlled oscillator (NCO). The waveform is further upconverted to microwave frequency using the radar's stable local oscillator (STALO), amplified by a solid-state power amplifier (SSPA), and transmitted. Received signals are downconverted to the intermediate frequency where they are directly sampled by an analog-to-digital converter (ADC). Digital samples are further filtered and downconverted to baseband in-phase and quadrature samples that are recorded or further processed by a conventional CPU. Pulse compression may be implemented within the DREX or it may be performed on the host computer CPU. Within the DREX, everything to the left of the DAC and ADC in Figure 3.3 may be implemented in an FPGA that is programmable using a language such as Very High Speed Integrated Circuit (VHSIC) Hardware Description Language (VHDL). For RFSoCs, the dividing line between analog and digital functions moves to the right, between the frequency conversion stage and the RF front-end.

For active phased-array radars, the transmit power is distributed among many lower power TR-modules, one at each element. For radars employing a reflector antenna, the only significant difference from a tube-based radar is the presence of a SSPA in place of the tube, as most modern weather radar systems now incorporate digital IF transceivers.

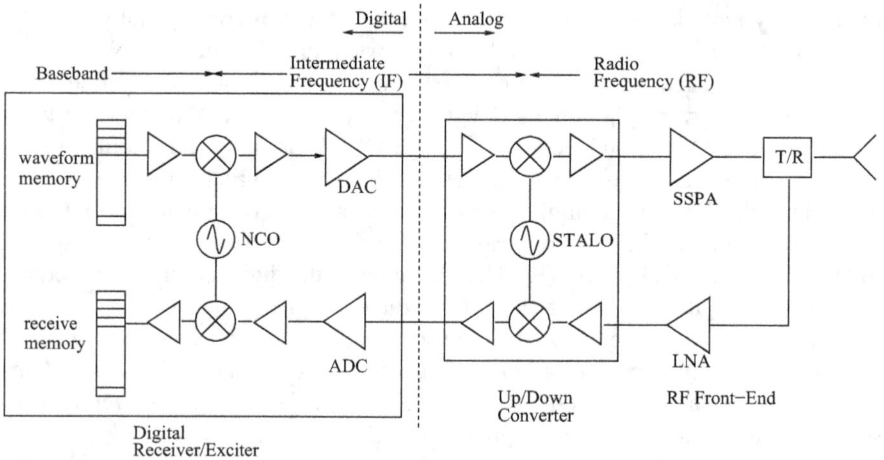

Figure 3.3 Principal components of a modern solid-state weather radar

3.3 Survey of solid-state weather radar

Table 3.1 shows a representative, though incomplete, list of solid-state weather radar systems as of this writing. Notably absent from the table are commercial aviation weather avoidance radars, as we focus here on ground-based and airborne research and operational radar systems. The systems listed are grouped according to the type of antenna or modulation used. The first grouping is all phased-array weather radar systems, a few of which are currently in development. At S-band, the Advanced Technology Demonstrator (ATD) is a dual-polarized phased-array test-bed developed by MIT Lincoln Laboratories and operated by the NSSL of NOAA [11]. A prototype all-digital S-band phased-array architecture is under development at the University of Oklahoma [27]. At C-band are a ground-based mobile phased-array [13] and an airborne phased array targeted for the National Science Foundation's C-130 aircraft [14]. Both of these systems are currently in development. At X-band, the University of Massachusetts developed a dual-polarized phased-array with electronic scanning in one dimension [12], and Raytheon has developed a two-dimensional scanning dual-polarized phased array [19]. Toshiba and Osaka University have also developed phased-array architectures in single and dual-polarization with electronic-scanning in the elevation plane [29].

In all of these phased-array architectures, the transmit power is distributed among many elements, each driven by a TR module. Each module contributes a small amount to the total transmitted power, from several milliWatts to a few Watts. These phased arrays are referred to as "active" or Active Electronically Scanned Arrays (AESAs). The active array stands in contrast to the passive array, where a single high-power amplifier drives an array consisting only of passive components (phase shifters and couplers). The predecessor to the Advanced Technology Demonstrator at the National Weather Radar Testbed, a modified

Table 3.1 A survey of solid-state weather radars (as of 2022)

System	Developer	Band	Status	Ref
Phased Arrays				
ATD[1]	NOAA-NSSL/MIT-LL	S	Research	[11,26]
HORUS	U. Oklahoma	S	In development	[27]
CPAIR[2]	”	C	”	[13]
APAR[3,a]	NCAR	C	”	[14]
CPPAR[4]	U. Oklahoma/NOAA-NSSL	C	Research	[28]
PTWR[5]	U. Massachusetts	X	”	[12]
Skyler	Raytheon/U. Massachusetts	X	Res./Comm.	[19,21]
PAWR[6]	Toshiba/Osaka U.	X	”	[29]
Reflectors				
PX-1000	U. Oklahoma	X	Research	[30]
IWRAP[7,a]	U. Massachusetts/NOAA	C,Ku	”	[31]
HIWRAP[8,a]	NASA-GSFC	Ku,Ka	”	[32]
CRS[9,a]	”	W	”	[23]
DPR[10,s]	JAXA (Japan)	Ku,Ka	”	[33]
RainCube[s]	NASA-JPL	Ka	”	[34]
Ranger	EEC	X	Commercial	[8]
SSWR	Toshiba (Japan)	C,X	”	[6]
GMWR-SST	Gamic (Germany)	X	”	[7]
Rainwatcher	JRC (Japan)	S,C,X	”	[35]
WRS400	Vaisala (Finland)	X	”	[36]
WRX-100	BIRM (China)	X	”	[37]
E700 PDR	EWR	X	”	[38]
WR2120	Furuno (Japan)	X	”	[39]
FMCW/QPC				
S-Band FMCW	U. Massachusetts	S	Research	[40,41]
IDRA[11]	Delft U. Tech. (Netherlands)	X	”	[42]
QX-120	MetaSensing (Netherlands)	X	Commercial	[43]
MRR[12]	METEK (Germany)	K	Res./Comm.	[44]
BASTA[13]	LATMOS/Meteomodem (France)	W	”	[45]
KPR[14,a]	U. Wyoming/ProSensing	Ka	”	[46]

[1]Advanced Technology Demonstrator, [2]C-band Polarimetric Atmospheric Imaging Radar, [3]Airborne Phased-Array Radar, [4]Cylindrical Polarimetric Phased-Array Radar, [5]Phase-Tilt Weather Radar, [6]Phased-Array Weather Radar, [7]Imaging Wind and Rain Airborne Profiler, [8]High-Altitude Wind and Rain Airborne Profiler, [9]Cloud Radar System, [10]Dual-frequency Precipitation Radar, [11]IRCTR Drizzle Radar, [12]Micro Rain Radar, [13]Bistatic Radar System for Atmospheric Studies, [14]Ka-band Precipitation Radar, *a* denotes airborne, *s* denotes spaceborne

SPY-1A radar, is an example of a passive phased array. While the passive array provides all the capabilities of rapid and arbitrary scanning, the high-power amplifier remains a single point of failure. Active arrays are more robust as they can tolerate the failure of a few to several elements and are said to degrade gracefully.

The second grouping in Table 3.1 all use reflector or flat-plate antennas and therefore employ much higher power solid-state transmitters than do the active phased-array systems. These systems employ a single transmitter typically of hundreds of Watts. The majority of commercially available solid-state weather radars fall into this category. The NASA radars are all airborne or spaceborne systems with nadir or near-nadir pointing and typically with tens of Watts of transmit power.

All of the radars in the first two groups are pulsed, and with the exception of the DPR radar, all of them employ pulse compression which we discuss in the next section. The final grouping is radars that are not pulsed in the usual sense but rather use frequency or phase modulation as a means to increase their average power. These are generally vertically profiling instruments, although the IDRA and QX-120 radars scan horizontally. We discuss frequency and phase modulation techniques later in this chapter.

3.4 Peak and average power considerations

To understand the practical operation of solid-state radars, it is necessary to understand the underlying sensitivity requirements. The sensitivity of a monostatic radar is often expressed in terms of the single-pulse signal-to-noise ratio (SNR),

$$SNR_1 = \frac{P_t G^2 \lambda^2 \sigma}{(4\pi)^3 R^4 [k(T_{ant} + T_{rec})B]} \tag{3.1}$$

where P_t is the peak transmitted power, G is the antenna gain (the same antenna is assumed for transmission and reception), λ is the radio wavelength, R is the range, and σ is the target's radar cross-section.* The term in square brackets in the denominator is the equivalent receiver noise referred to the antenna terminals, where k is Boltzmann's constant, T_{ant} is the radiometric temperature sensed by the antenna in Kelvins, T_{rec} is the radar receiver's noise temperature or the equivalent radiometric temperature produced by the receiver electronics, and B is the receiver bandwidth.

For rectangular pulses without any other modulation, the SNR is optimized if the receiver bandwidth is approximately the reciprocal of the transmitted pulse length, $B \approx 1/T$, where T is the pulse length. Employing this, (3.1) can be rewritten as

$$SNR_1 = \frac{(P_t T) G^2 \lambda^2 \sigma}{(4\pi)^3 R^4 k(T_{ant} + T_{rec})} \tag{3.2}$$

where the quantity $P_t T$ is the transmitted pulse energy, E_t. Thus, for a given antenna and receiver, it is the pulse energy that fundamentally dictates the single-pulse sensitivity of the radar.

*In this form, the equation implicitly assumes the target is a "point target." For volume targets such as hydrometeors, the radar cross-section incorporates both the volume reflectivity and the illuminated volume, $\sigma = \eta V$. For the purposes of this development, we will retain the point target form.

A single pulse is nearly useless in practice, due to the fluctuating nature of most radar targets. Typically, several pulses are combined or integrated to improve detectability. In the microwave range, radars require multiple pulses in order to measure the Doppler frequency shifts of moving targets. If the pulse returns are combined coherently (including both magnitude and phase information), then the SNR is improved at most by the number of pulses integrated, $SNR_N \leq N \cdot SNR_1$. The number of pulses available to integrate in this way is dictated by the dwell time, T_d, which itself may be determined by the radar antenna's scan rate or by the rate of fluctuation of the target. The number of samples available to integrate during the dwell time is $N = f_p T_d$, where f_p is the pulse rate of the radar. Incorporating these expressions, we obtain

$$SNR_N \leq \frac{[P_t(f_p T)T_d]\ G^2\lambda^2\sigma}{(4\pi)^3 R^4 k(T_{ant} + T_{rec})} \tag{3.3}$$

where now the term in square brackets in the numerator represents the total energy of the N pulses. The product $f_p T$ is the duty cycle of the transmitter or the fraction of time that the transmitter is on. Multiplying this by the peak power, one obtains the average power transmitted over time, $P_{ave} = P_t f_p T$. Thus, for a given dwell time, the SNR depends upon the average transmitted power, and further, for a given peak power, the SNR depends upon the duty cycle of the radar.

The duty cycle of a tube-based radar transmitting high-power unmodulated pulses is typically quite low. Often, the duty cycle is of the order of 0.1%. For example, a 25 kW peak-power magnetron with a 0.1% duty cycle has an average power of 25 W. To achieve a comparable average power, solid-state weather radars must operate with a substantially higher duty cycle, of the order of 10% or more. For example, a solid-state power amplifier with peak power of 250 W operating at 10% duty cycle also has an average power of 25 W.

To summarize, (3.2) indicates that single-pulse sensitivity is fundamentally dictated by pulse energy, while (3.3) indicates that N-pulse sensitivity is dictated by the total energy of the N pulses which may be expressed as average power times the dwell time. In the former case, sensitivity is improved by lengthening the transmitted pulse. In the latter case, sensitivity is improved either by lengthening the transmitted pulse or by increasing the pulse rate (or both). Most of the remainder of this chapter focuses on pulse compression, which is based on the lengthening of the pulse to increase the average power. At the end of the chapter, we consider an alternative method which is based upon increasing the pulse rate.

3.5 Pulse compression and the matched filter

Increasing the pulse energy implies lengthening the transmitted pulse. Doing so however results in a degradation of the radar's range resolution. The range resolution of a radar is usually characterized by its ability to discriminate between two closely spaced targets, and this is dictated by the pulse length. Because of the round-trip propagation of the pulse upon reflection by a target, the range resolution is half of the radar's pulse length as it propagates through space,

$$\Delta r = \frac{cT}{2}, \tag{3.4}$$

where Δr is range resolution, c is the speed of light, and T is the pulse length. A more general definition of the radar range resolution, however, is written in terms of the pulse bandwidth, B,

$$\Delta r = \frac{c}{2B}. \tag{3.5}$$

The idea of pulse compression is to transmit long, modulated pulses that occupy a desired bandwidth, and to process the received pulses in a way that exploits both the full energy of the pulse, which comes from the duration and the full bandwidth of the pulse, which defines the range resolution. The key to this is the so-called matched filter.

The matched filter originally attributed to North [47] is the filter that maximizes the instantaneous SNR at the filter output at a given instant in time. The derivation can be found in many texts [48] and involves the maximization at a desired time instant $\bar{t}$ of the ratio

$$\chi = \frac{|s_o(\bar{t})|^2}{P_n} \tag{3.6}$$

where $s_o(t)$ is the real signal at the filter output and P_n is the noise power also at the filter output. The numerator and denominator can both be expressed in terms of their Fourier transforms

$$\chi = \frac{|\int_{-\infty}^{\infty} S_i(f)H(f)e^{j2\pi f\bar{t}}df|^2}{N_0/2\int_{-\infty}^{\infty}|H(f)|^2 df} \tag{3.7}$$

where $S_i(f)$ is the Fourier transform of the signal at the filter input (assumed not affected by Doppler and therefore proportional to the transmitted signal), $H(f)$ is the filter transfer function yet to be determined, and $N_0/2$ is the (double sided) noise power spectral density. At this point, the derivation makes use of the Schwarz inequality applied to the numerator,

$$\left|\int_{-\infty}^{\infty} S_i(f)H(f)e^{j2\pi f\bar{t}}df\right|^2 \leq \int_{-\infty}^{\infty} |S_i(f)|^2|H(f)e^{j2\pi f\bar{t}}|^2 df \tag{3.8}$$

where the left-hand side is maximized (equality holds) when the filter transfer function is chosen to be $H(f) = S_i^*(f)e^{-j2\pi f\bar{t}}$. In the time domain, the filter's impulse response is the time reverse of the input waveform, $h(\tau) = s_i(\bar{t} - \tau)$. With this choice for $H(f)$, the ratio in (3.7) reduces to

$$\chi = \frac{\int_{-\infty}^{\infty}|S_i(f)|^2 df}{N_0/2}, \tag{3.9}$$

where the numerator is recognized as the signal energy applied to the filter input.

While the matched filter maximizes the SNR at the desired time instant, that is the only constraint in its design. The filter does not preserve the transmitted signal's shape in any way. In the time domain, the filter output is expressed as the convolution of the input signal and the filter's impulse response. The convolution operation itself involves the time reverse of either the input signal or the impulse response,

$$s_o(t) = \int_{-\infty}^{\infty} s_i(\tau)h(t-\tau)d\tau. \tag{3.10}$$

Because the impulse response of the matched filter is already the time reverse of the desired signal, the matched filter output becomes

$$s_o(t) = \int_{\infty}^{\infty} s_i(\tau)s_i(t+\tau)d\tau \tag{3.11}$$

which can be recognized as the deterministic autocorrelation function of the input waveform. Thus, convolution with a matched filter and correlation with a copy of the waveform are mathematically equivalent.

Herein lies the key aspect of pulse compression. A long-duration pulse is transmitted which has been modulated in a way to occupy the desired bandwidth. Upon reception, the matched filter "compresses" the long-duration pulse yielding an output waveform which is the autocorrelation of the transmitted waveform. The effective length of the compressed pulse is approximately the reciprocal of its bandwidth, $T_{eff} \approx 1/B$. Because energy is conserved, the energies of the input and output pulses are equal, so the peak power of the compressed pulse must increase. This increase in effective power is called the compression gain and is given by the ratio of the actual pulse width transmitted, T, to the effective pulse width, T_{eff}, output by the matched filter, $G_c = T/T_{eff} \approx BT$. Thus, the compression gain is approximately given by the time–bandwidth product of the transmitted waveform. Note that for unmodulated, simple pulse waveforms, the time–bandwidth product is approximately unity. The conservation of pulse energy can be written as

$$E_t = P_tT = P_tG_cT_{eff}. \tag{3.12}$$

The first expression is in terms of the actual transmitted pulse width. In the absence of pulse compression, the range resolution would be degraded with an increased pulse length. The second expression is in terms of the effective pulse width which retains range resolution and incorporates the compression gain.

Recognizing that the pulse compression gain is the product of the pulse length, T, and the pulse bandwidth, B, it is important to reiterate that the sensitivity improvement comes from the pulse length and not from the bandwidth. While increasing the pulse bandwidth does indeed increase the compression gain, it also reduces the compressed pulse length and requires that the radar receiver's noise bandwidth be increased as well. The result is an increase in receiver noise power and no net improvement in sensitivity. The real improvement comes from the increased pulse length, resulting in a higher duty cycle and higher average transmitted power.

A variety of modulation schemes may be employed to cause the pulse to occupy a desired bandwidth. By far the most common method is linear frequency modulation (LFM) or "chirp" modulation, where the instantaneous frequency of the pulse varies linearly during the pulse duration. The term "chirp" alludes to the sound one might hear if the waveform were an audible signal. This waveform also has the desirable property that its performance is robust in the presence of Doppler shift [48].†

3.6 Range sidelobes and their reduction

While pulse compression methods increase the average power without sacrificing range resolution, they have undesirable properties that must be addressed in a practical weather radar. The most serious of these is the presence of so-called time sidelobes or range sidelobes. These are most readily illustrated in the case of LFM waveforms with large time–bandwidth products. Figure 3.4 shows properties of a chirp waveform. This waveform has a rectangular envelope and a linearly varying frequency versus time

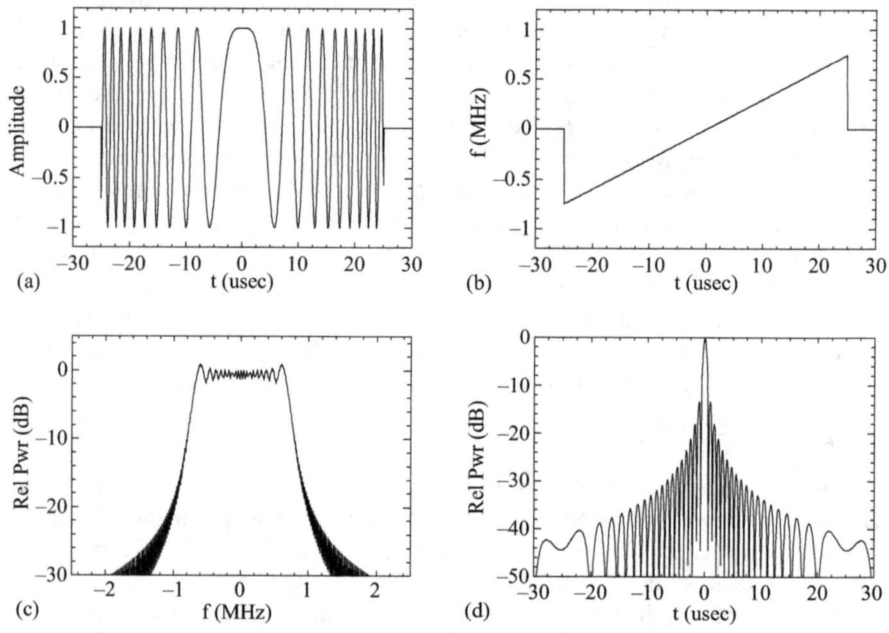

Figure 3.4 Properties of a LFM waveform with bandwidth 1.5 MHz and duration 50 μs: (a) real part of the chirp waveform, (b) instantaneous frequency versus time, (c) power spectrum, (d) autocorrelation function.

†Despite this fact, some of the early works on pulse compression for weather radar applications actually employed phase coding techniques [49,50], as binary phase modulation was more readily implemented on the weather radar hardware at the time.

during the pulse, as indicated by the two upper panels of the figure. The power spectrum of the chirp waveform is approximately rectangular as shown in the lower left. The matched filter output waveform, which is the autocorrelation function of the chirp, can be expressed as the inverse Fourier transform of the power spectrum. Due to the rectangular spectrum shape, the autocorrelation function will have the form of a sinc function, or $\sin(x)/x$, as illustrated in the lower right panel. The half-power width of the compressed pulse is approximately the reciprocal of the bandwidth as expected. However, there are sidelobes on either side of the peak, and in the case of a rectangular spectrum, the first sidelobe is at a relative level of about -13 dB. These sidelobes are undesirable, as they may mask other targets or features near the peak.

For weather radar applications, it is desired that the sidelobes be very low: typically well below -50 dB. Indeed, weather radar presents one of the most challenging applications of pulse compression. The weather signal has a high dynamic range, with reflectivity variations of 30–40 dB within 1 km being a frequent occurrence and sharp contrasts exceeding 55 dB on occasion. Very low sidelobes are also necessary for nadir-looking airborne or spaceborne precipitation and cloud radars, as near-surface atmospheric echoes compete with the sidelobes from the very strong surface return [23,34,51].

The challenge of low sidelobes is compounded by the fact that the time–bandwidth product for weather radars is often not particularly large (<100). So-called Fresnel ripples in the spectrum of the chirp waveform (evident in Figure 3.4c) limit the utility of conventional methods to reduce the sidelobes. In the following sections, we outline several methods to address range sidelobes.

3.6.1 Windowing

The unacceptably high sidelobes of the compressed waveform arise in part from the sharp gradient of the power spectrum at the band edges. The classical way to reduce these is to modify the spectrum in some way to remove the sharp edges. This can be done through a technique called windowing, where the matched filter is modified so as to taper the spectrum, removing the sharp edges. By virtue of the chirp waveform, the instantaneous frequency is linearly related to time within the pulse, so a time weighting is equivalent to a frequency weighting. The impulse response of the receiver filter can be modified to

$$h(\tau) = w(\tau)h_m(\tau) \tag{3.13}$$

where $h_m(\tau)$ is the original matched filter (the time reverse of the transmitted waveform) and $w(\tau)$ is a window function that varies smoothly from zero at the beginning to unity at the center and to zero at the end of the waveform. Technically, once amplitude weighting is applied, this filter is no longer the matched filter. However, in practice, it is not uncommon to refer to any impulse response that is based on the time reverse of the waveform as a matched filter.

Many window functions have been proposed including [52] triangular (Bartlett), Hamming, Hanning, Blackman, Kaiser–Bessel, Chebychev, and others commonly used in applications such as spectral estimation. Table 3.2 summarizes the properties of a few

Table 3.2 Properties of some common window functions

Window	Peak Sidelobe (dB)	Compressed Width ($\cdot B^{-1}$)	SNR loss (dB)
Rectangular	−13.2	0.88	0.0
Bartlett	−26.5	1.27	−1.25
Hanning (von Hann)	−31.5	1.44	−1.76
Hamming	−43.7	1.30	−1.34
Blackman	−58.2	1.64	−2.37
Blackman–Harris[1]	−74.4	1.70	−2.54

[1]−74dB 3-term.

common window functions. Each of these provides a certain level of sidelobe suppression as indicated by the peak sidelobe level (PSL). Because the amplitude weighting shapes the spectrum, generally reducing its width, the compressed pulse broadens slightly. The second column of the table expresses the half-power width of the compressed pulse relative to the reciprocal of the bandwidth of the linear FM pulse. Lower peak sidelobes generally come at the expense of increased broadening of the compressed pulse. Some windows such as the Chebychev and Kaiser–Bessel are parametric meaning they can be tuned to achieve particular sidelobe levels or main lobe widths.

The effect of windowing on the peak SNR at the output of the receiver filter is most easily described in the case of a discrete-time filter with a finite (fixed length) impulse response. Here, the loss in signal power relative to a rectangular window can be estimated as

$$L_s = \frac{1}{N_s^2}\left(\sum_{n=0}^{N_s-1} w(n)\right)^2 \tag{3.14}$$

where N_s is the length of the filter (i.e. the number of samples of the time-reversed waveform) and $w(n)$ are the filter weights. This expression is the ratio of the total energy obtained through coherent integration of the window (the square of the sum of the weights) compared to that of a rectangular window.

The filter will also reject some of the input noise power. The loss in noise power is given by

$$L_n = \frac{1}{N_s}\sum_{n=0}^{N_s-1} w^2(n) \tag{3.15}$$

which is the ratio of the total energy obtained through non-coherent integration of the window (the sum of the squares of the weights) compared to that of the rectangular window. The loss in peak SNR is just the ratio of these two,

$$L_{SNR} = \frac{\left(\sum_{n=0}^{N_s-1} w(n)\right)^2}{N_s \sum_{n=0}^{N_s-1} w^2(n)}. \tag{3.16}$$

The signal experiences a greater loss than does the noise, so the SNR is degraded by windowing. The last column of the table indicates the SNR loss of the window. In determining sensitivity, this is to be subtracted from the compression gain. As an example, a linear FM waveform with a time–bandwidth product of 100 has a compression gain of 20 dB. If a Blackman window is used upon reception, then the pulse compression gain is effectively reduced by 2.37 dB to 17.63 dB.

Shaping of the spectrum via amplitude weighting can be achieved in a few different ways. Because most radar transmitters operate in saturation, it is usually not feasible to shape the amplitude of the transmitted pulse. Therefore, it is more common to apply weighting in the receiver alone. It should be noted that in the case where weighting is applied both in transmission and reception, the window coefficients in the above expressions should be replaced by the combined weighting by transmitter and receiver, $w(n) = w_t(n)w_r(n)$.

The performance of the window functions indicated in Table 3.2 is predicated on the assumption of a rectangular input spectrum. For chirp waveforms, the use of amplitude weighting, especially when implemented in the receiver alone, is generally unsuccessful in achieving adequate sidelobes for typical weather radar applications. The problem is illustrated in Figure 3.5, where a Blackman window has been applied to the LFM waveform of Figure 3.4. While the near-in sidelobes close to the main lobe are suppressed to the expected level, the far-out sidelobes are only suppressed to a level approximately equal to 20log(BT) as indicated in the figure. The far-out sidelobes are observed to peak at a time lag of T/2 relative to the center of the pulse.

The sources of these far-out sidelobes are the remaining Fresnel ripples that exist throughout the spectrum. Some improvement is possible by applying

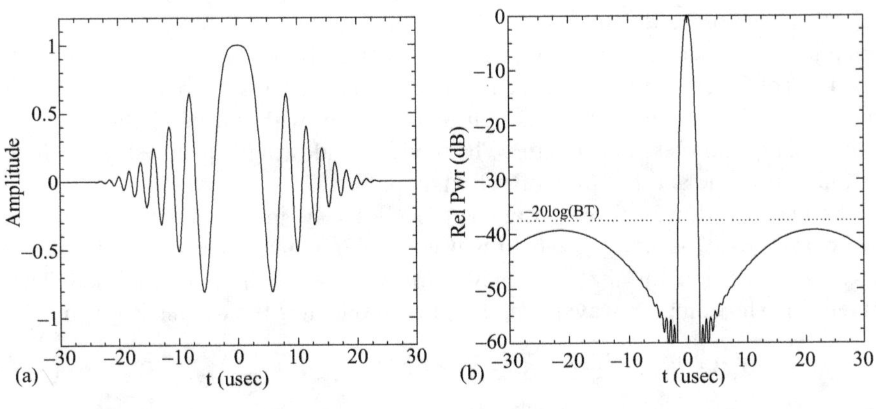

Figure 3.5 Effect of amplitude weighting alone implemented in the receiver on the LFM waveform of Figure 3.4 (left), and the resulting compressed pulse (right). A Blackman window has been applied. The sidelobes close to the peak are at the expected level, but the more distant sidelobes increase to approximately $-20log_{10}(BT)$.

weighting also in transmission. For example, Tanner *et al.* [51] demonstrated range sidelobes of better than −55 dB with an airborne, nadir-looking rain radar. In their case, they applied amplitude tapering also to the transmitted waveform, which required operating their transmitter in the linear portion of its dynamic range, thereby sacrificing some sensitivity.

3.6.2 Nonlinear frequency modulation

Because of the shortcomings of windowing, an alternative means of shaping the spectrum is through nonlinear frequency modulation (NLFM). In this scheme, the rate of change of the frequency, or the "chirp-rate," is varied during the pulse. It is rapid near the beginning and end of the waveform, corresponding to the band edges, and slower in the middle, corresponding to the band center. Because the waveform "spends more time" in the center of the bandwidth, more energy is contained there compared to the band edges. Thus, the spectrum shaping is achieved.

The earliest treatment of NLFM was given by Cook and Paolillo [53] who approached the problem as a predistortion of the LFM waveform presented to the transmitter of high-power radars. Their analysis suggested a modification of the instantaneous phase of the waveform would have a similar effect as an amplitude modulation. The resulting instantaneous frequency modulation appeared as a piecewise linear FM rate. Since then, a number of nonlinear waveforms have been proposed with chirp rates that are polynomial based [54–56], tangent-shaped [57], based on Fourier series [58,59], or based on Bezier curves [60]. An example of a NLFM waveform is shown in Figure 3.6, where the time-domain waveform, instantaneous frequency, power spectrum, and autocorrelation are shown [59].

The primary advantage of the NLFM waveform is its ability to shape the spectrum, thereby improving range sidelobes, without resorting to amplitude weighting. Despite this, most NLFM waveform designs still do resort to some form of amplitude weighting. In most cases, the amplitude weighting is in the form of a Tukey window, which is nearly rectangular but with a raised cosine-shaped taper at either end. The waveform of Figure 3.6 employs such a Tukey window. The incorporation of modest amplitude weighting aids in the design optimization, with only a slight penalty in SNR loss, typically less than 0.5 dB. The modest windowing is justified as it mimics the finite rise-time and fall-time of pulsed amplifiers.

Some NLFM waveform design approaches rely in part on the method of "stationary phase" [53,58,59] which is used to obtain analytical expressions relating to the time-domain and frequency-domain representations of the modulating waveform. The complex waveform, $u(t)$, that modulates the carrier is given by

$$u(t) = a(t)e^{j\phi(t)}, \quad -\frac{T}{2} \leq t \leq \frac{T}{2} \tag{3.17}$$

where $a(t)$ is the amplitude, $\phi(t)$ is the phase, and T is the duration of the pulse waveform. The autocorrelation function of $u(t)$ can be expressed as

$$R(\tau) = \int_{-\infty}^{\infty} |U(f)|^2 e^{j2\pi\tau} df, \tag{3.18}$$

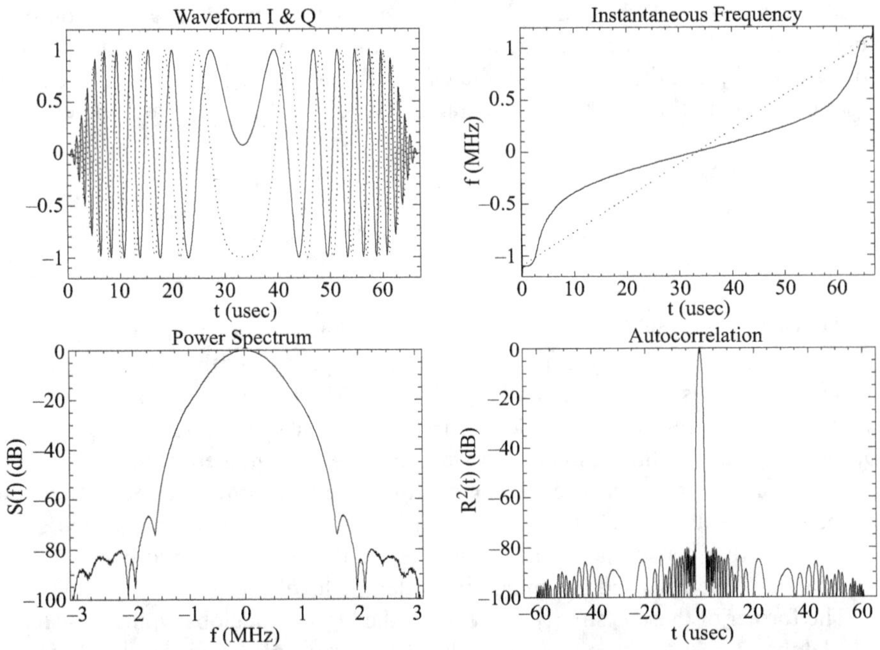

Figure 3.6 Properties of a NLFM waveform from [59] with swept bandwidth of 2.2 MHz and duration 67 μs: (a) real frequency and imaginary parts of the chirp waveform, (b) instantaneous freqency versus time (dotted line denotes LFM), (c) power spectrum, and (d) autocorrelation function.

where $U(f)$ is the Fourier transform of $u(t)$ and is given by

$$U(f) = U_m(f)e^{j\theta(f)}, \quad -\frac{B}{2} \leq f \leq \frac{B}{2}. \tag{3.19}$$

where $U_m(f)$ is the amplitude, $\theta(f)$ is the phase, and B is the swept bandwidth of the waveform. Note that in NLFM pulses, the swept bandwidth does not necessarily coincide with the spectral (i.e. equivalent rectangular) bandwidth, as is the case when using linear frequency modulated pulses. Similarly, BT may happen to be quite different from the compression gain T/T_{eff}. For the waveform of Figure 3.6, the equivalent rectangular bandwidth is 1.57 MHz, and the half-power width of the compressed pulse is $T_{eff} = 0.86\ \mu s$, implying a compression gain of 78.

The object of the stationary phase method is to relate the amplitude and phase of $u(t)$ given by (3.17) to the corresponding amplitude and phase of $U(f)$. For sufficiently large time–bandwidth products, an approximate relation can be found. The method of stationary phase purports that the main contribution to the integration of oscillatory functions occurs around "stationary points" that satisfy,

$$\frac{d}{df}\theta(f) = -2\pi t. \tag{3.20}$$

These are points where the phase of $U(f)$ is the conjugate of the complex exponential in (3.18). These define an approximate relation between the instantaneous frequency and the time within the waveform, $f = \gamma(t)$. With this relation, the magnitudes and phases of the Fourier transform pair $u(t) \leftrightarrow U(f)$ are related by

$$a(t) \approx \sqrt{2\pi} \frac{U_m(\gamma(t))}{\sqrt{|\theta'\prime(\gamma(t))|}} \tag{3.21}$$

$$\frac{d\phi(t)}{dt} \approx 2\pi\gamma(t). \tag{3.22}$$

As was the case with the performance of window functions, the quality of this approximation varies directly as the time–bandwidth product of the waveform, which for weather radars is not particularly large. For NLFM waveforms, the lack of a direct analytical relationship between time and frequency domain representations requires that the design and optimization of the waveform be done numerically.

The design of these waveforms is generally based on some sort of optimization scheme over the autocorrelation. Usually, it is minimizing the energy in the sidelobes subject to a constraint on the main-lobe width. Metrics used to constrain the design include the PSL and the integrated sidelobe level (ISL).

The former of these is simply the level of the highest sidelobe, which is usually the sidelobe closest to the main lobe. The ISL is ratio of the energy in all of the sidelobes relative to the main lobe,

$$ISL = 10\log_{10} \frac{\int_{-\infty}^{-a} |R(\tau)|^2 d\tau + \int_{a}^{\infty} |R(\tau)|^2 d\tau}{\int_{-a}^{a} |R(\tau)|^2 d\tau}, \tag{3.23}$$

where $\pm a$ denotes the first nulls about the main lobe. Optimization schemes used include least-mean-squares [56], genetic algorithms [60], simulated annealing [58], and eigenfilter and quadratic optimization methods [59].

Beginning with the stationary phase method, Pang *et al.* [58] arrived at an expression for the frequency versus time behavior of a NLFM waveform of the form

$$t = T\left[\frac{f}{B} + \frac{1}{2\pi}\sum_{n=1}^{N-1} \frac{a_n}{na_0} \sin\left(\frac{2\pi nf}{B}\right)\right] \tag{3.24}$$

which is seen to be a linear FM waveform perturbed with an odd function expressed as a sine series. The stationary phase method is not necessary to arrive at this expression *per se*, but it is necessary in order to relate it to the power spectrum. The N coefficients, a_n, also describe the shape of the power spectrum according to

$$U_m^2(f) = \frac{K}{2\pi}\sum_{n=0}^{N-1} a_n \cos\left(\frac{2\pi n}{B} f\right). \tag{3.25}$$

where K is a constant. The optimization over the coefficients a_n was then done via simulated annealing based on sidelobe levels of the autocorrelation. They found acceptable results with $N = 7$ terms, achieving sidelobes below −63 dB.

Argenti and Facheris [59] also began with the stationary phase method but found that the approximation inherent to the method precluded its practical use in a non-iterative design. They resorted to a nonlinear optimization to find the optimum filter also based on a power spectrum of the form in (3.25). The NLFM waveform of Figure 3.6, with sidelobes below −80 dB, was obtained using this iterative stationary phase technique and $N = 20$ terms.

Kurdzo *et al.* [60] specified the shape of the frequency versus time curve through 12 symmetrically placed tie points about the center of the waveform. They then used a genetic algorithm to optimally locate the (time, frequency) coordinates of the six unique tie points obtaining a waveform with sidelobes below −59 dB. Since the specification of the curve is based purely on geometry, there is no direct relation to the other methods.

3.6.3 Predistortion

While the aforementioned methods to developing efficient waveforms and their associated filters promise acceptably low range sidelobes, achieving these in practice is complicated as the radar electronics introduce distortion. The distortion comes principally from the power amplifier in the transmitter, and it generally raises the sidelobes above the designed level. Figure 3.7 illustrates the gain characteristic typical of a small microwave power amplifer that might be used, for example, in a phased-array TR module. A similar curve would also describe a high-power amplifier. This figure shows the output power versus the input power with both specified in dBm (deciBels relative to 1 milliWatt). For low levels of input power, the output is linearly related to input by the amplifier's small-signal gain of

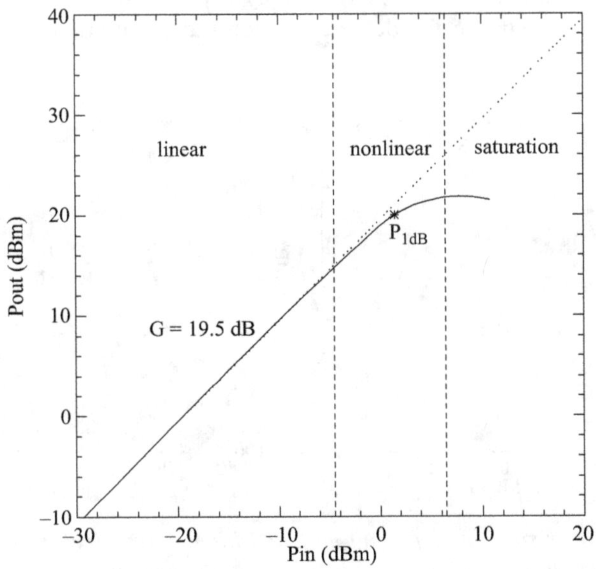

Figure 3.7 Gain characteristic of a typical amplifier illustrating the linear, nonlinear, and saturation regions of operation.

19.5 dB. The upper end of an amplifier's dynamic range is typically specified by its 1-dB compression point, P_{1dB}, or the point at which the gain is reduced by 1 dB. At power levels approaching this point, the amplifier's gain characteristic begins to deviate from linear, and a few dB beyond this point, the output power reaches its maximum. To maximize transmit power and hence radar sensitivity, most transmitters are driven beyond their linear range of amplification into saturation. Some amplifiers are intended for use only in saturation.

A consequence of nonlinear amplification is the generation of harmonics. When multiple frequencies are present at the input, these harmonics undergo intermodulation, resulting in the production of sum and difference frequencies. Harmonics themselves are usually well out of the band of interest, perhaps twice or three times the transmit frequency. However, when more than one frequency is present in the modulating signal, the intermodulation products (or frequencies) may lie within or just adjacent to the band of interest. For the case of two modulating frequencies present (i.e. two tones at f_1 and f_2) the most troublesome intermodulation products are third-order products occurring at frequencies $2f_1 - f_2$ and $2f_2 - f_1$. These products cause a spreading of the transmitted spectrum known as "spectral regrowth." The problem is illustrated in Figure 3.8, where the NLFM baseband waveform of Figure 3.6 is translated to an intermediate frequency of 40 MHz, passed through the amplifier of Figure 3.7 at various output levels, and translated back to baseband. When operated in the linear range, the effect of the amplifier on autocorrelation sidelobes is minimal. However, when driven into the nonlinear or saturation range, the growth of sidebands within and outside the signal power spectrum yields a substantial reduction in the autocorrelation sidelobes if not compensated.

A method to deal with this distortion is by adjusting the waveform input to the transmitter so as to produce the desired waveform at the transmitter output. This is

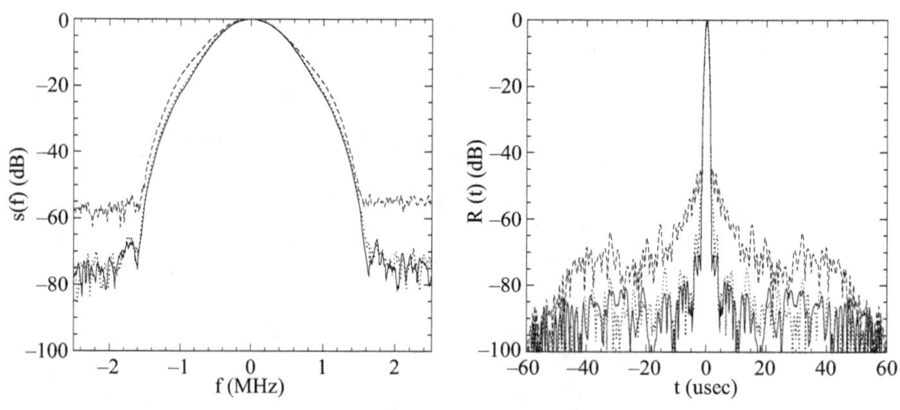

Figure 3.8 Power spectra (left) and matched filter output (right) for the NLFM waveform of Figure 3.6 transmitted through the amplifier of Figure 3.7 operating in the linear range (P_{1dB}−10 dB, solid line), at 1dB compression (P_{1dB}, dotted line), and in saturation.

referred to as predistortion, and it involves a careful characterization of the transmitter. With the advent of arbitrary waveform generation and software radio technology, it is relatively straightforward to implement and is referred to as digital predistortion (DPD).

Predistortion was built into the waveform design approach of Kurdzo *et al.* [60], as the sidelobes were optimized through evaluation through the actual transmit-receive system. In their case, the predistortion was applied as an inverse transfer function, $H^{-1}(f)$, applied in the frequency domain to the ideal input waveform's Fourier transform. The inverse transfer function was that which was necessary to obtain the input (desired) waveform from the output (measured) waveform. Figure 3.9 shows the autocorrelations for their ideal waveform, the waveform distorted by the transmitter, and their final predistorted waveform.

This straightforward approach was sufficient to reclaim about 10 dB of sidelobe suppression that was lost due to distortion. To achieve better results, more sophisticated modeling of the power amplifer is required to capture nonlinear effects. These may include so-called "memory effects" due to self-heating of the electronics and other causes. Various nonlinear models are available to characterize nonlinear effects including the Volterra, Hammerstein, Wiener, and MP models [61]. Of these, the MP model is perhaps the most popular. The MP model describes the system through

$$y(n) = \sum_{k=0}^{K-1} \sum_{m=0}^{M-1} h_{km} x(n-m) |x(n-m)|^k \tag{3.26}$$

where $y(n)$ is the current output, $x(n)$ is the current input, K is the order of nonlinearity ($k = 0$ is linear, $k = 1$ is quadratic, etc.), M is the memory depth, and h_{km} is a matrix of coefficients determined experimentally that describes the nonlinear behavior. The MP model is a simplification of the more general Volterra model, and because it is linear in the coefficients, h_{km}, it is relatively straightforward to determine them from measurements once the order of nonlinearity and memory depth are known. However, determining the forward model and its inverse can also be subject to numerical instabilities as the order of nonlinearity and/or memory depth increase [62].

The predistorted waveform may also be determined through adaptive approaches where feedback is used to repeatedly adjust the input waveform until the desired output waveform or autocorrelation function is obtained. Figure 3.10 illustrates the results of this approach which may be referred to as "open-loop adaptive digital predistortion with indirect learning" [61]. In this case, the ideal sidelobes are nearly recovered after several learning iterations. The predistorted input waveform is observed to have autocorrelation sidelobes similar but complementary to the original output waveform.

Though the primary source of distortion is the radar transmitter, another potential source of error in achieving desired sidelobes is the effect of quantization. This is an important consideration if the pulse compression filter is to be implemented in fixed-point arithmetic (usually in hardware) as quantization errors in the designed filter coefficients may adversely affect sidelobe performance. We do not address this issue here.

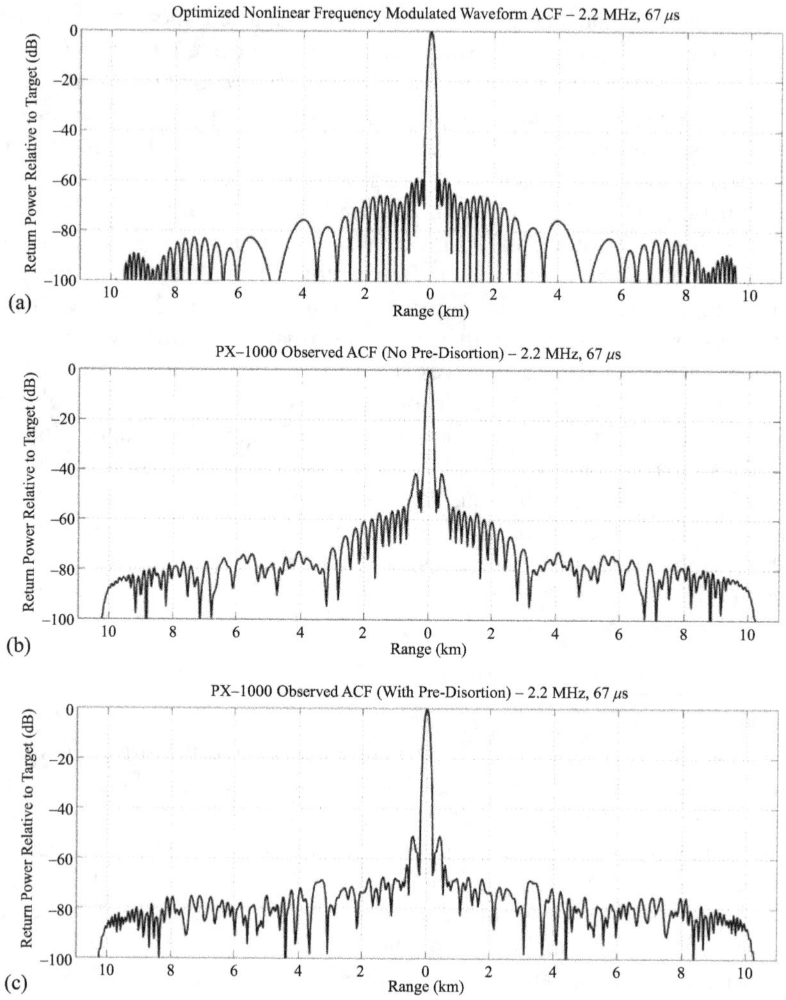

Figure 3.9 (a) Autocorrelation function of an optimized waveform with PSL of −59 dB. (b) Autocorrelation function of the waveform distorted by the transmitter with PSL of −42 dB. (c) Autocorrelation of an optimized waveform output by the transmitter with a predistorted input. PSL is −52 dB (from [60], © American Meteorological Society, used with permission).

3.6.4 Mismatched filters

Most of the literature addressing pulse compression and sidelobe reduction begins with the matched filter and adds various modifications such as amplitude tapering and phase or frequency adjustments. The resulting filters all sacrifice some resolution (i.e. bandwidth) and energy compared to the matched filter. Comparatively few authors have approached the problem differently.

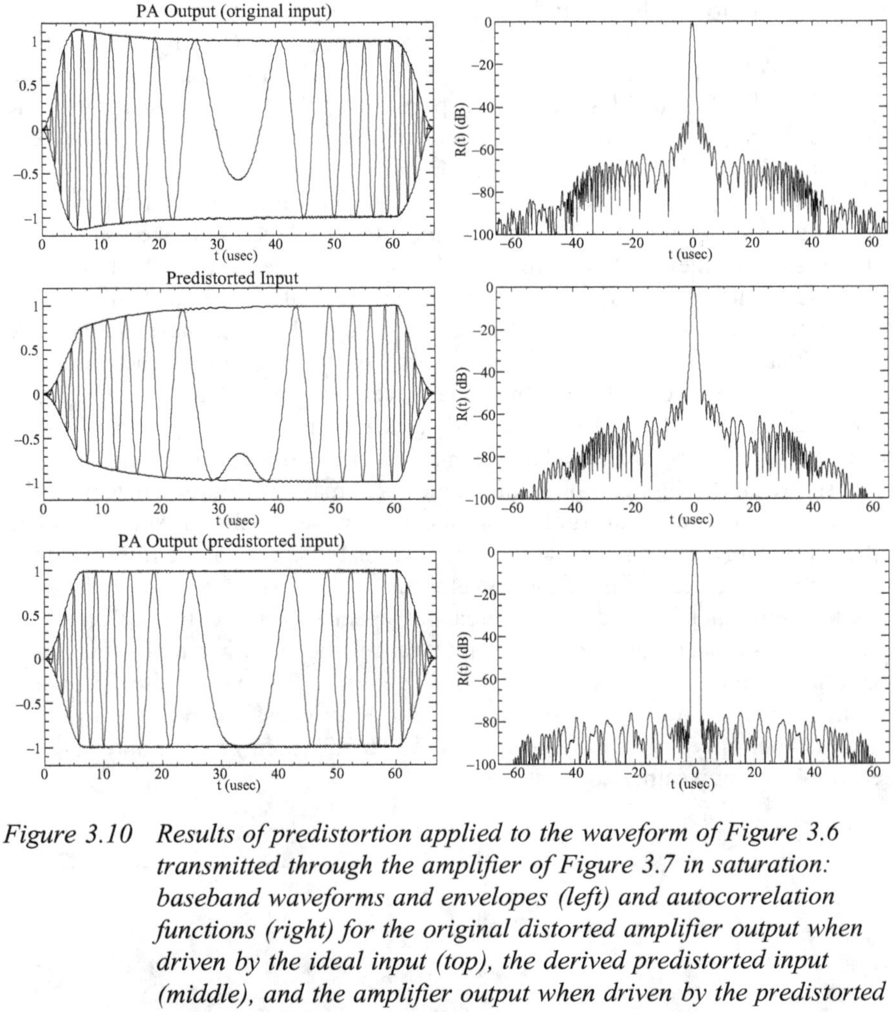

Figure 3.10 Results of predistortion applied to the waveform of Figure 3.6 transmitted through the amplifier of Figure 3.7 in saturation: baseband waveforms and envelopes (left) and autocorrelation functions (right) for the original distorted amplifier output when driven by the ideal input (top), the derived predistorted input (middle), and the amplifier output when driven by the predistorted input (bottom).

Mismatched filters are those whose design is not predicated initially on the matched filter. Keeler and Hwang [57] reported on the use of "inverse filters" using a design approach of Ashe *et al.* [63] for Barker-coded waveforms and for linear and nonlinear FM waveforms. These algorithms attempt to minimize the mean square error between the actual filter output and a desired output which is an impulse-like function, with a few points about the main peak left unconstrained. Cilliers and Smit [64] reported on mismatch filter design for linear FM waveforms based upon minimization of L_p-norms.

Another approach attempts to deal with the residual Fresnel ripples of the spectrum directly through the so-called "reciprocal ripple correction." In this scheme, the filter is designed to compensate the ripple in order to obtain a rectangular spectrum which can then be subject to a given window function. This technique was first applied in the design of SAW devices [65]. Figure 3.11 illustrates this with a filter designed by reciprocal ripple correction. The filter was designed by dividing the desired output spectrum by the Fourier transform of the input waveform. In this case, the desired output spectrum had the shape of a Blackman window. The figure shows the impulse response of the filter.

The filter output has sidelobe properties consistent with the Blackman window. The compression gain of this waveform and filter combination is the time–bandwidth product of the LFM waveform ($10\log_{10}(75)$ dB) less the SNR loss of the window (2.37 dB), or 16.4 dB. Were it a matched filter, the coefficients outside the dashed lines in Figure 3.11a would be identically zero. Because these methods are not specifically based upon the time reverse of the waveform, their impulse response is not limited to the length of the waveform, and in fact, it can be much longer than the waveform. This can lead to difficulties in their practical implementation. In this case, the impulse response extends well beyond the range of the x-axis shown in the figure, though the values are small. To limit their length, the filters can be truncated, but truncation tends to re-raise the far-out sidelobes. The dashed curve in Figure 3.11 shows the sidelobes that result when the filter length is truncated to twice the length of the waveform.

Other mismatch filter design approaches operate on the output of an initial matched (or similar) filter and attempt to cancel the resulting range sidelobes. Algorithms in this class include the CLEAN algorithm borrowed from astronomy and image processing [66], and adaptive pulse compression methods [67–69]. These methods appear generally geared toward "hard" (point) targets and have not been employed in weather applications.

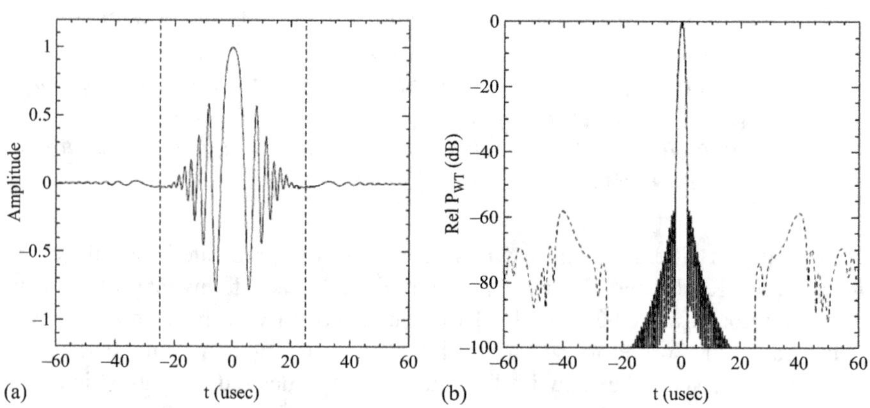

Figure 3.11 Mismatched filter design by ripple correction. (a) Filter impulse response for the LFM waveform of Figure 3.4 where dashed lines mark the duration of the transmitted waveform. (b) The mismatch filter output using the full impulse response (solid line) and truncated impulse response (at ±50 μs, dashed line).

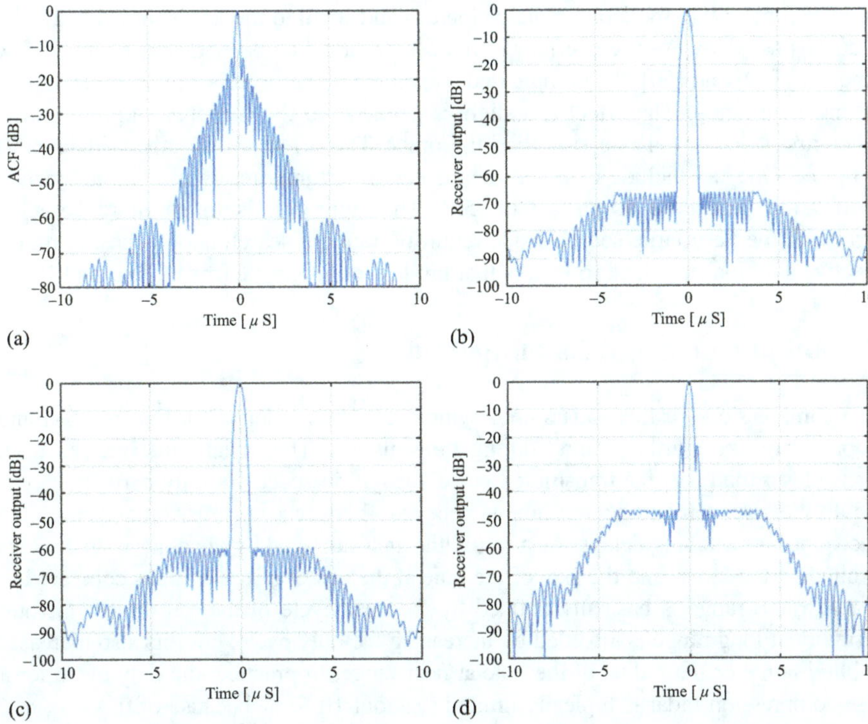

Figure 3.12 Results obtained with the mismatched receiver filter algorithms for a NLFM waveform of swept bandwidth 5 MHz and duration 50 μs: (a) Matched filter output (SNR loss = 0.36 dB), (b) MIMMF algorithm (SNR loss = 2.58 dB), and CMMF algorithm with (c) $\Delta_{loss} = 2$ *dB and (d)* $\Delta_{loss} = 1$ *dB (from [59] © IEEE, used with permission).*

A mismatch filter is proposed in [59] that permits a tradeoff between SNR loss and ISL. Figure 3.12b illustrates a "Minimum ISL Mismatched Filter" (MIMMF) implemented for a linear FM chirp with a Tukey window and no other optimizations. The sidelobe levels obtained are below −65 dB; however, the SNR loss is 2.58 dB which is a little higher than may be desirable. By adding a constraint on SNR loss, they can sacrifice ISL. The figure also shows two "Constrained Mismatched Filter" (CMMF) designs with SNR losses of 2 dB and 1 dB. When the MIMMF algorithm is applied to the already-optimized waveform of Figure 3.6, a slight improvement of 2.4 dB in PSL is reported.

3.6.5 Doppler tolerance

Most pulse compression filter designs, whether based upon the matched filter or not, assume that the received echo has zero Doppler shift. When present, a Doppler-shifted echo waveform is no longer "matched" to the original design criteria. In the case of weather, Doppler shifts are usually sufficiently small so as not to seriously degrade performance, particularly for FM waveforms which are known to exhibit Doppler tolerance for small to moderate radial velocities and small to moderate time–bandwidth products [48]. However, the effect is noticeable and has been discussed by a few authors. In [59] and [60], the compressed waveform in the presence of Doppler shift is presented explicitly. The effect is a slight asymmetry of the compressed pulse with a higher first sidelobe on one side than the other. The effect on all other sidelobes is negligible. One can include Doppler tolerance as a design criterion in a multiobjective optimization procedure as shown in [58]. This approach is much more complex because of the very large search space requiring random search approaches. A multi-objective solution may lead to results that are less optimal in terms of PSL or ISL.

3.7 Blind range and its mitigation

Pulse compression weather radars employing a common antenna for transmission and reception are necessarily "blind" during transmission. This means that echoes from targets closer than half the transmitted pulse's length in space are only partially recovered. In practice, some transition time is necessary between transmission and reception to account for switching delays. As a result, the area surveyed by the chirp is an annulus surrounding the radar, and the size of the hole at the center of the annulus compared to the maximum range is basically defined by the duty cycle of the transmitter. Greater sensitivity at long range is afforded by increasing the duty cycle, but this also increases the blind range or the radius of the hole at near range. In practice, the duty cycle for a pulse-compression radar is typically limited to about 10–20% because of this.

It is desirable therefore to have another means to monitor the closer ranges. The most common method is to transmit a short pulse immediately following the chirp prior to switching to receive mode. The short pulse is offset in frequency from the chirp so that it may be processed separately from the chirp echo in the receiver [30,70,71]. In [30], they followed a 67 μs chirp with a 2 μs short pulse. Because the energy of the shorter pulse is much lower, the sensitivity of the radar is also much lower within the blind range. This lack of sensitivity is offset by the much shorter range of interest; however, there remains a transition from lower sensitivity to higher sensitivity at the blind range as illustrated in Figure 3.13.

An alternative is to note that echoes from within the blind range but beyond zero range are partially present in the received signal, so partial compression of those signals is possible. Salazar *et al.* [72] refer to this as "progressive pulse compression." Within the blind range, this results in a variable range resolution that degrades with decreasing range, as progressively less chirp is available for compression. This also affects the absolute calibration, as the effective time–bandwidth product, and hence the compression gain, of the partially received chirp also varies directly with range.

3.8 Other techniques to increase average power

While to date most solid-state weather radars are pulsed radars employing pulse compression, there are alternative methods to increase the average power. The following sections describe two approaches.

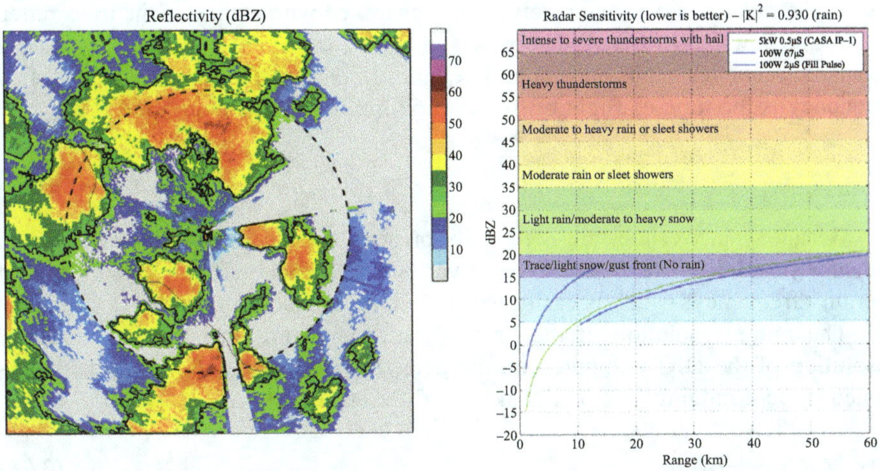

Figure 3.13 Example of the transition in sensitivity between short pulse mode (left, within dashed circle) and chirp mode (outside dashed circle), and minimum reflectivity versus range (right) for the PX-1000 radar developed by the University of Oklahoma [30] (© IEEE, reprinted with permission).

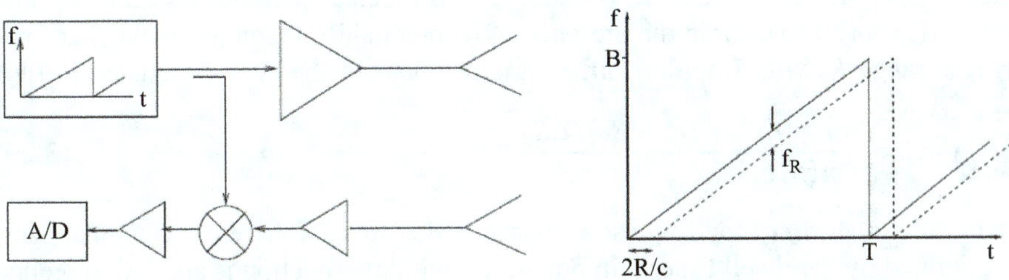

Figure 3.14 Simplified block diagram of an FMCW radar employing separate antennas for transmission and reception (left) and depiction of the FMCW transmitted waveform in time–frequency space (right, solid line) and an echo (dashed line) where the time-of-flight, 2R/c, is small compared to the sweep time, T. The detected beat frequency, f_R, indicates the range.

3.8.1 Frequency-modulated continuous-wave

The FMCW radar may be thought of as the extreme example of a pulse-compression radar that operates with a duty cycle of 100%, or nearly so. In this scheme, the radar simultaneously transmits and receives all the time. The radar transmits a linear FM waveform of duration, T, and bandwidth, B. However, unlike the traditional pulse-compression radar, the duration is longer compared to the round-trip time-of-flight of the radar echo, $T \ll 2R/c$, as illustrated in Figure 3.14.

In the radar receiver, the received echo is mixed with a copy of the transmitted signal and low-pass filtered. The resulting signal from a target at range, R, is a constant frequency sinusoid whose frequency indicates the range. The detected frequency is the time-of-flight of the echo multiplied by the chirp rate, B/T,

$$f_R = \frac{2R}{c}\frac{B}{T}. \tag{3.27}$$

Because many echoes are potentially present, the received signal will contain the superposition of many sinusoids of various frequencies. It is necessary to sort them by frequency, which is easily performed via a DFT.

The range resolution of the FMCW radar is determined by the frequency resolution of the DFT computed on the detected waveform, Δf_R. It can be determined by re-arranging (3.27) and taking a differential, $\Delta R/\Delta f_R$,

$$\Delta R = \frac{c}{2}\frac{T}{B}\Delta f_R. \tag{3.28}$$

Noting that the frequency resolution of the DFT is given by $1/T$, one obtains the expected result, namely $\Delta R = c/2B$.

Once the detected signal from a given frequency sweep is Fourier transformed, it may then be treated like any pulse echo waveform, with the FM sweep period, T, serving as the pulse interval. Doppler shift can be measured from pulse to pulse in the usual manner. Because the FMCW radar uses the detected frequency to determine range, it is subject to error in the presence of Doppler shift. It can be shown that for a target at range R_0 with Doppler shift f_D, the response of the FMCW radar is [40]

$$y(R) = \frac{\sin \pi(f_D T + (R - R_0)/\Delta R)}{\pi(f_D T + (R - R_0)/\Delta R)} \tag{3.29}$$

where the sinc form of the impulse response is due to the inherent rectangular window defined by the dwell time T. In practice, a window function is applied to reduce the range sidelobes to an acceptable level. We note that the problem of Fresnel ripples in the spectrum that was characteristic of the pulse-compression radar is not present here, as the detected signals are constant-frequency sinusoids. In the presence of Doppler shift, the apparent range of the target is offset from the true range

$$R_{app} = R_0 - f_D T \Delta R. \tag{3.30}$$

For most meteorological applications, this error is usually less than one range bin. The misregistration is less than one half of a range bin for unambiguously measured velocities [40].

FMCW radars are very sensitive, not only because of their high average power but also because their receiver noise bandwidth is usually very small. The detected frequencies are often in the audio frequency range, so the receiver's instantaneous bandwidth can be very small. As was the case for the pulse-compression radar, the improvement factor over the equivalent pulsed radar with the same transmit power is given by the time–bandwidth product of the waveform, BT. Because T is now quite long, this can be a very large improvement factor. In practice, T is limited by the coherence time of the microwave echo. That is, the FMCW technique assumes that the scatterers are essentially stationary or are translating together without re-arranging their relative positions. Relative motion amongst the scatterers results in a spread of radial velocities and spreads the echo in range. In practice, the sweep time is limited to a few tens of milliseconds at most.

Because of the relatively long FM sweep interval, the FMCW radar is best suited for relatively short ranges (i.e. $\lesssim$10 km) and slow-moving targets, ($|v_r| < \lambda/4T$). FMCW techniques are well suited to vertical profiling applications. One example is the MRR [44], a vertically pointed 24 GHz FMCW radar that is widely used for precipitation profiling. With a transmit power of only 50 mW, the MRR profiles up to 5 km altitude providing the fall velocity of hydrometeors obtained from the Doppler spectrum. Using expressions for terminal velocity as a function of drop diameter, the measured Doppler spectrum can be converted to a drop-size distribution (DSD).

A significant practical concern for FMCW radars is the adequate isolation between transmitter and receiver. It is important that the transmitted signal does not saturate the receiver since both functions must occur simultaneously. This concern often precludes single antenna operation, requiring separate antennas for transmission and reception. In the case of the MRR, the transmitter power is sufficiently low and the echo is sufficiently large that a single antenna is sufficient. The FMCW technique with dual antennas is employed in the UMass S-band FMCW radar (which was converted from a traveling-wave-tube transmitter to a solid-state GaN amplifier in 2017), the X-band IDRA and QX-120 scanning precipitation radars [42,43], and the W-band BASTA cloud profiling radar [45].

3.8.2 Quadratic phase coding

A recently developed scheme that is analogous to the FMCW radar is the use of quadratic phase coding from pulse to pulse and a very high pulse rate. In this scheme, described by Mead and Pazmany [46], the radar is pulsed at a rate of an integer factor higher than that which dictates the maximum or unambiguous range,

$$f_p = Mf_p^0 = M\frac{c}{2R_{max}}, \tag{3.31}$$

where f_p^0 is the pulse rate that defines maximum range, c is the speed of light, and R_{max} is the maximum range.

The transmitted pulses are themselves unmodulated, but each pulse is transmitted with a progressive starting phase given by

$$\phi(n) = \frac{\pi n^2}{M} \tag{3.32}$$

where M is related to the length of the base code. For M even, the code length is $L_0 = M$, and for M odd, the code length is $L_0 = 2M$. The phase code is symmetric, or palindromic, when observed modulo-2π. With this scheme, each range bin sampled by the receiver will contain up to M overlaid echoes from different (ambiguous) ranges. Upon reception, the echo signal is multiplied by the complex conjugate phase of the last transmitted pulse as illustrated in Figure 3.15. It can be shown that each range-overlaid echo will have a unique frequency shift

$$\Delta f_m = -mf_p^0, \qquad m = 0, 1, 2, \ldots M - 1. \tag{3.33}$$

where m denotes the echo-trip. Thus, the echo trips may be separated by frequency through a discrete Fourier transform.

The QPC radar has a duty cycle M times that of the nominal pulsed radar; hence, the improvement factor is equal to M. Doppler processing of the echoes for each range bin and echo trip occurs in the usual way. Indeed, a single, long DFT is sufficient to separate the echo trips, with a unique Doppler spectrum for each trip.

In the limit of the QPC technique, the pulse rate is increased until there is only a single range bin within the maximum range. At this point, the duty cycle of the radar is nearly 50% after accounting for some switching time. When the duty cycle is increased to this upper limit, the scheme is referred to as a phase-modulated square wave (PMSW) radar as illustrated in Figure 3.16.

An advantage of the QPC or PMSW technique is that there is no pulse compression. There are no range sidelobes to contend with. There is also no long blind range associated with transmitting a long pulse. Instead, there are multiple blind ranges associated with the transmit cycle that are interleaved between each of the blocks of overlaid range samples. For continuous volumetric signals such as weather, these interleaved missing ranges may not be a major concern as adjacent range bins are often correlated. Otherwise, they should be mitigated.

There is the necessity of a fast switch in PMSW mode. For a radar with 150 m range resolution, the switching speed would be 1 MHz. Such a switch is available for transmit powers of the order 1–10 W. However, with lesser duty cycles, the switching speed can be slower. If separate antennas for transmission and reception

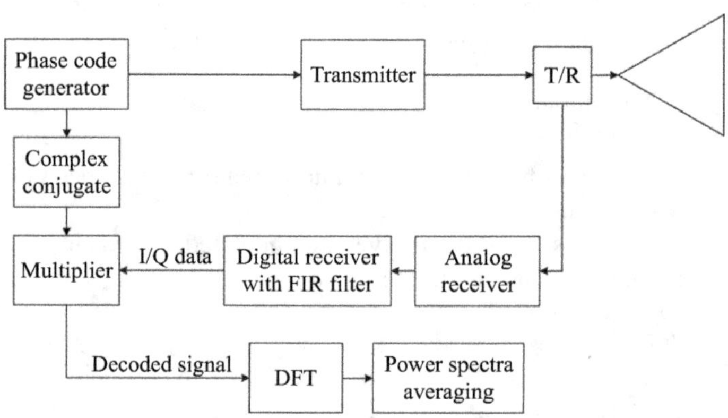

Figure 3.15 Simplified block diagram of a quadratic phase code (QPC) radar (from [46] © American Meteorological Society, used with permission).

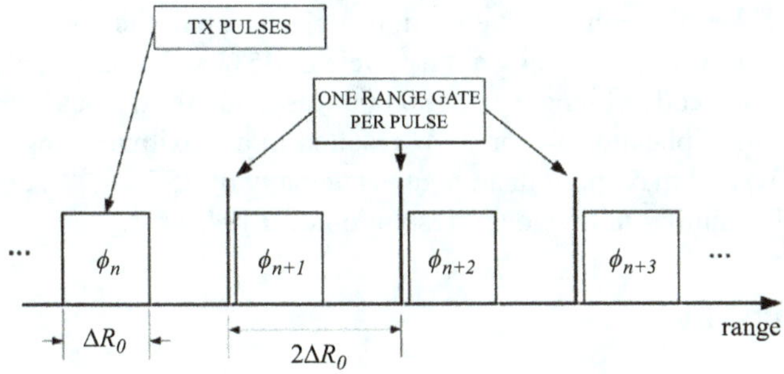

Figure 3.16 PMSW mode in which there is only a single range bin containing overlaid echoes from M trips (from [46] © American Meteorological Society, used with permission).

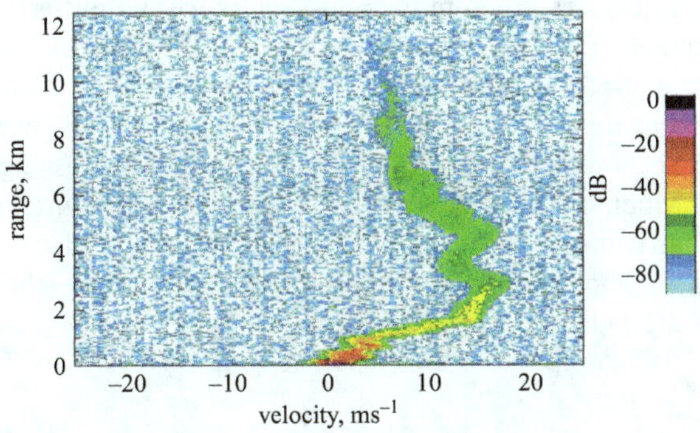

Figure 3.17 Range–Doppler plot in rain with mean noise subtraction using a composite code of length 130 000 and FFT length N = 65 000 with two power spectra averaged for a total dwell time of 43.3 ms. No spurious-induced range sidelobes are visible above the residual noise floor after mean noise subtraction (from [46] © American Meteorological Society, used with permission).

can be employed, then the *T*/*R* switch can be removed, the duty cycle can go to 100%, and the blind ranges can be eliminated.

Analogous to the problems encountered in pulse compression with non-linearities induced by the transmitter requiring predistortion, errors in the transmitted amplitude and phase may also occur in QPC radar. The effect of these errors is spurious peaks at zero Doppler shift for each range bin. Short of correcting these errors on transmission, it is shown that these can be suppressed by extending the length of the code by a large integer factor and by modifying the phase increment resulting in a "composite code."

The technique has been demonstrated on a Ka-band cloud radar employing a 10 W solid-state transmitter. Figure 3.17 shows the results of operating the radar in

PMSW mode observing rain at 45° elevation angle. The pulse length was 150 ns and the pulse rate was 3 MHz, yielding a duty cycle of 45% and a range gate interval of 50 m. A composite code of length $L = 130\,000$ was made from a base code of length $M = 250$. The multiplication factor of M resulted in a maximum range of 12.5 km (250 range bins) and maximum unambiguous velocity of 25.2 ms^{-1}. A 65 000-point FFT was used, resulting in a velocity resolution of 0.194 ms^{-1}.

3.9 Conclusion

In this chapter, we have reviewed the state of the art in solid-state weather radars. These radars achieve sensitivity by operating with a high-duty cycle compared to their tube-based counterparts. Advances in solid-state microwave power generation are now yielding practical and affordable transmitters. Advances in digital technology have enabled both the generation of sophisticated waveforms and compensation for nonlinearities inherent to solid-state power amplifiers. Looking to the future, we can expect further improvements in and broader availability of high-power solid-state sources as well as continuing sophistication of digital receivers and exciters. At the time of this writing, the state of the art in receiver/exciter technology is RFSoCs which blend RF and digital signal processing at frequencies through C-band. Much of this development is driven by the wireless communications industry, particularly 5G technology. The weather radar community will benefit as these technologies are further integrated pushing the digital/analog boundary within the radar systems ever closer to the antenna.

References

[1] McLaughlin D, Pepyne D, Chandrasekar V, *et al.* Short-wavelength technology and the potential for distributed networks of small radar systems. *Bull Amer Meteorol Soc*. 2009;90(12):1797–1818.

[2] Junyent F, Chandrasekar V, McLaughlin D, *et al.* The CASA integrated project 1 networked radar system. *J Atmos Ocean Technol*. 2010;27: 61–78.

[3] Brewster KA, Bajaj A, Philips BJ, *et al.* CASA Dallas-Fort Worth urban testbed operations: Network of networks at work. In: *Special Symposium on Meteorol. Obs. and Instr, 97th AMS Annual Meeting*. 22–26 January, Seattle, WA; 2017. p. 1027.

[4] Westrelin S, Diss S, Mériaux P, *et al.* Hydrometeorological risks in Mediterranean mountainous areas. In: *Proc Sixth European Conf. Radar in Meteorology and Hydrology*. Sibiu, Romania; 2010. Available from: https://www.erad2010.com/pdf/oral/thursday/hidrostu/05_ERAD2010_0363.pdf.

[5] Antonini A, Melani S, Corongiu M, *et al.* On the implementation of a regional X-band weather radar network. *Atmosphere*. 2017;8(2):25.

[6] Wada M, Horikomi J and Mizutani F. Development of solid-state weather radar. In: *2009 IEEE Radar Conf.* 4–8 May, Pasadena, CA; 2009. p. 1–4.

[7] Solid state X-band weather radar with dual polarization. GAMIC GmbH; [cited 2022 Feb 16]. Available from: https://www.gamic.com/radar-systems/gmwr-solid-state.

[8] Ranger X-band radar systems. EEC, Inc.; [cited 2022 Feb 16]. Available from: https://www.eecweathertech.com/ranger-x-band-radar-systems.php.

[9] Winslow G. Climavision pitches radar as a service. TVTech [serial on the Internet]; 2021 July [cited 2022 Feb 16]. Available from: https://www.tvtechnology.com/news/climavision-pitches-radar-as-a-service.

[10] Pressman A. Boston startup wants to combat the climate crisis by launching network of radar satellites. *Boston Globe*. 2021; 1. Available at: https://www.bostonglobe.com/2021/09/05/business/boston-startup-wants-combat-climate-crisis-by-launching-network-radar-satellites/.

[11] Hondl KD and Emmanuel MJ. The advanced technology demonstrator: An S-band, dual polarization, phased array radar at the National Weather Radar Testbed. In: *35th Conf. Environmental Information Processing Technol.* 6–10 January, Phoenix, AZ: AMS; 2019. p. 7A.1.

[12] Orzel KA and Frasier SJ. Weather observation by an electronically scanned dual-polarization phase-tilt radar. *IEEE Trans Geosci Remote Sens*. 2018;56 (5):2722–2734.

[13] Salazar JL, Yu TY, McCord M, *et al*. An ultra-fast scan C-band polarimetric atmospheric imaging radar (PAIR). In: *2019 IEEE Int Symp Phased Array Syst Technol (PAST)*; 2019. p. 1–5.

[14] Lussier LL, Lee WC and Grubisic V. Airborne phased array radar (APAR): The next generation of airborne polarimetric Doppler weather radar. In: *37th Conf Environ Inf Process Technol*. 10–15 Jan, Boston, MA: AMS; 2021. p. 11.8.

[15] Wang H, Huang TY, Sasikanth Mannem N, *et al.* Power amplifiers performance survey 2000-present; [updated 2020 Nov 19; cited 2022 Feb 17]. Available from: https://gems.ece.gatech.edu/PA_survey.html.

[16] Theeuwen SJCH and Qureshi JH. LDMOS technology for RF power amplifiers. *IEEE Trans Microw Theory Tech*. 2012;60(6):1755–1763.

[17] Berg H, Thiesies H, Hemmendorff E, *et al.* Silicon–germanium for phased array radars. In: *2008 Eur Microw Integr Circuit Conf*; 2008. p. 266–269.

[18] Puzella A and Alm R. Air-cooled, active transmit/receive panel array. In: *2008 IEEE Radar Conf*. 26–30 May, Rome, IT: IEEE; 2008.

[19] Skyler. Raytheon intelligence and space; [cited 2022 Feb 17]. Available from: https://www.raytheonintelligenceandspace.com/capabilities/products/skyler.

[20] Kollias P, McLaughlin DJ, Frasier S, *et al.* Advances and applications in low-power phased-array X-band weather radars. In: *Proc Radar Conf.* 23–27 April, Oklahoma City, OK: IEEE; 2018.

[21] Heberling W and Frasier SJ. On the projection of polarimetric variables observed by a planar phased-array radar at X-band. *IEEE Trans Geosci Remote Sens*. 2021;59(5):3891–3903.

[22] Drake P, Bourgeois J and McLaughlin D. X-band phased array: Current radar performance and plans for wake vortex experimentation. In: *WakeNet-Europe Workshop 2014*. 13–14 May, Bretigny-sur-Orge, France; 2014. p. 602.

[23] McLinden MLW, Li L, Heymsfield GM, *et al.* The NASA GSFC 94-GHz airborne solid-state cloud radar system (CRS). *J Atmos Ocean Technol.* 2021;38(5):1001–1017.

[24] XilinX/AMD. Zynq Ultrascale+ RFSoC; [cited 2022 May 3]. Available from: https://www.xilinx.com/products/silicon-devices/soc/rfsoc.html.

[25] Cooklev T, Normoyle RB and Clendenen DA. The VITA 49 analog RF-digital interface. *IEEE Circuits Syst. Mag.* 2012;12:21–32.

[26] Kowalski E, Conway D, Morris A, *et al.* Multifunction phased array radar advanced technology demonstrator (MPAR ATD) nearfield testing and fielding. In: *2019 IEEE Radar Conf (RadarConf)*; 2019. p. 1–4.

[27] Yeary M, Palmer R, Fulton C, *et al.* Recent advances on an S-band all-digital mobile phased array radar. In: *2019 IEEE Int. Symp Phased Array Syst Technol (PAST)*; 2019. p. 1–5.

[28] Fulton C, Salazar JL, Zhang Y, *et al.* Cylindrical polarimetric phased array radar: Beamforming and calibration for weather applications. *IEEE Trans Geosci Remote Sens.* 2017;55(5):2827–2841.

[29] Mizutani F, Ushio T, Yoshikawa E, *et al.* Fast-scanning phased-array weather radar with angular imaging technique. *IEEE Trans Geosci Remote Sens.* 2018;56(5):2664–2673.

[30] Cheong BL, Kelley R, Palmer RD, *et al.* PX-1000: A solid-state polarimetric X-band radar and time-frequency multiplexed waveform for blind range mitigation. *IEEE Trans Instrum Meas.* 2013;62:3064–3072.

[31] Fernandez DE, Kerr EM, Castells A, *et al.* IWRAP: The imaging wind and rain airborne profiler for remote sensing of the ocean and the atmospheric boundary layer within tropical cyclones. *IEEE Trans Geosci Remote Sens.* 2005;43(8):1775–1787.

[32] McLinden ML, Carswell J, Li L, *et al.* Utilizing versatile transmission waveforms to mitigate pulse-compression range sidelobes with the HIWRAP radar. *IEEE Geosci Remote Sens Lett.* 2013;10(6):1365–1368.

[33] Furukawa K, Hanado H, Hyakusoku Y, *et al.* Preliminary design of the spaceborne dual-frequency precipitation radar for the global precipitation measurement. In: *Int Geosci Remote Sens Symp.* IEEE; 2007. p. 3551–3554.

[34] Beauchamp RM, Tanelli S, Peral E, *et al.* Pulse compression waveform and filter optimization for spaceborne cloud and precipitation radar. *IEEE Trans Geosci Remote Sens.* 2017;55(2):915–931.

[35] Weather radar. Japan Radio Co., Ltd. (JRC); [cited 2022 Feb 16]. Available from: http://www.jrc.co.jp/eng/product/weather_disaster_prevention_rivers/weather_radar/index.html.

[36] X-band weather radar WRS400. Vaisala; [cited 2022 Jun 8]. Available from: https://www.vaisala.com/en/products/weather-environmental-sensors/x-band-weather-radar-wrs400-meteorology.

[37] WRX-100 X-band dual-polarized solid-state Doppler radar. BIRM; [cited 2022 Jun 8]. Available from: https://www.environmental-expert.com/products/birm-model-wrx-100-x-band-solid/state-dual-polarized-doppler-weather-radar-594914.

[38] E700 series portable Doppler radar. EWR Radar Systems; [cited 2022 Jun 8]. Available from: https://ewradar.com/radar/e700-series-portable-doppler-radar/.

[39] Compact dual-polarimetric X-band Doppler weather radar WR2120. Furuno; [cited 2022 Jun 8]. Available from: https://www.furuno.com/en/systems/meteorological-monitoring/WR2120.

[40] Ince T, Frasier SJ, Muschinski A, *et al.* An S-band frequency-modulated continuous-wave boundary layer profiler: Description and initial results. *Radio Sci*. 2003;38(4):1072.

[41] Waldinger J. Improvements to the UMass S-Band FMCW vertical wind profiling radar: System performance and data analysis [MS Thesis]. University of Massachusetts: Amherst, MA; 2018.

[42] Otto T and Russchenberg HWJ. High-resolution polarimetric X-band weather radar observations at the Cabauw experimental site for atmospheric research. *Geosci Data J.* 2013;1(1):7–12.

[43] MetaSensing X-band WR family. MetaSensing; [cited 2022 June 8]. Available from: https://www.weather.metasensing.com/products.

[44] METEK GmbH. Micro Rain Radar MRR; [cited 2022 Jan 13]. Available from: https://metek.de/product-group/micro-rain-radar-24-ghz/.

[45] Delanoe J, Protat A, Vinson JP, *et al.* BASTA: A 95-GHz FMCW Doppler radar for cloud and fog studies. *J Atmos Ocean Technol.* 2016;33(5):1023–1038.

[46] Mead JB and Pazmany AL. Quadratic phase coding for high duty cycle radar operation. *J Atmos Ocean Technol*. 2019;36(6):957–969.

[47] North DO. An analysis of the factors which determine signal/noise discrimination in pulsed-carrier systems. *Proc IEEE*. 1963;51(7):1016–1027.

[48] Skolnik MI. *Introduction to Radar Systems*. 3rd ed. New York, NY: McGraw-Hill; 2002.

[49] Bucci NJ, Owen HS, Woodward KA, *et al.* Validation of pulse compression techniques for meteorological functions. *IEEE Trans Geosci Remote Sens*. 1997;35(3):507–523.

[50] Mudukutore AS, Chandrasekar V and Keeler RJ. Pulse compression for weather radars. *IEEE Trans Geosci Remote Sens*. 1998;36(1):125–142.

[51] Tanner A, Durden SL, Denning R, *et al.* Pulse compression with very low sidelobes in an airborne rain mapping radar. *IEEE Trans Geosci Remote Sens*. 1994;32(1):211–213.

[52] Doerry AW. *Catalog of Window Taper Functions for Sidelobe Control*. Albuquerque, NM and Livermore CA: Sandia National Laboratories; 2017. Report No.: SAND2017-4042.

[53] Cook CE and Paolillo J. A pulse compression predistortion function for efficient sidelobe reduction in a high-power radar. *Proc IEEE*. 1964;52 (4):377–389.

[54] Griffiths HD and Vinagre L. Design of low-sidelobe pulse compression waveforms. *Electron Lett*. 1994;30(12):1004–1005.

[55] De Witte E and Griffiths HD. Improved ultra-low range sidelobe pulse compression waveform design. *Electron Lett*. 2004;40(22):1448–1450.

[56] Bharadwaj N and Chandrasekar V. Wideband waveform design principles for solid-state weather radars. *J Atmos Ocean Technol.* 2012;29(1):14–31.

[57] Keeler RJ and Hwang CA. Pulse compression for weather radar. In: *Proc Int Radar Conf.* 8–11 May, Alexandria, VA: IEEE; 1995.

[58] Pang C, Hoogeboom P, Chevalier FL, *et al.* A pulse compression waveform for weather radars with solid-state transmitters. *IEEE Geosci Remote Sens Lett.* 2015;12(10):2026–2030.

[59] Argenti F and Facheris L. Radar pulse compression methods based on nonlinear and quadratic optimization. *IEEE Trans Geosci Remote Sens.* 2020;59 (5):3904–3916.

[60] Kurdzo JM, Cheong BL, Palmer RD, *et al.* A pulse compression waveform for improved-sensitivity weather radar observations. *J Atmos Ocean Technol.* 2014;31(12):2713–2731.

[61] Ghannouchi FM, Hammi O and Helaoui M. *Behavioral Modeling and Predistortion of Wideband Wireless Transmitters.* West Sussex, UK: Wiley; 2015.

[62] Dunn Z, Yeary M, Fulton C, *et al.* Wideband digital predistortion of solid-state radar amplifiers. *IEEE Trans Aerosp Electron Syst.* 2016;52(5):2452–2466.

[63] Ashe JM, Nevin RL, Urkowitz H, *et al.* Range sidelobe suppression of expanded/compressed pulses with droop. In: *Proc Natl Radar Conf.* 29–31 Mar, Atlanta, GA: IEEE; 1994.

[64] Cilliers JE and Smit JC. Pulse compression sidelobe reduction by minimization of L_p-norms. *IEEE Trans Aerosp Electron Syst.* 2007;43(3):1238–1247.

[65] Judd GW. Technique for realizing low time sidelobe levels in small compression ratio chirp waveforms. In: *1973 Ultrasonics Symposium* IEEE; 1973. p. 478–481.

[66] Kulpa K. The CLEAN type algorithms for radar signal processing. In: *Proc. Microwaves, Radar and Remote Sensing Symp.* 22–24 Sept, Kiev, Ukraine: IEEE; 2008. p. 152–157.

[67] Blunt SD, Smith KJ and Gerlach K. Doppler-compensated adaptive pulse compression. *IEEE Trans Aerosp Electron Syst.* 2006;42:114–119.

[68] Blunt SD and Gerlach K. Adaptive pulse compression via MMSE estimation. *IEEE Trans Aerosp Electron Syst.* 2006;42(2):572–584.

[69] Gerlach K and Blunt SD. Radar pulse compression repair. *IEEE Trans Aerosp Electron Syst.* 2007;43(3):1188–1195.

[70] George J, Mishra KV, Nguyen CM, *et al.* Implementation of blind zone and range-velocity ambiguity mitigation for solid-state weather radar. In: *Proc IEEE Radar Conf.* 10–14 May, Arlington, VA; 2010.

[71] Mishra KV. Frequency diversity wideband digital receiver and signal processor for solid-state dual-polarimetric weather radars [M.S. Thesis]. Colorado State University; 2012.

[72] Salazar Aquino CM, Cheong B and Palmer RD. Progressive pulse compression: A novel technique for blind range recovery for solid-state radars. *J Atmos Ocean Technol.* 2021;38(9):1599–1611.

Chapter 4
Quality of polarimetric data in the WSR-88D system

Valery M. Melnikov[1,4] *and Dušan S. Zrnić*[1,2,3]

The US weather radar network consists of 160+ S-band Weather Surveillance Doppler Radars (WSR-88D, aka NEXRAD) serving two National Weather Service's (NWS) missions: monitoring severe weather and measuring precipitation. Severe weather events include tornadoes, squall lines, microbursts, strong winds, hail, freezing rain, and blizzard [1–6]. The amounts of rainfall or snowfall are measured with the WSR-88D network in real time. Heavy rain causing flash floods is also a severe weather event. Weather radars can quantify the attributes of some impactful events like winds and turbulence via the Doppler shift. Due to dual polarization, the WSR-88D delivers data that can improve significantly quantitative precipitation estimates (QPE) and hail sizing. Nonetheless, quantitative measurements require data with insignificant bias and small variance of estimates. To serve well its mission, the NWS is continually engaged in improving the WSR-88Ds and the "quality" of its data. Herein we describe calibration procedures used to reach this goal.

4.1 Introduction

Shortly after World War II, the USA Weather Bureau, predecessor to the National Weather Service (NWS), developed the first civilian network of weather radars. These modified military radars [5] displayed weather images on the position plan indicators (PPI). Over time, continuing improvements of the systems and their data quality followed.

The first comprehensive network of US weather radars was designed in 1957; hence, it is designated as WSR-57. It operated at the 10 cm wavelength and had a logarithmic and a linear receiver. A step attenuator was added for comparing (calibrating) the returned powers against a known standard. Further enhancement

[1]NOAA National Severe Storms Laboratory, Norman, OK, USA
[2]School of Meteorology, University of Oklahoma, Norman, OK, USA
[3]School of Electrical and Computer Engineering, University of Oklahoma, Norman, OK, USA
[4]Cooperative Institute for Severe and High-Impact Weather Research and Operations, Norman, OK, USA

was the Video Integrator and Processor (VPI) that reduced statistical errors in power estimates. Its analog circuitry averaged in range and time the returns from both receivers [6]. Seven categories of precipitation intensity were then encoded for display on the PPI. One category indicated insignificant signal (threshold above which the signals are displayed); hence, the actual number of displayed categories (levels) was six and corresponded to reflectivity values (dBZ): 0–(30); 30–(41); 41–(46); 46–(50); 50–(57); ≥ 57. Clearly, the reflectivity, Z, estimates had a very coarse resolution. A much-improved resolution of 1 dB was achieved in the 1980s with the Digital Video Integrator Processor [7] installed on some WSR-74 radars [7,8]. These radars gradually replaced the WSR-57s and had two models, one operating at the 10 cm wavelength and the other at 5 cm wavelength filled the gaps.

The fully digital integration reduced the quantization of Z but averaging the logarithms of powers was the remaining deficiency. It biased the Z estimates and unbiasing increased the standard deviations [9]. Direct averaging of power produces the estimates with lowest variance. Even the WSR-88D (NEXRAD), the successor radar to the WSR-74, averaged the logarithms until its upgrade (circa 2005) with a digital receiver and programmable signal processor. Prior to this improvement, the WSR-88D had a sliding dynamic range of the linear receiver controlled with the output from the logarithmic receiver. This was essential for high-quality Doppler measurements. The mean Doppler and spectrum width were computed from the output of the linear receiver. The radars had automatic calibration of reflectivity and estimation of noise level at the end of each volume scans; both features remained in subsequent upgrades. The NWS specified volume scans, including spacing in azimuth and elevation, rotation rate, and associated error bounds for reflectivity and Doppler estimates. Addition of dual polarization was completed in 2013 and is thus far the most significant upgrade to the network. The error specs for reflectivity and Doppler measurements carried over from the single-polarized WSR-88Ds while the required error limits for the polarimetric variables were chosen to be compatible with the ones for the spectral moments.

Starting in 1964, sun observations were part of absolute calibration of the azimuth pointing direction. By 1982, this was applied to the WSR-74s. The procedure became much more precise on the WSR-88Ds, thanks to digital control of the positioner and high-resolution measurement of Z. It was thus possible to calibrate the receiver using the solar flux. Further refinement was made after upgrade to dual polarization and the sun flux became one standard for calibrating the difference in gains between the receiver for horizontally (H) polarized waves and the one for vertically (V) polarized waves. Other calibration improvements and additions followed. Radial by radial noise estimation was adopted. Previous signal-to-noise thresholds for Doppler and reflectivity were augmented by including correlation in time and correlation between the H and V signals. In addition to automatic calibration of reflectivity, an automatic control of differential gain between the H and V signals was built into the system. For offline absolute calibration of differential reflectivity returns from light rain, snow aggregates and Bragg scatterers (turbulent eddies) are used. Moreover, polarimetric variables help separate meteorological from nonmeteorological returns.

The WSR-88D wavelength of about 10 cm enables observing precipitation up to 460 km at tolerable attenuation of signals and provides better resolution of the unambiguous range and Doppler velocity compared to shorter wavelengths of many weather surveillance radars. The transmitted power of 700 kW and the 8.4 m antenna diameter enable collecting data with sufficient spatial resolution up to a specified distance of 300 km within which QPE are produced and characteristic features of severe weather are detected (e.g., tornado vortices and microburst locations). Moreover, data with such spatial resolution are becoming suitable for assimilation into high-resolution numerical models. Beyond about 300 km, the radar resolution volume is large and radar data are used in semi-quantitative manner for nowcasting severe weather.

The spatial distribution of storms and the characteristics of the return signals impose challenging demands on weather radars. The dynamic range of weather signals is very large (can be about 100 dB). The signals fluctuate in time because they are formed by a superposition of the backscattered returns from the relative motion of scatterers in the radar resolution volume. Therefore, averaging of returns is needed to compute the Doppler spectral moments and the polarimetric variables [10,11]. The averaging time (dwell time) is specified so that the error requirements at some specific Doppler spectrum width and signal-to-noise ratio (SNR) are met. Because severe weather develops fast, real-time monitoring with an update time of a few minutes is needed. The WSR-88D's update time for severe weather observations is about 5 minutes. The tradeoff between the update time and measurement time to produce one radial of data is one of the main challenges in radar meteorology. How this is achieved and the functional description of the WSR-88D is in [12].

Accuracy of radar measurements depends on the hydrometeors' properties and quality of radar hardware. Accurate calibration is also required to achieve good data quality. This chapter describes required accuracy of radar measurements (Section 4.2), quality of WSR-88D's radar hardware and its calibration (Section 4.3), and current methods to assess calibration accuracy (Section 4.4). Some possibilities to enhance radar measurements related to calibration are discussed in Section 4.5.

4.2 Required accuracy for WSR-88D measurements

The WSR-88D network has been upgraded to dual polarization and the implemented mode is termed as Simultaneous transmission and reception of Horizontally and Vertically (SHV) polarized electromagnetic waves. A high-power pulse is transmitted through a 20–30 m waveguide (which depends on the height of radar tower) to the antenna. The radar transmitter and signal processor are located in a shelter on the ground. All other components (Figure 4.1) are mounted on the back of the antenna. The power divider splits the transmission path into two channels: one for the horizontal and the other for the vertical polarization. The Antenna Mounted Electronics (AME) includes a built-in radio frequency (RF) continuous wave generator, circuits to supply generator signals for hardware calibration,

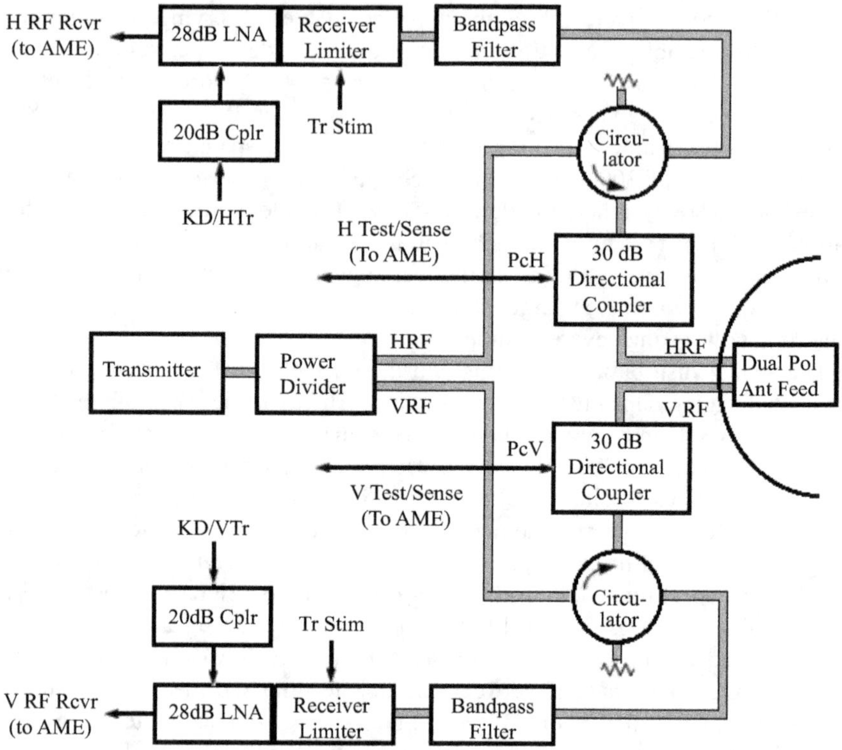

Figure 4.1 Simplified diagram of the RF part of the WSR-88D. The shaded thick lines indicate waveguides and the single lines are coaxial RF cables.

circuits for measuring the transmitter power, mixers for downconverting RF signals to intermediate frequency, and auxiliary electronics. The AME is housed in a climate-controlled box. The signals at the intermediate frequency are supplied to the digital receiver located in a cabinet inside the shelter. All these hardware components with corresponding software comprise the Radar Data Acquisition (RDA) subsystem. The received signals from each transmitted pulse are designated as level I data. The RDA produces six variables at each range location along radials: reflectivity (Z), Doppler velocity (v), Doppler spectrum width (σ_v), differential reflectivity (Z_{DR}), correlation coefficient between signals in the horizontal and vertical channels (ρ_{hv}), and differential phase between those signals (Φ_{DP}). Collectively these are referred to as level II data.

The backscattered signals are transmitted to the Radar Product Generator (RPG) which may be located several kilometers away from the RDA. The RPG transfers the radar variables as level II data to the NWS's storage facility and generates the meteorological products (level III data). The data are available to users free of charge. The real-time level II data are made available to vendors who add value (geographic displays, looping, extrapolating, etc.) and distribute these to

Table 4.1 Specified uncertainties of the radar variables obtained with WSR-88D

Variable	**Minimal SNR, dB**	**Minimal, ρ_{hv}**	**Uncertainty**
Reflectivity, Z	10		1 dB
Doppler velocity, v	8		1 m s^{-1}
Spectrum width, σ_v	10		0.5 m s^{-1}
Differential reflectivity, Z_{DR}	20	0.99	0.1 dB at Z_{DR} <1 dB 0.1 Z_{DR} at Z_{DR} > 1 dB
Differential phase, Φ_{DP}	20	0.99	0.033 deg
Correlation coefficient, ρ_{hv}	20	0.99	0.006

the public. Of all US government generated data, the ones from the WSR-88D are the second most used by the public. In the first place are the geographic positioning system (GPS) data.

Uncertainty of radar measurements depends on properties of weather signals and specifics of radar hardware. The latter includes the antenna gain at 3-dB beamwidth, receiver noise figure, stability of electronic circuits, transmitter spectral purity, crosstalk between channels for horizontally and vertically polarized waves, and other system attributes. Radar signals from range resolution volumes fluctuate in time. To achieve needed measurement uncertainties, these returns must be averaged. The needed dwell time (proportional to the number of averaged samples M) depends on the spectrum width and SNR. These factors control measurement uncertainties of Z, v, σ_v, Z_{DR}, Φ_{DP}, and ρ_{hv}. In addition, the errors of the three polarimetric variables depend on ρ_{hv}. To summarize, the four parameters determining measurement uncertainties of dual-polarization variables are M, σ_v, SNR, and ρ_{hv}. The statistical errors of the radar variables are analyzed in [10,11,13].

To verify some of the specified accuracies, theoretical expressions and/or simulations are used. The pulse repetition time T_s and the number of samples M for a specific Volume Coverage Pattern (VCP) have been chosen and entered in the formulas for the standard deviation containing specified signal parameters (i.e., the SNR, σ_v, and ρ_{hv}) and/or the same parameters are used in simulation experiments on the Level 1 data. The requirement is met if the resulting errors are smaller than the specified ones (Table 4.1). Finding biases in Z, Z_{DR} and Φ_{DP} is done via special calibration (Sections 4.3.3, 4.3.5, and 4.4).

4.3 Quality of the WSR-88D dual-polarization measurements

Whereas ultimate interest of users is in QPE and winds, we consider the polarimetric variables and the Doppler spectral moments as these directly affect any of the derived products. Henceforth, quality refers to the bias and variance of these measurements. It depends on radar hardware and accuracy of calibration. Some other metrics of data quality are also in use on weather radars. For example, the Signal Quality Index (SQI) is the estimated modulus of the autocorrelation coefficient of the weather signal at lag 1. It is uniquely related to the Doppler spectrum width and SNR. Its maximum of 1 indicates a sinusoidal signal and the minimum of 0 corresponds to white noise. Indiscriminate use of this index is not advised because it would eliminate signals with large spectrum widths like those from tornado circulations. A generalized quality index is in use on the UK weather radars. It consists of measurement-specific information that affects data [14], for example, errors in derived rain rate, or beam height above ground and other. These are scaled linearly between the highest value of 1 and the lowest of 0.

A simplified diagram of the RF hardware components critical for calibration is shown in Figure 4.1. High power pulse from the transmitter is supplied to the power divider to form two channels (waveguides for the H RF and V RF) connected to the antenna feedhorn through the circulators. The 30 dB directional couplers sample the powers in the waveguides so that these can be used for calibration. The feedhorn transmits two orthogonally polarized waves that combine into an elliptically polarized wave. The returned wave is received by the antenna and feedhorn and separated by the orthomode coupler (not shown) into H RF and V RF and these signals follow the path through the circulators, bandpass filters, receiver limiters, Low Noise Amplifiers (LNA), and AME to the processor. The system calibration front end is at the 30 dB directional couplers (Figure 4.1). Calibrated signals from the built-in RF generator are supplied to the couplers.

4.3.1 Isolation of the dual-polarization channels

Channel isolation depends on various hardware components. In the upgrade to dual polarization, the antenna dish did not change, and the number of antenna struts remained three; the lower two support waveguides for the polarization channels. In the back of the antenna feedhorn is an orthomode coupler designed to combine for transmission and separate on reception the two orthogonally polarized waves. The feedhorn must be accurately oriented in the antenna focal point to transmit and receive horizontally and vertically polarized waves in the polarization plane [15].

The Radar Operations Center (ROC) and the National Severe Storms Laboratory (NSSL) have specified various performance parameters for the WSR-88Ds upgrade to dual polarization. The important ones are the detectability for the SHV mode as well as the accuracy of the polarimetric variables. The team took specifications for the WSR-88D [10,16], i.e., error bounds for specific dwell times, pulse repetition times, SNR, M, σ_v, and ρ_{hv}. Then they computed the corresponding errors in the polarimetric variables and relaxed these slightly before including into the requirements. Doing so guaranteed that the upgrade to dual polarization would not affect the errors in the spectral moments. Moreover, it was argued that spatial averaging of the polarimetric variables would significantly reduce the statistical errors, consequently improving QPE. For other applications, similar averaging of polarimetric variables and spectral moment is not advised as it would smear important severe weather indicators like the mesocyclone and tornado vortex signatures. Nonetheless, researchers suggested that range oversampling combined with whitening techniques significantly improves the estimates of the polarimetric variables at high SNRs [17]. More recent study proposed adaptive pseudo-whitening to maintain performance of whitening at high SNR and equal or exceed the performance of the matched digital filter at low SNRs [18]. The authors [17] applied this approach to weather data and validated the expected improvements that can be realized operationally.

The specified accuracies of WSR-88D variables are listed in Table 4.1 [16]. In this table, minimal SNR, σ_v, and minimal ρ_{hv} are those for which the accuracies are determined. It is very challenging to achieve Z_{DR} accuracy of 0.1 dB; hence, the interim goal is set at 0.2 dB. Early studies of radar rain rate estimates suggested that Z_{DR} accuracy should be at least 0.2 dB [19]. To avoid errors in Z_{DR} estimates, the rain measurements with WSR-88D are transitioning to the use of specific differential phase (K_{DP}) and specific attenuation A [20]. The desired accuracy of Z_{DR} measurements in snow is 0.1 dB as the span of Z_{DR} values is frequently not more than 1 dB [21].

Seavey Engineering has measured the polarization isolation of the WSR-88D's antenna on an antenna range and reported isolation values of 34 dB. Similar isolation level was measured in 2007 on the WSR-88D KOUN located in Norman, OK, USA. In 2007–2009, KOUN was equipped with a waveguide assembly which could bypass the power divider so that the total transmitted power passes through the channel for horizontal polarization. On reception, the signal carrying copolar (horizontal polarization) information and the one carrying the cross-polar information were recorded and processed. One of the processing results is the linear depolarization ratio $L_{DR.}$ defined as the ratio of cross-polar to copolar powers.

The channel isolation measurements consist of two steps: (1) measurement of isolation in hardware below the antenna and (2) total isolation measurement. Then, the isolation level of the antenna is the difference between these two measurements. The first step is accomplished by bypassing the power divider, injecting the calibrated signal P_{CH} (Figure 4.1) through the 30 dB directional coupler, and measuring powers in both channels. The ratio of the coupled power in the vertical channel to the power in the horizontal channel defines the isolation. The results from KOUN (Figure 4.2) indicate that the channel isolation below the antenna is about 60 dB.

The second step is L_{DR} measurements from light rain. Figure 4.3a presents the reflectivity field collected with KOUN on June 26, 2007. One can see an area of reflectivities lower than 15 dBZ below the melting layer between 64 and 90 km in range. The reflectivities there were caused by light rain. Panel (b) presents the L_{DR} field obtained simultaneously with the reflectivity field. Panel (c) depicts a distribution of L_{DR} values from the area of light rain. The median L_{DR} is −34 dB and is consistent with the measurement by Seavey Engineering although the feedhorn on the KOUN was different from the feedhorn on the antenna tested by the Seavey Engineering. Because our measurement (−34 dB) was produced by rain, we conclude that the total isolation is at least 34 dB. Considering that the channel isolation

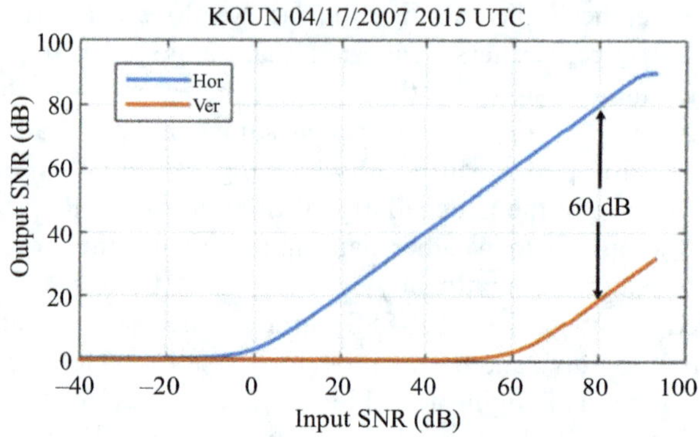

Figure 4.2 *The power of the signal injected into the horizontal channel (the blue curve) and the power coupled into the vertical channel (the red curve) measured on KOUN, March 17, 2007 at 2015 UTC.*

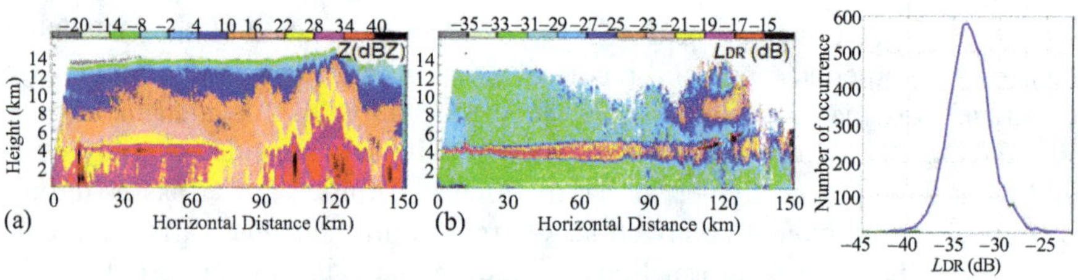

Figure 4.3 *Vertical cross-sections of (a) reflectivity and (b) L_{DR} collected with KOUN June 26, 2007 at 1205 UTC at azimuth of 209°. (c) Distribution of L_{DR} from areas of light rain.*

below the antenna is 60 dB, it follows that the antenna has isolation better than 34 dB. This isolation is about 4 dB less than obtained from similar data on the CSU-CHILL radar with a dual-offset Gregorian antenna [22, Figure 12]. Therefore, we submit that the WSR-88D antenna has sufficient isolation to take accurate polarimetric measurements.

4.3.2 WSR-88D's dynamic range

Reflectivity of atmospheric scatterers, observed with S-band radars, spans an interval from smaller than −20 dBZ to over 80 dBZ. The lower end of this interval is from non-precipitating clouds and Bragg scatter in clear air. The high reflectivities are from hail and ground clutter. Obviously to cover this wide range without saturating the receiver, it is important to have a dynamic range of about 100 dB. Because this interval is very large (ten orders of magnitude!) and the uncertainty of reflectivity measurements is 1 dB, the receiver's transfer functions over this interval should be as linear as possible. The WSR-88D dynamic range is specified at 93 dB [16, Section 5.4.7].

The receiver's dynamic range can be measured by injecting signals from the built-in RF generator into the 30 dB coupler (Figure 4.1) and measuring the output signals with the signal processor in the RDA. The results (Figure 4.4) show that the linear dynamic range of KOUN is from SNR = 0 dB to SNR = 91 dB, i.e., 91 dB. The deviations of the curves from a straight line are smaller than 1 dB and thus satisfy the specified accuracy for the reflectivity measurements (Table 4.1). The red and magenta curves (Figure 4.4) are the measured dependences in the two polarization channels. The blue and green curves have been obtained by subtracting the mean noise powers from the measured powers. It is seen that by subtracting noise power, the linear dependence can be extended down to the SNR of about −10 dB. This is useful for measurements in regions of very weak scatterers. Clearly, the variables dependent on the power (Z, σ_v, Z_{DR}, and ρ_{hv}) should be calculated after subtracting noise. The nonlinear dependence at the very strong signals (Figure 4.4) can be measured and accounted for if the weather signal powers are slightly below the saturating power for the continuous wave signal. This "linearization" extends the dynamic range by about 3 dB; with this extension and noise subtraction, the dynamic range on the KOUN is 104 dB.

So far, we have analyzed the effects of the dynamic range on reflectivity measurements involving only the channel for horizontal polarization (by definition). Z_{DR}, ρ_{hv}, and Φ_{dp} are obtained using signals from both channels and are therefore sensitive to small differences between the transfer curves of the two channels. The variables on the WSR-88Ds are typically computed for $SNR_h \geq 2$ dB (this threshold can be altered by operators). This is because the detectability is high so that precipitation from most storms produces larger SNRs. Moreover,

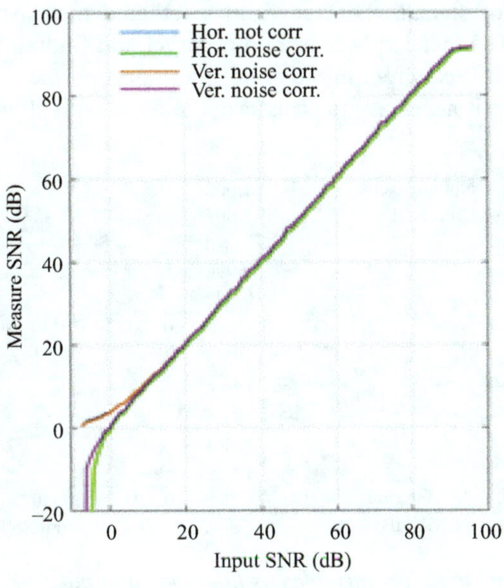

Figure 4.4 Dynamic range of KOUN measured on May 14, 2018, at 2015 UTC.

obtaining reliable estimates at lower SNR requires a long dwell time, hence is impractical for operations.

Ideally, the polarization variables would be independent of SNRs larger than 2 dB. This, however, is not the case as can be seen in the dependencies of Z_{DR}, ρ_{hv}, and Φ_{DP} on the input SNR (Figure 4.5). The SNR threshold of 2 dB is shown with the vertical dash magenta line. The blue curve (Figure 4.5a) represents the noise biased Z_{DR} and the unbiased one is the red curve. The ρ_{hv} values (Figure 4.5b) have been corrected for noise and Φ_{DP} is not biased by noise.

Noise-corrected Z_{DR} exhibits a flat dependence in the interval from about 20 to 90 dB. Clearly, the receiver nonlinearities should be compensated for at SNR < 20 dB and at SNR > 90 dB, where the deviations of Z_{DR} exceed specified measurement accuracy of 0.1 dB. The receiver nonlinearities do not affect the Φ_{DP} and ρ_{hv} values at low SNR (Figure 4.5b, c); the deviations of these values from the mean do not exceed specified accuracies (Table 4.1). Receiver nonlinearities at SNR > 90 dB begin to affect these variables. Nonetheless, such large SNRs would be caused by giant hail or flood-producing rain/hail mixtures that would not require polarimetric variables for diagnosis. In other words, the corrected variables at SNR > 90 dB may be redundant.

4.3.3 Reflectivity calibration

There are several methods to calibrate weather radars and a historical review is in [23]. We describe two popular ones next. In one, a calibrated signal is injected through the waveguide coupler ahead of the LNA, and the power is recorded at the receiver output. In a self-calibration procedure of the WSR-88D, a signal from the built-in generator is injected at the end of each VCP. The second method is based on measurements of signals reflected from a tethered metal sphere of a known diameter [23,24]. The metal sphere is a point target and finding its location within the beam is difficult. Scatterers in the atmosphere occupy the whole radar resolution volume making it necessary to transform the data (with all uncertainties) from

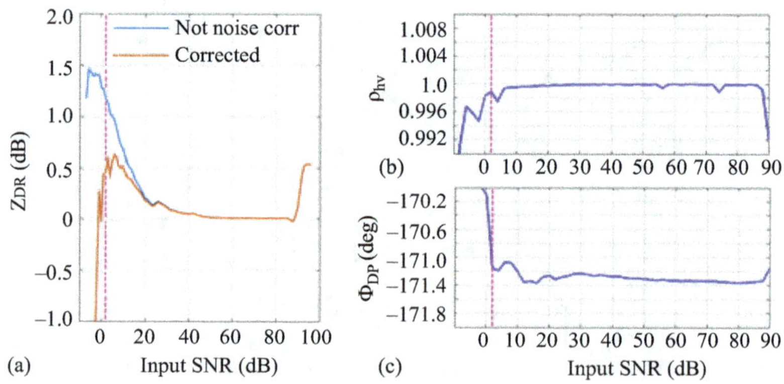

Figure 4.5 Dual-polarization variables as functions of input SNR measured on June 12, 2013 on KOUN. The red vertical dash line indicates SNR = 2 dB.

the point measurement to the volume measurement. The tethered sphere method is not practical for calibrating network radars as these require automation and real-time monitoring. The WSR-88Ds have a dedicated hardware and a "baseline" procedure for calibrating the reflectivity factor via the weather radar equation (e.g., section 4.4.5 in [10], section 6.3 in [11]):

$$Z_e = 10\log\left(\frac{P_{rh} - N_h}{N_h}\right) + 20\log(r) + ar + dBZo \text{ (dBZ)}, \tag{4.1}$$

where Z_e is equivalent reflectivity factor (reflectivity Z in short), P_{rh} is the received power in the horizontal channel, N_h is the noise power in the channel, r is the distance to the radar resolution volume, a is the atmospheric attenuation factor, and $dBZo$ combines relevant radar parameters. The value of $dBZo$ is Z_e measured at a distance of 1 km at the SNR_h = 0 dB, i.e., at $P_{rh} = 2N_h$, and is calculated as

$$dBZo = 10\log\left(\frac{2^{10}10^{21}\ln 2\, \lambda^2 N_h}{\pi^3 c\tau G^2\theta^2 |K|^2 g P_{Th} L_t L_r L_d}\right) \text{ (dB)}. \tag{4.2}$$

We define the variables in (4.2) and indicate their values for KOUN in parentheses as follows. The radar wavelength is λ (11.1 cm), c is the speed of light, τ is the radar pulse width (1.54 μs for the short radar pulse and 4.5 μs for the long pulse), G^2 is the two-way antenna gain (91.50 dB), θ is the beamwidth (0.95° for azimuthal and elevation directions), $|K|^2$ depends on the dielectric permittivity of water (0.93 at S-band), g is the receiver gain (35.75 dB), P_{Th} (350 kW) is the transmitted power in the horizontal channel. $L_t, L_r,$ and L_d are waveguides losses from the transmitter to the power divider (2.25 dB), receiver waveguides (0.62 dB), and receiver detection loss (1.5 dB). All the parameters in (4.2) except G and P_{Th} are measured by supplying calibrated signals to the 30 dB directional coupler (Figure 4.1). The antenna gain including radome is measured using the solar flux by comparing the radar-measured power from the Sun with the calibrated power recorded at the Dominion Radio Astrophysical Observatory in Canada [25]. P_{Th} is obtained by measuring the transmitted power via the 30 dB coupler. The atmospheric attenuation factor a has been calculated as the mean per 1 km for the standard atmosphere and for each antenna elevation angle. Actual temperature and humidity height profiles are currently not used for estimating attenuation.

Accuracy of the reflectivity measurements depends on the accuracy of $dBZo$. The estimate made by the ROC engineers is 0.56 dB. This value is the standard deviation of $dBZo$ caused by the variations in its parameters and it satisfies the specification for reflectivity (Table 4.1). $dBZo$ is adjusted in each radial in accord with the measured noise power. It is also adjusted at the end of each VCP by updating P_{Th}. For KOUN, the mean $dBZo = -43.3$ dB. The updated $dBZo$ is used on the data from the next VCP.

4.3.4 Radar detectability

Minimum detectable reflectivity (detectability for short) is an important parameter quantifying the radar capability to observe distant precipitation and weak echoes from its edges, non-precipitating clouds, and clear air. It is defined by the value of reflectivity (in dBZ) from a certain range that produces reflected power of a single pulse at the input of the receiver that equals the noise power. Detectability of −7.5 dBZ at a distance of 50 km is specified for single polarization WSR-88D ([6] Table 4.2). The achieved value on the network was about 5 dB better (i.e., −12.5 dBZ). Therefore, for the polarimetric WSR-88D, the government specified detectability of −9.5 dBZ at 50 km [16]. This is because, in the NWS's chosen SHV mode, the power divider reduces the SNR per polarization channel by 3 dB. Detectability depends on many radar parameters entering the main radar equation (4.2); therefore, it exhibits some variations among the WSR-88Ds.

The transmitted power is one of the parameters affecting detectability. Dual-polarization WSR-88D retained its 700 kW transmitter. Therefore, splitting the power between polarization channels in the polarimetric WSR-88D cuts the power in the horizontal channel by 3 dB increasing *dBZo* according to (4.2). Relocating the receiver hardware to the back of the antenna and using the modern receiver components reduced the losses, thus lowering the receiver noise power. This reduction in the receiver noise power turned out to be significantly larger than the 3 dB loss in the transmitted power. The net result is an increase in the detectability. The *dBZo* is calculated at a distance of 1 km and for KOUN it is −43.3 dBZ. At 50 km its detectability is 20log(50) = 34 dB larger, i.e., −9.3 dBZ very close to the specified −9.5 dBZ [16].

A distribution of radar detectability over the network (Figure 4.6) shows that two-third of sites exhibit detectability equal or better than the specified −9.5 dBZ at 50 km (short radar pulse). Detectability with the long transmitted pulse is about 9.5 dB better than the one with the short pulse.

The *dBZo* is a figure of merit handy for comparison between radars and for predicting detection of weak signals by various integration techniques. Although these techniques enhance meteorological signal compared to noise, there are often large regions where only noise is present. These may confuse forecasters and are detrimental if assimilated into models. To eliminate these pesky regions, the WSR-88D operators use SNR thresholds. A typical one is at the SNR of 2 dB. If this threshold is reduced, noise speckles appear in the data fields, but can be removed with a spatial filter. A simple despeckling 2D filter [26] removes noise speckles enabling weather observations. For example, in the case of non-precipitating clouds

Table 4.2 Coupling coefficients of the 30 dB couplers installed on the KOUN

Coupling coefficient	Loss (dB)
H coupler in	−30.20
H coupler out	−30.14
V coupler in	−30.36
V coupler out	−30.31

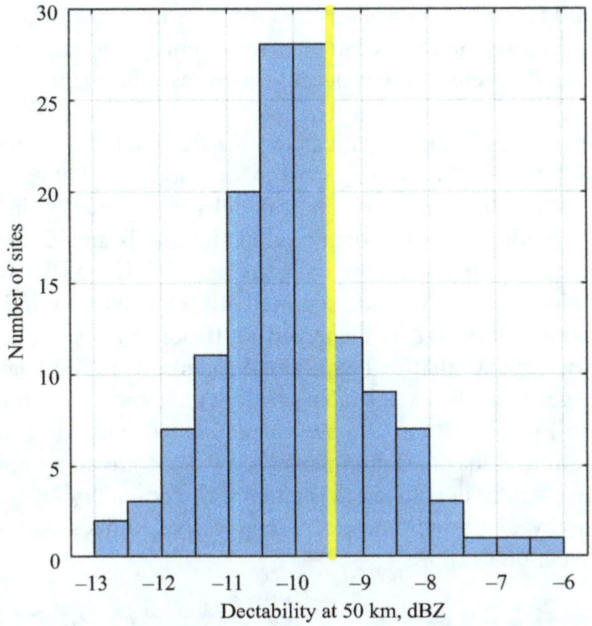

Figure 4.6 Distribution of the detectability across the WSR-88D network for the short pulse mode in February 2021. The vertical yellow line is the specified detectability for the mode.

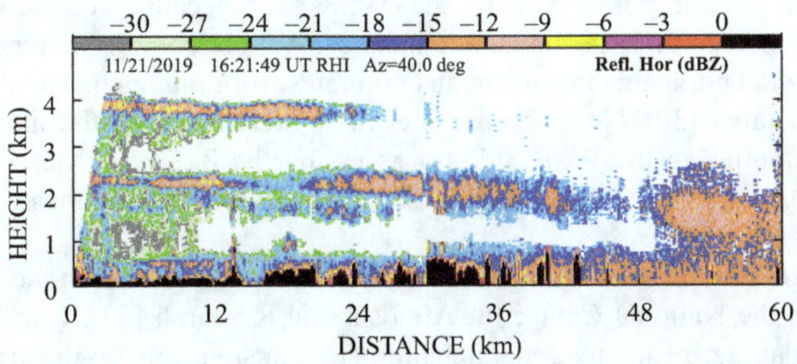

Figure 4.7 Vertical cross-sections of reflectivity collected with KOUN on November 21, 2019 at 1621 UTC at the azimuth of 40°. The lowest layer of reflectivity stronger than 0 dBZ is contaminated by the ground clutter residue after filtering.

(Figure 4.7), reflectivity of −18 dBZ is observed at 50 km, which is 8.7 dB lower than −9.3 dBZ obtained from the radar parameters at SNR = 0 dB. Note that measurements at these low SNRs require longer dwell times. The number of samples used to generate Figure 4.7 is 256 and because the pulse repetition frequency is 1000 Hz, the dwell time is 256 ms. The time to complete this vertical scan is 15 s. Similar measurements can be made in non-precipitating clouds or clear air in the absence of severe weather because then fast volume coverage would not be needed.

Information about clouds is critical to climate studies and the observations with the WSR-88Ds could fill existing gaps in knowledge and coverage. The radars are located in various climate regions and can deliver information in real time. The Doppler velocity could be used to obtain winds in clouds and Z_{DR} to estimate the shape of cloud particles. It is useful to compare detectability of the WSR-88D with detectability of cloud radars typically reported at 10 km. We see in Figure 4.7 that reflectivity of about −30 dBZ is displayed at 10 km and beyond. This level of detectability is better than that of many cloud radars [27]. The beamwidth of the WSR-88D is larger than the cloud radars beamwidth, but many fine field details remain visible in Figure 4.7 and can be complemented with the Doppler velocity and Z_{DR} information. Radar measurements at low SNR are also important in clear air to sense Bragg scatter developed at the top of the boundary layer. Observations in clear air can reveal the parameters of convective plumes that determine the intensity of convection [28].

4.3.5 Z_{DR} calibration

Z_{DR} can be calibrated by engineering methods or by measurements from scatterers isotropic in the polarization plane (i.e., producing statistically the same scattering of the copolar component regardless of the impinging E field orientation). The simplest of the latter ones is by sensing rain at vertical incidence [11,29]. That measurement is not possible with the WSR-88D because it has mechanical obstructions to keep its large antenna in balance. A more complex and possibly less accurate method compares instantaneous rain rates from rain gauges with the radar-derived rain rates [30,31]. Measurements in light rain are in the arsenal of the procedures applied to the WSR-88D for assessing the Z_{DR} bias. Measurements in snow aggregates or regions of Bragg scatter are also part of these procedures (Section 4.4.1).

A method based on scattering reciprocity [32] has been used on the S-pol radar belonging to the National Center for Atmospheric Research [33]. It requires more complex radar hardware to generate alternate transmission of the orthogonally polarized electric fields. That is not possible on the WSR-88D but could be implemented on polarimetric phased array radars. These are under development and consideration for replacing the current WSR-88Ds at the end of their designed lifetime (in about 2040).

WSR-88D is calibrated using built-in hardware and dedicated software; the procedure is referred to as baseline calibration. The initial calibration procedure for the WSR-88D was suggested in [34] and became the foundation for the upgraded network to dual polarization. The procedure has been enhanced by the ROC [35,36] and some proposed simplifications are in [37]. Next, we describe the current baseline calibration on the WSR-88D.

On the WSR-88D, the imbalances in the transmitted powers, the receiver gains, and the antenna are measured and accounted for. These differences are called Z_{DR} biases in transmission (TXB is the notation used in many publications), reception (RCB), and antenna (SMB Sun Measurement Bias). TXB is the difference between the transmitted powers in the waveguide paths from the power divider to the 30 dB couplers (Figure 4.1). The power difference in the paths from the 30 dB couplers through the antenna and radome is SMB. That is, the Z_{DR} bias upon transmission equals to TXB + SMB (in dB). The difference in received powers in the signal paths from the antenna to the 30 dB couplers is SMB and the difference in gains in the signal path from the 30 dB couplers to the signal processor is RCB. Thus, the Z_{DR} bias upon reception equals SMB + RCB (in dB) and the system Z_{DR} is the sum

$$Z_{DRsys} = \mathrm{TXB} + \mathrm{RCB} + 2\mathrm{SMB}(\mathrm{dB}). \tag{4.3}$$

To measure the contributors in (4.3), first an offline calibration is accomplished. Attenuation in the signal paths from the transmitter to the 30 dB couplers is measured as follows. The test/sense cables from the couplers' ports are disconnected and a power meter is sequentially connected to these ports through a 20 dB attenuator (Figure 4.1). These measurements deliver the calibrated difference of transmitted powers. The out-coupling coefficients of the 30 dB couplers must be known. These coefficients are measured in the factory. After the measurements of the transmitted powers with a power meter, the test/sense cables are reconnected back to the couplers' ports and measurements of the transmitted powers are taken at the output of the AME by the signal processor. This produces the loss difference in the measurement lines. Once this manual calibration is completed, TXB is ready to be measured during radar operations at the end of each VCP. Accuracy of the TXB measurements is determined by the accuracy of the power meter measurements, i.e., 0.02 dB that can be considered as the standard deviation in the measurements. Because TXB is the difference of two measurements in the channels each of which consisted of the measurements of power and coupling coefficient, the total standard deviation is $\sqrt{4}\ 0.02 = 0.04$ dB.

The RCB is measured by injecting calibrated signals P_{CH} and P_{CV} into the 30 dB couplers and computing the received powers by the signal processor (Figure 4.1). The signal source is the built-in continuous-wave RF generator. The test/sense cables are disconnected from the 30 dB couplers and the powers P_{CH} and P_{CV} are measured with a power meter at the outputs of the cables. Then, the cables are reconnected back to the ports, the built-in signal generator is turned on, and the powers are measured with the signal processor. These powers depend on losses in the couplers and losses/amplifications in the path from the couplers to the processor. These coupling coefficients are also measured in the factory. Table 4.2 lists the coefficients of the KOUN's couplers. Because the RCB is the difference of measurements, its standard deviation is $\sqrt{4}\ 0.02 = 0.04$ dB same as for the TXB.

TXB and RCB do not include Z_{DR} bias from the parts of waveguides above the couplers, i.e., in the circulators, feedhorn, antenna, and radome. That bias constitutes the SMB and is measured using the solar flux. Because the solar flux has a broad band, the measurement consists of two steps. First, the test/sense cables are disconnected from the ports of the 30 dB couplers (Figure 4.1), the built-in noise source is turned on, and its power is sequentially measured at the test/sense cables in the channels. This calibrates the powers injected into the 30 dB couplers, i.e., P_{CH} and P_{CV} for broad band signals. Then, the test/sense cables are reconnected back to the couplers' ports, the noise source is turned on, and the received powers are measured with the signal processor. That way the difference between channel amplifications in the paths from the 30 dB couplers to the signal processor is measured. Let this difference be ΔA, i.e., the difference in paths amplification. In the second step, the radar antenna is directed at the center of the solar disk and the powers in the channels are measured with the signal processor. Assuming that the solar radiation is unpolarized, i.e., the input flux powers at orthogonal polarizations are equal, the power difference measured from the Sun is ΔS: $\Delta S = \Delta A + \mathrm{SMB}$. Because ΔA has been measured, SMB is determined. Uncertainty in these measurements can be related to uncertainty of ΔA, i.e., the standard deviation of 0.04 dB which is same as for the TXB measurements.

Summing up variances in the TXB, RCB, and SMB measurements, the standard deviation in Z_{DRsys} is estimated to be $\sqrt{4}\,0.04 = 0.08$ dB. This value is close to 0.1 dB specified for ultimate Z_{DR} accuracy (Table 4.1) and less than 0.2 dB that is set as goal for the present WSR-88D. We highlight that TXB, RCB, and SMB require multiple measurements by trained technicians. To achieve Z_{DR} accuracy of 0.1 dB, the uncertainties in TXB, RCB, and SMB need to be reduced possibly by utilizing more accurate modern instruments. Alternatively, a simpler method proposed in [37] may be sufficient.

Another uncertainty in SMB measurements comes from the polarization properties of solar flux at the 10 cm wavelength. It is not known if the powers of the two orthogonal components are within 0.1 dB of each other. Solar physics teaches that radiation by solar spots can contain linearly polarized flux [38]. Therefore, the assumption about $Z_{DR} = 0$ dB in the solar radiation at times of Sun's active stages may not hold.

Z_{DRsys} is automatically adjusted at the end of each VCP, when the antenna is at an elevation of 19.5°. At this time, test calibrated signals are injected into the 30 dB couplers and current RCB and TXB are measured. From these measurements and known SMB, Z_{DRsys} is computed and used in the observations during the next VCP as follows:

$$Z_{DR} = Z_{DRm} - Z_{DRsys} \tag{4.4}$$

4.3.6 Differential phases upon transmission and reception

In the microwave transmission circuits, the signal paths for the H and V polarizations differ causing a phase shift β between the two signals. Similar differential phase γ occurs upon reception. The sum of these phases is the system differential phase $\Phi_{DPsys} = \beta + \gamma$. The Φ_{DPsys} is estimated by comparing the phases of returns from the edges of light precipitation closest to the radar. The receiver phase can be changed in the signal processor by multiplying V channel voltage by exp($j\psi$) with ψ being the introduced phase shift. Then, the phase upon reception becomes $\gamma + \psi$, so that $\Phi_{DPsys} = \beta + \gamma + \psi$ can be set at any value in the interval 0–359°. Return signals from the edges of precipitation are weak, hence exhibit large fluctuations. If Φ_{DPsys} is close to 0 degree, the fluctuating signals may alias the phases. Because the Φ_{DPsys} estimate is an average of instantaneous phases, it may be biased and the corresponding K_{DP} in the precipitation edges would be wrong. To avoid this possibility, Φ_{DPsys} on all WSR-88Ds is set at 60°. Radar observations show that this phase offset is sufficient to prevent aliasing of the differential phases.

On the WSR-88Ds, there is no control of the transmitted differential phase β. The length of waveguides from the power splitter to the feedhorn (Figure 4.1) is several meters and a difference in the waveguide lengths of 1 cm leads to a phase difference of 20.8° on the KOUN radar. The phase β can be measured using a signal from external RF source. Measurements of KOUN's β in 2021 were conducted by placing a standard horn antenna on a tower 321 m from the radar. The horn antenna transmits a linearly polarized plane wave. By adjusting the horn in the polarization plane so that its edges are at $\pm 45^{\circ}$ with respect to the ground, a linearly polarized wave acquires the same orientation. Thus, the horizontally and vertically polarized waves with zero differential phase are transmitted to the receiving KOUN's antenna. The differential phase upon reception of these signals is γ. The phase $\Phi_{DPsys} = \beta + \gamma$ is measured from precipitation; hence, β can be readily computed. Such measurement taken in 2021 showed that β = 112.5°, i.e., KOUN transmits elliptically polarized waves.

The phase β affects the measurements of the scattering properties of frozen hydrometeors. This has been noticed in icy parts of thunderstorms, where crystals become oriented by electric fields [39]. The phase β strongly impacts the estimates of the polarization variables from insects making the Z, Z_{DR}, Φ_{DP}, and ρ_{hv} fields asymmetric [40]. Influence of β on the polarimetric measurements in the melting layers and ice clouds has also been observed [41]. Examples of the predicted β-impact on ρ_{hv} and the backscatter differential phase δ in hail are plotted in Figure 4.8. The figure shows simulation results for wet hailstones of various diameters D, the axis ratio (b/a) of 0.6, and a 20° standard deviation of canting angles. Z_{DR} does not depend on β because the projection of the hailstone's symmetry axis on the horizontal plane is assumed to be uniformly distributed. The impact on ρ_{hv} is very strong and may affect the WSR-88D hail detection algorithm because it uses Z, Z_{DR}, and ρ_{hv} [42]. The dependencies of ρ_{hv} and δ on β hinder interpretations of radar data if β is not known. On the other hand, if β is known (measured), analysis of ρ_{hv} and δ could yield additional information about the scattering media.

4.3.7 Monitoring radar hardware

To monitor radar hardware, the WSR-88Ds are equipped with various devices and supporting software. In place are three major calibration procedures. One is after every VCP, one after every eight hours, and one every 28 days. The following radar

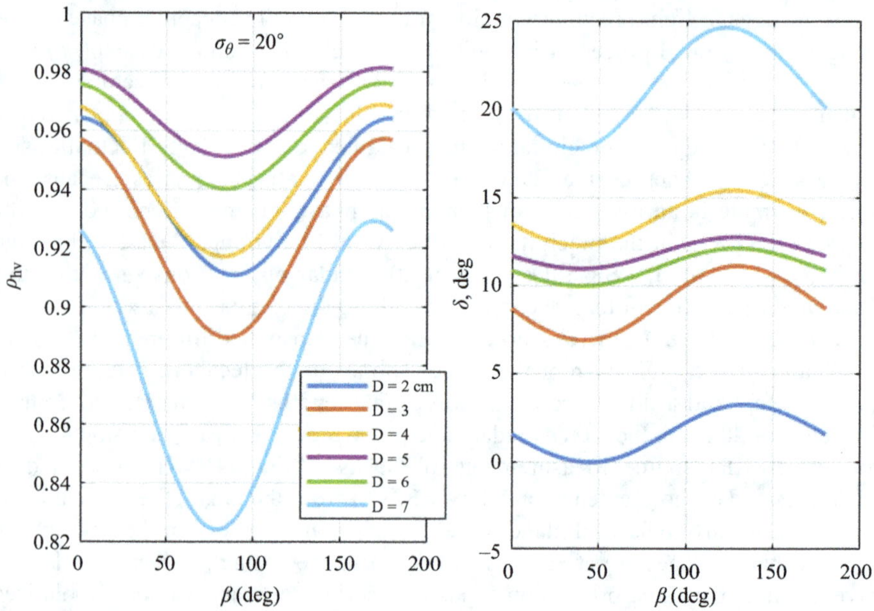

Figure 4.8 The correlation coefficient ρ_{hv} and backscatter differential phase δ from wet hailstones of various sizes D as functions of the incident differential phase β. The standard deviation of canting angles is 20°.

parameters are monitored at the end of each VCP during the antenna retrace: *DBZo*, receiver linearity, minimum detectable signal, dynamic range, noise levels, system Z_{DR} offset, differential phase, and velocity and spectrum width check. This is done with signals from the built-in CW and noise generators. These tests except noise measurements are performed by injecting CW signal via the 30-dB directional couplers (Figure 4.1). The differential phase, velocity, and spectrum width checks involve the built-in phase shifter. The noise levels are measured using signals of the built-in noise source. TXB and RCB parts of the system Z_{DR} offset are monitored during this test every 5–10 minutes depending on the longevity of the VCP. The third contribution to the Z_{DR} offset, i.e., SMB, is measured using the solar flux.

If weather permits, the eight-hour test is conducted. It includes all the procedures from the previous tests plus routines to monitor the transmitter limiters, power sensors (Figure 4.1), phase shifter, and delayed klystron chains. The clutter suppression test is also done at this time.

The solar flux measurements are taken every 28 days if there is no precipitation in a radar area. A boxcar scanning is made over the Sun position and maximal solar signal is measured and compared against the one from the Dominion Radio Astrophysical Observatory in Canada [25]. This enables monitoring waveguide attenuation and measuring the SMB contribution to the system Z_{DR} offset (4.3). The solar radiation at the 10 cm wavelength has a wide band. To correctly measure the power of the signal, the built-in noise source is used. The monitoring results are stored in the Level II data and in the system log file. The latter also contains more than 800 messages, events, and possible alarms from various radar devices and software routines.

Monitoring radar receivers with built-in generators is a common procedure on weather radars. This is used on the German and Switzerland networks [31,43]. The French network utilizes a calibration period of 72 hours [44]. The time interval of 5–10 minutes on the WSR-88Ds insures quick response to hardware malfunctions, compensation of temperature dependencies, and is handy for real-time monitoring of various radar components.

4.4 Estimating Z_{DRsys} using weather observations

Z_{DR} analysis of meteorological scatterers has a long history since it has been proposed in 1976 [19]. The values vary between about −3 dB and 10 dB at the 10 cm wavelength. Graupel [45] and giant resonating hailstones [37,46] produce negative Z_{DR}, whereas the highest observed values are from thin plate-like ice particles [41,47]. Wet oblate hailstones with a toroidal water coating can raise Z_{DR} to about 5–6 dB [46].

The Z_{DRsys} values measured on the WSR-88Ds are mainly in the −2 to 2 dB interval. Radars with Z_{DRsys} out of this range require engineering attention. Estimating the calibration accuracy of Z_{DRsys} can be made by comparisons with values from scatterers that have Z_{DR} close to 0 dB. If Z_{DR} computed in (4.4) agrees with the expected values from meteorological scatterers, then Z_{DRsys} is assumed to be correct. Light rain consisting of spherical drops, dry snow aggregates, and turbulent clear air producing Bragg scattering are used on WSR-88D to assess Z_{DRsys} accuracy.

4.4.1 *Measurements in light rain*

Theoretical results of scattering by various raindrop sizes, data from rain gauges and optical disdrometers, as well as radar observations of rain at S-band have been used to establish a relation between Z and Z_{DR} in light rain. The procedure is as follows. The reflectivity intervals in 2 dBZ steps from 19 to 30.5 dBZ are mapped to the following Z_{DR} intervals: 0.23, 0.27, 0.32, 0.38, 0.46, and 0.55 dB [48–50]. To assess correctness of Z_{DRsys}, collected data are stratified into the indicated Z intervals. The measured Z_{DRm} are then distributed into these Z intervals, and true Z_{DR} values are obtained according to (4.4). Radar data are collected under the following conditions: antenna elevations are below 1.8° and distances are beyond 10 km to avoid possible ground clutter contamination. Three filters are applied to the data: SNR $\geq$ 20 dB, $\rho_{hv} \geq$ 0.98, and types of rain. The ρ_{hv} threshold separates the meteorological data from ground clutter and atmospheric biota. Data from stratiform rain are only analyzed in this method and the convective rain is rejected using an existing separation algorithm. To avoid contaminations from the melting layer, data are analyzed from ranges at which the beam center is at least 1 km below the bottom of the melting layer. The melting layer is located with the WSR-88D's detection algorithm. The number of data collected in the indicated intervals should exceed 200. Subtracting the median radar variable value in each category (interval) from the climatological mean leaves a residue. If that residue is within the ±0.2 dB interval, it is assumed that Z_{DRsys} is correct. The operational VCPs are used in these automated measurements, and the residue is produced when appropriate data are available.

Possible problems with this method are the presence of large drops and contamination from wet snowflakes originating in the melting layer. A small number of large drops can be present in light tropical rain [50]. Calculations indicate that such drops can bias positively the climatological Z_{DR} values. Wet snowflakes can also increase the Z_{DRm} and their presence cannot be excluded with the current filters. Because of these issues, additional methods are used to assess the accuracy of Z_{DRsys}.

4.4.2 Measurements in dry snow aggregates

Dry snow aggregates in areas above the melting layer produce intrinsic Z_{DR} close to 0 dB. The temperature there is few degrees below 0°C and crystal aggregation frequently takes place. The aggregates decrease Z_{DR} due to random orientation and/or quasi-spherical shape and reduction in the dielectric constant. ROC meteorologists have found the 0.2 dB for Z_{DR} estimates from such areas, which is similar to the value from light rain [49]. Therefore, Z_{DR} = 0.2 dB is used as intrinsic Z_{DR} of aggregated dry snow. In the example of stratiform precipitation (Figure 4.9), the area of aggregated snow is depicted with a thin layer of close to zero Z_{DR} just above 4 km (i.e., above the top of the melting layer).

In this method, data are collected at antenna elevations higher than 1° and from areas where reflectivities are in the interval of 15–25 dBZ. The following three thresholds are applied to the data: SNR $\geq$ 20 dB, $\rho_{hv} \geq 0.98$, and $\Phi_{DP} < 100°$ (to avoid signals attenuated by precipitation). The data should be taken from heights smaller than 1 km above the melting layer. The top of the melting layer is determined by a WSR-88D's algorithm. The number of data should exceed 500 to compute a stable mean Z_{DR} value. The data from 12 consecutive VCPs are averaged. If this mean does not deviate from 0.2 dB by more than 0.2 dB, Z_{DRsys} is assumed to be correct.

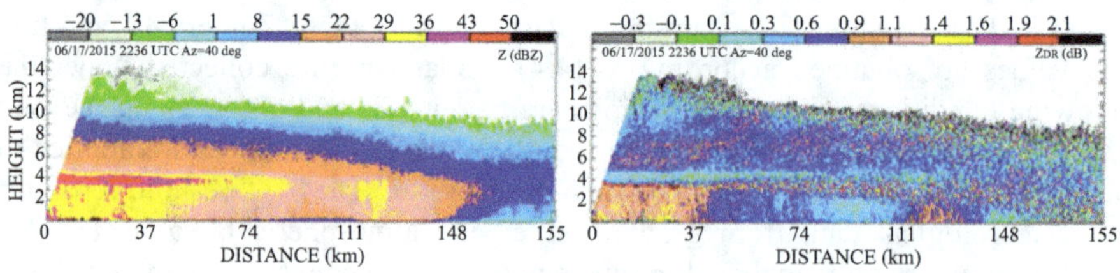

Figure 4.9 Vertical cross-sections of (left) reflectivity and (right) Z_{DR} collected with KOUN on June 17, 2015 at 2236 UTC and azimuth of 40°.

Incomplete knowledge about the aggregation process limits applicability of this method. Some radar observations exhibit no obvious decrease in Z_{DR} above the melting layer. In the case in Figure 4.10, the melting layer is clearly identified but above it, there are no pronounced areas of Z_{DR} smaller than 0.4 dB. It may be that the Z_{DR} values of aggregated snowflakes depend on the intensity of aggregation, shapes of the aggregates, and intensity of their fluttering. Therefore, significant variations in Z_{DR} from such areas are possible and intrinsic Z_{DR} = 0.2 dB may not be the median. Research to better detect areas of dry snow is ongoing.

4.4.3 Measurements in regions of Bragg scatter

S-band radars can observe reflections from clear turbulent air [10]. Turbulent eddies of half the size of radar wavelength (i.e., about 5 cm) scatter S-band radiation and create various structures of echoes [51]. Small randomly orientated eddies reflect equal power of linearly polarized incident waves. Therefore, the mean Z_{DR} of Bragg scatterers equals 0 dB [52]. The example in Figure 4.11 depicts a layer with ρ_{hv} larger than 0.98 and Z_{DR} close to zero indicating that it contains Bragg scatterers. The layer's increasing thickness and decreasing SNR with range suggest the beam smearing is broadening it and that it is about 200 meters thick. These types of layers are suitable for verifying accuracy of the Z_{DRsys} measurements.

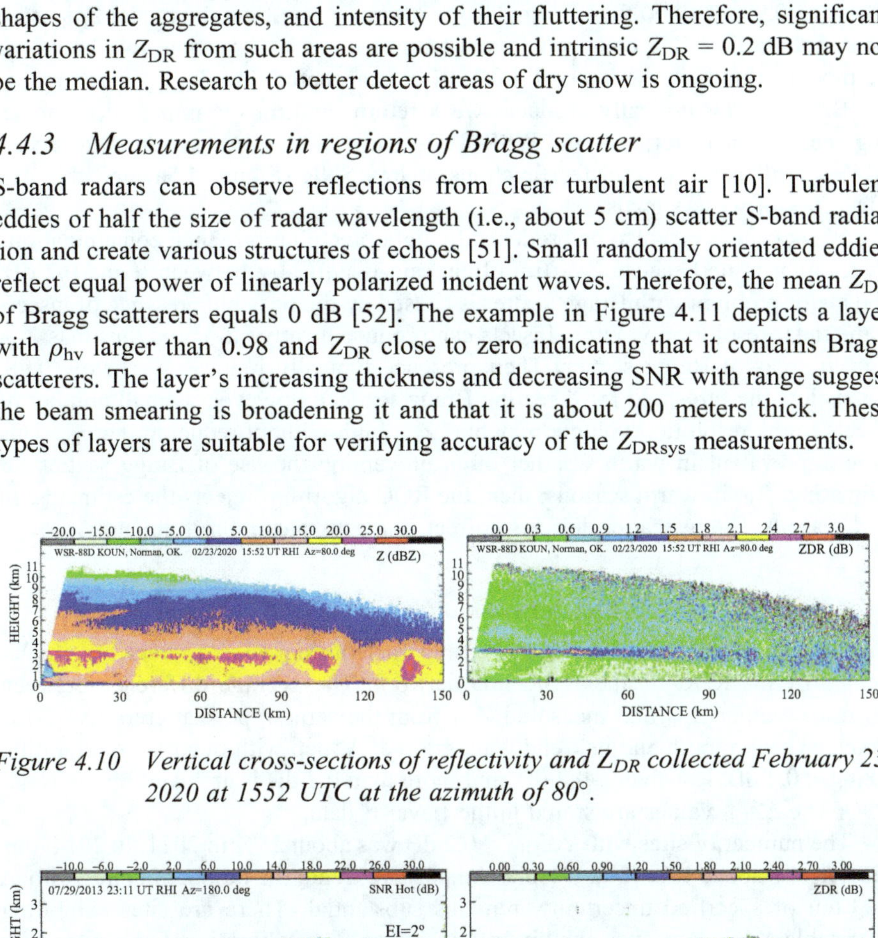

Figure 4.10 Vertical cross-sections of reflectivity and Z_{DR} collected February 23, 2020 at 1552 UTC at the azimuth of 80°.

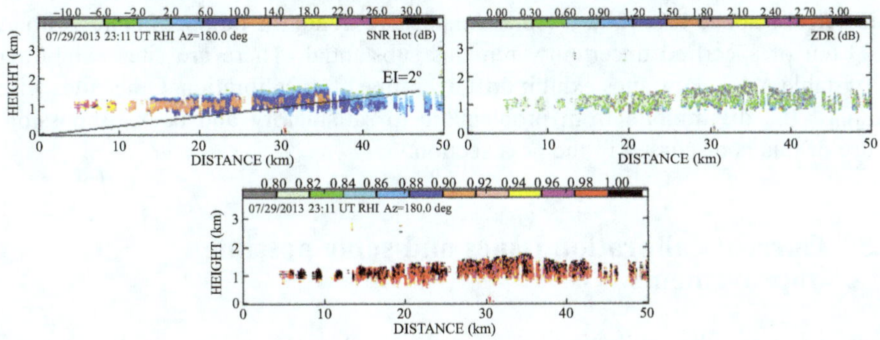

Figure 4.11 Vertical cross-sections of SNR, Z_{DR}, and ρ_{hv} collected with KOUN in clear air on July 29, 2013 at 2311 UTS and the azimuth of 180°.

To verify Z_{DR} calibration on WSR-88Ds, ROC personnel has designed a Bragg scatter detection procedure and applied it in real time as follows [47,53,54]. (1) Collect data from antenna elevations between 2.4° and 4.5° in clear-air VCPs (VCPs 21 and 32). (2) Reject data likely to be contaminated by precipitation. Select data with $Z \leq -3$ dBZ at the 90th percentile. (3) Apply thresholds on SNR < 15 dB, $\rho_{hv} > 0.98$, $\sigma_v \geq 0.5$ m s^{-1}, and $|v| > 2.0$ m s^{-1}. (4) Find the modal Z_{DR} using the distribution of Z_{DR} collected from 12 consecutive VCPs (about two hours). If the modal value is in the interval ± 0.2 dB, it is accepted that Z_{DRsys} is correct.

Bragg scatter typically produces weak returns and thus is impacted by noise. Engineering measurements on WSR-88Ds show that some systems' receivers exhibit nonlinear Z_{DR} transfer functions at low SNR (Figure 4.5a) and this can affect Z_{DR} of weak signals.

Weak echoes can also be produced by light precipitation. Such contaminations are a minor issue because Z_{DR} from light rain is typically between 0 and 0.2 dB. The major problem with Bragg scatter is caused by the possible presence of insects in the radar resolution volume. Insects can produce positive Z_{DR} and thus bias Z_{DR} high in areas of Bragg scatter. Their presence typically reduces ρ_{hv} below 0.98, (i.e., below the threshold for detecting Bragg scatter); however, a small number of insects in the resolution volume may bias Z_{DR} high without reducing the ρ_{hv}. This issue is prevalent in warm weather often preventing the use of Bragg scatter for estimating Z_{DR} in warm seasons; then, the ROC algorithm rejects the estimates. In cold seasons, Bragg scatter delivers correct Z_{DR} estimates more frequently.

4.4.4 Z_{DRsys} across the network

The three methods to assess Z_{DRsys} using meteorological scatterers are implemented by the ROC on the WSR-88D network. The average difference between intrinsic weather Z_{DR} and measured Z_{DR} from the same type scatterers (ΔZ_{DR}) is produced each month and its trend is monitored. A map with radar sites exhibiting $|\Delta Z_{DR}| < 0.2$ dB, less than -0.2 dB, and more than 0.2 dB is archived by the NWS ROC. The ΔZ_{DR} values are stored in the Level II data.

The number of sites with $|\Delta Z_{DR}| > 0.2$ dB was about 25% in 2021. In 2014, this number was about 40% [36]. Despite the positive trend, the number of sites with $|\Delta Z_{DR}|$ out of specified uncertainty remains substantial. There are sites exhibiting very stable ΔZ_{DR}. Few sites exhibit drift over time after calibration. Other sites with $|\Delta Z_{DR}| > 0.2$ dB migrate from problematic to satisfactory and back. A possible cause of this is discussed in the next section.

4.5 Current calibration issues and some possible improvements

Next, we review the calibration issues on the WSR-88D. Then, we highlight some specific processing on the WSR-88D and contrast it to the ones on other national weather radars.

4.5.1 Current WSR-88D calibration issues

Most of the 25% sites exhibiting $|\Delta Z_{DR}|$ out of the specified tolerance (0.2 dB) have fluctuating $|\Delta Z_{DR}|$ between less than −0.2 dB and more than 0.2 dB. This could be caused by receiver nonlinearity, and we use Figure 4.5(a) to illustrate. Suppose Z_{DRsys} is measured from Bragg scatter at SNR of about 15 dB. The measured Z_{DR} is 0.5 dB which equals Z_{DRsys} because intrinsic $Z_{DR} = 0$ dB. At some other time, light rain produces SNR > 40 dB (close to the radar); this measured Z_{DRm} equals the intrinsic Z_{DR} of 0.2 dB suggesting that Z_{DRsys} is 0 dB. The difference between the first and second measurements is 0.5 dB. Therefore, receiver nonlinearity could cause differences in Z_{DRsys} from meteorological scatterers with different SNRs.

A similar conclusion can be drawn from Figure 4.12 in which the transfer function, Z_{DR} versus SNR, differs from the one in Figure 4.5a. These figures indicate that the Z_{DR} transfer function should be measured using SNR steps of no more than 1 dB across the whole dynamic range. Then, these transfer functions can be used to "linearize" the responses.

Another calibration problem could be caused by the power divider. WSR-88Ds are equipped with a variable-ratio power divider [55]. It splits the transmitted pulse and balances the output powers by shifting the phases of the split signals with a mechanical plunger. The plunger operates automatically, and practice indicates that these dividers are affected by temperature causing uncertainty in performance. The plunger also changes the output phases and consequently the differential phase upon transmission. Furthermore, the variable-ratio dividers are difficult to maintain. Contrast these to the simpler passive ones which utilizes the "magic T." Although the output powers of passive dividers may not be equal (to within 0.1 dB), these dividers are more stable, significantly simplifying hardware in the transmission path, but still require measurements of the output powers as do the variables ratio ones. Another improvement could be made in the power measurements by replacing the power meters with network analyzers.

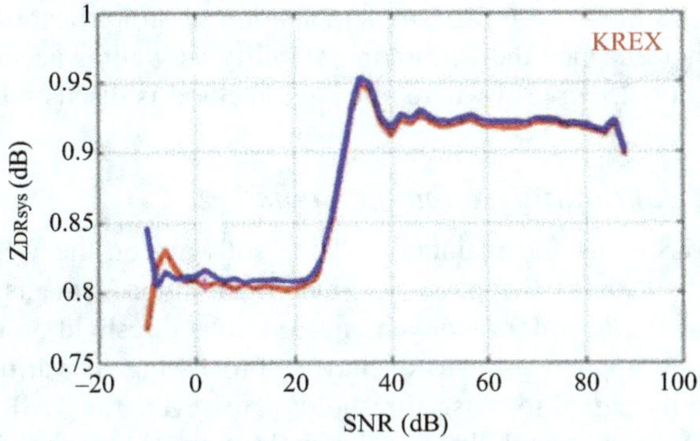

Figure 4.12 Z_{DR} receiver transfer function of the WSR-88D KREX, located in Norman, OK. The blue and red curves have been obtained at different times 93 min apart.

The calculations of Z_{DR} and ρ_{hv} require precise measurements of the noise powers in the polarization channels. Currently, noise is determined in each radial [56] except in radials completely filled with weather returns. For these, it is assumed that the noise estimates are same as in the adjacent radials. The use of the lag 1 estimators would avoid this issue [13] because these are not biased by noise. Moreover, the lag 1 estimators could be combined with the lag 0 (dependent on powers and cross correlation) to reduce the statistical uncertainties.

The SMB measurements are currently taken with the transmitter switched off. This state deviates from the operational state when the transmitter is turned on. There is evidence that the solar measurements while the transmitter is turned on differs from the one while it is off. Furthermore, it is assumed that the Sun radiates randomly polarized waves, hence its intrinsic Z_{DR} = 0 dB. Measurements of solar radiation at cm wavelengths prove this for the Sun's quiet states. This does not necessarily hold in the Sun's active states, i.e., sunspots.

The WSR-88D's hail detection algorithm uses ρ_{hv} for determining hail size. But the ρ_{hv} dependence on the incident differential phase (Section 4.3.6) can confuse the algorithm. To improve hail sizing, knowledge of the incident differential phase is desirable. The contribution to it by the propagation differential phase is routinely measured, but the transmitted differential phase (β) is not measured. A routine robust method to measure β on the network is not available but would be useful to aid in hail sizing.

Improvement over the current tracing of the radar beam's position in height is possible. That can be done by using the profiles of humidity and temperature measured by rawinsondes or generated by numerical models rather than the profiles obtained from the standard atmosphere. Better adjustment is also possible in regions with rain by assuming a 100% humidity. The greatest benefit would be at long distances where the beam position is highest. The projections of precipitation from the beam height to the ground would improve the estimates. In addition, the attenuation estimates would be better and would lead to improved reflectivity estimates.

Some changes in the hardware implementation or simplification of the procedure could further enhance the calibration stability as well as accuracy. We refer readers to section 6.2.1 in [37] where one such method is discussed.

4.5.2 Other data quality enhancements

Because the NWS owns and maintains all the software on the WSR-88D, it can make changes to address the most cost-beneficial improvements. A couple of examples follow. For operators' convenience, various threshold on the field of the radar variables are applied prior to display or processing by various algorithms. These are typically signal-to-noise thresholds. However, the WSR-88D data are subject to semi-coherent threshold. The threshold is an additive combination of the SNR, the cross-correlation between the H and V components, as well as the autocorrelation at lag 1 of the returned signals [57]. This way the weak but semi-coherent signals are pulled out of the noise increasing the radar spatial coverage.

Another specific processing is the estimation of the spectrum width called the "hybrid" estimator [58]. It has a smaller bias and standard deviation than traditional pulse pair estimator [10]. The estimator chooses one of three estimates depending on the spectrum width. At small, medium, and large widths, the estimates are from the autocorrelation at lags 0 and 1, 1 and 2, and 1 and 3.

The French Meteo France has also software control. Long-term evaluation of their polarimetric QPE indicates that the stability of Z_{DR} is ± 0.4 dB [44]. As the study considers radars at C band, the evaluation of polarimetric data quality calls for correcting attenuated *Z*.

The German Weather Service's polarimetric radars operate at the 5 cm wavelength. The antenna can be pointed to zenith; therefore, the calibration procedure includes Z_{DR} measurements at vertical incidence every 5 min when it is raining overhead. These measurements and analysis of solar returns during operational scanning enable constraining the Z_{DR} uncertainty to $\pm$ 0.1 dB [59].

A very important for operations are the composite fields of the various variables and the derived products. For the WSR-88D network, this is done with the Multi-Radar Multi-Sensor (MRMS) platform [60]. As the name implies, it gathers data from many sensors (satellite, surface observations, and others) and applies these to censor radar data. Several layers of quality control exist (too many to list in this short overview) to free the composite from artifacts. MRMS also takes data from the Canadian weather radars and the Terminal Doppler Weather Radars (TDWRs). The US government operates (and maintains) the two networks; hence, all the parameters pertinent to data quality are known. The Canadian government shares its radar data and the corresponding quality parameters (dwell times, SNR, PRT, etc.). This facilitates tremendously the blending of data and uniform quality control. The platform is implemented at the NWS National Center for Environmental Prediction and its latest operating version includes polarimetric data for quality control [61].

European countries use a similar concept in their Operational Program on the Exchange of Weather Radar Information (OPERA), [62]. It produces Pan-European radar composites and many applications including data assimilation, flood warnings, and monitoring of animal migration. Several approaches homogenize the disparate raw data and indicate the reliability of its products. A small number of OPERA countries are using dual-pol attenuation correction. Much of the dual-polarization quality control is implemented in the radar signal processor as part of the proprietary software. Therefore, it can only be applied to the radar during measurements. Presently, the main benefit of dual-polarization radars is the quality improvement of reflectivity and Doppler moments achieved by removing interferences and nonmeteorological returns.

4.6 Concluding remarks

Dual-polarization S-band WSR-88D is an outstanding radar for monitoring severe weather, measuring precipitation, and classifying echoes into weather and nonmeteorological. Acceptable attenuation of radar waves in precipitation at S-band, its large dynamic range, and exceptional detectability enable quantitative measurements at distances up to 300 km. The 8.4 m prime focus parabolic reflector antenna inherited from the single polarization system provides isolation of at least 34 dB between polarization channels that allows quantitative measurements of dual-polarization variables. Although observations are made to the range of 460 km, quantitative interpretations are hindered by the large transverse dimension of the beam and its increased height above the ground.

The WSR-88D's built-in calibration subsystem meets almost all specified requirements for good weather radar measurements. Only measurements of Z_{DR} on about a quarter of the WSR-88Ds do not reach the desirable ±0.2 dB uncertainty range. Nevertheless, about a quarter of these radars exhibit the uncertainty range of about ±0.1 dB or better. Accurate Z_{DR} is needed for measurements in snow and clouds, for the hydrometeor classification algorithms, for hail and graupel detection, and for measurements in clear air to name a few.

We have described the current calibration procedures on the WSR-88D and the achieved accuracies in reflectivity and dual-polarization variables. The lifetime of the WSR-88Ds has been extended beyond the year 2030. Over these years, the number of users will grow and so will their demands on the quality of the radar data. This will exert further pressure to improve calibration. Automating and improving most of the calibration procedures on such large number of radars require expertise in engineering, physics, and meteorology.

Some additional potentials for weather observations with WSR-88D are close to operational use. Among these are more accurate measurements of the rain and snow rates, measurements of convection parameters in pre-storm environment, and observations of the boundary layer. These new capabilities require more accurate radar variables. Resolving current calibration issues will enable more accurate measurements and for transitioning this expertise to phased array radars which may be the next generation of weather radars.

List of abbreviations

2D	Two dimensional
AME	Antenna Mounted Electronics
CSU-CHILL	Colorado State University–Chicago Illinois
LNA	Low Noise Amplifier
NSSL	National Severe Storms
NWS	National Weather Service
QPE	Quantitative precipitation estimates
RCB	Receiver Bias
RDA	Radar Data Acquisition
RF	Radio frequency
ROC	Radar Operations Center

RPG	Radar Product Generator
SHV	Simultaneous Horizontal and Vertical
SQI	Signal Quality Index
SMB	Sun Measurement Bias
SNR	Signal-to-noise ratio
TDWR	Terminal Doppler Weather Radar
TXB	Transmitter Bias
VCP	Volume Coverage Pattern
VPI	Video Integrator and Processor
WSR-88D	Weather Surveillance Radar, 1988, Doppler

References

[1] Crum T.D. and Alberty, R.L. The WSR-88D and the WSR88D operational support facility. *Bull. Amer. Meteorol. Soc.* 1993;74(9):1669–87.

[2] Klazura G.E. and Imy D. A description of the initial set of analysis products available from the NEXRAD WSR-88D System. *Bull. Amer. Meteorol. Soc.* 1993;74(7):1293–311.

[3] Ortega K.L., Smith T.M., Manross K.L., *et al.* The severe hazards analysis and verification experiment. *Bull. Amer. Meteorol. Soc.* 2009;90(10):1519–30.

[4] Smith T.M., Lakshmanan V., Stumpf G.J., *et al.* Multi-radar multi-sensor (MRMS) severe weather and aviation products: Initial operating capabilities. *Bull. Amer. Meteorol. Soc.* 2016;97(9):1617–30.

[5] Whiton, R.C., Smith P.L., Bigler S.G., Wilk K.E. and Harbuck A.C. History of the operational use of weather radar by U.S. weather services. Part I: The Pre-NEXRAD era. *Weather. Forecast.* 1998;13(2): 219–43.

[6] Shreeve, K.H. and Erdahl, R.J. A weather radar video integrator and processor. *IEEE Trans. Geosci. Electron.* 1968;6(3):152–5.

[7] Shreeve, K.H. Design and operation of a digital video integrator processor (DVIP). *NOAA Tech. Memo. 1974; NWS EDL.* 1974;13:129

[8] Klazura, G.E. and Aman, R. Quality control of archiving digital data gathered by the Bureau of Reclamation's weather radar. *J. Appl. Meteorol.* 1989;28(12):1336–40.

[9] Doviak R.J. and Zrnić D.S. *Doppler Radar and Weather Observations*. San Diego: Academic Press; 1984. p. 458.

[10] Doviak R.J. and Zrnić D.S. *Doppler Radar and Weather Observations.* New York: Dover Publications; 2006. p. 562.

[11] Bringi V.N. and Chandrasekar V. *Polarimetric Doppler Weather Radar. Principles and Applications.* Cambridge: Cambridge University Press; 2001. p. 636.

[12] Zrnic D.S. Doppler radar for USA weather surveillance. In *Doppler Radar Observations – Weather Radar, Wind Profiler, Ionospheric Radar, and*

Other Advanced Applications. J. Bech (ed.). Rijeka, Croatia: InTech-Open Access Publisher; 2012; ISBN 978-953-51-0496-4.

[13] Melnikov V.M. and Zrnić D.S. Autocorrelation and cross-correlation estimators of polarimetric variables. *J. Atmos. Ocean. Technol.* 2007;24 (8):1337–50.

[14] Sandford, C. and Gaussiat, N. Use of a radar quality index to mitigate the effects of attenuation at C-band in the UK composite. *ERAD 2012, Seventh Eur. Conf. Radar Meteorol. Hydrol.*

[15] Doviak R.J., Bringi V., Ryzhkov A., Zahrai A. and Zrnić D.S. Considerations for polarimetric upgrades of operational WSR-88D radars. *J. Atmos. Ocean. Technol.* 2000;17(3):257–78.

[16] NOAA/NWS Radar Functional Requirements. 2015. https://www.roc.noaa.gov/WSR88D/PublicDocs/NOAA_Radar_Functional_Requirements_Final_Sept%202015.pdf.

[17] Torres S.M. and Zrnic D.S. Whitening of signals in range to improve estimates of polarimetric variables. *J. Atmos. Ocean. Technol.* 2003;20(12):1776–89.

[18] Curtis C. and Torres S.M. Adaptive range oversampling to improve estimates of polarimetric variables on weather radars. *J. Atmos. Ocean. Technol.* 2014;31(9):1853–66.

[19] Seliga T.A. and Bringi V.N. Potential use of radar differential reflectivity measurements at orthogonal polarizations for measuring precipitation. *J. Appl. Meteorol.* 1976;15(1):69–76.

[20] Zhang J., Tang L., Cocks S., *et al.* A dual-polarization radar synthetic QPE for operations. *J. Hydrometeorol.* 2020;21(11):2507–21.

[21] Ryzhkov A.V. and Zrnic D.S. Discrimination between rain and snow with a polarimetric radar. *J. Appl. Meteorol.* 1998;37(10):1228–40.

[22] Bringi, V.N., Hoferer R., Brunkov D.A., *et al.* Antenna design and performance characteristics of the new 8.5-m dual-offset Gregorian antenna for the CSU–CHILL radar. *J. Atmos. Ocean. Technol.* 2011;28(7):907–20.

[23] Atlas D. Radar calibration. *Bull. Amer. Meteorol. Soc.* 2002;83(9):1313–6.

[24] Williams E., Hood K., Smalley D., *et al.* End-to-end calibration of Nexrad differential reflectivity with metal spheres. *36th Conf. Radar Meteorol. AMS*, 2013. https://ams.confex.com/ams/36Radar/webprogram/Paper228796.html.

[25] Dominion Radio Astrophysical Observatory. https://nrc.canada.ca/en/research-development/nrc-facilities/dominion-radio-astrophysical-observatory-research-facility.

[26] Melnikov V. and Schlatter T. Enhancing sensitivity on the polarimetric WSR-88D. *91st Annu. Meet.* Amer. Meteorol. Soc. Seattle, WA, USA, 2011. Available from https://ams.confex.com/ams/91Annual/webprogram/Paper178856.html.

[27] Kollias P., Bharadwaj N., Clothiaux E.E., *et al.* The ARM radar network: At the leading edge of cloud and precipitation observations. *Bull. Amer. Meteorol. Soc.* 2020;101(5):E588–607.

[28] Melnikov V. and Zrnić D.S. Observations of convective thermals with weather radar. *J. Atmos. Ocean. Technol.* 2017;34(7):1585–90.

[29] Gorgucci E., Scarchilli G. and Chandrasekar V. A procedure to calibrate multiparameter weather radar using properties of the rain medium. *IEEE Trans. Geosci. Remote Sens.* 1999;37(1):269–76.

[30] Gourley J.J., Tabary P. and Parent du Chatelet J. Data quality of the Meteo-France C-band polarimetric radar. *J. Atmos. Ocean. Technol.* 2006;23(10):1340–56.

[31] Frech M., Hagen M. and Mammen T. Monitoring the absolute calibration of a polarimetric weather radar. *J. Atmos. Ocean. Technol.* 2017;34(3):599–615.

[32] Tragl K. Polarimetric radar backscattering from reciprocal random targets. *IEEE Trans. Geosci. Remote Sens.* 1990;8:856–64.

[33] Hubbert J.C., Bringi V.N. and Brunkow D. Studies of the polarimetric covariance matrix. Part I: Calibration methodology. *J. Atmos. Ocean. Technol.* 2003;20(5):696–706.

[34] Zrnić D.S., Melnikov V.M. and Carter J.K. Calibrating differential reflectivity on the WSR-88D. *J. Atmos. Ocean. Technol.* 2006;23(7):944–51.

[35] Cunningham J.G., Zittel W.D., Lee R.R., Ice R.L. and Hoban N.P. Methods for identifying systematic differential reflectivity (ZDR) biases on the operational WSR-88D network. In *36th Conf. Radar Meteorol.* Albuquerque, NM:AMS; 2013.

[36] Ice R.L., Heck A.K., Cunningham J.G. and Zittel W.D. Challenges of polarimetric weather radar calibration. In *8th Eur. Conf. Radar Meteorol. Hydrol.* Garmisch-Partenkirchen, Germany; 2014. www.roc.noaa.gov/WSR88D/PublicDocs/Publications/117_Ice.pdf.

[37] Ryzhkov A.V. and Zrnić D.S. *Radar Polarimetry for Weather Observations*. Cham, Switzerland: Springer, Atmospheric Sciences; 2019. p. 486.

[38] Kundu M.R. *Solar Radio Astronomy*. New York: Wiley; 1965. p. 660.

[39] Ryzhkov A.V. and Zrnić D.S. Depolarization in ice crystals and its effect on radar polarimetric measurements. *J. Atmos. Ocean. Technol.* 2007;24(7):1256–67.

[40] Melnikov V.M., Istok M.J. and Westbrook J.K. Asymmetric radar echo patterns from insects. *J. Atmos. Ocean. Technol.* 2015;32(4):659–74.

[41] Melnikov V.M. Impacts of the phase shift between incident radar waves on the polarization variables from ice cloud particles. *J. Atmos. Ocean. Technol.* 2020;37(7):1423–36.

[42] Park H.S., Ryzhkov A.V., Zrnić D.S. and Kim K.E. The hydrometeor classification for polarimetric WSR-88D; description and application to an MCS. *Weather Forecast.* 2009;24(3):730–48.

[43] Boscacci M., Gabella M., Clementi L., Sartori M. and Germann U. Short journey into the best practices of monitoring and calibration of the weather radar receiver at Meteoswiss during the last 25 years. ERAD-2022. https://www.erad2022.ch/_files/ugd/25a7b1_23b223c508be440ca402c5a29bba166e.pdf.

[44] Figueras i Ventura J., Boumahmoud A-A, Fradon B, Dupuy P., and Tabary P. Long-term monitoring of French polarimetric radar data quality and

evaluation of several polarimetric quantitative precipitation estimators in ideal conditions for operational implementation at C-band. *Q. J. R. Meteorol. Soc.* 2012;138:2212–28.

[45] Bringi V.N., Kennedy P.C., Huang G.J., Kleinkort C., Thurai M. and Notaros B.M. Dual-polarized radar and surface observations of a winter graupel shower with negative Z_{dr} column. *J. Appl. Meteorol. Climatol.* 2017;56(2):455–70.

[46] Ryzhkov A.V., Kumjian M.R. and Ganson S.M. Polarimetric radar characteristics of melting hail. Part I: Theoretical simulations using spectral microphysical modeling. *J. Appl. Meteorol. Climatol.* 2013;52(12):2849–70.

[47] Westbrook C.D. Rayleigh scattering by hexagonal ice crystals and the interpretation of dual-polarization radar measurements. *Q. J. R. Meteorol. Soc.* 2014;140(683):2090–96.

[48] Cao Q., Zhang G., Brandes E., Schuur T., Ryzhkov A. and Ikeda K. Analysis of video disdrometer and polarimetric radar data to characterize rain microphysics in Oklahoma. *J. Appl. Meteorol. Climatol.* 2008;47(8): 2238–55.

[49] Zittel W.D., Cunningham J.G., Lee R.R., Richardson L.M, Ice R.L. and Melnikov V. Use of hydrometeors, Bragg scatter, and sun spikes to determine system ZDR biases in the WSR -88D fleet. In *8th Eur. Conf. Radar Meteorol. Hydrol.* 2014. www.roc.noaa.gov/WSR88D/ PublicDocs/ Publications/132_Zittel.pdf].

[50] Bringi V.N., Chandrasekar V., Hubbert J., Gorgucci E., Randeu W.L. and Schoenhuber M. Raindrop size distribution in different climatic regimes from disdrometer and dual-polarized radar analysis. *J. Atmos. Sci.* 2003;60(2):354–65.

[51] Melnikov V.M., Doviak R.J., Zrnić D.S. and Stensrud D.J. Structures of Bragg scatter observed with the polarimetric WSR-88D. *J. Atmos. Ocean. Technol.* 2013;30(7):1253–58.

[52] Melnikov V.M., Doviak R.J., Zrnić D.S. and Stensrud D.J. Mapping Bragg scatter with a polarimetric WSR-88D. *J. Atmos. Ocean. Technol.* 2011;28 (10):1273–85.

[53] Richardson L.M., Cunningham J.G., Zittel W.D., *et al.* Bragg scatter detection by the WSR-88D. Part I: Algorithm development. *J. Atmos. Ocean. Technol.* 2017;34(3):465–78.

[54] Richardson L.M., Zittel W.D., Lee R.R., Melnikov V.M., Ice R.L. and Cunningham J.G. Bragg scatter detection by the WSR88D. Part II: Assessment of ZDR bias estimation. *J. Atmos. Ocean. Technol.* 2017;34 (3):479–93.

[55] Teeter W.L. and Bushore K.R. A variable-ratio microwave power divider and multiplexer. *IRE Trans. Microw. Theory Tech.* 1957;5(4):227–9.

[56] Ivić, I.R., Curtis C. and Torres S.M. Radial-based noise power estimation for weather radars. *J. Atmos. Ocean. Technol.* 2013;30(12):2737–53.

[57] Ivic, I.R., Zrnic, D.S. and Yu, T.-Y. The use of coherency to improve signal detection in dual-polarization weather radars. *J. Atmos. Ocean. Technol.* 2009;26(11):2474–87.

[58] Meymaris G., Williams J. and Hubbert J. Performance of a proposed hybrid spectrum width estimator for the NEXRAD ORDA. In *25th Int. Conf. Interact. Inf. Process. Syst. Meteorol. Oceanogr. Hydrol.* Phoenix, AZ, Amer. Meteor. Soc. 2009; 11B.1. [Available online athttps://ams.confex.com/ams/89annual/techprogram/paper_145958.htm.]

[59] Frech, M. and Hubbert, J. Monitoring the differential reflectivity and receiver calibration of the German polarimetric weather radar network. *Atmos. Meas. Tech.* 2020;13(3):1051–69.

[60] Zhang J., Howard K., Langston C., *et al.* Multi-radar multi-sensor (MRMS) quantitative precipitation estimations: Initial operating capabilities. *Bull. Amer. Meteorol. Soc.* 2016;97(4):621–38.

[61] Tang L., Zhang J., Simpson M., *et al.* Updates on the radar data quality control in the MRMS quantitative precipitation estimation system. *J. Atmos. Ocean. Technol.* 2020;37(9):1521–37.

[62] Saltikoff, E., Haase G., Delobbe L., *et al.* OPERA the radar project. *Atmosphere*. 2019;10(6):1–13. https://doi.org/10.3390/atmos10060320

Chapter 5

Improvement of GPM dual-frequency precipitation radar algorithms for Version 07

Shinta Seto[1]

This chapter describes the standard algorithms to retrieve the vertical profile of precipitation rate from the measurements of the Dual-frequency Precipitation Radar (DPR) on the core satellite of the Global Precipitation Measurement (GPM) mission. At first, in Section 5.1, the DPR sensor and the standard algorithms are overviewed. Next, in Section 5.2, as theoretical basis for the DPR standard algorithms, the relations between drop size distribution (DSD) parameters and other variables are derived, then some retrieval methods for DSD parameters are introduced. In Section 5.3, the standard algorithm for Precipitation Radar (PR) on the Tropical Rainfall Measurement Mission (TRMM) satellite is reviewed and the history of the DPR standard algorithm development (until version 06) is summarized. In Section 5.4, the latest version (Version 07) of the DPR standard algorithms is explained focusing on the improvement from the previous version. In Section 5.5, remaining tasks for the next version are noted. Finally, Section 5.6 is given for the summary of this chapter.

5.1 Introduction

5.1.1 Overview of the DPR

The first space-borne precipitation radar, PR on the TRMM satellite, was operated from December 1997 to April 2015 [1]. As the successor of PR, DPR started its measurement in March 2014 [2]. DPR is composed of two radars called KuPR and KaPR. KuPR emits and receives microwave radiation at 13.6 GHz (Ku-band), while PR does at 13.8 GHz. As shown in Figure 5.1(a), KuPR makes cross-track scans, the width of which is approximately 245 km. Forty-nine field of views (FOVs or pixels in short) with the diameter of 5.0 km are included in a scan. The FOV size is nearly same as that of PR after the orbit boost in August 2001. On the other hand, KaPR emits and receives microwave radiation at 35.5 GHz (Ka-band). For several years after the launch, KaPR made a normal scan and an interleaved scan alternatively. The

[1]Graduate School of Engineering, Nagasaki University, Nagasaki, Japan

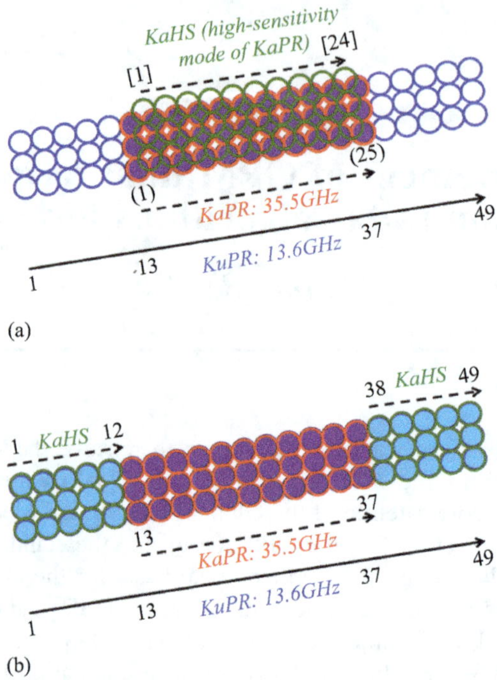

Figure 5.1 Scan pattern of the DPR (a) before May 2018 and (b) after May 2018

normal scan has the width of 125 km and includes 25 FOVs, while the interleaved scan has the width of 120 km and includes 24 FOVs. The normal scan of KaPR was overlapped with the middle part of KuPR's scan (called inner swath). An FOV in the inner swath was observed by both KuPR and KaPR, or the dual-frequency measurement was realized there. The interleaved scan was measured by KaPR's high-sensitivity mode, where the range bin interval is 250 m and the range resolution is 500 m. These values are double of those in the normal scan (range bin interval is 125 m and the vertical resolution is 250 m). The high-sensitivity mode is expected to detect light precipitation including solid precipitation. Though the swath width of KaPR is nearly half of that of KuPR, the spatial density of FOVs is higher for KaPR than for KuPR. The high spatial density was designed to investigate the non-uniformity of precipitation within an FOV. In May 2018, the scan pattern of KaPR has been changed so that the high-sensitivity mode has been moved to the outer part of the normal scan (called outer swath), and the interleaved scan has been abandoned as shown in Figure 5.1(b). After the scan pattern change, all FOVs are observed by both KuPR and KaPR.

5.1.2 Overview of DPR algorithms

The standard algorithms for DPR are categorized into three levels. At level-1 algorithms, geometric information is given to each FOV and engineering variables such as received echo power are calculated. Separately for KuPR and KaPR, level-1 standard products are output. At level-2 algorithms, physical variables such as precipitation rate are calculated. Three kinds of level-2 standard products are output: KuPR product, KaPR product, and the dual-frequency product. At level-3 algorithm, monthly/daily gridded values are produced by processing level-2 standard products. In this chapter, the focus is on level-2 algorithms.

5.1.2.1 DPR level-2 algorithms

DPR level-2 algorithms are composed of three algorithms (Figure 5.2). Each algorithm produces one product file per one orbit. The first one is KuPR algorithm, which uses KuPR measurement but no KaPR measurement. The output is called KuPR product. The second one is KaPR algorithm, which uses KaPR measurement but no KuPR measurement. The output is called KaPR product. The third one is the dual-frequency algorithm, which uses both KuPR and KaPR measurements. The output is called the dual-frequency product. The dual-frequency algorithm (product) is sometimes called as "DPR algorithm (product)," but not in this chapter. Here, "DPR algorithms (products)" is used as the general name of the three algorithms (products). KuPR algorithm (product) and KaPR algorithm (product) are also called the single-frequency algorithms (products). By using measurements by KuPR and KaPR, the dual-frequency algorithm can apply methods such as dual-frequency surface reference technique (SRT) which are not available in single-frequency algorithms. Although the dual-frequency algorithm is expected to be more accurate than the single-frequency algorithms, the single-frequency algorithms are prepared. It is partly because KuPR product is required as an extension of the PR standard product. It is also because even if one radar of DPR is out of order and the dual-frequency algorithm is not available, a single-frequency product can be made by another radar. The latest KuPR algorithm is applied to TRMM/PR measurements (the details are given in Section 5.1.2.5).

Before the scan pattern change, KuPR algorithm was not applied for the interleaved scan and KaPR algorithm was not applied for the outer swath. The dual-frequency algorithm is applied at all FOVs, but in outer swath, it works almost same as KuPR algorithm, and in interleaved scan, it works almost same as KaPR algorithm. In this chapter, the dual-frequency algorithm is explained only for the case that it is applied for an FOV with dual-frequency measurements.

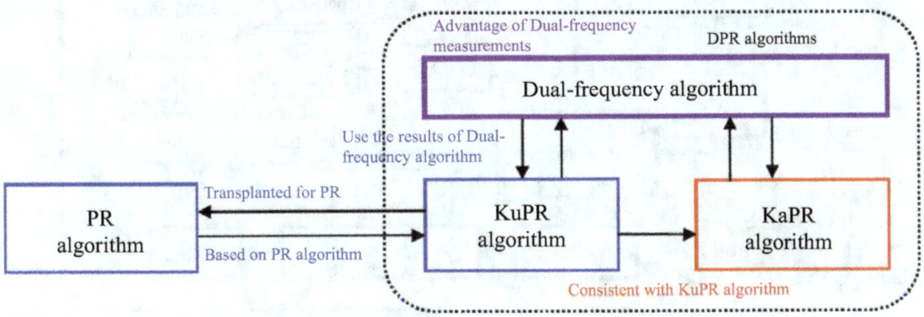

Figure 5.2 Relation among DPR algorithms and PR standard algorithm

5.1.2.2 Modules

Each of the DPR level-2 algorithms has the structure shown in Figure 5.3. Each algorithm is composed of modules: Preparation module, Vertical profile module, Classification module, SRT module, DSD module, Solver module, and Trigger module.

Main functions of Preparation module are clutter removal [3], determination of precipitation/no-precipitation, calculation of radar reflectivity factors, and surface backscattering cross-sections [4]. Main functions of Vertical profile module are the estimation of attenuation caused by non-precipitating particles (such as water vapor, oxygen molecules, and cloud water [5]) and the determination of atmospheric variables such as air temperature and air pressure based on global analysis dataset. Function of Classification module is to detect a bright band and to judge precipitation types [6]. SRT module is to estimate path integrated attenuation (PIA) by SRT [7]. Main functions of DSD module is to assume the characteristics of precipitation particles and the relations among DSD parameters and their related variables. Solver module is to estimate precipitation rates and related variables such as DSD parameters. Trigger module has functions such as detection of multiple scattering. This module is included only in the dual-frequency algorithm and the outputs of this module do not affect the other modules.

In each of the single-frequency algorithms, six modules are executed sequentially (Preparation module, Vertical profile module, Classification module, SRT module, DSD module, Solver module) twice. The second iteration is necessary

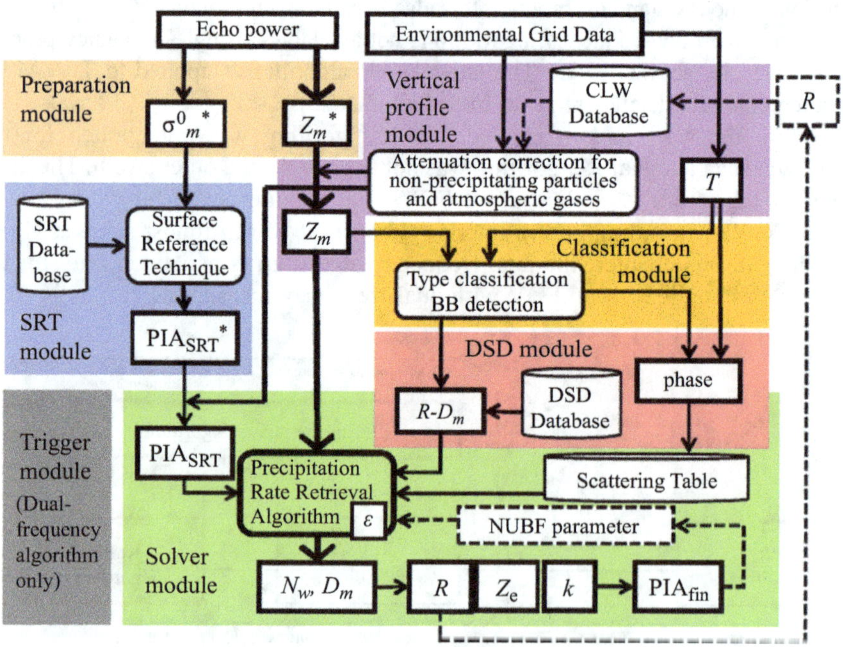

Figure 5.3 Modules in DPR algorithms

because some outputs by a module executed later are required as inputs for another module executed earlier. For example, Vertical profile module assumes that cloud liquid water depends on the surface precipitation rate. Hence, the surface precipitation rate estimated by Solver module in the first iteration is used in Vertical profile module in the second iteration. In the dual-frequency algorithm, seven modules including Trigger module are executed one time only. As the dual-frequency algorithm can refer KuPR and KaPR level-2 products, no iterations are necessary.

5.1.2.3 Strategies for the algorithm development

In order to keep the continuity between PR product and KuPR product, DSD and Solver modules of the KuPR algorithm were developed based on the part of level-2 PR algorithm called 2A25 [8,9]. Other parts of the level-2 PR algorithm are 2A21 and 2A23, and they are taken over by SRT module and Classification module, respectively.

In order to maintain the consistency among the three DPR algorithms, common assumptions are used for different algorithms as much as possible. For example, because the falling speed of precipitation particles cannot be measured by the DPR, common equations for the falling speed are applied to the three DPR algorithms.

The dual-frequency algorithm is expected to be more accurate than the single-frequency algorithms. To clarify the advantage of the dual-frequency algorithm, while the dual-frequency algorithm has basically the same structure with the single-frequency algorithm, some optional methods are added to the dual-frequency algorithm.

5.1.2.4 Brief history of the algorithm development

In March 2014, the DPR has started its operation and the DPR algorithms Version 03 (V03) were used to produce the DPR products, which were released in September 2014. The DPR algorithms were updated to Version 04 (V04) in March 2016, Version 05 (V05) in May 2017, Version 06 (V06) in October 2018, and Version 07 (V07) in December 2021. After a new-version algorithm is released, the algorithms in older versions are no longer used for the processing of new measurement. The new-version algorithm is applied to measurements from the start of the mission to new measurements.

Although the scan pattern was changed in May 2018, the DPR algorithms V06 were not adapted to the new scan pattern. In V06, KaPR's high-sensitivity mode was not processed and KaPR's interleaved scan always had missing values after the scan pattern change. As an experimental product, the DPR algorithms Version 06X (V06X) were released in June 2020. V06X is applied to the measurement only after the scan pattern change. V06X uses basically the same algorithm with V06, but can be applied to all FOVs including high-sensitivity mode.

5.1.2.5 Application of KuPR algorithm to PR measurement

For continuity of PR and KuPR products, KuPR algorithm V06 was applied to PR measurements. KuPR algorithm requires a level-1 product. From PR measurements, level-1 PR product is produced in the same format with level-1 KuPR product. As the sampling interval was different between PR (250 m) and KuPR (125 m), PR's vertical profile of echo power is interpolated to 125 m. KuPR algorithm with technically minor changes is applied to level-1 PR product and the output is called PR product Version 8 (V8), as the PR standard products were released up to Version 7. In the same way, KuPR algorithm V07 is applied to PR measurements and the product is called PR Version 9 (V9), which is released in May 2022.

5.2 Basic theory

5.2.1 Drop size distribution

In the DPR algorithms, all the precipitation particles are assumed to be spherical and the diameter is denoted by D [mm]. The drop size distribution function is denoted by $N(D)$ [mm^{-1} m^{-3}] and it is defined as follows. The value of integral in (5.1) represents the number of precipitation particles with the diameter between D_1 and D_2 in the atmosphere of a unit volume (1 m^3).

$$\int_{D_1}^{D_2} N(D)dD. \tag{5.1}$$

The following integral in (5.2) gives the number of all precipitation particles in the atmosphere of a unit volume.

$$\int_{0}^{\infty} N(D)dD. \tag{5.2}$$

One of the main purposes of the DPR algorithms is to estimate the drop size distribution $N(D)$. Once $N(D)$ is determined, the precipitation rate R [mm h^{-1}] can be calculated by (5.3).

$$R = 0.6\pi \times 10^{-3} \int_{0}^{\infty} V(D)D^3N(D)dD, \tag{5.3}$$

where $V(D)$ is the falling speed of precipitation particle [m s^{-1}]. Radar variables such as radar reflectivity factor Z_e [mm^6 m^{-3}] and attenuation coefficient k [dB km^{-1}] are calculated in (5.4) and (5.5), respectively.

$$Z_e = \frac{\lambda^4}{\pi^5|K_w|^2}\int_{0}^{\infty} \sigma_b(D)N(D)dD, \tag{5.4}$$

$$k = \frac{0.01}{\ln 10}\int_{0}^{\infty} \sigma_e(D)N(D)dD, \tag{5.5}$$

where σ_b is backscattering cross-section [mm^2], σ_e is extinction cross-section [mm^2], λ is wavelength [mm]. $|K_w|^2$ is a function of refractivity index, and the value is 0.9255 for KuPR and 0.8989 for KaPR. If $N(D)$ has higher degree of freedom

(DoF) than the radar measurement has, $N(D)$ cannot be uniquely determined. Generally, DoF of the single-frequency measurement is 1 per range bin, and DoF of the dual-frequency measurement is 2 per range bin. In the DPR algorithms, $N(D)$ is assumed to follow a Gamma distribution as shown in (5.6).

$$N(D) = N_0 D^{\mu} \exp(-\Lambda D), \tag{5.6}$$

where $N_0[\mathrm{mm}^{-1-\mu}\,\mathrm{m}^{-3}]$, μ, and $\Lambda[\mathrm{mm}^{-1}]$ are parameters (called DSD parameters). $N(D)$ converges to 0, when D approaches to positive infinity. It represents that larger particles are more difficult to survive as they tend to be broken up to multiple small particles. If μ is zero, $N(D)$ converges to N_0 when D approaches to 0 and $N(D)$ is a monotonic decreasing function. If μ is positive, $N(D)$ converges to 0 when D approaches to 0. This represents the fact that tiny particles are subjected to evaporation.

The next equation calculates the mass-weighted mean particle size which is denoted by D_m [mm].

$$D_m = \int_0^{\infty} N(D) D^4 dD \Big/ \int_0^{\infty} N(D) D^3 dD = \frac{\mu + 4}{\Lambda}. \tag{5.7}$$

In the derivation of (5.7), mathematical formula in (5.8) is used.

$$\int_0^{\infty} D^q \exp(-pD) dD = \frac{\Gamma(q+1)}{p^{q+1}}, \tag{5.8}$$

where p and q are positive constants, and Γ is complete gamma function. Equation (5.6) is rewritten to (5.9) by using D_m.

$$N(D) = N_0 D^{\mu} \exp\left[-(\mu + 4)\frac{D}{D_m}\right]. \tag{5.9}$$

Mass of water in the atmosphere of a unit volume is denoted by W [g m^{-3}] and is calculated as follows:

$$W = \rho_w \times 10^{-3} \int_0^{\infty} \frac{4}{3}\pi \left(\frac{D}{2}\right)^3 N(D) dD = \frac{\pi}{6} \times 10^{-3} \rho_w N_0 {D_m}^{\mu+4} \frac{\Gamma(\mu + 4)}{(\mu + 4)^{\mu+4}}, \tag{5.10}$$

where ρ_w is the density of water [g cm^{-3}]. A new DSD parameter N_w is defined as N_0 of the $N(D)$ with $\mu = 0$ and the same W. When μ=0, W is calculated in (5.11).

$$W = \frac{\pi}{4^4} \times 10^{-3} \rho_w N_0 {D_m}^4. \tag{5.11}$$

Therefore, N_w satisfies the following equation:

$$\frac{\pi}{4^4} \times 10^{-3} \rho_w N_w {D_m}^4 = \frac{\pi}{6} \times 10^{-3} \rho_w N_0 {D_m}^{\mu+4} \frac{\Gamma(\mu + 4)}{(\mu + 4)^{\mu+4}} \tag{5.12}$$

From (5.12), N_w is derived as follows:

$$N_w = \frac{4^4}{6} D_m{}^{\mu} \frac{\Gamma(\mu+4)}{(\mu+4)^{\mu+4}} N_0. \tag{5.13}$$

Equation (5.9) is rewritten to (5.14) by using N_w.

$$N(D) = N_w \frac{6(4+\mu)^{4+\mu}}{\Gamma(4+\mu)4^4} \left(\frac{D}{D_m}\right)^{\mu} \exp\left[-(\mu+4)\frac{D}{D_m}\right]. \tag{5.14}$$

Here, $N(D)$ has three unknown parameters N_w, D_m, and μ. As DoF should be 2 or fewer in the DPR algorithms, μ is fixed to 3. $N(D)$ becomes (5.15) when $\mu = 3$.

$$N(D) = \frac{7^7}{5!4^4} N_w \left(\frac{D}{D_m}\right)^3 \exp\left(-7\frac{D}{D_m}\right). \tag{5.15}$$

For simple notation, (5.15) is rewritten to (5.16).

$$N(D) = N_w n(D; D_m), \tag{5.16}$$

where

$$n(D; D_m) = \frac{7^7}{5!4^4} \left(\frac{D}{D_m}\right)^3 \exp\left(-7\frac{D}{D_m}\right). \tag{5.17}$$

In similar way, (5.3)–(5.5) are rewritten to (5.18)–(5.20).

$$R = N_w f_R(D_m), \tag{5.18}$$

$$Z_e = N_w f_Z(D_m), \tag{5.19}$$

$$k = N_w f_k(D_m), \tag{5.20}$$

where

$$f_R(D_m) = 0.6\pi \times 10^{-3} \int_0^{\infty} V(D) D^3 n(D; D_m) dD, \tag{5.21}$$

$$f_Z(D_m) = \frac{\lambda^4}{\pi^5 |K_w|^2} \int_0^{\infty} \sigma_b(D) n(D; D_m) dD, \tag{5.22}$$

$$f_k(D_m) = \frac{0.01}{\ln 10} \int_0^{\infty} \sigma_e(D) n(D; D_m) dD. \tag{5.23}$$

These equations mean N_w is proportional to R, Z_e, and k. In other words, R/N_w, Z_e/N_w, and k/N_w are the functions of D_m.

5.2.2 Retrieval of DSD parameters from unattenuated observations

The retrieval method of DSD parameters is explained. At first, for simplicity, unattenuated Z_e is assumed to be obtained by radar measurements. In the case of dual-frequency measurements, two Z_e's are obtained per range bin and it may be possible to retrieve two DSD parameters N_w and D_m. Here, Z_e measured by KuPR is denoted by $Z_{e,u}$ and Z_e measured by KaPR is denoted by $Z_{e,a}$. Using these notations, (5.19) is rewritten to (5.24) and (5.25) for KuPR and KaPR, respectively.

$$Z_{e,u} = N_w f_{Z,u}(D_m), \tag{5.24}$$

$$Z_{e,a} = N_w f_{Z,a}(D_m), \tag{5.25}$$

where

$$f_{Z,u}(D_m) = \frac{\lambda^4}{\pi^5 |K_{w,u}|^2} \int_0^\infty \sigma_{b,u}(D) n(D; D_m) dD, \tag{5.26}$$

$$f_{Z,a}(D_m) = \frac{\lambda^4}{\pi^5 |K_{w,a}|^2} \int_0^\infty \sigma_{b,a}(D) n(D; D_m) dD, \tag{5.27}$$

where subscripts u and a represent that the values are for KuPR and KaPR, respectively.

The ratio of $Z_{e,a}$ to $Z_{e,u}$ is calculated in (5.28) and the value is called dual-frequency ratio (DFR).

$$\frac{Z_{e,a}}{Z_{e,u}} = \frac{f_{Z,a}(D_m)}{f_{Z,u}(D_m)}. \tag{5.28}$$

DFR does not depend on N_w and it is a function of D_m. As an example, for liquid precipitation with the temperature of 0 °C, DFR is calculated for different D_m (Figure 5.4). In this case, as DFR is not a monotonic increasing/decreasing function of D_m, D_m may not be determined uniquely by a given DFR. Under the conditions that D_m is larger than D_{m0}, where DFR takes the local maximum, D_m can be determined uniquely. Once D_m is determined, N_w can be calculated by (5.24) or (5.25). Finally, R is derived by (5.18).

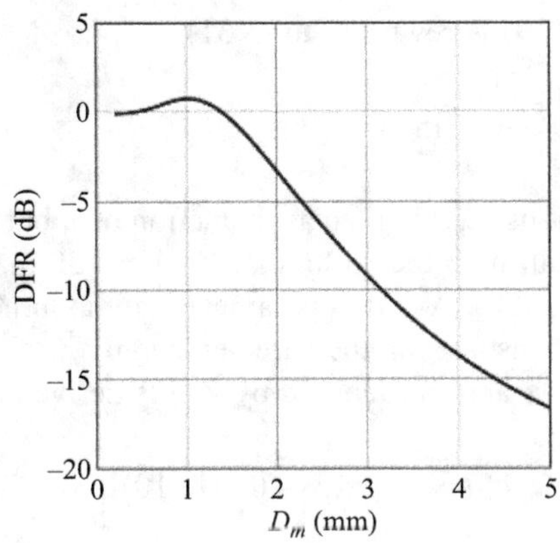

Figure 5.4 Relation between DFR and D_m for liquid precipitation of 0 °C

In single-frequency algorithms, as DoF of measurement is one per range bin, it is not possible to retrieve two unknown parameters. Additional conditions are needed to retrieve DSD parameters. If N_w or D_m is fixed, another DSD parameter may be determined, but this condition is not realistic. Relation between N_w and D_m can be used as a condition to make the retrieval possible. The relation is not necessarily expressed by N_w and D_m, but it can be expressed by the variables, which can be calculated from N_w and D_m. For example, relation between Rayleigh's Z and R (Z-R relation) is often used for conventional radar measurement. In the PR algorithm, relation between Z_e and R is used after Z_e is obtained by the attenuation correction. In the DPR algorithms, relation between R and D_m is used.

5.2.3 Retrieval of DSD parameters from attenuated observations

The spaceborne precipitation radars (PR and DPR) emit/receive microwave of Ku-band or Ka-band, the frequency of which is higher than the frequency of microwave usually used by conventional precipitation radars (S-band, C-band, or X-band). It is because the distance between the spaceborne radar and precipitation is longer and the size of antenna carried on the spacecraft is limited. Microwave at higher frequency is subject to attenuation by precipitation particles. Attenuation correction is necessary for the PR and DPR algorithms. Measured radar reflectivity factor is denoted by Z_m, and it is related to Z_e and k by (5.29).

$$10\log_{10}Z_m(r) = 10\log_{10}Z_e(r) - 2\int_0^r k(s)ds, \tag{5.29}$$

where r axis is taken along the beam and r is set to be zero at the top of precipitation as illustrated in Figure 5.5. As convention, $10\log_{10}X$ is denoted by dBX and (5.29) is rewritten to (5.30).

$$\mathrm{dB}Z_m(r) = \mathrm{dB}Z_e(r) - 2\int_0^r k(s)ds. \tag{5.30}$$

In another way, (5.29) is rewritten to (5.31).

$$Z_m(r) = Z_e(r)\exp\left[-0.2(\ln 10)\int_0^r k(s)ds\right]. \tag{5.31}$$

In the above equations, Z_m is given as a function of r, but actually measured Z_m is given at each range bin with the width of 125 m or 250 m. Consider a range bin where r is between r_1 and r_2. As DSD parameters are assumed to be constant in a range bin, Z_e and k are constant and they are denoted by $\overline{Z_e}$ and $\overline{k}$, respectively. The average of Z_m between r_1 and r_2 (denoted by $\overline{Z_m}$) is derived as follows:

$$\overline{Z_m} = \frac{1}{r_2 - r_1}\int_{r_1}^{r_2} Z_m(r)dr = \overline{Z_e}\exp\left[-0.2(\ln 10)\int_0^{r_1} k(s)ds\right]\frac{1-\exp(-u)}{u}, \tag{5.32}$$

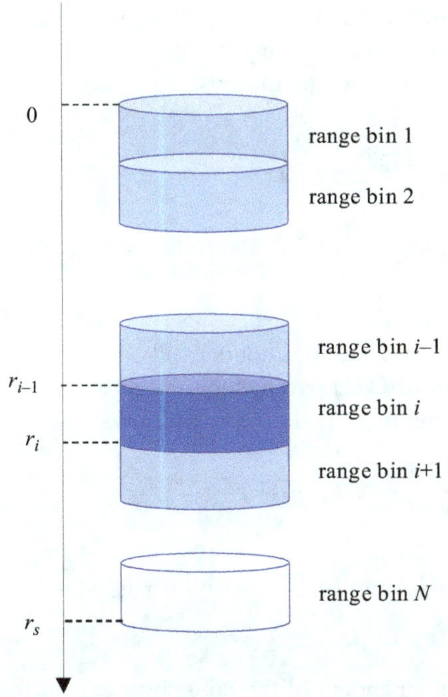

Figure 5.5 Schematic figure to explain range bins

where $u = 0.2(\ln 10)\overline{k}(r_2 - r_1)$. If u is small enough, the following approximation will hold.

$$\frac{1-\exp(-u)}{u} \sim \exp\left(-\frac{u}{2}\right) \tag{5.33}$$

By using (5.33), (5.32) is rewritten to (5.34).

$$\overline{Z_m} \approx \overline{Z_e}\exp\left\{-0.2(\ln 10)\left[\int_0^{r_1} k(s)ds + \frac{r_2 - r_1}{2}\overline{k}\right]\right\}. \tag{5.34}$$

All the precipitation particles in the measured range bin contribute to attenuation, but the further the range from the radar, the contribution is smaller. As a result, the attenuation is calculated as if it is caused by half of the precipitation particles.

As illustrated in Figure 5.5, N range bins exist; then, the nearest range bin from the radar is called range bin 1 and the farthest range bin from the radar is called range bin N. Z_m for range bin i (i = 1, …, N) is given as follows:

$$\mathrm{dB}Z_m[i] = \mathrm{dB}Z_e[i] - 2\sum_{j=1}^{i-1} k[j]L - k[i]L, \tag{5.35}$$

where the variable with $[i]$ means that its value is for range bin i, j is a dummy variable of i, and L is the width of range bin.

A new variable Z_f is defined as follows:

$$\mathrm{dB}Z_f[i] \equiv \mathrm{dB}Z_m[i] + 2\sum_{j=1}^{i-1} k[j]L. \tag{5.36}$$

By substituting (5.35) into (5.36), the following equation is derived:

$$\mathrm{dB}Z_f[i] = \mathrm{dB}Z_e[i] - k[i]L. \tag{5.37}$$

This equation means that Z_f does not depend on the range bins except for range bin i, or Z_f is a function of DSD parameters of range bin i.

In similar way, variable Z_b is defined as follows:

$$\mathrm{dB}Z_b[i] \equiv \mathrm{dB}Z_m[i] + 2\sum_{j=1}^{i} k[j]L. \tag{5.38}$$

By substituting (5.35) into (5.38), the following equation is derived:

$$\mathrm{dB}Z_b[i] = \mathrm{dB}Z_e[i] + k[i]L. \tag{5.39}$$

Same as Z_f, Z_b is independent of the range bins except for range bin i, or it can be calculated by the DSD parameters of range bin i.

Retrieval methods to solve DSD parameters from Z_m's at multiple range bins are classified into forward retrieval method and backward retrieval method. The forward retrieval method starts at the nearest range bin (range bin 1), goes downward, and ends at the farthest range bin (range bin N).

- At range bin 1, as (5.36) becomes $\mathrm{dB}Z_f[1] = \mathrm{dB}Z_m[1]$, Z_f is obtained by measurements. As Z_f is a function of N_w and D_m, it may be possible that the DSD parameters are estimated by dual-frequency Z_f's. Otherwise, additional condition is given to make solution unique. In single-frequency algorithms, additional condition is necessary. Once N_w and D_m are determined, $k[1]$ can be calculated by (5.20).
- At range bin 2, as (5.36) becomes $\mathrm{dB}Z_f[2] = \mathrm{dB}Z_m[2] + 2k[1]L$, Z_f can be calculated. In the same way with range bin 1, DSD parameters at range bin 2 are estimated, and $k[2]$ is calculated.
- The same procedure is applied sequentially at the following range bins, until DSD parameters are obtained at all range bins.

  On the other hand, the backward retrieval method starts at the farthest range bin (range bin N), goes upward, and ends at the nearest range bin (range bin 1).

- At range bin N, (5.38) becomes $\mathrm{dB}Z_b[N] = \mathrm{dB}Z_m[N] + 2\sum_{j=1}^{N} k[j]L$. $2\sum_{j=1}^{N} k[j]L$ is equal to the PIA. If PIA is estimated by SRT, Z_b can be calculated. As Z_b is a function of N_w and D_m, it may be possible to retrieve DSD parameters from dual-frequency measurement. If necessary, additional condition is given. Once DSD parameters are obtained, $k[N]$ can be calculated.

- At range bin $N-1$, (5.38) becomes $\mathrm{dB}Z_b[N-1] = \mathrm{dB}Z_m[N-1] + \mathrm{PIA} - 2k[N]L$. As PIA and $k[N]$ are known, $Z_b[N-1]$ can be calculated. In the same way, DSD parameters at range bin $N-1$ are retrieved and $k[N-1]$ is calculated.
- The same procedure is applied at the other range bins, until DSD parameters are obtained at all range bins.

Note that the forward retrieval method and the backward retrieval method have no essential difference. They should give the same solution if the same conditions are given. Actually, when different conditions are given, the results are different. For example, in the forward retrieval method, to solve D_m from DFR, additional condition (i.e. $D_m > D_{m0}$) is used. In the backward retrieval method, PIA needs to be assumed.

5.3 History of the algorithm development

In this section, history of the algorithm development is introduced. PR algorithm is explained first, then KuPR algorithm, KaPR algorithm, and the dual-frequency algorithm are explained.

5.3.1 *PR algorithm*

PR algorithm applies a hybrid method of the Hitschfeld–Bordan method (HB method) and SRT. The HB method assumes the relation of k and Z_e in (5.40)

$$k(r) = \alpha(r) Z_e(r)^{\beta}, \tag{5.40}$$

where α depends on r and β is independent of r. By substituting (5.40), (5.29) becomes (5.41).

$$10\log_{10} Z_m(r) = 10\log_{10} Z_e(r) - 2\int_0^r \alpha(s) Z_e(s)^{\beta} ds. \tag{5.41}$$

Equation (5.41) can be rewritten by using natural logarithm to (5.42)

$$\ln Z_m(r) = \ln Z_e(r) - 0.2(\ln 10)\int_0^r \alpha(s) Z_e(s)^{\beta} ds \tag{5.42}$$

By differentiating both the sides of (5.42) in terms of r, (5.43) is obtained.

$$\frac{1}{Z_m(r)}\frac{d}{dr}Z_m(r) = \frac{1}{Z_e(r)}\frac{d}{dr}Z_e(r) - 0.2(\ln 10)\alpha(r) Z_e(r)^{\beta}. \tag{5.43}$$

$\beta \frac{Z_m(r)^{\beta}}{Z_e(r)^{\beta}}$ is multiplied to both the sides of (5.43); then, the following equations are derived:

$$\frac{\beta Z_m(r)^{\beta-1}}{Z_e(r)^{\beta}}\frac{d}{dr}Z_m(r) = \frac{\beta Z_m(r)^{\beta}}{Z_e(r)^{\beta+1}}\frac{d}{dr}Z_e(r) - 0.2(\ln 10)\alpha(r)\beta Z_m(r)^{\beta}. \tag{5.44}$$

$$\frac{1}{Z_e(r)^\beta}\frac{d}{dr}\left[Z_m(r)^\beta\right] = -Z_m(r)^\beta \frac{d}{dr}\left(\frac{1}{Z_e(r)^\beta}\right) - 0.2(\ln 10)\alpha(r)\beta Z_m(r)^\beta. \tag{5.45}$$

$$\frac{d}{dr}\left[\frac{Z_m(r)^\beta}{Z_e(r)^\beta}\right] = -0.2(\ln 10)\alpha(r)\beta Z_m(r)^\beta. \tag{5.46}$$

$$\frac{Z_m(r)^\beta}{Z_e(r)^\beta} = \int_0^r -0.2(\ln 10)\alpha(s)\beta Z_m(s)^\beta ds + C, \tag{5.47}$$

where C is a constant, and it is determined that $C = 1$ by substituting $r = 0$ into (5.47). Therefore, Z_e is given as follows:

$$Z_e(r) = Z_m(r)\left[1 - \int_0^r 0.2(\ln 10)\alpha(s)\beta Z_m(s)^\beta ds\right]^{-\frac{1}{\beta}}. \tag{5.48}$$

It can be rewritten to

$$Z_e(r) = Z_m(r)[1 - \zeta(r)]^{-\frac{1}{\beta}}, \tag{5.49}$$

where

$$\zeta(r) = \int_0^r 0.2(\ln 10)\alpha(s)\beta Z_m(s)^\beta ds. \tag{5.50}$$

PIA is calculated as follows:

$$\mathrm{PIA} = -10\log_{10}\frac{Z_m(r_s)}{Z_e(r_s)} = -\frac{10}{\beta}\log_{10}[1 - \zeta(r_s)], \tag{5.51}$$

where r_s is the value of r at the Earth surface. In this way, the HB method gives analytical solutions of Z_e and PIA. However, as Z_m cannot be measured continuously, discretization of (5.50) is required to calculate ζ.

PIA is estimated by SRT independently. The estimate is denoted by PIA(SRT). Generally, PIA(SRT) and PIA calculated by (5.51) are not equal. Possible reasons of the difference are inappropriate assumption of k–Z_e relation in (5.40) and the error of SRT. For the former, (5.40) is modified as follows:

$$k(r) = \varepsilon\alpha(r)Z_e(r)^\beta, \tag{5.52}$$

where ε is an adjustment factor independent of r. By the modified k–Z_e relation, (5.49) and (5.51) are modified, respectively, as follows:

$$Z_e(r) = Z_m(r)[1 - \varepsilon\zeta(r)]^{-\frac{1}{\beta}}, \tag{5.53}$$

$$\mathrm{PIA} = -10\log_{10}\frac{Z_m(r_s)}{Z_e(r_s)} = -\frac{10}{\beta}\log_{10}[1 - \varepsilon\zeta(r_s)]. \tag{5.54}$$

It is assumed that ε follows a log-normal distribution or $\log_{10}\varepsilon$ follows a normal distribution. The average of $\log_{10}\varepsilon$ is assumed to μ_x and the standard deviation of $\log_{10}\varepsilon$ is assumed to σ_x. The error of PIA(SRT) is assumed to follow a normal distribution and the average of the error is assumed to be zero and the standard deviation of the error is assumed to be σ_{SRT}. According to the maximum likelihood method, ε and PIA which satisfy (5.54) and minimize the value of (5.55) are selected.

$$\left\{\frac{\mathrm{PIA(SRT)} - \mathrm{PIA}}{\sigma_{SRT}}\right\}^2 + \left(\frac{\log_{10}\varepsilon - \mu_x}{\sigma_x}\right)^2. \tag{5.55}$$

Finally, Z_e's are calculated by (5.53) with the selected ε. R is estimated by R–Z_e relation, which can be modified by ε. In the above procedure, the DSD does not appear explicitly and DSD parameters are not estimated. However, PR algorithm assumes that DSD function is given as a gamma function with $\mu = 3$. Then, k–Z_e relation and R–Z_e relations are derived with the assumed DSD function [10].

5.3.2 KuPR and KaPR algorithms

KuPR algorithm applies the forward retrieval method instead of the HB method and outputs DSD parameters N_w and D_m. In KuPR algorithm V03, k–Z_e relation was used as in PR algorithm. By substituting (5.52) into (5,37), (5.56) is obtained.

$$dBZ_f[i] = dBZ_e[i] - \varepsilon\alpha[i]Z_e[i]^{\beta}L. \tag{5.56}$$

From (5.56), Z_e is determined; then k is calculated by (5.52). The ratio of k to Z_e is given by (5.19) and (5.20).

$$\frac{k}{Z_e} = \frac{f_k(D_m)}{f_Z(D_m)}. \tag{5.57}$$

The ratio of k to Z_e depends only on D_m, not on N_w. The righthand side of (5.57) is generally a monotonic function of D_m, and D_m can be determined from the ratio. Once D_m is determined, N_w is calculated by (5.19). Then, R is calculated by (5.18) from N_w and D_m. After DSD parameters are retrieved at all range bins, PIA is calculated. In the same way with PR algorithm, ε and PIA are selected to minimize the value of (5.55). For this process, different ε's are tested from $\varepsilon = 0.2$ to $\varepsilon = 5.0$.

In KaPR algorithm V03, the same method was applied to KaPR measurement, but k–Z_e relation should be different between KuPR and KaPR, as both k and Z_e are dependent on the frequency. In V03, the coefficient α for KaPR is set to be eight times as large as α for KuPR, while β is set to be the same for KuPR and KaPR. In V03, KaPR algorithm underestimates heavy precipitation. It is partly caused by k–Z_e relation used for KaPR is inappropriate. Seto *et al.* [11] showed that it is not possible to represent k–Z_e relation of KaPR in the form of (5.40).

While the HB method requires that the k–Z_e relation is given in the form of (5.40) to get the analytical solutions of Z_e and PIA, the forward retrieval method numerically solves the DSD parameters and does not require the form of (5.40). In

V04, instead of k–Z_e relation, R–D_m relation is used. As R and D_m are independent of the frequency, the same equation can be used for KuPR and KaPR. In the forward retrieval method of KuPR and KaPR algorithms, the following procedure is applied to retrieve DSD parameters from Z_f and R–D_m relation. The range of D_m is from 0.5 mm to 5.0 mm for KuPR algorithm and it is from 0.5 mm to 3.0 mm for KaPR algorithm to avoid overestimation of D_m.

- D_m is assumed and R is calculated by R–D_m relation.
- N_w is calculated by substituting R and D_m into (5.18).
- Z_e and k are calculated by substituting D_m and N_w into (5.19) and (5.20), respectively.
- Z_f is calculated by substituting Z_e and k into (5.37)
- Calculated Z_f and given Z_f are compared. To make the difference between the two values of Z_f smallest, D_m is selected in the following steps. First, D_m between 0.5 mm and 5.0 mm (3.0 mm for KaPR algorithm) is tested with the step of 0.1 mm, and the best D_m is selected (denoted by $D_{m,1}$). Second, D_m between $D_{m,1}-0.1$ mm and $D_{m,1}+0.1$ mm is tested with the step of 0.01 mm, and the best D_m is selected (denoted by $D_{m,2}$). Finally, D_m between $D_{m,2}-0.01$ mm and $D_{m,2}+0.01$ mm is tested with the step of 0.001 mm, and the best D_m is selected.

5.3.3 Dual-frequency algorithm

In the dual-frequency algorithm, it may be possible that N_w and D_m are retrieved from dual-frequency Z_f's. For example, the following procedure is considered.

- In (5.37), assume $k = 0$, or $Z_f = Z_e$.
- From the dual-frequency Z_e's, DFR is calculated.
- From (5.28), D_m is retrieved. Once D_m is determined, N_w and k are calculated.
- Using the calculated k, Z_e is recalculated in (5.37).
- Repeat the above procedure until the estimates are converged.

This method has some problems. There are sometimes multiple solutions in solving D_m from DFR. As noted earlier, in the case of Figure 5.4, if D_m is assumed to be larger than D_{m0}, D_m can be determined uniquely. However, with this assumption ($D_m > D_{m0}$), there can be multiple solutions in solving DSD parameters from dual-frequency Z_f's. KuPR's Z_f (denoted by $Z_{f,u}$) and KaPR's Z_f (denoted by $Z_{f,a}$) are written as follows:

$$dBZ_{f,u} = dBZ_{e,u} - k_u L = dBN_w + dBf_{Z,u}(D_m) - N_w f_{k,u}(D_m)L, \tag{5.58}$$

$$dBZ_{f,a} = dBZ_{e,a} - k_a L = dBN_w + dBf_{Z,a}(D_m) - N_w f_{k,a}(D_m)L, \tag{5.59}$$

where subscripts u and a mean that the variables are for KuPR and KaPR, respectively. By subtracting (5.58) from (5.59), (5.60) is obtained.

$$dBZ_{f,\delta} = dBZ_{e,\delta} - k_\delta L = dBf_{Z,\delta}(D_m) - N_w f_{k,\delta}(D_m)L, \tag{5.60}$$

where subscript δ means that the value is for the difference between KuPR and KaPR. Equation (5.60) shows the relation between $dBZ_{f,\delta}$ and DSD parameters. In

Figure 5.6, the horizontal axis is D_m and the vertical axis is $dBZ_{f,\delta}$. The solid black line is DFR–D_m curve same as drawn in Figure 5.4. The same condition with Figure 5.4 (liquid precipitation with air temperature of 0 °C) is assumed for Figure 5.6. For given D_m and $dBZ_{f,\delta}$, N_w can be calculated by (5.60), then $dBZ_{f,u}$ and $dBZ_{f,a}$ are calculated by (5.58) and (5.59). Figure 5.6(a) shows the contour of $dBZ_{f,u}$ and Figure 5.6(b) shows the contour of $dBZ_{f,a}$. As N_w is positive, $dBZ_{f,\delta}$ is smaller than $dBf_{z,\delta}(D_m)$ or DFR; therefore, the contours always exist below the DFR–D_m curve. Figure 5.6(c) shows some examples; if $dBZ_{f,u}$ is 43 dBZ and $dBZ_{f,a}$ is 40 dBZ, the contours cross with each other at $D_m = 0.9$ mm and $D_m = 1.8$ mm. One solution has D_m smaller than D_{m0} (= 1.0 mm) and another solution has D_m larger than D_{m0}. However, if $dBZ_{f,u}$ is 49 dBZ and $dBZ_{f,a}$ is 40 dBZ, the contours cross with each other at $D_m = 1.1$ mm and $D_m = 2.8$ mm, both of which are larger than D_{m0}. The problem in solving N_w and D_m from dual-frequency Z_f's has multiple solutions and the condition that D_m is larger than D_{m0} does not necessarily make solution unique.

In V04 and later, R–D_m relation is used in the dual-frequency algorithm. If Z_m is available both for KuPR and KaPR, KuPR's Z_m is used because KuPR's Z_m has smaller effects of attenuation and Mie scattering and is easier to be used than KaPR's Z_m. In the same way with KuPR algorithm, DSD parameters are derived in the forward retrieval method. It means that the dual-frequency algorithm works the same with KuPR algorithm as long as KuPR is available. However, for the selection of the best ε, KaPR's measurements are used in the dual-frequency algorithm. At a range bin where both KuPR's Z_m and KaPR's Z_m are available, KaPR's Z_f is calculated in two ways. First, by (5.36), Z_f for KaPR is calculated. In the forward retrieval method, k for KaPR can be calculated by the retrieved DSD parameters at upper range bins and Z_f can be calculated from them. This Z_f is denoted as Z_{f1}. Second, by (5.37), Z_f for KaPR is calculated from the retrieved DSD parameters at the current range bin. This Z_f is denoted as Z_{f2}. Ideally, Z_{f1} and Z_{f2} are equal, but usually they are different. The difference between Z_{f1} and Z_{f2} is considered to evaluate ε, which is common to KuPR and KaPR. The details of this method are written in Section 5.4.3 and in [11]. In this way, KaPR's Z_m is not directly used for the retrieval of DSD parameters, but it affects the final results through the selection of ε.

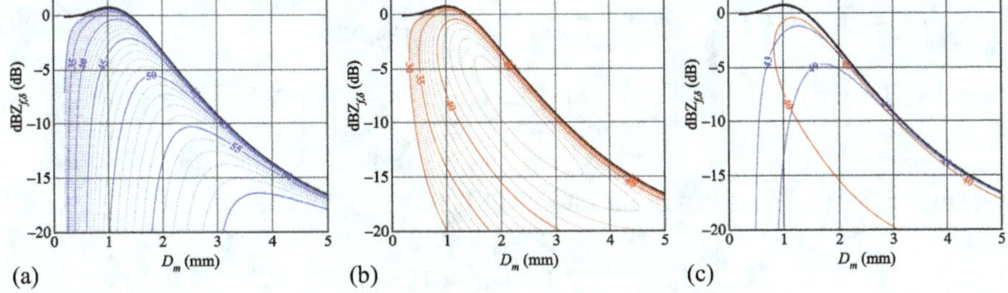

Figure 5.6 On DFR$_f$–D_m plain, (a) the contour of $dBZ_{f,u}$ and (b) the contour of $dBZ_{f,a}$ are shown. (c) Examples shown.

5.4 Improvement in Version 07

5.4.1 R–D_m relation

As noted earlier, the DPR algorithms assume the R–D_m relation to constrain DSD parameters. As R and D_m are not dependent on the frequency, the same equation can be used for KuPR and KaPR. For the continuity between PR algorithm and DPR algorithms, the R–D_m relation is designed to work as constraints similar to the k–Z_e relation. The k–Z_e relation was originally derived from Z–R relation in [9], which is

$$Z = aR^b, \tag{5.61}$$

where a and b are constants; $a = 298.84$ and $b = 1.38$ for the stratiform precipitation. $a = 184.20$ and $b = 1.43$ for the convective precipitation. Z is the radar reflectivity factor calculated by the Rayleigh approximation in (5.62).

$$Z = \int_0^\infty D^6 N(D) dD. \tag{5.62}$$

By substituting (5.6) into (5.62), (5.63) is derived

$$Z = N_0 \int_0^\infty D^{\mu+6} \exp(-\Lambda D) dD = N_0 \frac{\Gamma(\mu+7)}{\Lambda^{\mu+7}}. \tag{5.63}$$

Falling speed $V(D)$ is assumed by (5.64).

$$V = mD^n, \tag{5.64}$$

where m and n are constants. By substituting (5.6) and (5.64) into (5.3), (5.65) is derived.

$$R = 0.6\pi \times 10^{-3} m N_0 \frac{\Gamma(\mu+n+4)}{\Lambda^{\mu+n+4}}. \tag{5.65}$$

By substituting (5.63) and (5.65) into (5.61), the following equations are derived:

$$N_0 = a^{\frac{1}{1-b}} \left[0.6\pi \times 10^{-3} m \frac{\Gamma(\mu+n+4)}{\Lambda^{\mu+n+4}} \right]^{\frac{b}{1-b}} \left[\frac{\Lambda^{\mu+7}}{\Gamma(\mu+7)} \right]^{\frac{1}{1-b}} \tag{5.66}$$

$$R = \left[a \times 0.6\pi \times 10^{-3} \times m \times \frac{\Gamma(\mu+n+4)}{\Gamma(\mu+7)} \times \Lambda^{3-n} \right]^{\frac{1}{1-b}} \tag{5.67}$$

By using (5.7), (5.67) is rewritten into (5.68)

$$R = \left[a \times 0.6\pi \times 10^{-3} \times m \times \frac{\Gamma(\mu+n+4)}{\Gamma(\mu+7)} \times \left(\frac{D_m}{\mu+4} \right)^{n-3} \right]^{\frac{1}{1-b}} \tag{5.68}$$

It can be summarized as follows:

$$R = pD_m^{\,q}, \tag{5.69}$$

where

$$p = \left(\frac{a \times 0.6\pi \times 10^{-3} \times m \times \Gamma(\mu + n + 4)}{(\mu + 4)^{n-3} \times \Gamma(\mu + 6)} \right)^{\frac{1}{1-b}} \tag{5.70}$$

$$q = \frac{n-3}{1-b} \tag{5.71}$$

Kozu *et al.* [10] assumed that $m = 3.78$ and $n = 0.67$ [12]. And, $\mu = 3$ is assumed in PR and DPR algorithms. These values as well as a and b in (5.61) are substituted into (5.70) and (5.71) to obtain values for p and q. For stratiform precipitation, $p = 0.392$ and $q = 6.131$. For convective precipitation, $p = 1.348$ and $q = 5.418$. These values are used in the DPR algorithms until V06.

R–D_m relation is also modified by ε as k–Z_e relation is. With (5.18)–(5.20), (5.52) is rewritten as follows:

$$R = \left\{ \frac{\varepsilon\alpha[f_Z(D_m)]^{\beta}}{f_k(D_m)} \right\}^{\frac{1}{1-\beta}} f_R(D_m) \tag{5.72}$$

When $\varepsilon = 1$, (5.72) shows an R–D_m relation and it is well agreed with (5.69) [11]. Therefore, (5.72) is written as follows:

$$R = \varepsilon^{r} pD_m^{\,q}, \tag{5.73}$$

where

$$r = \frac{1}{1-\beta} \tag{5.74}$$

As β is 0.7923, r is 4.815 for stratiform precipitation. And as β is 0.7713, r is 4.373 for convective precipitation.

According to [10], different equations are used for different precipitation types in k–Z_e relation in PR algorithm and R–D_m relation in DPR algorithms until V06. Hence, the adjustment factor ε has different definitions for different precipitation types. In some studies, ε is statistically analyzed and it is tried to find the relation between ε and climatological characteristics. However, ε cannot be compared between stratiform and convective precipitation. In V07, the same values of p, q, and r are used for stratiform precipitation and convective precipitation so that ε can be compared between different precipitation types. From V06 to V07, R–D_m relation with $\varepsilon = 1$ changes and it may significantly change precipitation rate estimates for convective precipitation. $\varepsilon = 1$ is used as the default value of ε, as the average of $\log_{10}\varepsilon$ (μ_x) is set to be 0 in the dual-frequency algorithm until V06. In V07, μ_x is set to be $\log_{10}(1.25)$ for convective precipitation, which means $\varepsilon = 1.25$ is used as a new default value. As shown in Figure 5.7,

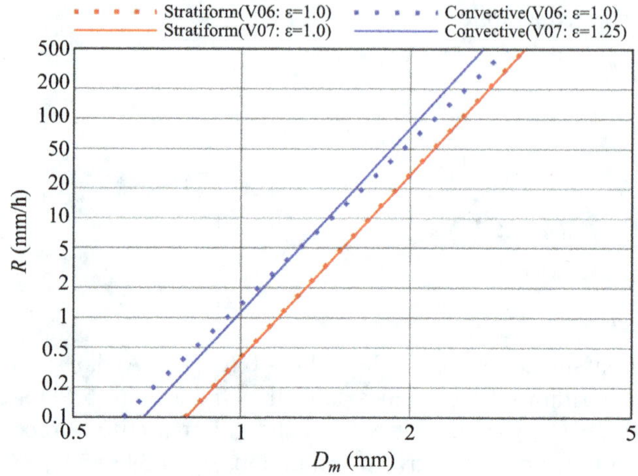

Figure 5.7 R–D_m relation. Solid (dotted) line is for V07 (V06) and red (blue) line is for stratiform (convective) precipitation. ε = 1.25 is assumed for convective V07, while ε = 1.0 is assumed for others.

R–D_m relation with $\varepsilon = 1$ in V06 and that with $\varepsilon = 1.25$ in V07 are nearly the same. For stratiform precipitation, μ_x is kept to be zero in V07. For single-frequency algorithms, μ_x and the standard deviation of $\log_{10}\varepsilon$ (σ_x) are statistically calculated from the retrieved ε in the dual-frequency algorithm for each month and each 5-by-5-degree grid. The values of μ_x and σ_x for stratiform and convective precipitation in July 2018 are shown in Figure 5.8. For stratiform precipitation, μ_x is generally negative. For convective precipitation, μ_x is negative over land while it is positive over ocean. The values are calculated with the dual-frequency products in July 2018, and they are used for July every year.

5.4.2 Correction of SRT considering the soil moisture effect

5.4.2.1 Principles of SRT

SRT estimates PIA by using surface backscattering cross-sections. At a pixel with precipitation (called R-pixel), measured surface backscattering cross-section is denoted by σ^0_m(R), and attenuation corrected surface backscattering cross-section is denoted by σ^0_e(R). Equation (5.75) shows the relation among these values.

$$\sigma^0_e(\mathrm{R}) = \sigma^0_m(\mathrm{R}) + \mathrm{PIA}, \tag{5.75}$$

At a pixel without precipitation (called NR-pixel), measured surface backscattering cross-section is denoted by σ^0_m(NR). As no attenuation is assumed at an NR-pixel, σ^0_e(NR) should be equal to σ^0_m(NR) as shown in (5.76).

$$\sigma^0_e(\mathrm{NR}) = \sigma^0_m(\mathrm{NR}) \tag{5.76}$$

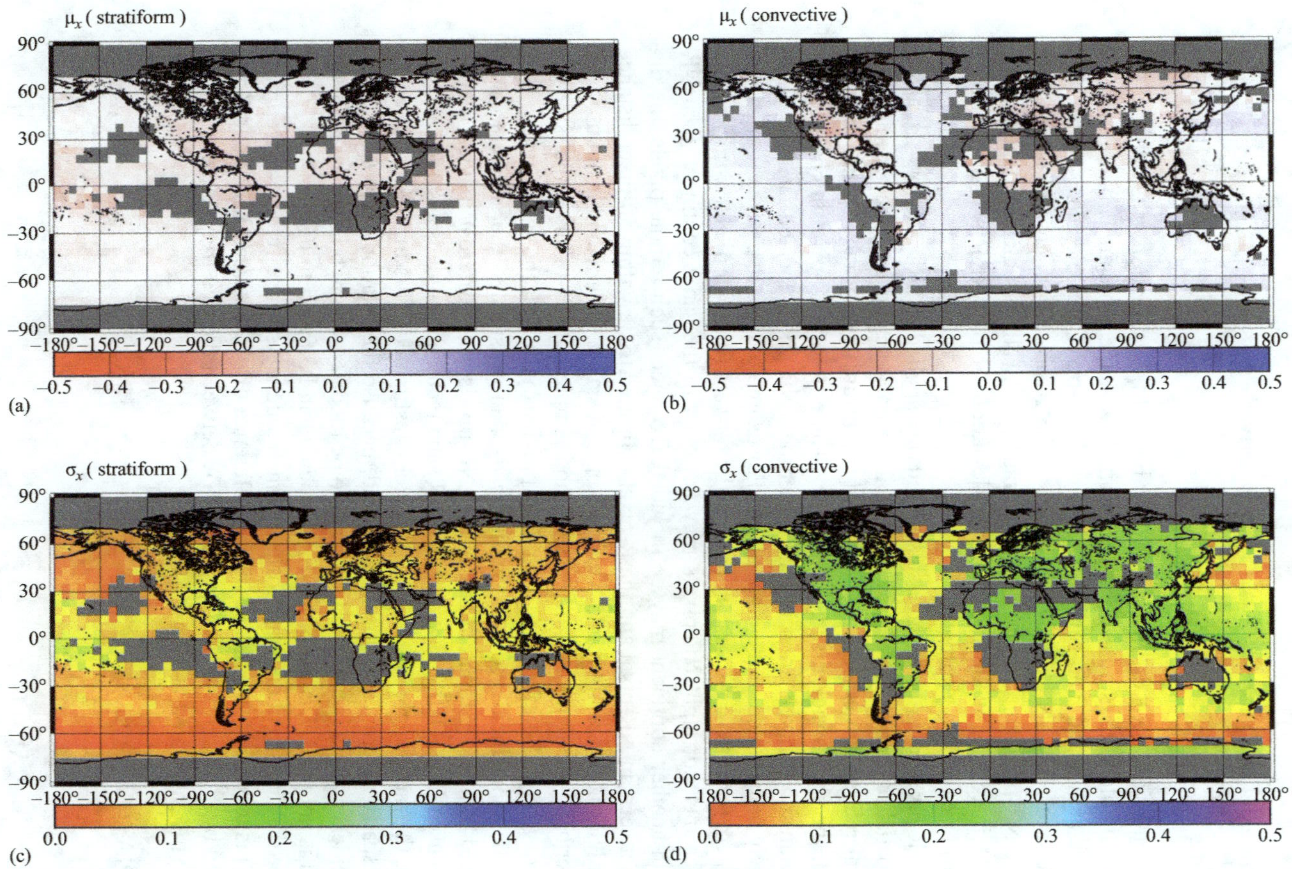

Figure 5.8 *The values of (a, b) μ_x and (c, d) σ_x for (a, c) stratiform and (b, d) convective precipitation calculated with the dual-frequency products in July 2018*

In SRT, to estimate the PIA at an R-pixel, some NR-pixels are referenced by assuming

$$\sigma_e^0(\mathrm{NR}) = \sigma_e^0(\mathrm{R}) \tag{5.77}$$

By combining (5.75) and (5.77), (5.78) is derived

$$\mathrm{PIA} = \sigma_m^0(\mathrm{NR}) - \sigma_m^0(\mathrm{R}) \tag{5.78}$$

By this equation, SRT estimates PIA. For better PIA estimation, it is required that (5.77) will hold with good accuracy. NR-pixels with the same angle bin number (or incidence angle) and land surface type with the R-pixel are selected. Additionally, it is desirable that NR-pixels are close to the R-pixel both in space and time. Multiple NR-pixels are taken, and the average of $\sigma^0{}_m$(NR) is calculated and used as $\sigma^0{}_m$(NR) in (5.78), while the standard deviation of $\sigma^0{}_m$(NR) is used to be the standard error of PIA estimates.

For R-pixels over land, three submethods are applied in SRT: the forward along-track spatial reference method, backward along-track spatial reference method, and temporal reference method. In the forward (backward) along-track spatial reference method, the along-track distance from the R-pixel to an NR-pixel is checked, and eight nearest NR-pixels are selected. Sometimes, NR-pixels located far from the R-pixel may be included in the eight pixels. For example, if an R-pixel is located near the western (eastern) coast of continent, in the forward (backward) along-track spatial reference method, eight NR-pixels may not exist in the same continent, then NR-pixels in another continent may be included. To reject such remote pixels, the distance between an NR-pixel and the R-pixel is limited to be within 50-pixel distance. If eight NR pixels are not found within 50-pixel distance, the method is not used. On the other hand, the temporal reference method samples all NR-pixels in the same 0.5-by-0.5-degree grid, in the same month (of previous years), at the same surface type, and at the same incidence angle with the R-pixel. Results of the submethods are combined into one estimate. Over ocean, NR-pixels with different incidence angle are also used. The relation between $\sigma^0{}_m$(NR) and incidence angle is assumed and $\sigma^0{}_m$ (NR) at the same angle bin with the R-pixel is estimated. This method is called the hybrid spatial reference method. Over land, because the relation between $\sigma^0{}_m$(NR) and incidence angle is not simple enough to be modeled, the hybrid spatial reference method is not used. Details of the SRT module can be found in [7].

For the dual-frequency algorithm, dual-frequency surface reference technique (DSRT) is used. $\sigma^0{}_m$ is replaced by the difference of $\sigma^0{}_m$ between KuPR and KaPR (denoted by $\sigma^0{}_{m,\delta}$) and the difference of PIA between the two frequencies (denoted by PIA_δ) is calculated as shown in (5.79).

$$\mathrm{PIA}_\delta = \sigma_{m,\delta}^0(\mathrm{NR}) - \sigma_{m,\delta}^0(\mathrm{R}) \tag{5.79}$$

where

$$\mathrm{PIA}_\delta = \mathrm{PIA}_a - \mathrm{PIA}_u \tag{5.80}$$

$$\sigma_{m,\delta}^0 = \sigma_{m,a}^0 - \sigma_{m,u}^0 \tag{5.81}$$

As the variations in $\sigma^0{}_m$ between the R-pixel and NR-pixels are partly cancelled out by taking the difference of $\sigma^0{}_m$, PIA_δ is expected to have better accuracy than single-frequency PIA.

5.4.2.2 Soil moisture effect

Seto and Iguchi [13] analyzed surface backscattering cross-sections in PR V6. NR-pixels located within eight pixels from an R-pixel along the track were taken and they are denoted as NR* pixels. $\sigma^0{}_m$ at NR* pixels (denoted by $\sigma^0{}_m$(NR*)) were compared with the average of $\sigma^0{}_m$(NR) in the same grid, in the same month, with the same angle bin, and at the same surface type; then it was found that $\sigma^0{}_m$(NR*) is likely to be larger than the average of $\sigma^0{}_m$(NR) in the land area except for dense forest such as Amazonia and desert such as Sahara. NR* pixels are classified into ones located west to precipitation area (measured just before an R-pixel) and ones located east to precipitation area (measured just after an R-pixel). In midlatitude, the former shows higher $\sigma^0{}_m$ (NR*) than the latter, but the reverse is true in the Sahel of Africa. These results are explained by temporal increase in surface soil moisture by precipitation (soil moisture effect). Surface backscattering cross-section at soil surface increases with the increase in surface soil moisture. In Amazonia, soil surface is densely covered by forest and the soil moisture effect does not appear clearly. At an NR* pixel, it is more likely that precipitation existed just before the measurement than at other NR-pixels; then $\sigma^0{}_m$ (NR*) becomes higher than the average of $\sigma^0{}_m$(NR). In midlatitude, as precipitation system generally moves from west to east, NR* pixels located west to precipitation are expected to show soil moisture effect. On the other hand, as precipitation system generally moves from east to west in the Sahel region, the reverse is true.

When the soil moisture effect exists, (5.77) is not hold and $\Delta\sigma^0{}_e$ defined in (5.82) becomes positive.

$$\Delta\sigma_e^0 = \sigma_e^0(\mathrm{R}) - \sigma_e^0(\mathrm{NR}) \tag{5.82}$$

With (5.75), (5.76), and (5.82), (5.83) is derived.

$$\mathrm{PIA} - \Delta\sigma_e^0 = \sigma_m^0(\mathrm{NR}) - \sigma_m^0(\mathrm{R}) \tag{5.83}$$

This means that the SRT underestimates PIA by $\Delta\sigma^0{}_e$. In the along-track spatial reference method, NR* pixels are generally sampled and they may have some soil moisture effect. $\Delta\sigma^0{}_e$ calculated in (5.84) may be smaller than that in (5.82).

$$\Delta\sigma_e^0 = \sigma_e^0(\mathrm{R}) - \sigma_e^0(\mathrm{NR}*) \tag{5.84}$$

In the temporal reference method, sampled NR-pixels are not expected to have soil moisture effect. $\Delta\sigma^0{}_e$ is considered to be larger for the temporal reference method than for the along-track spatial reference method.

In PR V6, the backward along-track spatial reference method was not used. The forward along-track spatial reference method and temporal reference method are combined. This resulted in underestimation of PIA in the Sahel of Africa. In PR V7, the backward along-track spatial reference method is included and an offset of 0.5 dB is added to PIA estimates before the precipitation rate retrieval at all R-pixels over land.

5.4.2.3 Correction of soil moisture effect in DPR algorithms

In DPR algorithms, soil moisture effect was not considered until V06. Seto *et al.* [14] analyzed the surface backscattering cross-sections in KuPR and KaPR products in V06 and got similar results about soil moisture effect with [13]. KuPR shows the soil moisture effect as well as PR does. For KaPR, uncertainty of the attenuation correction for non-precipitation particles makes the quantitative analysis of soil moisture effect difficult, but it is concluded that the soil moisture effect exists. For R-pixels, it is suggested that $\sigma^0{}_e$(R) increases with the surface precipitation rate. With heavier precipitation, surface soil moisture and $\sigma^0{}_e$(R) become higher.

Based on the above analysis, in KuPR and KaPR algorithms V07, PIA estimates by SRT are corrected in Solver module of the second iteration. The offset of PIA estimates depends on the area, angle bin, and surface precipitation rate estimated in the first loop of Solver module. The offset becomes around 5 dB in some cases. In KuPR (KaPR) algorithm, surface precipitation rate estimates increase by approximately 20% (15%) over land from V06 to V07 (Figure 5.9). It is mainly caused by the correction of PIA.

$\sigma^0{}_e(R)$ is calculated by substituting the final estimates of PIA into (5.75) and it is denoted by $\sigma^0{}_e$(SLV), where SLV stands for Solver. The anomaly of $\sigma^0{}_e$(SLV) from the average of $\sigma^0{}_m$(NR) in the same month, in the same grid, with the same surface type, at the same angle bin is calculated and is denoted by $\Delta\sigma^0{}_e$(SLV). The relation between $\Delta\sigma^0{}_e$(SLV) and surface precipitation rate estimates (by KuPR algorithm) is shown in Figure 5.10. In V06, KuPR's $\Delta\sigma^0{}_e$(SLV) is positive, which

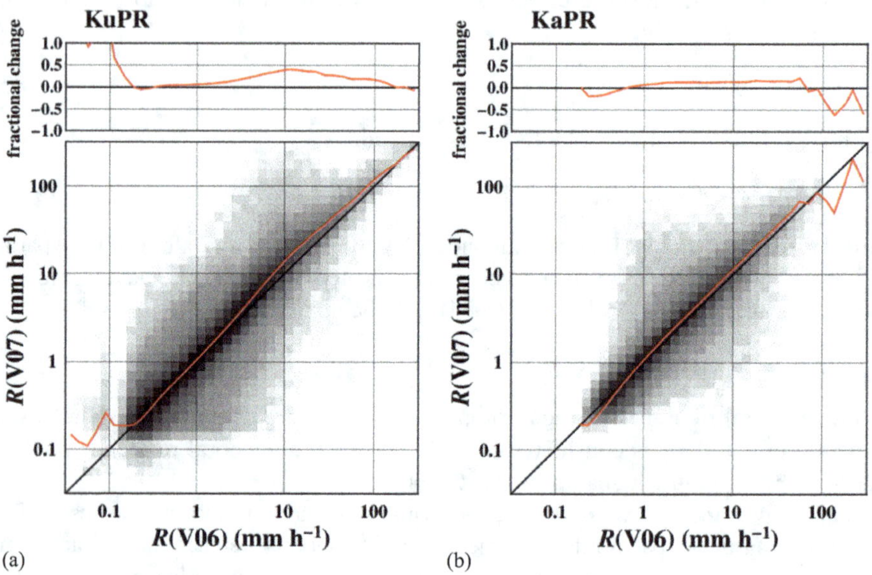

Figure 5.9 Scatterplots of surface precipitation rate estimates between V06 and V07 for (a) KuPR and (b) KaPR algorithms

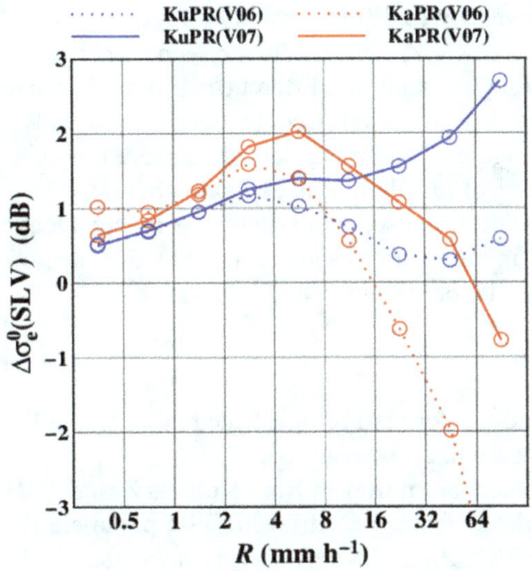

Figure 5.10 $\Delta\sigma^0{}_e$*(SLV) for the category of precipitation rate estimates. Blue (red) curves are for KuPR (KaPR), and dotted (solid) curves are for V06 (V07).*

suggests that soil moisture effect is represented. However, there is no significant correlation between the surface precipitation rate estimates and $\Delta\sigma^0{}_e$(SLV). In V07, KuPR's $\Delta\sigma^0{}_e$(SLV) increases with the increase in the surface precipitation rate estimates. KaPR's $\Delta\sigma^0{}_e$(SLV) is negative when the surface precipitation rate estimate is higher than 16 mm/h in V06. Soil moisture effect is not represented under heavy precipitation. KaPR's $\Delta\sigma^0{}_e$(SLV) increases from V06 to V07, but no significant correlation between the surface precipitation rate estimate and KaPR's $\Delta\sigma^0{}_e$(SLV) is seen in V07 as well as in V06. These results mean that soil moisture effect is not well represented in KaPR algorithm and heavy precipitation is underestimated.

In the dual-frequency algorithm, soil moisture effect is not considered and no correction is applied for PIA estimates by SRT. If $\Delta\sigma^0{}_e$(R) is the same for KuPR and KaPR, they are cancelled out when PIA_δ is calculated by (5.79). In [14], KaPR's soil moisture effect is shown to be smaller than KuPR's soil moisture effect, but this result is quantitatively uncertain because the attenuation correction for non-precipitation particles is not very accurate. If the soil moisture effect is different between KuPR and KaPR, the correction for PIA_δ is necessary in the dual-frequency algorithm. More accurate analysis of soil moisture effect for KaPR is required.

5.4.3 Estimation of the vertical profile of ε

As explained in Section 5.3.3, the dual-frequency algorithm V06 mainly uses KuPR's Z_m, while KaPR's Z_m is used to evaluate ε. KaPR's Z_{f1} derived by (5.36) and KaPR's Z_{f2} derived by (5.37) are usually not equal. ε is selected to make the difference between KaPR's Z_{f1} and KaPR's Z_{f2} smaller, but as ε is vertically constant, it is not possible to make the difference zero at all range bins. In the dual-frequency algorithm V07, ε is allowed to change vertically. The difference is evaluated at each range bin, and ε is selected to make the difference smaller. The estimation of the vertical profile of ε is explained below.

The value of ε is decomposed into ε_S which is independent of range bin and ε_D which is dependent on range bin as shown in (5.85). ε_S is set to be between 0.4 and 2.5, while ε_D is set to be between 0.5 and 2.0.

$$\varepsilon = \varepsilon_S \times \varepsilon_D[i] \tag{5.85}$$

1) ε_S is assumed, and the forward retrieval method is applied.
   1.1) At each range bin, ε_D is assumed.
   1.2) DSD parameters are retrieved from KuPR's Z_f and R–D_m relation. KaPR's Z_{f2} is calculated from the retrieved DSD parameters, and the difference between KaPR's Z_{f1} and KaPR's Z_{f2} is calculated.
   1.3) ε_D is selected to make the value of (5.86) smallest.

$$\left(\frac{\log_{10}\varepsilon_D}{0.1}\right)^2 + \left[w\left(\mathrm{dB}Z_{f2} - \mathrm{dB}Z_{f1}\right)\right]^2, \tag{5.86}$$

   where w is a weighting factor. ε_D is assumed to follow a log-normal distribution and the difference between Z_{f1} and Z_{f2} is assumed to follow a normal distribution. If w is set to be large enough, the solution with $Z_{f1} = Z_{f2}$ may be selected as in the DFR method. However, as the DFR method is generally unstable, w is set to be moderately small. In V07, w is set to be 0.5.

2) After DSD parameters are retrieved at all range bins, KuPR's PIA and KaPR's PIA are calculated from DSD. ε_S is selected to minimize the value of the following equation in the single-frequency algorithms:

$$\left(\frac{\log_{10}\varepsilon_S}{0.1}\right)^2 + \left[\frac{\mathrm{PIA(SRT)} - \mathrm{PIA}}{\sigma_{\mathrm{SRT}}}\right]^2. \tag{5.87}$$

In the dual-frequency algorithm, if DSRT is available, (5.87) is replaced by (5.88)

$$\left(\frac{\log_{10}\varepsilon_S}{0.1}\right)^2 + \left[\frac{\mathrm{PIA}_\delta\mathrm{(DSRT)} - \mathrm{PIA}_\delta}{\sigma_{\mathrm{DSRT}}}\right]^2. \tag{5.88}$$

Figures 5.11 and 5.12 show retrieval results in V06 and V07, respectively, as a case study (orbit number 12826, scan number 1071, angle bin number 19). In V06, the vertical profile of ε is constant (Figure 5.11(a)). The difference between Z_{f1} and Z_{f2} is large particularly at higher altitude (Figure 5.12(a)). In V07, ε changes

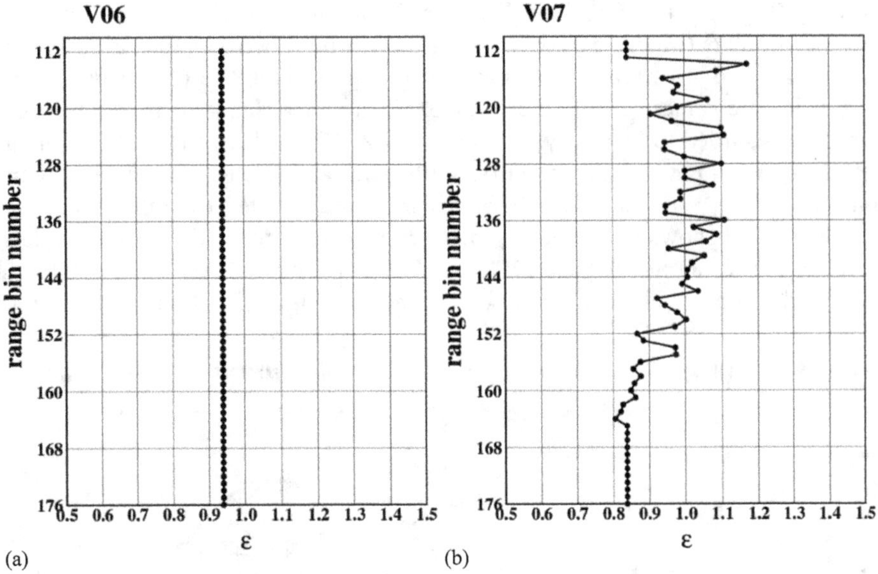

Figure 5.11 The vertical profile of ε in (a) V06 and (b) V07

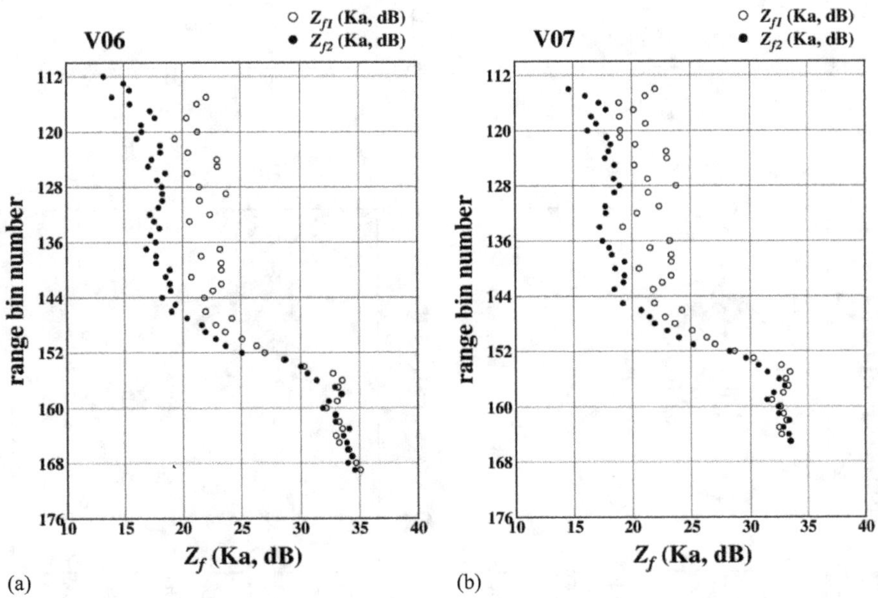

Figure 5.12 The vertical profiles of KaPR's Z_{f1} and Z_{f2} in (a) V06 and (b) V07

vertically (Figure 5.11(b)) and the difference between Z_{f1} and Z_{f2} is slightly smaller than that in V06 (Figure 5.12(b)). If w is set to be much larger, the difference between Z_{f1} and Z_{f2} becomes smaller, but the vertical profile of ε may become unrealistic. The values of Z_{f1} and Z_{f2} are not output in the standard product V07, but Z_{f2} can be regarded to be nearly the same with Z_e. The value of KaPR's Z_{f1} minus KuPR's Z_{f1} is output as DFRforward1. KuPR's Z_{f1} and KuPR's Z_{f2} are usually the same. Therefore, KaPR's Z_{f1} can be approximately estimated from the outputs.

Figure 5.13 shows the scatterplot between DFRforward1 and D_m (by blue dots) as well as the scatterplot between DFR and D_m (by red dots). Pixels without a bright band in orbit number 12826 are used for this analysis. Figure 5.13(a) is for range bins

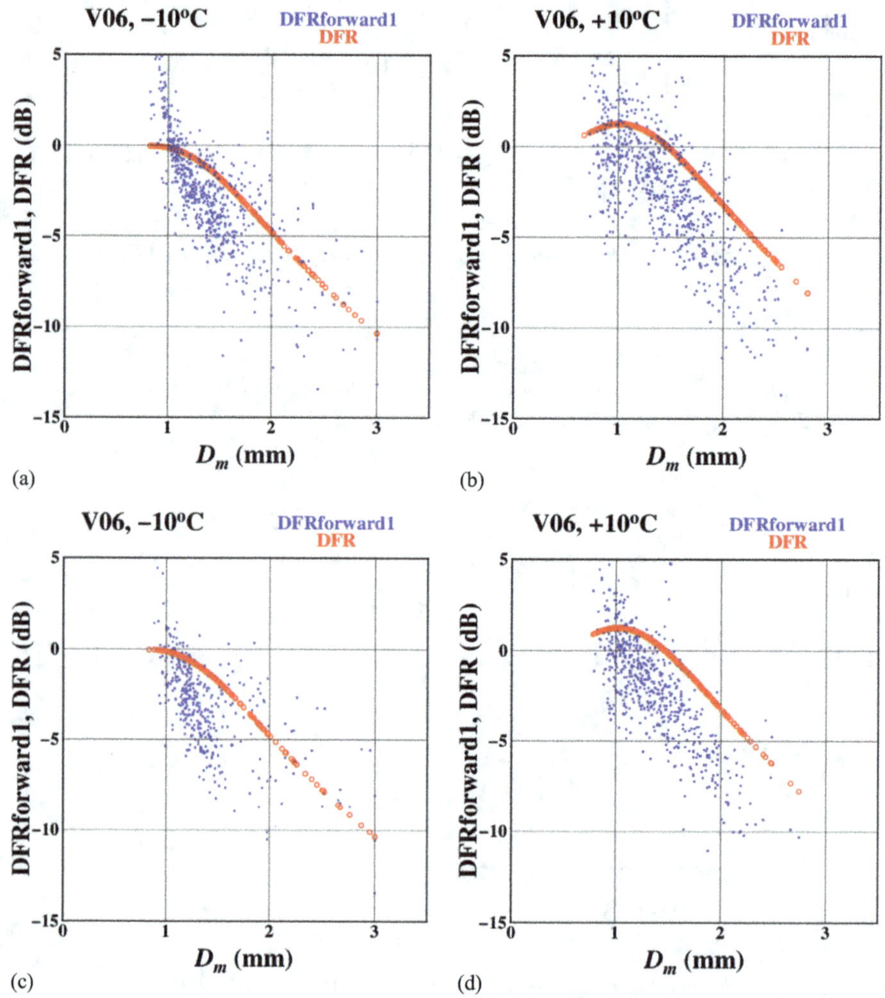

Figure 5.13 Scatterplot between DFRforward1 vs D_m and DFR and D_m in (a, b) V06 and (c ,d) V07 at (a, c) −10 °C and (b, d) at 10 °C

with the air temperature of −10 °C and Figure 5.13(b) is for range bins with the air temperature of 10 °C. In both figures, the relation between DFR and D_m agrees with theoretical relationship, but it is not surprising because KaPR's Z_e is calculated from the retrieved D_m and N_w. If $Z_{f1} = Z_{f2}$, the relation between DFRforward1 and D_m is closer to the DFR–D_m curve, but actually they are different, which means that Z_{f1} is different from Z_{f2} at many pixels. Figure 5.13(c) and 5.13(d) are the same with Figure 5.13(a) and 5.13(b), respectively, but for V06. Though DFRforward1 was not output in the standard product of V06, it is calculated for this study. Compared with the results of V06, it is suggested that DFRforward1–D_m relation is closer to DFR–D_m curve or the difference between Z_{f1} and Z_{f2} becomes smaller in V07.

5.4.4 *Correction of scattering tables for PR*

As noted in Section 5.1.2.5, KuPR algorithm V06 is applied for PR measurements and PR product V8 is produced. Seto [15] statistically compared KuPR V06 and PR V8 for the period from April to September 2014 and in the region between 35 degrees south and north. The frequency of the surface precipitation rate estimates over 10 mm/h is higher in PR V8 than in KuPR V06. To investigate the reasons of the difference, related variables such as Z_m and PIA estimates by SRT of KuPR V06 and PR V8 are analyzed. Z_m of PR V8 is found to be slightly smaller than that of KuPR V06 in the case of heavy precipitation, while the difference of Z_m is not clearly seen for light precipitation. It is considered to be caused by the difference in attenuation. PR's frequency is 13.8 GHz, which is slightly higher than KuPR's frequency of 13.6 GHz. The difference of 0.2 GHz does not affect the value of Z_e significantly, but changes k by approximately 3%. PIA estimates by SRT in PR V8 are statistically larger than those in KuPR V06. It can be also explained by the difference in frequency. In PR V8, smaller Z_m and larger PIA make ε larger. Statistical analysis shows ε is larger for PR V8 than KuPR V06. Z_e is not significantly different in PR V8 and KuPR V06 as the differences in Z_m and PIA are cancelled. Finally, larger ε results in larger precipitation rate estimates in PR V8.

As PR V8 is produced with the same algorithm of KuPR V06, the difference in frequency is not considered. In PR V9, KuPR algorithm V07 is applied, but the scattering table is modified for PR's frequency. $Z_e(D_m)/N_w$ and $k(D_m)/N_w$ are calculated for 13.8 GHz. Figure 5.14 shows the comparison of the results of KuPR V07, PR V9 (with scattering table for 13.6 GHz), and PR V9 (with scattering table for 13.8 GHz). Figure 5.14(a) shows the average value of ε, where the horizontal axis is the exceedance probability of the surface precipitation rate. If the exceedance probability is smaller (larger), the surface precipitation rate is higher (lower). ε between KuPR V07 and PR V9 (with the scattering table for 13.6 GHz) is different. With the scattering table for 13.8 GHz, PR V9's histogram moves closer to KuPR V07's histogram. Figure 5.14(b) shows the surface precipitation rate estimates. The difference among three products is not large, but PR V9 (with the scattering table for 13.8 GHz) shows the highest estimates and KuPR V07 shows the lowest estimates. PR V9 (with the scattering table for 13.8GHz)'s result is located between the two. By modifying the scattering table for PR, PR's estimates become closer to KuPR's estimates.

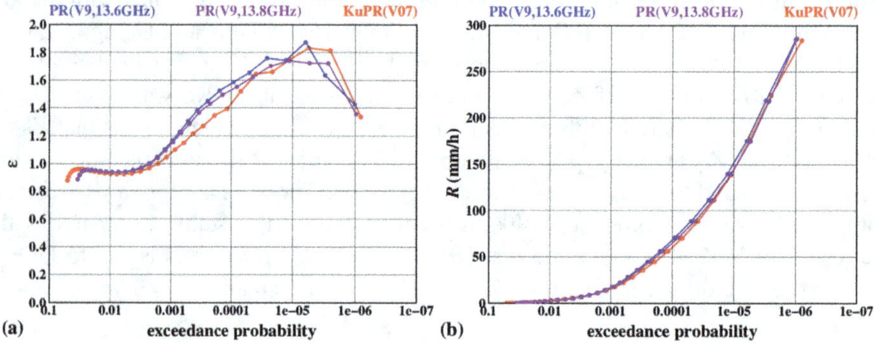

Figure 5.14 *Statistics of (a) ε and (b) surface precipitation rate. Horizontal axis is the exceedance probability of surface precipitation rate.*

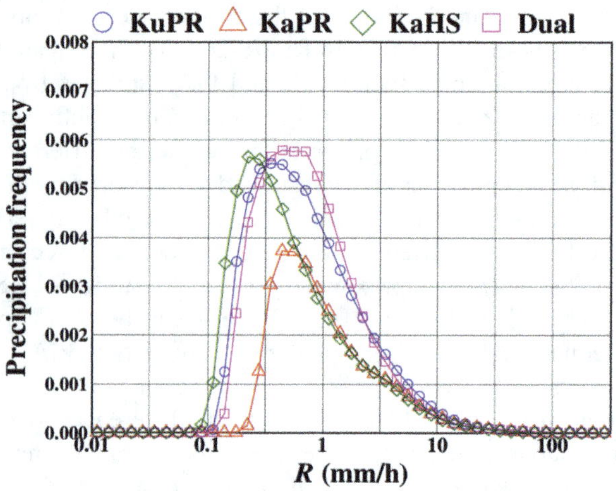

Figure 5.15 *Histogram of surface precipitation rate estimates*

5.4.5 *Effects of the change in the precipitation detection method*

In Preparation module, each pixel is judged to have precipitation or not. This method is modified in V07 and the ratio of precipitation pixel increases. Figure 5.15 shows the histogram of surface precipitation rate in V07. Lines are for KuPR product, KaPR product (normal scan), KaPR product (high-sensitivity mode; denoted by KaHS), and the dual-frequency product. All data measured in June 2016 are used. One may compare this figure (for V07) with Fig. 11 of [11] for V06. Both in V06 and V07, KuPR shows higher sensitivity than KaPR (normal scan), but KaHS has slightly higher sensitivity than KuPR.

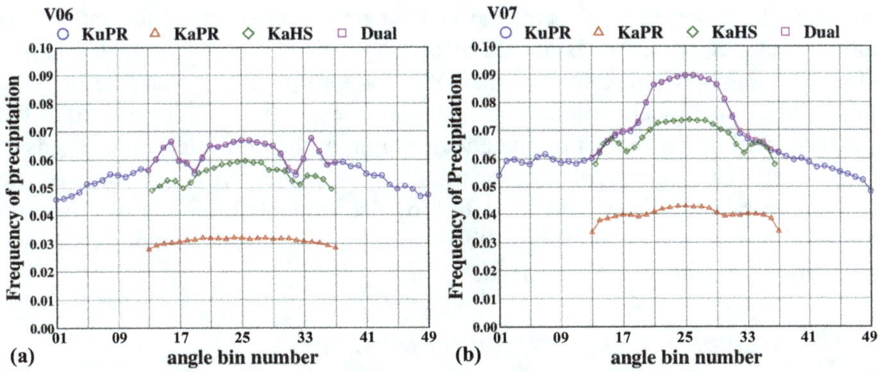

Figure 5.16 Angle bin dependence of frequency of precipitation (a) in V06 and (b) in V07

Figure 5.16 shows angle bin dependence of the frequency of precipitation. As in the previous figure, different lines are for four products and all measurement data in June 2016 are used. Line for the dual-frequency algorithm and line for KuPR algorithm are mostly overlapped. Figure 5.16(a) is for V06 and Figure 5.16(b) is for V07. The increase in the frequency of precipitation from V06 to V07 is clearly seen. In V06, the frequency of precipitation is lower at angle bins 17–20 and 30–33 for KuPR and the dual-frequency products because of the side-lobe clutter routine. In V07, the side-lobe clutter routine is improved as well as precipitation detection method, and the depressions disappear. At near nadir (around angle bin 25) in KuPR and the dual-frequency products, the frequency of precipitation becomes much higher.

5.5 Improvement plan for the next version

For the next version (Version 08), further improvements of DPR algorithms are planned.

5.5.1 Extrapolation of the vertical profile in the main-lobe clutter region

Because of the main-lobe clutter, radar reflectivity factors cannot be used for the precipitation retrieval near the Earth's surface. For the main-lobe clutter region, the vertical profile of Z_e or R is extrapolated. In the DPR algorithms, Z_e is assumed to be constant in the main-lobe clutter region. In the forward retrieval method, Z_e at the previous range bin is copied if Z_m is not available. DSD parameters are retrieved from Z_e and R–D_m relation. R becomes almost constant in the main-lobe clutter region, but it slightly decreases downward because of the increase in air temperature and air density. In PR algorithm, Z_e is assumed to decrease by 0.5 dBZ per km for stratiform precipitation over land to represent the effect of evaporation from falling raindrops, while Z_e is assumed to be constant for convective precipitation or over ocean. Hirose *et al.* [16] analyzed KuPR products in V06 and produced a database for the vertical profile of precipitation rates based on measurements at near nadir. By using this database, the vertical profile of precipitation rate in the main-lobe clutter region is estimated at off-nadir. This method estimates higher surface precipitation rate estimates than V06 by approximately 5%. The increase ratio is larger in higher latitude and at higher elevation. This suggests that the estimation method for the main-lobe clutter region in the DPR algorithms needs to be improved. The improvement may mitigate the angle bin dependence of the precipitation rate estimates. Hirose *et al.* [16] also evaluated missing of precipitation with lower storm height. These effects result in the underestimation of the total precipitation amount by 8% over land.

5.5.2 Assumptions of precipitation particles

In DPR algorithms, precipitation particles are assumed to be a mixture of solid ice, liquid water, and air in a bright band or above 0 °C height. When a bright band exists, the volumetric ratio of ice, water, and the density of particle is given at some specific heights: −50 °C (node A), the top of bright band (node B), the peak of bright band (node C), the bottom of bright band (node D), between node B and node C, and between node C and node D. At these range bins, refractivity index is calculated by the mixing rule. The shape of precipitation particle is assumed to be sphere, and $f_Z(D_m)/N_w$ and $f_k(D_m)/N_w$ are calculated with Mie scattering theory. Between node A and node B, $\log[f_Z(D_m)/N_w]$ and $f_k(D_m)/N_w$ are linearly interpolated with air temperature. Over node A (air temperature is under –50 °C), $f_Z(D_m)/N_w$ and $f_k(D_m)/N_w$ are the same with those at node A. When no bright band exists, 0 °C level is at node D, then $\log[f_Z(D_m)/N_w]$ and $f_k(D_m)/N_w$ are linearly interpolated between nodes A and D with air temperature.

At node A, the volumetric ratio of liquid water is set to zero. It is assumed that some liquid water exists where air temperature is higher than −50 °C. In PR algorithm, no liquid precipitation is assumed where air temperature is −15 °C or lower. This difference affects the continuity of PR and DPR products. Linear interpolation between nodes makes the physical assumptions for range bins between the nodes unclear. Non-spherical shape of solid precipitation particles should be considered.

5.5.3 Differences between PR algorithm and KuPR algorithm

Although KuPR algorithm has been developed based on PR algorithm, there are some differences between them.

5.5.3.1 R–Z_e relation

PR algorithm uses R–Z_e relation after Z_e is obtained by attenuation correction. The R–Z_e relation is modified by ε when k–Z_e relation is modified. On the other hand, in KuPR algorithm, DSD parameters are retrieved by the forward retrieval method and R is calculated by DSD parameters. R–Z_e relation is not explicitly given but it can be calculated numerically. Figure 5.17 shows the comparison of the R–Z_e relation in PR V7 with that in KuPR V06 for liquid precipitation with the

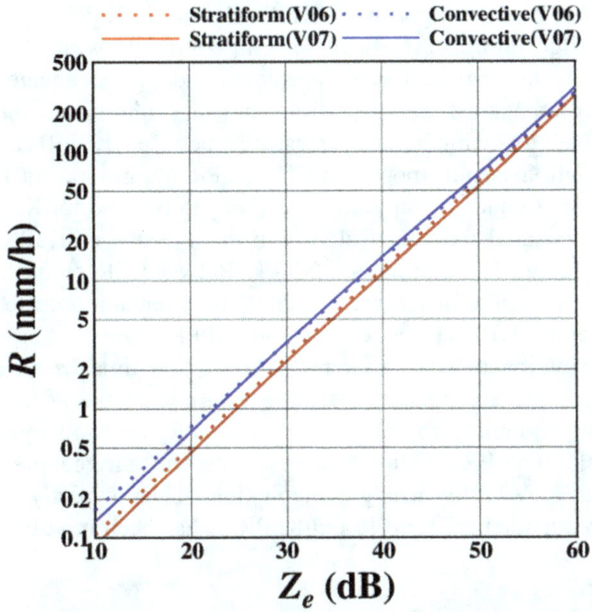

Figure 5.17 R–Z_e relation in KuPR V06 (solid lines) and PR V7 (dashed lines) for stratiform (red lines) and convective (blue lines) precipitation

temperature of 0 °C. When ε = 1.0, *R*–Z_e relations of the two algorithms are nearly the same. In PR algorithm, if ε is lower than 0.7, ε = 0.7 is used for *R*–Z_e relation and if ε is higher than 1.428, ε = 1.428 is used. In these cases, *R*–Z_e relations in the two algorithms differ significantly. PR algorithm tries to avoid severe overestimation or underestimation caused by extreme value of ε, but *k*–Z_e relation and *R*–Z_e relation become inconsistent.

5.5.3.2 Non-uniform beam filling correction

Non-uniform beam filling (NUBF) needs to be considered as the size of FOV is as large as 5 km for PR and DPR. In both PR and KuPR algorithms, NUBF correction is applied. PR algorithm assumes that *k* follows gamma distribution and ε is constant, KuPR algorithm assumes that N_w follows gamma distribution and D_m is constant. Each assumption is suited for each algorithm, which means that it may be difficult to apply the assumption of PR algorithm in the forward retrieval method of KuPR algorithm and vice versa. In PR algorithm, *R*–Z_e relation is affected by NUBF correction, but *R*–D_m relation in KuPR algorithm is not. The details of NUBF in KuPR algorithm are written in [11].

5.6 Summary

In this chapter, the basic theory, development history, improvements in the latest version, and some remaining tasks of DPR algorithms are explained. DPR algorithms consist of three algorithms (KuPR algorithm, KaPR algorithm, and the dual-frequency algorithm). In the development, continuity with PR standard algorithm, consistency among the three algorithms, and the advantages of the dual-frequency algorithm against the single-frequency algorithms are considered. R–D_m relation is applied to maintain the consistency between KuPR and KaPR algorithms. k–Z_e relation is not appropriate for KaPR algorithm. As R and D_m are independent of the frequency, the same equation can be used for KuPR and KaPR. Advantages of the dual-frequency algorithm are the use of KaPR's Z_m for the evaluation of ε as well as the use of DSRT. Between PR algorithm and KuPR algorithm, there are some differences such as the attenuation correction method, R–Z_e relation, and NUBF correction method. KuPR algorithm assumes that the probability distribution of ε is different by space and time based on the results of the dual-frequency algorithm. This is an example of the advantage of DPR. KuPR algorithm is applied to PR measurement. PR V8 shows good agreement with KuPR V06, but PR V8 tends to estimate the surface precipitation rate higher than KuPR V06 for heavy precipitation. This is partly caused by the difference of frequency between PR and KuPR. It has been modified in PR V9.

Acronyms

DFR dual-frequency ratio
DoF degree of freedom
DSRT dual-frequency surface reference technique
FOV field of view
DPR Dual-frequency Precipitation Radar
DSD drop size distribution
GPM Global Precipitation Measurement
HB Hitschfeld-Bordan
NR no rain
NUBF non-uniform beam filling
PIA path integrated attenuation
PR Precipitation Radar
SLV solver
SRT surface reference technique
TRMM Tropical Rainfall Measuring Mission

References

[1] Kummerow, C., W. Barnes, T. Kozu, J. Shiue, and J. Simpson. The tropical rainfall measuring mission (TRMM) sensor package. *J Atmos Ocean Technol.* 1998;15:809–17.

[2] Iguchi T. Dual-frequency precipitation radar (DPR) on the global precipitation measurement (GPM) mission's core observatory. *Satellite precipitation measurement. Vol. 1. Advances in Global Change Research Book Series vol. 67*. Cham: Springer, 183–92.

[3] Kubota T, Iguchi T, Kojima M, *et al.* A statistical method for the reducing sidelobe clutter for the Ku-band precipitation radar on board the GPM core observatory. *J Atmos Ocean Technol.* 2016;33:1413–28.

[4] Masaki T, Iguchi T, Kanemaru K, *et al.* Calibration of the dual-frequency precipitation radar (DPR) onboard the global precipitation measurement (GPM) core observatory. *IEEE Trans Geosci Remote Sens.* 2022;60:5100116.

[5] Kubota T, Seto S, Satoh M, *et al.* Cloud assumption of precipitation retrieval algorithms for the dual-frequency precipitation radar. *J Atmos Ocean Technol.* 2020;37:2015–31.

[6] Awaka J, Le M, Brodzik T, *et al.* Improvements of rain type classification algorithms for a full scan mode of GPM dual-frequency precipitation radar. *J Meteorol Soc Japan.* 2021;99:1253–70.

[7] Meneghini R, Kim H, Liao L, Kwiatkowski J, and Iguchi T. Path attenuation estimates for the GPM dual-frequency precipitation radar (DPR). *J Meteorol Soc Japan.* 2021;99:181–200.

[8] Iguchi T, Kozu T, Meneghini R, Awaka J, and Okamoto K. Rain-profiling algorithm for the TRMM precipitation radar. *J Appl Meteorol.* 2000;39:2038–52.

[9] Iguchi T, Kozu T, Kwiatkowski J, Meneghini R, Awaka J, and Okamoto K. Uncertainties in the rain profiling algorithm for the TRMM precipitation radar. *J Meteorol Soc Japan.* 2009;87A:1–30.

[10] Kozu T, Iguchi T, Kubota T, *et al.* Feasibility of raindrop size distribution parameter estimation with TRMM precipitation radar. *J Meteorol Soc Japan.* 2009;87A:53–66.

[11] Seto S, Iguchi T, Meneghini R, *et al.* The precipitation rate retrieval algorithms for the GPM dual-frequency precipitation radar. *J Meteorol Soc Japan.* 2021;99(2):205–37.

[12] Atlas D, and Ulbrich C.W. Path- and area-integrated rainfall measurement by microwave attenuation in the 1–3 cm band. *J Appl Meteorol.* 1977;16:1322–31.

[13] Seto S, and Iguchi T. Rainfall-induced changes in actual surface backscattering cross sections and effects on rain-rate estimates by spaceborne precipitation radar. *J Atmos Ocean Technol.* 2007;24:1693–709.

[14] Seto S, Iguchi T, and Meneghini R. Correction of path-integrated attenuation estimates considering the soil moisture effect for the GPM dual-frequency precipitation radar. *J Atmos Ocean Technol.* 2022;39:803–21.

[15] Seto S. Examining the consistency of precipitation rate estimates between the TRMM and GPM Ku-band radars. *SOLA.* 2022;18:53–7.

[16] Hirose M, Shige S, Kubota T, Furuzawa A, Minda H, and Masunaga H. Refinement of surface precipitation estimates for the dual-frequency precipitation radar on the GPM core observatory using near-nadir measurements. *J Meteorol Soc Japan.* 2021;99:1231–52.

Chapter 6

The NASA Polarimetric (NPOL) weather radar facility and some applications

David B. Wolff[1], *David A. Marks*[1,2], *Charanjit S. Pabla*[1,2], *Jason L. Pippitt*[2,3], *Ali Tokay*[3,4,5], *Jianxin Wang*[2,3] *and Michael Watson*[1,6]

The NASA POLarimetric (NPOL) Radar is based at NASA's Goddard Space Flight Center

(GSFC) Wallops Flight Facility (WFF) in Wallops Island, Virginia, and is physically located about 38 km northeast in Newark, MD (38.263N, 75.342W). NPOL is NASA's flagship weather radar and provides well-calibrated high-quality data for the Global Precipitation Measurement (GPM) Ground Validation (GV) program. NPOL is an S-band, Doppler, dual-polarimetric radar that measures reflectivity, radial velocity, and spectrum width, as well as differential reflectivity, differential phase, and co-polar correlation. Subsequent data processing provides retrievals of rain rate, specific differential phase, and particle size distribution parameters.

A discussion of the NPOL system is provided in Section 6.1. The various means of both engineering and data-based calibration of the radar are given in Section 6.2. A summary of the software tools that have been developed and used by GPM GV personnel is provided in Section 6.3. A brief review of the NPOL-based GPM field campaigns, as well as NPOL's contributions to the WFF Precipitation Research Facility, is given in Section 6.4. Finally, Section 6.5 provides some examples that utilize NPOL for validation.

6.1 The NPOL system

The current NASA Polarimetric (NPOL) radar is a transportable, S-band, dual polarimetric research radar first developed at NASA's Wallops Flight Facility

[1]NASA Wallops Flight Facility, Wallops Island, VA, USA
[2]Science Systems & Applications, Inc., Lanham, MD, USA
[3]NASA Goddard Space Flight Center, Greenbelt, MD, USA
[4]Goddard Earth Sciences Technology and Research (GESTAR-II), Baltimore, MD, USA
[5]University of Maryland, Baltimore County, Baltimore, MD, USA
[6]Peraton Inc., Reston, VA, USA

(WFF). NPOL was redesigned in 2011 with a state-of-the-art center-fed parabolic reflector in preparation for several pre- and post-launch GPM field campaigns, thereby providing improved research quality volumetric measurements of precipitation. NPOL consists of six sea-containers with dimensions 2.44 m × 2.44 m × 6.1 m. Each sea-container weighs between 5,455 and 15,000 kg totaling 60 tons when fully loaded for shipping. The containers form the foundation of the stable base on which a steel platform is placed for mounting the antenna. The stable base configuration comprises a footprint of approximately 30 m^2. In this configuration, the center line of the antenna is 8.5 m above the ground. Figure 6.1 provides a photograph of NPOL as deployed in Iowa for the Iowa Floods Studies (IFloodS) field campaign in 2013.

6.1.1 Transmitter

The NPOL radar transmitter is an Enterprise Electronics Corporation (EEC) DWSR-8500 commissioned in 2000 and upgraded in 2018 with a Pulse System Incorporated (PSI) TR-1157. The main transmitter components are the High Voltage Power Supply, the Modulator, and the Magnetron. The NPOL transmitter specifications are provided in Table 6.1.

6.1.2 Receiver

The receiver was designed and built by WFF Engineers in 2000 and upgraded to the RVP900 in 2009. The receiver consists of an analog front end and an Intermediate Frequency Digital Receiver (IFDR). The receiver is a dual-independent channel design with one channel processing the horizontal (H) signal and the other processing the vertical (V) signal. The NPOL receiver specifications are provided in Table 6.1.

Figure 6.1 NPOL deployed in Iowa for the GPM field campaign "IFloodS" in 2013.

Table 6.1 NPOL transmitter, receiver, pedestal and reflector specifications

Transmitter	
Transmitter type	Coaxial magnetron
Magnetron	CPI VMS-1197B
Modulator type	Solid state
Rise time	200 nsec
Fall time	350 nsec
Pulse jitter	5 nsec maximum
Stability	0.1dB pulse to pulse
Operating frequency	2.700 to 2.900 GHz tunable
Spectrum	12dB down on side bands
Polarization	Simultaneous, horizontal, vertical, alternating
Peak pulse power output simultaneous	400 kW H, 400 kW V
Peak pulse power output horizontal, vertical, alternating	800 kW
Pulse width	0.5, 0.8, 1.0 and 2.0 μs, selectable
Pulse repetition frequency	250 to 1200 p/s, duty cycle limited
Duty cycle	0.001 maximum
Receiver	
Operating frequency	2.700 GHz to 2.900 GHz
Mixers	Image reject
Intermediate frequency	30 MHz
Automatic frequency control	Digital
Digital receiver	Vaisala RVP901
Pedestal	
Pedestal type	Elevation over Azimuth
Azimuth/elevation mode	Computer controlled
Types of scan patterns	PPI, RHI, full volume, sector
Azimuth angular travel	Continuous
Azimuth angular acceleration	20 °/s^2
Azimuth angular velocity	1 °/s to 20 °/s
Azimuth position accuracy	+0.1°
Elevation angular travel	−2° to +91° operational, −3° to + 95° mechanical
Elevation angular acceleration	20 °/s^2
Elevation angular velocity	1 °/s to 20 °/s
Elevation position accuracy	+0.1°
Reflector	
Reflector type	Prime-focus parabolic
Reflector size	8.5 m
Feed type	Dual-polarized OMT horn
Focal distance	3.20 m
Polarization	Linear horizontal, linear vertical

(Continues)

Table 6.1 (Continued)

	Reflector
Gain	44.87 dBi H, 45.82 dBi V
Gain flatness	0.65 dB H, 0.63 dB V
Beamwidth	≤ 0.93° at ±3 dB H, ≤ 0.94° at ±3 dB V
Beam symmetry	0.02 Δ max at ±3 dB, 0.07, Δ max at ±10 dB, 0.09 Δ max at ±20 dB
Sidelobe levels	≤ −26 dB at ± 2° H & V, ≤ Straight-line connection −26 dB at ±2° to −38 dB at ±10° H & V, ≤ Straight-line connection −38 dB at ±10° to −42 dB at ±180° H & V
Collimation error	≤ 0.1 Δ (H_{peak}−V_{peak})
Cross-polarization	< −41.5 dB on axis H, < −37.6 dB on axis V, < −30 .0 dB ± 2° H, < −30 .0 dB ± 2° V, < −40 .0 dB ± 5° H , < −40 .0 dB ± 5° V
Port isolation	52.5 dB
VSWR	1.30 H, 1.29 V

6.1.3 Antenna

The primary antenna components are the pedestal and the center-fed parabolic reflector (see Figure 6.1). The Antenna Control Unit (ACU), located in a sea-container, controls the positioning of the antenna. The pedestal and ACU were manufactured by LEONARDO Germany GmbH. The reflector was manufactured by ASC Signal Corporation. The NPOL pedestal and reflector specifications are provided in Table 6.1.

6.2 Radar calibration and quality control

Accurate calibration of weather radars directly impacts measurement quality and is critical for all applications. The NPOL radar team employs numerous engineering measurements and data analysis techniques to ensure accurate NPOL calibration. The following measurements and techniques are separated into four categories: transmitter, receiver, antenna, and data as a calibration tool. Polarimetric calibration measurements are made in both the horizontal and vertical channels.

6.2.1 Transmitter, receiver, and antenna calibration

Steps describing the transmitter, receiver, and antenna calibration are summarized in Figure 6.2.

The receiver calibration is performed using the Vaisala Z-Auto Utility as shown in Figure 6.3. Known signal and noise data are injected into the receiver, and measurements are retrieved. The measurements above the system noise level determine relevant calibration parameters for received power and reflectivity

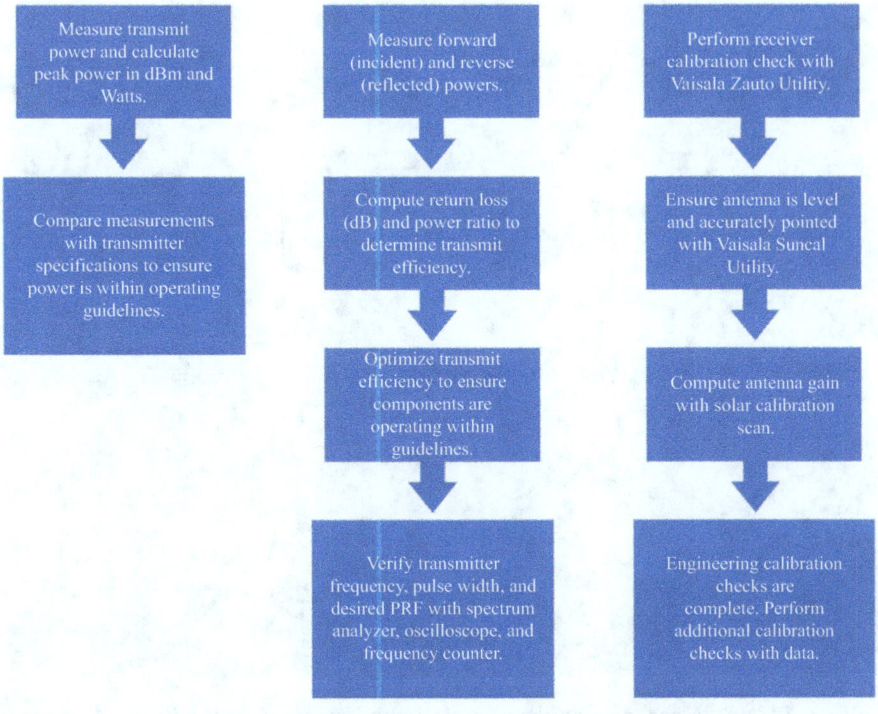

Figure 6.2 A summary flowchart of the transmitter, receiver, and antenna calibration steps.

calculations. Z-Auto is performed monthly, and the parameters are analyzed for consistency and updated as appropriate.

Cross-levels installed in the pedestal and elevation box are monitored to ensure the antenna is level. For pointing accuracy, the sun serves as an external radiation source for calibration of the radar system. The sun's position can be calculated provided that accurate time, latitude, and longitude information are known. The Vaisala Suncal Utility performs a raster scan of the sun and generates a cross-section beam pattern image showing range-averaged solar microwave power in azimuth and elevation. The difference between the predicted position and the observed position of the sun is recorded as azimuth and elevation position errors.

6.2.2 Calibration techniques using observed data

Calibration checks of NPOL observed reflectivity (Z_H) and differential reflectivity (Z_{DR}) are routinely performed with several tools employed by the GPM GV group. Among these tools are the Relative Calibration Adjustment (RCA) [1,2], the self-consistency of polarimetric properties within the rain medium [3–7] comparisons with independently computed Z_H from disdrometers, and vertical profiles of Z_{DR} in light rain.

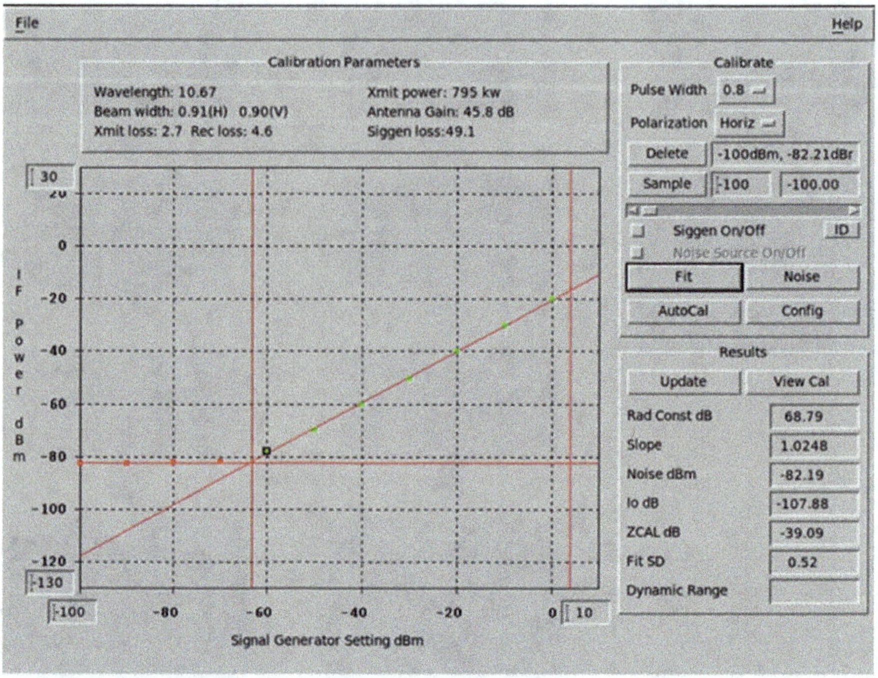

Figure 6.3 NPOL receiver calibration with the Vaisala Z-Auto Utility. A series of calibrated Radio Frequency (RF) signals are injected into the receiver and then sampled to compute the corresponding Intermediate Frequency (IF) levels. The green points show computed IF power levels for the injected RF signals. The slope of the diagonal line provides a best-fit quantitative measure of their agreement and is expected to be near 1.0. This example shows a slope of 1.02 and indicates excellent agreement between the injected RF signal and the computed IF power. The receiver is well calibrated to the setup parameters.

6.2.2.1 Self-consistency

One method of independent calibration that is routinely applied to NPOL data is the use of a self-consistency approach from [7]. The stability of radar data calibration is of paramount importance for evaluating differences in reflectivity and rain rate estimates between space-borne and ground-based platforms. The absolute reflectivity calibration of the NPOL radar is evaluated and monitored using the principle of self-consistency. The consistency of polarimetric properties within the rain medium was quantified by Gorgucci *et al.* [3,4], Scarchilli *et al.* [5], Vivekanandan *et al.* [6], and Ryzhkov *et al.* [7]. Precipitation sources in the coastal mid-Atlantic area include mid-latitude synoptic systems, tropical systems, and sea-breeze convergence zones. This area is ideal for self-consistency calibration because of the common events producing light-to-moderate stratiform rain. Following the method of Ryzhkov *et al.* [7], a self-consistency equation describing the relationship among Z_H, Z_{DR}, and specific K_{DP} was derived using DSD data obtained from disdrometers. Six two-dimensional video disdrometers (2DVD) sited at WFF from 2013 to 2014 provided raindrop size distribution (DSD) data to regress a relationship in the following form:

$$K_{DP} = AZ_H^b Z_{DR}^C \tag{6.1}$$

where Z_H is in mm^6 m^{-3}; Z_{DR} is in dB, and K_{DP} is in deg km^{-1}.

As described in [8], the one-minute DSD observations were used to calculate the polarimetric parameters for an S-band radar (10.7 cm wavelength) at a temperature of 20°C with observed drop shapes. The mean axis ratios in [9] were adopted for drops less than 4 mm in diameter and equilibrium drop shapes [10] for larger drops. For the fall velocity, the terminal fall velocity drop diameter relation given by Beard and Chuang [11] is used. The coefficient and exponents were derived via a linear, least squares fit regression ($a = 2.056 \times 10^{-5}$, $b = 0.9927$, and $c = -0.3722$).

The following steps are taken to determine the NPOL absolute Z_H bias.

1) The Z_{DR} bias is determined first with vertical profiles in light rain via [4].
2) Approximately NPOL data with one hour are selected for self-consistency, based on event echo coverage and intensity.
3) The selected data are quality-controlled via the GV radar method as described in [12].
4) Following [7], the triplet data (Z_H, Z_{DR}, and K_{DP}) are initially segregated within user-defined spatial and temporal domain. The data are then filtered to select triplet data within the Z_H range of 30–48 dBZ. The internal K_{DP} field and theoretical K_{DP} field (as a function of Z_H and Z_{DR}) are integrated over the Z_H range in 1 dB increments, and $n(Z)$ is the number of gates with reflectivity Z.

$$I_1 = \sum_{Z_{min}}^{Z_{max}} \overline{K_{DP}(Z)} n(Z) \Delta Z \tag{6.2}$$

$$I_2 = A \sum_{Z_{min}}^{Z_{max}} Z_H^b \overline{Z_{DR}(Z)^c} n(Z) \Delta Z \tag{6.3}$$

where $n(Z)$ is the number of gates with reflectivity Z.

5) The absolute Z_H bias is determined by the ratio as defined in [6]

$$Z_{Bias}(dB) = 10\log\left(\frac{I_2}{I_1}\right) \tag{6.4}$$

The iterative Z_{Bias} is then added or subtracted to the Z_H data and new triplet data are obtained. The iterative process continues until the Z_{Bias} is within 0.1 dB. The final Z_{Bias} value represents the absolute Z_H calibration adjustment needed for consistency between the polarimetric variables. Convergence typically occurs within three iterations.

6.2.2.2 Reflectivity comparison with disdrometers

As an independent calibration check, NPOL reflectivity is frequently compared to 2DVD and Parsivel disdrometers using the following method. The equilibrium drops shapes (Beard and Chuang 1987) distribution with Raleigh scattering assumption is used to calculate the disdrometer-based reflectivity with one-minute temporal resolution. Following [8], the polarimetric parameters are calculated for an S-band radar (10.7 cm wavelength) at a temperature of 20° C. For the radar dataset, high-resolution Range Height Indicator (RHI) scans from NPOL are performed at the relevant azimuths directly over the disdrometer locations. Based on the RHI scan timestamp, the disdrometer data are selected one-minute later to allow drops to fall to near surface. To reduce or mitigate the effects of blockage, ground clutter, and sidelobe issues, the 0.9° elevation is used for comparison.

The disdrometer-based comparison method was employed during the GPM IFloodS field campaign to independently assess results from self-consistency calibration [2]. During IFloodS, four 2DVDs operated nearly continuously at ranges from 15 to 69 km from NPOL from May 3 to June 15, 2013. Combining rain events from all disdrometers and NPOL RHI scans during this period resulted in a dataset of 6,440 entries for comparison. Figure 6.4 shows a histogram of reflectivity difference (disdrometer minus NPOL) from the entire dataset. The Gaussian fit shows

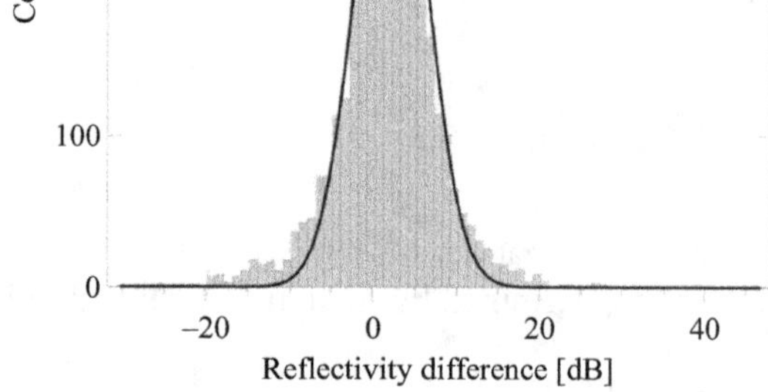

Figure 6.4 Histogram showing the reflectivity difference (2DVD minus NPOL) from four-2DVD instruments combined and NPOL RHI scans above the disdrometers at the 0.9° elevation from the IFloodS field campaign from May 3 to June 15, 2013. Data indicated that NPOL reflectivity values were low by approximately 2.15 dB.

a difference of +2.15 dB indicating that NPOL reflectivity was lower than the disdrometers. Individual disdrometer comparisons (from all rain events combined) indicated that NPOL reflectivity was low within the range of 1.8–2.6 dB. These findings agreed well with the independent self-consistency approach [2] and provided further confidence in required calibration adjustments. This method continues to be used for GPM ground validation efforts with NPOL using an array of 2DVD and Parsivel^2 disdrometers on the Maryland and Virginia Eastern Shore.

6.2.2.3 Relative calibration adjustment

NASA also uses the probability distribution of clutter area reflectivity to provide estimates of the relative calibration of reflectivity in near-real-time. The Relative Calibration Adjustment (RCA) technique was first presented by Silberstein *et al.* [1] for oceanic radar data over Kwajalein Atoll, Republic of the Marshall Island (RMI), which is loosely based on the method in [13]. Wolff *et al.* generalized the technique for application to NPOL and other S-band radars over land or water. As explained in [2], the method provides a tool for maintaining consistent relative calibration with uncertainty of +/−0.5 dB or better and is simply explained as a calibration adjustment applied to reflectivity data to obtain agreement to an established baseline. The adjustment is computed by comparison of the 95th percentile of the clutter area reflectivity to the baseline. The probability distribution of clutter area reflectivity from a specified elevation can vary significantly; however, the 95th percentile of the clutter distribution is remarkably stable and allows the detection of relative calibration changes.

As detailed in [2], a series of steps were taken to develop a 5 km range RCA clutter map showing locations where reflectivity values exceeded a 55 dBZ threshold more than 50% of the time. Figure 6.5 shows an example of the technique as applied to the NPOL radar data from the GPM IFloodS field campaign.

The next step is to determine the baseline for comparison to relative calculations from other periods. The cumulative distribution function (CDF) from the clutter map dataset is computed and the 95th percentile is used as the baseline. The RCA value for any given day is then defined by the following equation:

$$\mathrm{RCA} = \left(\mathrm{dBZ95}_{\mathrm{baseline}} - \mathrm{dBZ95}_{\mathrm{daily}}\right)$$

where $\mathrm{dBZ95}_{\mathrm{baseline}}$ is the 95th percentile baseline reflectivity and $\mathrm{dBZ95}_{\mathrm{daily}}$ is a specific daily 95th percentile clutter area reflectivity.

As an example of application from NPOL during the IFloodS field campaign, the hourly and daily probability density function (PDF) and CDF were computed on May 4, 2013 with significant precipitation occurring. Figure 6.6 shows the PDF and CDF curves along with the hourly RCA values (gray-scaled text on left). Considerable spread in the hourly curves is due to precipitation; however, the curves converge prior to the 95th percentile (intersection of dashed horizontal and vertical lines). In this example, the daily computed RCA value is −0.11 dB. This application was executed each day during the IFloodS field campaign to monitor the relative calibration.

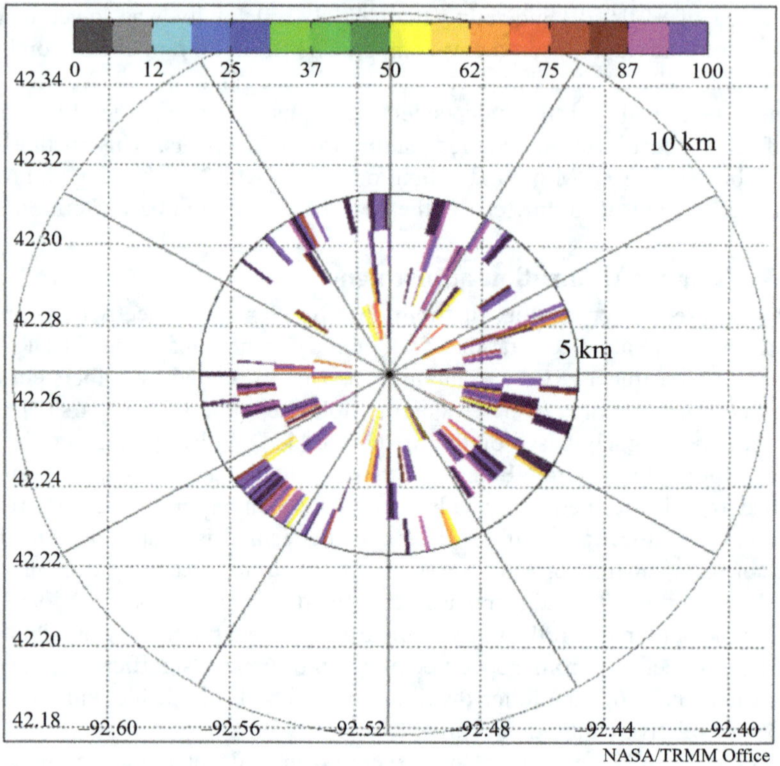

Figure 6.5 Clutter map for the NPOL radar from the IFloodS field campaign. Colors represent the percentage of time that 1 km × 1° gate reflectivity values exceeded an arbitrary threshold of 55 dBZ.

In an operational sense, the RCA technique is applied to NPOL on most days to monitor for clutter area reflectivity changes that exceed +/−0.5 dB. A change by this magnitude or higher is indicative of a system change to be investigated. To adjust the actual calibration, an accurate estimate of the absolute calibration must be made by independent means first (see Sections 2.21 and 2.2.2 in this chapter). When the absolute calibration value is known, the RCA baseline value can be set to zero, so subsequent changes in the absolute calibration are detectable by a 1:1 change in the RCA value.

6.2.2.4 Differential reflectivity calibration: birdbath scans

Vertical profile (or birdbath) scans are unique Plan Position Indicator (PPI) scans at 90° elevation angle. NPOL uses these scans primarily for a Z_{DR} calibration check as detailed in [4]. NPOL performs two complete 360° vertical rotations as a routine component of its scanning strategies. Figure 6.7 shows a four-panel plot of Z_H, Z_{DR}, ρ_{HV}, and mean Z_{DR} by azimuth from July 26, 2022. For Z_{DR} calibration, we choose those scans with a steady Z profile of light rain in the low-to-mid 20 dBZ range

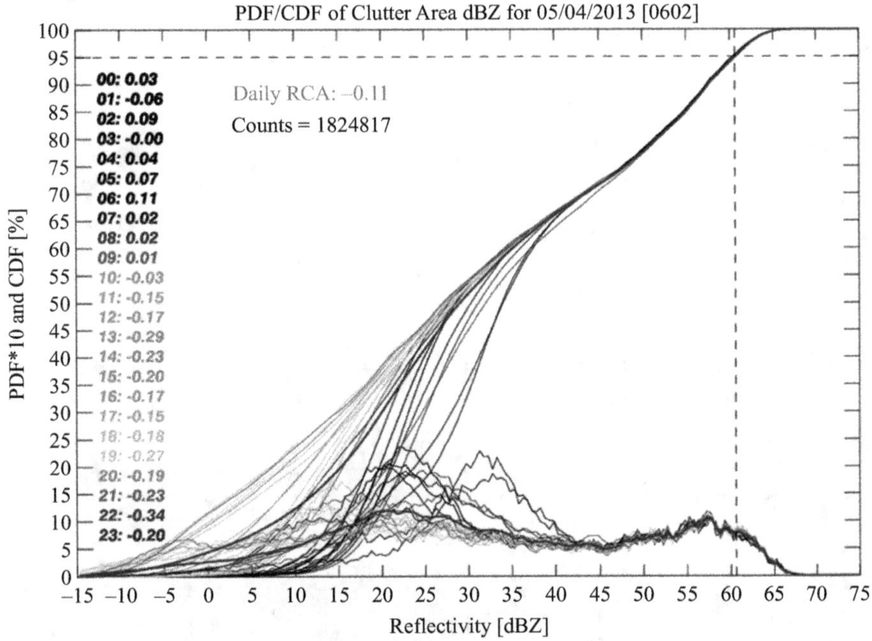

Figure 6.6 Hourly clutter area PDF/CDF on May 4, 2013.The 95th percentile reflectivity which Silberstein et al. and Wolff et al. showed is quite stable when calibration or engineering changes are not encountered.

(Panel A), the Z_{DR} profile (Panel B) will be uniform with height, ρ_{HV} will be near 1.0 (Panel C), and Z_{DR} by azimuth should show low standard deviation (Panel D). In this example, the mean Z_{DR} bias over 720° azimuth (two rotations) is approximately 0.011 dB. Note the bright band associated with the melting layer is about 4.9 km in height. The level chosen for Z_{DR} calibration in this case is 3.5 km and represents the level of highest correlation between the H and V channels that is at least 1 km below the freezing level.

6.2.3 *Quality control*

NASA's GPM GV team developed a Python-based radar data processing system referred to as GVradar. GVradar consists of two modules: Dual Polarimetric Quality Control (DPQC) and dual polarimetric precipitation product generation (dp_products). GVradar also utilizes both the Python Atmospheric Radiation Measurement (ARM) Radar Toolkit (Py-ART) [14], and Colorado State University's Radar Tools (CSU Radar Tools; [15]).

Dual polarimetric (DP) radars allow GPM-GV to develop improved understanding of precipitation microphysics and provide essential input for development of precipitation retrieval algorithms. The quality control (QC) of these datasets is a critical first step in this process. GPM-GV developed an algorithm based on [16],

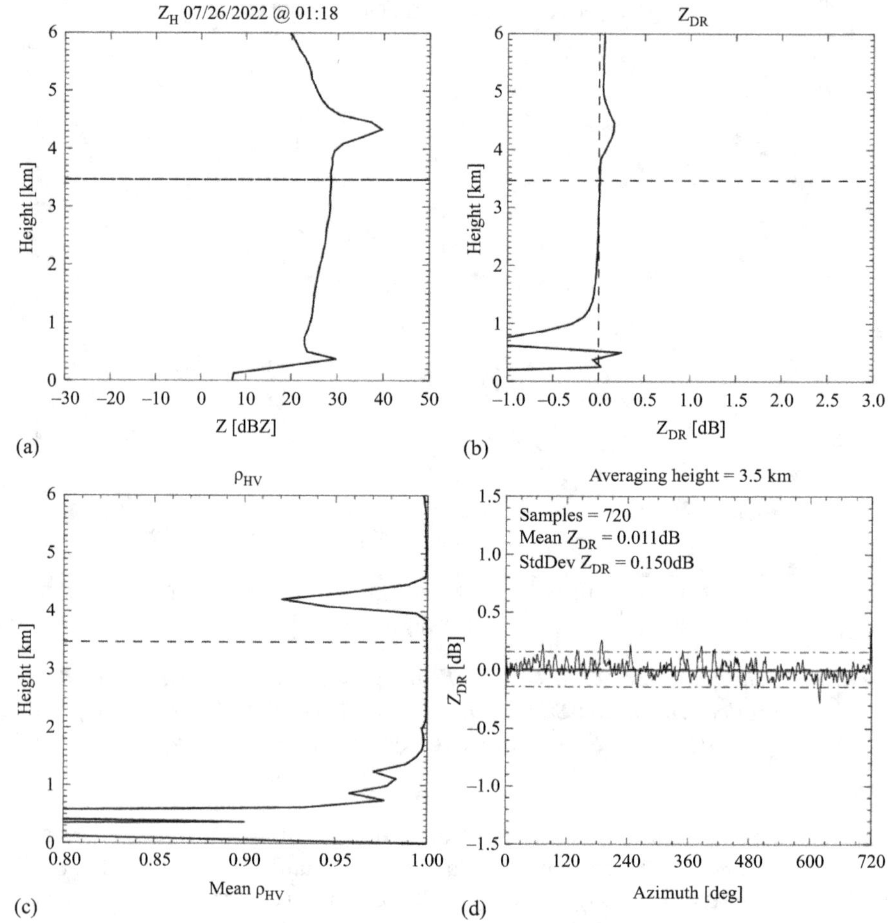

Figure 6.7 Sample output image depicting vertical profiles of reflectivity (Z_H), differential reflectivity (Z_{DR}), and copular correlation (ρ_{HV}) in panels A, B, and C, respectively. The dashed horizontal line shows the level of maximum ρ_{HV}. Panel D shows the mean and standard deviation of Z_{DR} over two antenna rotations. The Z_{DR} offset is then calculated as the overall mean Z_{DR} over the two rotations. The freezing level at this time was at about 4,800 m.

that uses DP parameters to QC radar data (DPQC) [12,17]. QC algorithms based on DP parameters have been successful in the identification of nonprecipitating echoes [16,18,19].Tunable DP parameter threshold gate filters are utilized to identify and remove non-precipitating echoes. Py-ART allows the user to easily view and manipulate the radar structure by volume, sweep, ray, and range. There are various fields within the radar structure and DPQC threshold gate filters are dependent on the values associated with these fields. When the value of a gate falls outside one of the thresholds, a missing data mask will be applied to that specific gate for all fields. DP values commonly fall outside the established QC thresholds near the melting level, resulting in unwarranted data removal. To account for this problem, DPQC calculates the height of the radar beam and only applies QC below a user-defined height or sounding retrieved height. In an operational environment, hourly model soundings are retrieved from the Rapid Refresh (RAP) model to estimate the melting level, and QC is applied to data 1 km below this level. Additionally, thresholds can be applied based on a user-defined sector. This allows the user to target a troublesome area of false echo without affecting the entire radar scope. Figure 6.8 provides a flowchart of the DPQC algorithm.

A function within DPQC determines if a differential phase (Φ_{DP}) gate is aliased. An aliased gate is identified when the absolute value of the phase difference between the consecutive gates exceeds 149° [20], and the gate is not considered to be noise or clutter. When an aliased Φ_{DP} gate is detected DPQC will correct the data by adding 180° to the aliased gate. DPQC utilizes functions within CSU Radar Tools to calculate Specific Differential Phase (K_{DP}) [21] and the standard deviation of differential phase $\sigma(\Phi_{DP})$. Velocity de-aliasing based on [22] and application of calibration offsets are additional features of DPQC.

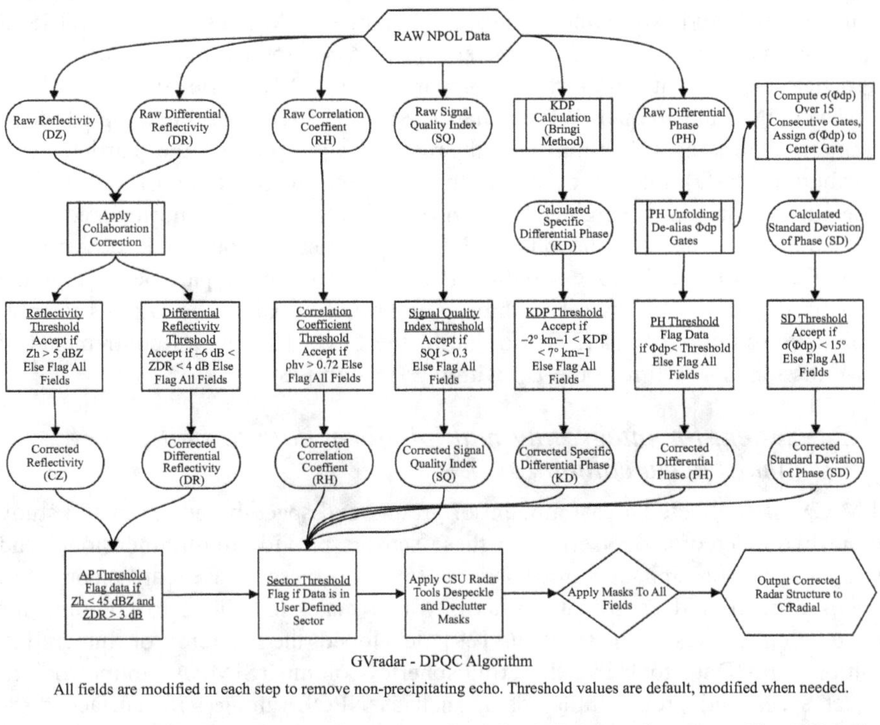

Figure 6.8 Flowchart for quality control of the NPOL radar data. The DPQC system uses a mixture of in-house and open-source python libraries and is publicly available from github.com.

GVradar will run with default settings; however, it is recommended to utilize a user-designated parameter dictionary. The use of a dictionary allows the user to control QC thresholds, specify fields to generate, select output options, and set plotting preferences.

6.3 Tools for validation

Several value-added NPOL products are generated, including specific differential phase, rain rates, and select DSD parameters. Each of these procedures is performed with publicly available python modules and the applications are available on https://github.com/GPM-GV.

6.3.1 Polarimetric radar retrievals

DP-based rain rate, DSD, and other products are generated from DPQC output and are selected within a Python dictionary. Available products include rain rate estimates, hydrometeor classification, water mass, ice mass, and DSD retrievals.

Precipitation products generated with CSU Radar Tools include HIDRO Rain Rate (RC) [23], Hydrometeor Identification (FH) [24], and Ice and Liquid mass (MI, MW) [19]. Additional products include GPM-GV's Mass Weighted Mean Diameter (D_m) and Normalized Intercept Parameter (N_w). D_m and N_w retrieval equations were derived from disdrometer Z_H and Z_{DR} computations obtained during GPM field experiments and directly compared with NPOL retrievals [8,25].

The GPM Dual-frequency Precipitation Radar (DPR) algorithm adopted three-parameter normalized gamma size distribution to represent the Particle Size Distribution (PSD) that is observed through conical scanning of GPM Core Observatory. The mass-weighted mean drop diameter ($\mathrm{D_m}$) and normalized intercept parameter ($\mathrm{N_W}$) are the retrieved PSD parameters from DPR observables through a complicated process outlined in [26]. The shape parameter μ of the normalized gamma PSD is set to three. The primary goal of the DPR is to infer precipitation rate and PSD [27] and one of the Level I Science Requirements of GPM mission is to estimate the D_m within ± 0.5 mm [28].

6.3.2 System for integrating multiplatform data to build the atmospheric column

GPM GV has collected a vast amount of ground and space-based sensors to study falling hydrometeors. Datasets from these sensors provide multidimensional and multiparameter precipitation information. However, these datasets are recorded in various formats and coordinate systems which make it tedious to combine for precipitation studies. Wingo *et al.* [29] developed the System for Integrating Multiplatform Data to Build the Atmospheric Column (SIMBA) framework to project space and ground radar data, such as NPOL, along with surface disdrometers and rain gauges into a common coordinate system. SIMBA ultimately unifies different precipitation platforms and data formats in a common 3D coordinate system. The SIMBA system is a modular design with different modules for different platforms. For example, ground radar data module such as NPOL is dual-polarimetric quality controlled via Pippitt *et al.* [12] while rainfall properties such as DSD is based on Tokay *et al.* [8,25] and rainrate based on Cifelli *et al.*'s [19] algorithms. Ground radar data are inserted into SIMBA grid via NCAR's Radx gridding software (https://github.com/nsf-lrose/lrose-topaz) that takes advantage of Sorted Position Radar INTerpolation (SPRINT). All other platform data are assigned closest to the SIMBA grid point depending on platform location. For example, SIMBA does no horizontal interpolation of space-based data. Pixels from the original data structure are assigned to the SIMBA grid. However, linear interpolation is done by converting original bin levels to SIMBA gird in the vertical level. The current SIMBA system is written in the commercial Interactive Data Language (IDL) but is being converted to python for easier access to an open software system.

Figure 6.9 illustrates the key aspects of the SIMBA framework. More details are available in Wingo *et al.* (2018) [29] and Pabla *et al.* (2022) [30].

6.4 In situ validation

Validation of spaceborne precipitation is a difficult task due to challenges of validating a global precipitation product with ground-based data over limited area or regions. The goal of the GPM GV program is to provide multiple validation approaches that provide the most robust estimates possible to satellite algorithm developers. Hou *et al.* [31] stated that one of the key lessons learned from the Tropical Rainfall Measuring Mission (TRMM; [32]) was that those simple comparisons between ground-based observations

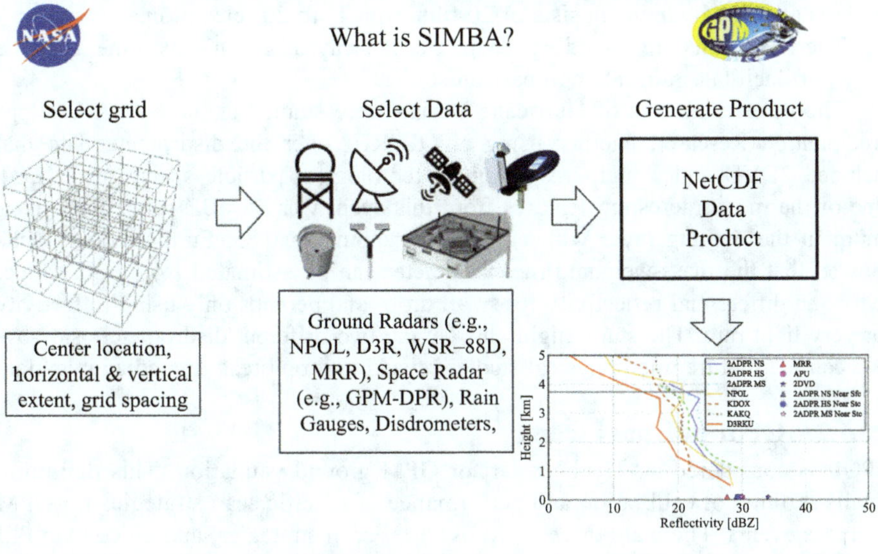

Figure 6.9 Illustration of the SIMBA framework for merging various GPM precipitation measuring platforms into a combined data structure.

from instruments such as radars and rain gauges, while useful, do not provide sufficient information to provide direct feedback to satellite algorithm developers so that they can improve the underlying physics of their algorithms. To address this, the GPM GV approach was threefold: (1) direct statistical validation at the surface using national networks and other areas of interest; (2) physical validation of precipitation properties in a vertical column through field experiments; and (3) integrated validation over space and time via hydrometeorological applications [31]. NPOL has served as the key instrument in several field campaigns which are discussed later in this section. Section 4.1 will discuss NPOL's ongoing contributions at its home location at NASA WFF. Section 4.2 provides brief summaries of the GPM field campaigns in which NPOL was deployed.

6.4.1 The Wallops GPM Precipitation Research Facility

The Wallops Precipitation Research Facility (WPRF) is located on the Eastern Shore of Virginia and is considered an active and ongoing data collection site for GPM GV. The location of NPOL near Newark, MD is the base of operations when not deployed on field campaigns. Figure 6.10 shows the location of NPOL, the WPRF, and numerous ground instrument locations.

In support of the GPM ground validation program and microphysical studies, NPOL runs numerous scanning strategies.

6.4.1.1 Newark scanning strategies

The "Newark" strategy consists of a combination of PPI sweeps and RHI scans designed for low-level rain detection and near-surface rain rate estimation for comparison with ground instrumentation. Scan details are shown in Table 6.2. The vertical profile (birdbath) scans are for Z_{DR} calibration per [4]. A full volume-scan mode is also used which consists of 20-tilts from 1° to 20° elevation.

The RHI scans are directly over the dense array of ground instruments at the WPRF to facilitate rain rate comparison studies.

The outer rainbands of Hurricane Dorian were studied at the NASA Wallops Precipitation Research Facility using NPOL RHI over the disdrometer site that included 2DVD and a high- resolution meteorological particle spectrometer [33]. One of the most interesting features from this event was 55 dBZ reflectivity maximum in the melting layer with correlation dipping to 0.85. Furthermore, results showed that the mass-weighted mean diameter can be estimated from both reflectivity and differential reflectivity for small drops and perhaps only using reflectivity for very light rain. The study highlights having two different disdrometers side-by-side can help future microphysical studies related to drop break up and aggregation.

6.4.1.2 GPM overpass scans

NPOL is considered a "Tier1" radar for GPM ground validation. This definition entails scrutiny of calibration and performance of specific scan strategies for GPM overpass events. There are three GPM overpass scan strategies, and all contain PPI and RHI components. Table 6.3 describes the "GMI," "MEDIUM," and "SHORT" overpass scan strategies as defined by nadir distance range. Scan timing is set so

Figure 6.10 GoogleEarth® overview of the GPM Precipitation Research Facility, highlighting the Wallops Flight Facility instrument network, a high-density rain gauge network in Pocomoke City, MD, and the location of the NPOL radar in Newark, MD. Green circles indicate locations of active rain gauges, green "P" symbols represent sites with Parsivel disdrometers deployed. NPOL is about 38 km NNE of Wallops Flight Facility. Image courtesy of GoogleEarth®.

Table 6.2 NPOL Newark scanning strategy

Newark	Elevation (°)	Azimuth (°)	Samples	PRF
PPI	0.7, 1.2, 1.7	0–360	70	1100
RHI	0–25	195, 197, 199	70	1100
Vertical Profile	90	0–360 (two revolutions)	128	1100

that RHI scans are being performed when the overpass occurs for optimal profile comparisons. The NPOL overpass scans are quality-controlled and sent to the GPM Validation Network (VN) for statistical validation. Figure 6.11 illustrates a near-nadir GPM overpass at WFF. This overpass occurred on July 25, 2022 at 2309 UTC. NPOL captured convection and stratiform rain in the GMI and DPR swaths.

Table 6.3 Scanning strategy for capturing GPM overpasses over the Delaware, Maryland, and Virginia (Delmarva) region

Task	Nadir dist (km)	PPI sweep elev (°)	RHI azimuth (0°–25° elev)
"GMI"	> 120	0.7, 1.2, 1.8	Three cross-track toward nadir Two over WPRF ground instrument pad
"MEDIUM"	60–120	0.7, 1.6, 2.8, 3.9, 5.1, 6.3, 7.8, 9.5, 11.7, 14.3, 17.5, 21.2, 25.5	One along-track on GPM approach angle One along-track on GPM recede angle Three cross-track toward nadir Two over WPRF ground instrument pad
"SHORT"	< 60	0.7, 1.6, 2.8, 3.9, 5.1, 6.3, 7.8, 9.5, 11.7, 14.3, 17.5, 21.2, 25.5	Two along-track on GPM approach angle Two along-track on GPM recede angle Three cross-track toward nadir Two over WPRF ground instrument pad

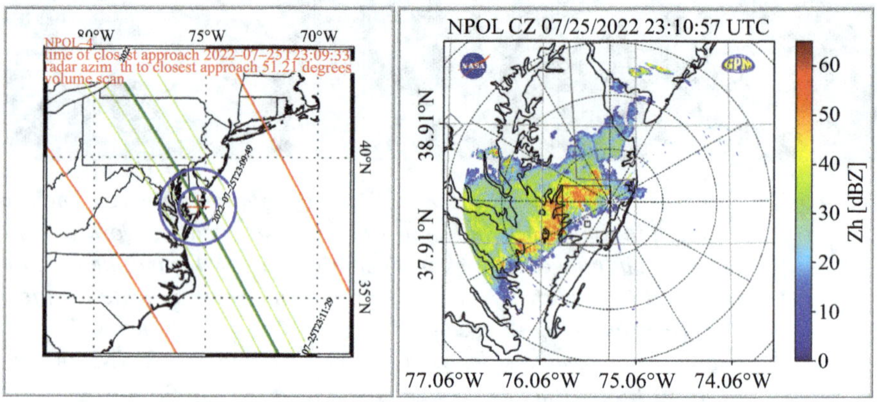

Figure 6.11 GPM track during near-nadir overpass of WFF on July 25, 2022 at 2309 UTC (left). NPOL PPI Image showing convective system that brought heavy rain and strong winds to the area.

NPOL GPM overpass data have been used extensively within the VN for ground validation of rain DSD [34]. NPOL data were combined with other S-band radars around the world to compare DSD parameters in stratiform and convection with those retrieved from GPM Core retrievals. Results show that mass-weighted mean diameter is overestimated by 0.2 mm for all rain types. The bias is near 0.5 mm for convective rainfall. In addition, the normalized intercept parameter is underestimated in the DPR algorithm leading to biases in rainfall rate algorithms. This finding can help algorithm developers improve assumptions when forming retrievals for future algorithms.

6.4.2 GPM remote field campaigns

GPM conducted several pre- and post-launch field campaigns for which NPOL was the lynchpin. All of GPM field campaign data are publicly available from the Global Hydrometeorology Resource Center (GHRC; https://ghrc.nsstc.nasa.gov/home).

6.4.2.1 Midlatitude Continental Convective Clouds Experiment (MC3E)

The NPOL radar provided near-continuous data collection near Lamont, Oklahoma from April 11 to June 3, 2011 supporting the Midlatitude Continental Convective Clouds Experiment (MC3E) field campaign. The experiment was a collaborative effort between the U.S. Department of Energy (DOE) Atmospheric Radiation Measurement (ARM) Climate Research Facility and NASA's GPM mission GV program. The overarching goal was to provide characterization of convective cloud systems, precipitation, and the environment in the development of model constraints for cumulus parameterizations and space-based rainfall retrieval algorithms over land. The NPOL scanning strategy emphasized vertical structure sampling and included three-tilt elevation surveillance PPI rain scans, narrow sector-volume scans, and high resolution (0.01°). RHI scans provide measurements of precipitation in liquid, mixed, and ice phases. Rapid DSD scans combined both PPI and RHI tasks with 40 sec resolution when precipitation was over the ARM Southern Great Plains Central Facility. Vertical profile PPI scans (MC3E BB) were performed for melting level (bright band) characterization and differential reflectivity (Z_{DR}) calibration. Data were processed with the CSU Hydrometeor Identification Algorithm (HID) providing classification and spatial distribution of hydrometeor types (e.g., rain, drizzle, large drops, hail, ice crystals, wet or dry snow, and graupel) (Table 6.4).

To achieve MC3E objectives on 3D convective cloud formation, microphysical parameterization in numerical weather prediction, and improve our

Table 6.4 MC3E scanning strategy

Task name	Scan type	Elevation (°)	Azimuth (°)	Samples	PRF
Surveillance – 0	PPI	0.5, 1.0, 1.5	0.0–360.0	60	500
Volume – 1A	PPI sector	20 sweeps from 0.5 to 19.5	Variable (30–120 deg spread)	60	1000
Volume – 1B	PPI sector	20 sweeps from 0.5 to 15.3	Variable (120 deg spread)	60	1000
RHI – 2	RHI	0.0–40.0	3 (variable)	60	1000
RHI – 3	RHI	0.0–40.0	6 (variable)	60	1000
DSD – 4A	PPI	1.0	245.0–325.0	60	1000
DSD – 4B	RHI	0.0–20.0	283.5	60	1000
MC3E BB	PPI	90.0	0.0–360.0	70	1000

understanding of satellite-based precipitation retrieval, scientists used a host of precipitation instrumentations, including NPOL to observe the environment [35]. There are several research experiments that have benefited from MC3E dataset.

A recent study used NPOL RHI in conjunction with aircraft and various ground-based data from the MC3E field campaign to thoroughly investigate a precipitation pattern in a Mesoscale Convective System (MCS) observed on May 11, 2011 that had not been studied thoroughly in literature. Under the influence of a relatively cold air mass, the MCS begins with a parallel stratiform rain type before transitioning to the more common trailing stratiform [36]. In addition, microphysical characteristics such as hydrometeor number concentration and total water content decreased with height. There is evidence of ice sublimation and aggregation in parallel stratiform region of MCS.

Another study evaluated the spatial correlation of DSD parameters and rain rate using NPOL and 2DVDs from two MC3E precipitation events [37]. Azimuthal dependence was found for the median drop diameter with increased variability in convection as opposed to stratiform. Keeping in mind the small sample size in the study, the authors suggest dual-polarization radar can be used to analyze spatial correlations of DSD parameters as long as the scan time is kept smaller than 1 min. The spatial correlations can be used for other radar applications such as nonuniform beam filling corrections and satellite retrieval development.

6.4.2.2 Iowa Flood Studies (IFloodS)

NASA conducted a field experiment in northeastern Iowa referred to as the Iowa Flood Studies (IFloodS) in collaboration with other government agencies and members of the U.S. academic research community, principally the University of Iowa. The main goal of the experiment was to support GV program activities of the GPM mission by specifically focusing on evaluating satellite rainfall data in flood forecasting. This entailed the collection of precipitation measurements using ground instruments and advanced weather radars while simultaneously collecting data from GPM and constellation satellite overpasses. NPOL employed numerous scanning strategies that were dependent on precipitation range, location, and echo height as detailed in Table 6.5.

The University of Iowa deployed a network of X-band polarimetric (XPOL) radars during the IFloodS field campaign within NPOL range for performance evaluation [38]. A PPI qualitative comparison between XPOL and NPOL shows similar reflectivity patterns. However, the two are not identical due to frequency and range resolution differences. In addition, both radars have different sampling volumes. For example, XPOL resembles a finer detail in the reflectivity pattern compared to NPOL due to its high-range sampling.

NPOL was collocated with the multifrequency D3R radar during IFloodS as part of the GPM GV program [39]. Results show D3R Ku-band reflectivity matches well with NPOL S-band reflectivity. In addition, Z_{DR} compares well in light to moderate precipitation, but D3R Ku overestimates relative to NPOL in heavy rain due to overcorrection of differential attenuation. This is likely due to the assumption of Rayleigh scattering when developing empirical relationship between

Table 6.5 IFloodS scanning strategy

Task name	Scan type	Elevation (°)	Azimuth (°)	Samples	PRF
D3Rnear30	PPI sector	22 sweeps from 0.7 to 21.3	110.0–140.0	60	1100
RHI sector	RHI	0.0–40.0	180.0, 182.0, 184.0, 186.0	72	1100
Rain volume	PPI full	9 sweeps from 0.0 to 4.0	0.0–360.0	60	1000
Rapid Rain A	PPI full	0.7, 1.4	0.0–360.0	60	1000
Rapid Rain B	RHI	0.0–40.0	83.0	72	1100
IFloodS BB	PPI full	90.0	0.0–360.0	128	1000
Top 8km@60km	PPI sector	12 sweeps from 0.7 to 7.9	338.0–70.0	60	1100
Top 6km@60km	PPI sector	12 sweeps from 0.7 to 5.3	338.0–70.0	60	1100

attenuation at horizontal and vertical polarization from K_{DP} measurements using disdrometer data.

Also, during IFloodS, a series of Micro Rain Radars (MRR) and disdrometers were deployed with increasing range from NPOL at a specific azimuth for an excellent comparative study to analyze effects of radar beam broadening [40]. MRR retrievals were compared with both disdrometers and NPOL during both stratiform and convective rainfall. Reflectivity comparisons show excellent agreement when considering all rainfall types with a bias < 1 dB while differential reflectivity bias is on the order of 0.20 dB. The study comments that MRR is a great tool for precipitation estimation in a vertical profile that can help fill gaps where point measurements and areal radar coverage can be lacking.

An experiment was performed that compared IFloodS datasets in assessing single- and dual-polarimetric radar-based rain rate retrievals to gauge corrected rainfall products with a network of gauges and disdrometers over Iowa in the context of flood forecasting [41]. Analyses were performed at different time scales to account for spatial and temporal variability. The normalized root mean square error decreased with increasing time up to 24 hours for all the algorithms tested. The study highlighted that gauge-based rainfall corrections are needed for radar-based rainfall estimation in the context of hydrological applications.

NPOL has used to derive a polarimetric radar rainfall algorithm using data from IFloodS experiment [42]. The algorithm is based on a regional hydrometeor classification that includes a temperature profile from a sounding. Results show that radar-estimated rainfall performance with ground disdrometers degrade with increasing radar range, an issue with radars called beam broadening. In addition, the new algorithm is superior to traditional single- and -dual polarization algorithms with root mean square errors closer to 0. This S-band rainfall algorithm can be widely used around the world, especially over the United States within the National Weather Service S-band radars.

6.4.2.3 Integrated Precipitation and Hydrology Experiment (IPHEX)

The Integrated Precipitation and Hydrology Experiment (IPHEx) was a NASA field campaign in the summer of 2014 to support the GPM GV program. The IPHEx field experiment was the first post-launch campaign since the launch of GPM satellite mission to study orographic precipitation over the Great Smokey Mountains of North Carolina [43]. NPOL provided high-quality polarimetric data, microphysical retrievals, and rain mapping to satisfy GPM GV efforts. The study area was in the Southern Appalachians of North Carolina in the southeastern United States. The goal of the campaign was to gain further understanding of how mountainous areas influence and interact with summertime precipitation before, during, and after rainfall. To accomplish these goals, NPOL, rain gauges, disdrometers, and other instrumentation were deployed throughout the mountainous region and in the Piedmont and Coastal Plain regions. Simultaneous data were collected using instruments on scientific research aircraft and satellites, including NASA's DC-8 and ER-2 aircraft. IPHEx was the first post-launch GPM GV field campaign. NPOL observations, coincident with NASA aircraft and the GPM satellite helped improve algorithms used to retrieve rainfall data from space. NPOL was located near Forest City, NC and collected data from April 27 to June 16, 2014. NPOL scanning strategies are shown in Table 6.6. NPOL rain rate estimates were computed over the Upper Tennessee, Catawba-Santee, Pigeon, Yadkin-Pee Dee, and Savannah River basins. Specific RHI scans were performed directly over rain gauge and disdrometer locations for DSD studies. For example, Tokay *et al.*'s study used the IPHEx disdrometer dataset with other GPM GV field campaign measurements to develop DSD parameter empirical relations that can be used for ground scanning radar retrievals to validate GPM Core products.

An experiment was conducted to evaluate synoptic patterns that are responsible for flooding rains in the Southern Appalachian Mountains using the IPHEx dataset [44]. The results show half of the heavy rain events were associated with large synoptic-scale atmospheric river phenomenon.

NASA's D3R radar was deployed for the IPHEx field experiment to demonstrate precipitation estimation at Ku- and Ka-band frequencies [45]. D3R has the same frequencies as GPM DPR; however, D3R can provide more information on

Table 6.6 IPHEx scanning strategy

Task name	Scan type	Elevation (°)	Azimuth (°)	Samples	PRF
FAR – 45	PPI sector	14 angles from 1.0 to 8.5	280.0–325.0	60	1100
NEAR – 45	PPI sector	19 angles from 1.0 to 30.9	40.0–85.0	60	1100
NEAR – 90	PPI sector	19 angles from 1.0 to 30.9	0.0–90.0	60	1100
Rain	PPI full	1.0, 1.5, 2.0	0.0–360.0	60	1000
Single RHI	RHI	0.0–60.0	291.5	72	1100
IPHEX BB	PPI full	90.0	0.0–360.0	128	1000
DSD	RHI	1.0–60.0	292.0, 298.0, 299.0	72	1000

precipitation microphysics due to dual-polarization capability. Operating with collocated NPOL radar, vast amount of data was collected at three different frequencies. Reflectivity comparison at an approximate azimuth of 300° shows good agreement between D3R-Ku and NPOL-S. Data from D3R can be crossvalidated with GPM.

Light rain measurements are critical due to their high occurrence over the globe [46]. Much of the literature focuses on S, C, and X-band radar frequencies when it comes to rainfall estimation. This study used D3R Ku-band frequency to estimate light rain. Data from IFloodS and IPHEx show the normalized standard error is 36.9% between D3R and APU comparison and 48.5% between D3R and gauge measurements. The error is higher between D3R and gauge due to sampling time for gauges.

6.4.2.4 Olympic Mountain Experiment (OLYMPEx)

The Olympic Mountains Experiment (OLYMPEx) was conducted in northwest Washington on the Olympic Peninsula from November 2015 through February 2016. OLYMPEx was the last GPM GV field campaign and followed the science guidelines in [47]. A variety of ground instruments and aircraft recorded data from storm systems as they approached and passed through the complex terrain of the Peninsula and Olympic Mountains. NPOL was positioned at a coastal site near Taholah, WA, and performed RHI sectors interspersed with low-level PPIs in a repeating cycle of 20 min (Table 6.7). A vertical profile "birdbath" scan was performed at the top of each hour. NPOL RHI coverage was over ocean sectors (210°–326°) and the Quinault River Valley (30°–60°). The scanning strategy covered the Chehalis rain gauge network, the University of Washington tipping bucket network, and numerous ground observation sites.

NPOL was deployed near the Olympic Mountains of Washington State to study terrain effects on precipitation [48]. During the full campaign, NPOL measured nearly 2,000 mm of rainfall. NPOL data provided insights on cloud structure and microphysical characteristics that resulted in the enhanced precipitation. The microphysical insights provided from this rich dataset help understand and eventually improve precipitation retrievals from space.

NPOL data have shown a secondary reflectivity maximum above the melting level indicating enhanced ice-phased processes during the OLYMPEX field campaign [49]. Results also indicate higher rain rates and larger drops measured at the ground on the windward side of the mountain when a secondary reflectivity

Table 6.7 OLYMPEx scanning strategy

Task name	Scan type	Elevation (°)	Azimuth (°)	Samples	PRF
RHI_A	RHI	0.0–45.0	40 angles from 210.0 to 326.0	72	1100
RHI_B	RHI	0.0–45.0	16 angles from 30.0 to 60.0	72	1100
PPI	PPI full	0.5, 1.5, 2.5	0.0–360.0	70	1100
BB	PPI full	90.0	0.0–360.0	128	1100

maximum was present. This finding can help understand precipitation at other mountainous regions around the world.

The OLYMPEX field experiment provided sufficient data to study the micro-physical characteristics associated with land-falling mid-latitude cyclones [50].

NPOL data near the mountains show DSD in stratiform rain within atmo-spheric river events showed variability due to the complex terrain. In addition, the dataset suggests the rainfall over the Olympic Mountains is a combination of warm-growth and ice-phased processes.

The unique dataset from the OLYMPEX experiment was used to demonstrate the utility of the SIMBA software [29]. A GPM Core Observatory overpass on December 3, 2015 measurements were compared with various ground platforms, including NPOL. Profiles of reflectivity, precipitation rate and mass-weighted mean diameter were compared between GPM Core products and NPOL. Results indicate profiles of reflectivity compare well for regions where there are no terrain effects, while discrepancies arise in complex terrain.

6.5 Validating GPM precipitation retrievals

NASA set science requirements for the GPM Mission in the pre-launch era for instrument performance, measurement accuracy, and science data [28].

One of the key requirements is for DPR to estimate the precipitation particle size distribution (such as D_m) to within +/− 0.5 mm with respect to ground-based retrieval. Another key requirement is for the bias (and random) error in instan-taneous rain rate retrieval to be within 50% at 1 mm h^{-1} and 25% at 10 mm h^{-1}. In order to confirm these requirements are met, ground validation (GV) is essential.

6.5.1 Using SIMBA to validate that GPM Level I requirements are met

The GPM sensor D_m retrieval accuracy requirement led an important study to develop ground-based empirical relationships using 2DVD and NPOL from several GPM GV field campaigns [8,25]. Ground "truth" validation with 2DVD and NPOL must be tested for accuracy before comparing to space-based retrieval due to assumptions and uncertainties [51,31]. NPOL was used to validate D_m retrieval with the GPM DPR and DPR/GMI combined algorithm from GPM overpasses from 2014 to 2020 over the eastern shore of Maryland [30]. Figure 6.12 illustrates how NPOL is used at the Wallops PRF to help validate DPR Level I Requirements. Each symbol represents a GPM overpass data value compared to reference GV datasets. The top row shows comparisons between the V06A 2ADPR [52] and 2BCMB [53] retrieved D_m versus disdrometer-measured D_m at Wallops, with correlations of 0.5–0.9 and biases of less than 0.15 mm, well within the Level I requirements. The middle panel shows satellite comparisons of reflectivity with correlations of 0.7–0.9 and biases of 1–5 dB which is generally within error bounds from the TRMM era [54]. The bottom panel shows comparisons of the retrieved

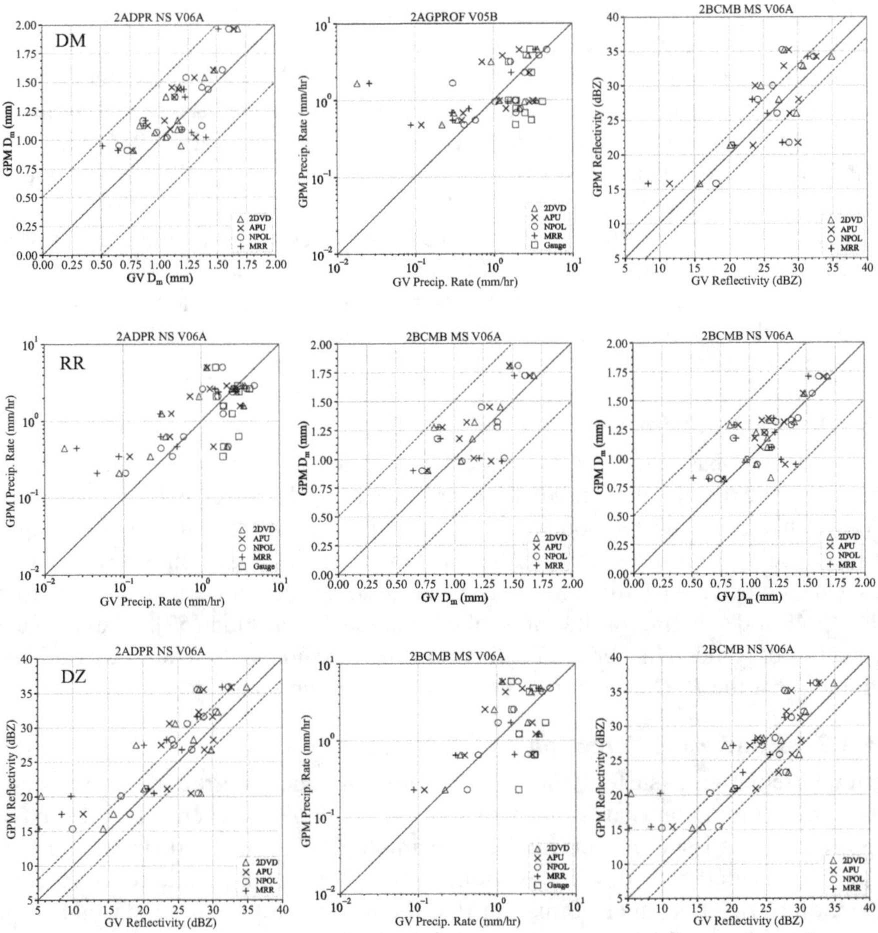

Figure 6.12 Comparisons of GPM DPR, and CMB products to ground-based measurements at the Wallops PRF from 2014 to 2020 GPM overpasses. Figure adapted from [30].

rain rates for satellite- and GV-measured rain rate, with correlation up to 0.75 and normalized biases within $\pm$ 40%, although there is tendency for the satellite estimates to overestimate light rain.

NASA's PMM Drop-Size Distribution Working Group (DSDWG) was initially established to investigate rainfall retrieval assumptions associated with DSD constraints using field campaign data from the TRMM era [55]. The group involved NASA PMM Science Team, algorithm developers, and ground validation scientists. This group evolved with time and during the GPM era was named the Particle Size Distribution Working Group (PSDWG) to include snowfall precipitation. The main goal of this group is to bridge the gap between ground validation and satellite algorithm development work to enhance GPM retrievals and products. The central theme is to improve PSD because it plays a critical role in precipitation (both liquid and solid) algorithm retrievals. To accurately retrieve rainfall (or snowfall) rate, PSD assumptions are required that involve information of the mass, count, volume, phase, and fall speed of falling particles. In addition, retrieval of PSD and precipitation rate profiles from GPM-DPR sensor are lacking near the surface due to ground clutter contamination.

Currently, there are three main goals that the PSDWG is attempting to investigate: (a) PSD parameter retrieval; (b) "blind zone" detection; and (c) phase discrimination. NASA WFF and hence NPOL play a critical part in topics (a) and (b); however, other more northern sites and platforms are being utilized to study topic (c).

6.5.1.1 PSD parameter retrieval

Accurate space-based radar retrievals are crucial to the success of the GPM Core mission. For example, the DSD parameter D_m is strongly correlated to rainfall rate R [56]. This important relation was first introduced in GPM DPR Version 4 algorithm. Results from a research study that looked at the R–D_m empirical relation using disdrometer data for rain show mean absolute percent error ranging from 29% to 47% depending on the prescribed relation in question [57]. The absolute percent error is much higher at 77% for snowfall measurements. The study findings highlight that the R–D_m relation is not adequate for snowfall.

6.5.1.2 Blind zone detection

Space-based radars suffer from the blind zone due to ground clutter contaminating the radar gates near the surface [58,59]. Several studies have reported on this issue in literature. For example, DPR estimated stratiform rain is underestimated over the Andes Mountains, which is likely due to blind zone detection [60]. This paper points out that the lowest clutter-free radar gate that can be used for analysis ranges up to 2 km above ground level and sometimes extending well above the melting layer up to 5 km depending on orbit track and cross-track scanning angle. In another study, stratiform profiles from two GPM overpasses over the Delmarva Peninsula are shown to illustrate key information DPR misses below the lowest clutter-free radar gate [30]. To help blind zone detection algorithm developers, these two cases illustrate how data from ground radars and disdrometers help fill in the blind zone from the surface up to the clutter-free gate.

Two stratiform vertical profiles show slightly different results compared to GV dataset. The lighter rain case (top) shows profiling algorithms generally agree while the strong BB case shows mismatches below the melting level. The surface algorithms overestimate Z, R, and D_m in both cases. In the first and third panel of Figure 6.13, NPOL, MRR, and surface platforms fill the gap below 1.0 km in both cases where DPR is blind due to ground clutter. More cases can provide precipitation and reflectivity patterns within the blind zone to develop a parameterization technique for future GPM satellite algorithms.

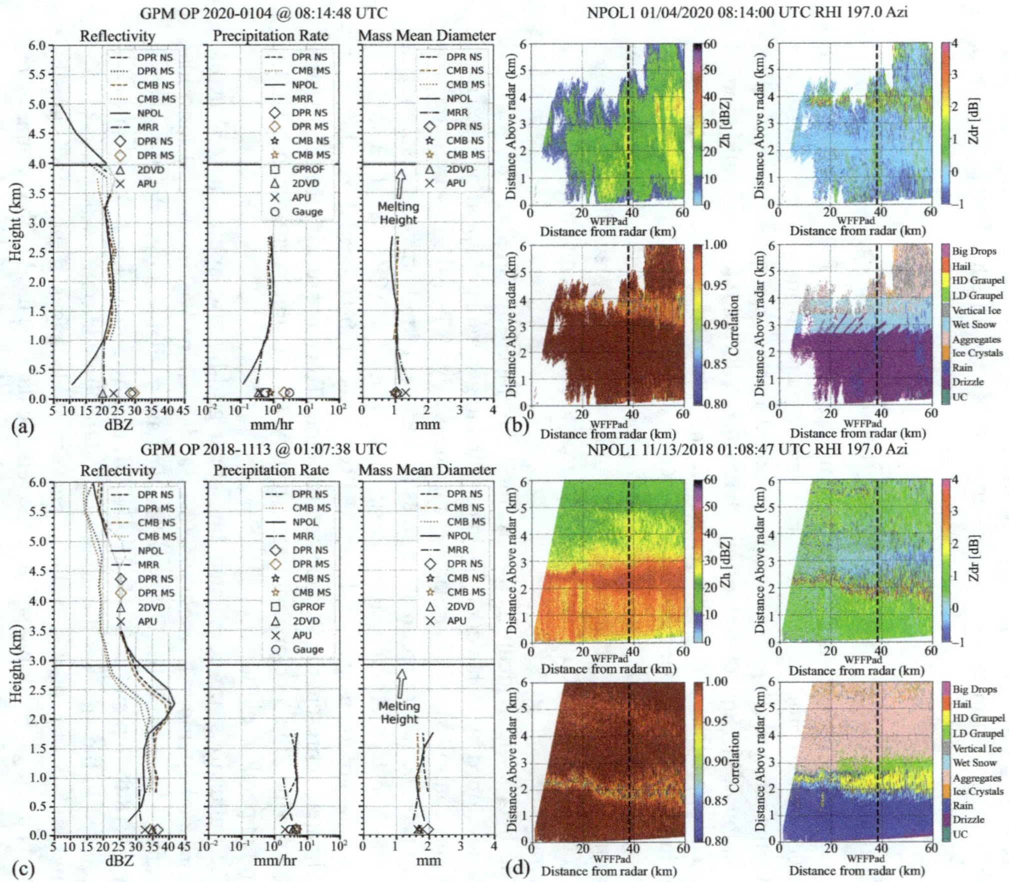

Figure 6.13 Mean vertical profiles from two different GPM overpasses along with NPOL RHI cross-sections. Vertical line in RHI plots shows the WFF Pad instrument site. Figure extracted from [30].

6.5.1.3 Phase discrimination

The third key goal of the PSDWG is to help improve precipitation phase discrimination for both passive and active algorithms [42,61]. According to the 2022 PSDWG Charter, there are two key research areas: (1) to understand model parameterization assumptions that drive phase transitions; and (2) to explore relationships between vertical profiles of environmental humidity and wet bulb temperature to guide model parameterizations. Utilize ground-based profiles from remote-sensing instruments (e.g., profiling radars) and in situ observations (e.g., radiosondes).

6.5.2 Validation of IMERG precipitation using NPOL

The Integrated Multi-satellitE Retrievals for GPM (IMERG) [62] is a global gridded precipitation product that unifies measurements from passive microwave (PMW) and infrared (IR)-based sensors onboard a network of satellites in the GPM constellation [59]. Substantial efforts have been devoted for the validation of IMERG precipitation estimates against ground observations such as radars and gauges in different areas of the world. The GPM GV program at NASA is currently underway to independently verify IMERG products. One of the direct statistical validation studies at the surface is carried out by comparing precipitation estimates from Version 06B IMERG Final Run with ground measurements from the NASA's NPOL radar.

The hydrometeor identification method developed by Cifelli *et al.* [23] and further developed by Dolan *et al.* [24] is applied to generate precipitation rates from the quality-controlled NPOL radar measurements. PyART [14] software is utilized to interpolate the radar precipitation rates to 1 km $\times$ 1 km horizontal resolution at the 1 km Constant Altitude Plan Position Indicator (CAPPI) level. The 1-km gridded data are then averaged to IMERG grids at 0.1° spatial resolution to approximately represent the surface precipitation. To compare with the half-hourly IMERG product, the precipitation rates from NPOL volume scans are averaged to 30 min temporal resolution.

The validations are conducted through a grid-by-grid approach [63] at the common spatial-temporal resolution (0.1° and 30 min) within 15–100 km radar scan radius when both IMERG and NPOL data are available during the period from January 2018 to September 2021. The NPOL data within 15 km radar scan radius are excluded in the analysis due to ground clutter. In addition, two sectors (155°–170° and 310°–330°) are also excluded because of blockages of the radar beams by trees and buildings. By doing so, only the reliable data are used in the comparisons.

IMERG performance can be traced back to each sensor (Table 6.8) onboard a satellite, or a retrieval method (e.g. morphing only, IR estimating only, or mixture of morphing and IR estimating) by utilizing the IMERG ancillary variables. Table 6.8 lists all sensors used in IMERG V06B over the NPOL radar scan area during the study period (January 2018 to September 2021). "Index" in Table 6.8 is the source number for each sensor provided in IMERG ancillary variable HQprecipSource. AMSR, SSMIS, and GMI are imagers, whereas MHS and ATMS are sounders.

Table 6.8 Sensors used in IMERG V06B over NPOL scan area from January 2018 to September 2021

Index	Sensor acronym	Sensor name	Sensor type	Satellite
3	AMSR	Advanced Microwave Scanning Radiometer Version 2	Imager	GCOM-W1
5	SSMIS	Special Sensor Microwave Imager/ Sounder	Imager	DMSP-F16-19
9	GMI	GPM Microwave Imager	Imager	GPM
7	MHS	Microwave Humidity Sounder	Sounder	NOAA-18,19; MetOp-A, B
11	ATMS	Advanced Technology Microwave Sounder	Sounder	Suomi-NPP

The statistical metrics such as Pearson Correlation Coefficient (CC), Relative Bias (Bias), Normalized Mean Absolute Error (NMAE), and Normalized Root Mean Square Error (NRMSE) are used to quantitatively compare the performance of IMERG product. To evaluate the precipitation detection capability, three widely applied categorical statistical metrics, i.e., Probability of Detection (POD), False Alarm Ratio (FAR), and Heidke Skill Score (HSS) are also used in the analysis. The equations of these metrics are not listed here as interested readers can refer to a textbook [64]. Furthermore, a volumetric error decomposition approach [65,66] is adapted to decompose the total error (E) of precipitation volume into three independent components: (1) hit bias (H)—precipitation is detected, but the detected precipitation amount is higher or lower than the reference precipitation amount; (2) bias due to missed precipitation (−M)—precipitation is missed, which results in a negative bias; and (3) bias due to false precipitation detection (F)—precipitation is mistakenly detected while there is no precipitation, which results a positive bias. The relationship of these components is governed by $E = H - M + F$. Each component could be possibly larger than the total error as one component could cancel another, resulting in a smaller total error. In the calculation of these statistics, the precipitation or no-precipitation threshold is set to 0.1 mm h^{-1}.

Figure 6.14 shows the comparisons of statistical metrics for IMERG sources. The sample size from each source is very different. The source with larger sample size certainly contributes more toward to IMERG generation. The morph-only algorithm has the largest sample size, followed by IR+morph, whereas IR-only has the smallest sample size (0.1%). IMERG uses the direct IR estimate only when the PMW estimates and associated morphing estimates are poor, mainly over ice- or snow-covered surfaces during winters. IR category shows the worst performance among all sources in terms of correlation, POD, and HSS. In contrast, GMI outperforms all other sources with best correlation, NMAE, NRMSE, FAR, and HSS values, which could be attributed to GMI's finest footprint and full spectrum of frequencies. The comparison among GMI and other radiometers against surface radars over land [67] and ocean [68] also showed that GMI has the highest correlation and best HSS score.

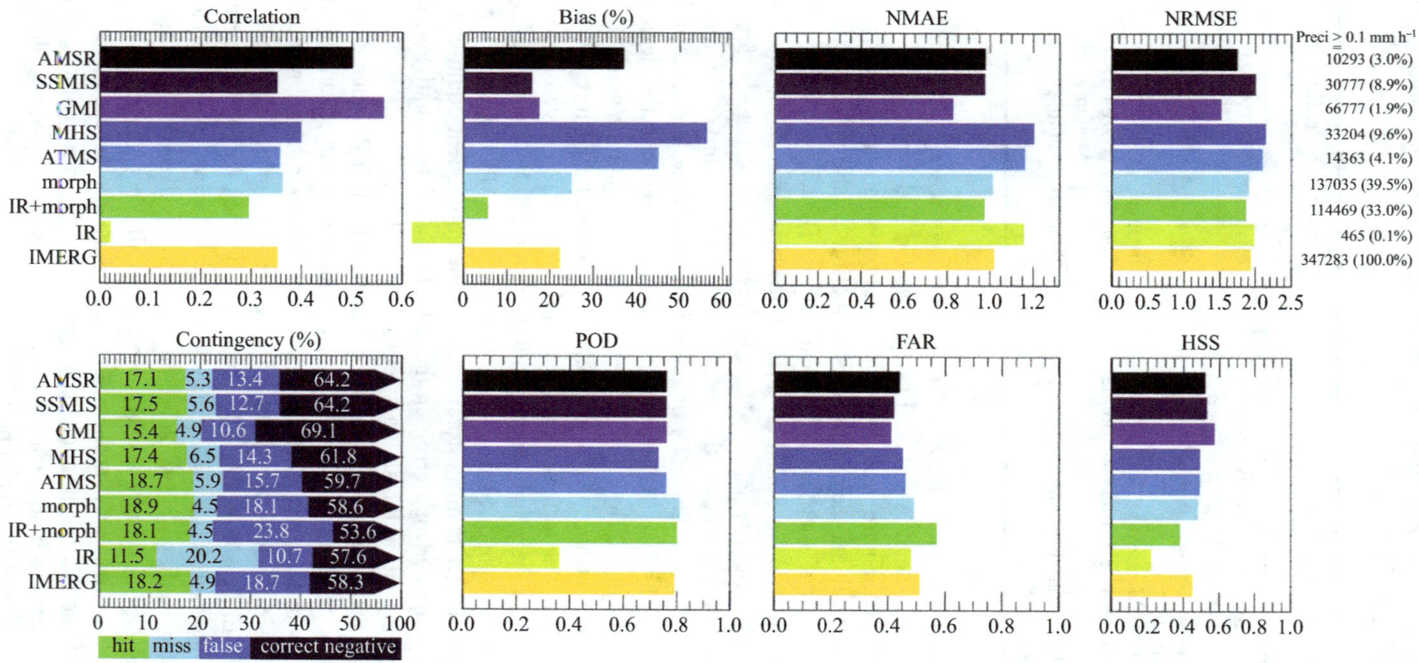

Figure 6.14 *Pearson correlation coefficient (Correlation), Relative Bias (Bias), Normalized Mean Absolute Error (NMAE), Normalized Root Mean Square Error (NRMSE), Hits, misses, false alarms, and correct negatives, Probability of detection (POD), false alarm ratio (FAR), and Heidke Skill Score (HSS) for V06B IMERG relative to NPOL. The numbers at the right in the upper panel are the sample sizes and their percentages of the total samples for precipitation rates at least 0.1 mm* h^{-1}*. The bottom rows in upper and lower panels are for IMERG regardless of its sources.*

The contingency table in Figure 6.14 shows that more than 50% of the occurrences in each source are correct negatives. Each source has fewer miss rates except for IR category, and considerable hit and false rates.

Comparison between two sounders shows that ATMS appears to be slightly better than MHS in terms of Bias, NMAE, NRMSE, POD, and FAR (Figure 6.14), but the correlation for ATMS is slightly lower than that for MHS. For overall detection performance, the difference between ATMS and MHS is neglectable as indicated by their very close HSS scores.

For HSS scores in Figure 6.14, all imagers (AMSR, SSMIS, and GMI) perform better than all sounders (MHS and ATMS). Among imagers, GMI is the best followed by SSMIS, then AMSR. Morph-based algorithms (morph and IR+morph) only perform better than IR-only. Morph-only performs comparably to sounders, and better than IR+morph. IR-only shows the worst performance among all sources.

The IMERG total bias for each source can be decomposed into hit, missed-precipitation, and false-precipitation biases (Figure 6.15). Only IR category has a negative hit bias and negative total bias. The total negative bias for IR is mainly due to large amount of missing precipitation. This is also reflected by the large miss rate and poor POD score in Figure 6.15. Figure 6.15 also shows that GMI has the least amount of missing and false precipitation, and relatively lower hit bias; as a result, the total bias for GMI is low. Surprisingly, MHS has the largest hit, false, and total biases among all imagers and sounders. The total bias for IMERG (regardless of sources) is positive, which is mainly due to large false precipitation.

To investigate the distribution of precipitation rates, joint distributions of half-hourly precipitation rates at 0.1° resolution are constructed for each IMERG source relative to NPOL for all matched data during the study period. The precipitation threshold is 0.1 mm h^{-1}. The black dotted line overlayed on each panel of Figure 6.16 is the quantile–quantile (Q–Q) plot for the precipitation rates from IMERG and NPOL. The numbers on the bottom right in each panel are the sample sizes (number of hits), ratio (IMERG/NPOL), and correlation. In comparison with all other categories, GMI has the lowest random error, which is demonstrated by its tightest scatters and highest correlations. The benefit of IR estimates in the IMERG V06B development over the NPOL scan area seems to be questionable as shown in

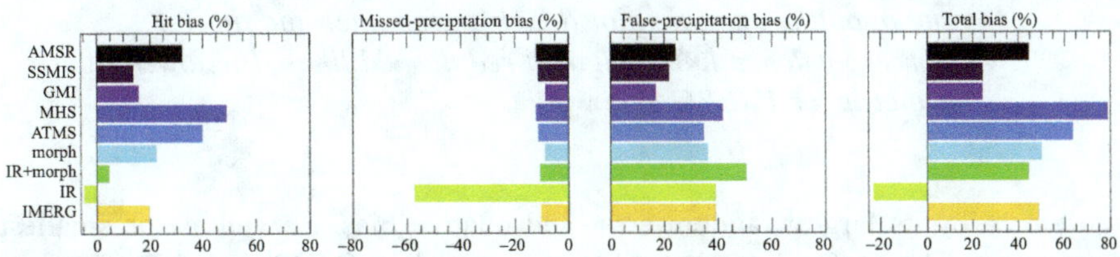

Figure 6.15 Error components of V06B IMERG sources: total bias (E), hit bias (H), biases due to missing precipitation (−M), and false precipitation (F). The components are related by E = H − M + F. The bottom row is for IMERG regardless of sources.

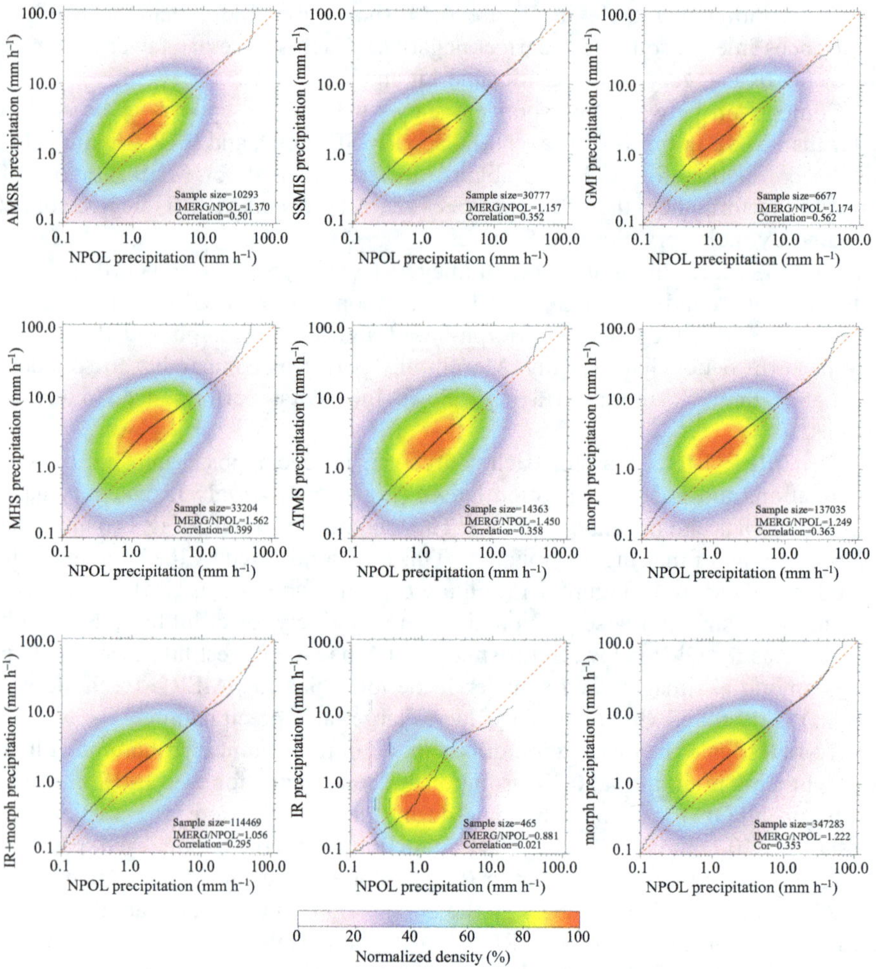

Figure 6.16 Joint distribution of half-hourly precipitation rates at 0.1° resolution for each IMERG source against NPOL conditional on both precipitation rates at least 0.1 mm h^{-1}. Overlayed on each panel are the quantile–quantile plot (black dotted line) and the 1:1 correspondence line (diagonal red dashed line). The lower right panel is for IMERG as a whole.

Figure 6.16 by its largest random error and relative bias, as well as the smallest sample size. The Q–Q plots show that two sounders (MHS and ATMS) overestimated precipitation at all intensities, especially for very heavy precipitation. The overestimation for imagers (AMSR, SSMIS and GMI) is relatively lower than sounders. The bias in the morphing estimation is comparable to those from PMW. IR+morph has the lowest relative bias but with the worst correlation among all sources except for IR category. IMERG overestimates precipitation rates with relative bias of 22%.

In summary, IMERG V06B was evaluated against NPOL precipitation product near NASA's Wallops Flight Facility in Newark, Maryland in the mid-Atlantic region of the United States. IMERG performance was traced to the individual sensors and retrieval techniques used in the algorithm. In general, IMERG estimates tend to overestimate precipitation with considerable random error. The overestimation is mainly due to large false precipitation. All imagers (AMSR, SSMIS and GMI) perform better than all sounders (MHS and ATMS) in terms of HSS score and other metrics. Among imagers, GMI is the best, followed by SSMIS, then AMSR. Between two sounders, ATMS appears to be slightly better than MHS in terms of Bias, NMAE, NRMSE, POD, and FAR, but the correlation for ATMS is slightly lower than that for MHS, whereas the overall detection performance is very similar. Morph-only algorithm performs comparably to sounders, and better than the mixture of morphing and IR. IR shows the worst performance among all IMERG sources.

It should be noted that this study only uses measurements from one ground radar as a reference. The precipitation product derived from a single radar over a 100 km radar scan area certainly cannot represent the entire precipitation distribution over a large area with complicated climate features. In addition, the intermittent operation of NPOL results in fewer samples available for our analysis and thus limits the robustness of our conclusions.

6.5.3 Validation studies using NPOL at Wallops

NPOL was used in conjunction with 2DVD data from several GPM GV field experiments to analyze horizontal and differential reflectivity, along with DSD parameter D_m and N_w empirical equations that are used for validation of GPM DPR retrievals ([25]—need to correct). Results show NPOL measured reflectivity and differential reflectivity comparisons with 2DVD improved when removing up to 15% of data that were outside of 2DVD estimated. In addition, the absolute biases dropped for reflectivity, differential reflectivity, mass-weighted mean diameter, and normalized intercept parameter.

Hurricanes exhibit a complex structure of precipitation phase and related size distribution. Thurai *et al.* [33] investigated the characteristics of raindrop size distribution in outer rainbands of Hurricane Dorian where NPOL's RHI scans showed bright band peaks of 55 dBZ. In a separate study, Bringi *et al.* [69] used NPOL and two different disdrometer data from the Wallops PRF to analyze a particle-based Monte Carlo microphysical model (McSnow) to simulate the outer rain bands of Hurricane Dorian and concluded that the collisional break-up plays an important role in shaping the DSD at rain rates as high as 10 mm h^{-1}.

Stratiform and convective rain represent distinct microphysical properties and provide unique profiles of latent heat release. Thurai *et al.* [70] used NPOL data from a widespread event to compare two different methods of convective and stratiform rain (C/S) partitioning, and these methods showed disagreement in only 12% of the radar pixels. In a separate study, Thurai *et al.* [33] tested a disdrometer-based

$N_W - D_m$ C/S partitioning algorithm where D_m is the mass-weighted mean diameter and N_W is normalized intercept parameter. NPOL RHI scans over the disdrometer sites revealed that the separation criteria from previous studies in other climate regimes (e.g. mid-latitude continental, semi-arid, and sub-tropical continental) required no modification at this coastal mid-latitude site, Wallops Island, Virginia.

Thurai *et al.* [71] focused on the retrieval of moments of the size distribution from GPM DPR and indirect comparison with the disdrometer-derived size distribution parameters.

They used NPOL quasi-vertical profiles to determine precipitation type during the passage of remnants of Hurricane Sally and concluded that the stratiform rain dominated the precipitation during GPM overpass.

6.6 Conclusion

NPOL has served the radar community well and is the flagship weather radar for NASA's Precipitation Measurement Missions (PMM). NPOL has and will continue to provide high-quality precipitation measurements and its transportability will allow for continued sampling in different regions. The key to maintaining these measurements is to rely on advances in weather radar data processing to provide accurate calibration, microphysical retrievals, and of course high-quality precipitation measurements.

List of acronyms

2ADPR	GPM Level 2A DPR product
2AGPROF	GPM Level 2A GMI product
2ACMB	GPM Level 2A Combined product
2DVD	Two-dimensional Video Disdrometer
AMSR	Advanced Microwave Scanning Radiometer Version 2
ARM	Atmospheric Radiation Measurement
ATMS	Advanced Technology Microwave Sounder
BB	Birdbath or 90° radar scan"
C/S	Convective/Stratiform
CAPPI	Constant Altitude Plan Position Indicator
CC	Pearson Correlation Coefficient
CDF	Cumulative Distribution Function
CMB	GPM Combined instrument product
DM	Mass-weight median drop diameter
DOE	Department of Energy
DP	Dual Polarization
DPQC	Dual-Polarization radar Quality Control

DPR	Dual-frequency Radar
DSD	Drop Size Distribution
EEC	Enterprise Electronics Corporation
FAR	False Alarm
GMI	GPM Microwave Imager
GPM	Global Precipitation Measurement mission
GV	Ground Validation
GVradar	Python toolkit to perform QC on radar data
HID	Hydrometeor Identification Algorithm
HSS	Heidke Skill Score
IFDR	Intermediate Frequency Digital Receiver
IFloodS	Iowa Floods Studies
IMERG	Integrated Multi-satellitE Retrievals for GPM
IPHEx	Integrated Precipitation and Hydrology Experiment
IR	Infrared radiation
KDP	Specific Differential Phase
MC3E	Midlatitude Continental Convective Clouds Experiment
MCS	Mesoscale Convective System
McSnow	Monte Carlo microphysical model
MHS	Microwave Humidity Sounder
MRR	Micro Rain Radar
NMAE	Normalized Mean Absolute Error
NPOL	NASA polarimetric S-band radar
NRMSE	Normalized Root Mean Square Error
NW	Normalized intercept parameter
OLYMPEx	Olympic Mountain Experiment
PDF	Probability Density Function
PMM	Precipitation Measurement Missions
PMW	Passive Microwave
POD	Probability of Detection
PSDWG	Particle Size Distribution Working Group
PSI	Pulse System Incorporated
PyART	Python Atmospheric Radiation Measurement (ARM) Radar Toolkit
Q–Q	Quantile–Quantile plot
QC	Quality Control
RAP	Rapid Refresh model
RCA	Relative Calibration Adjustment

RHI	Range Height Indicator
SIMBA	System for Integrating Multiplatform Data to Build the Atmospheric Column
SPRINT	Sorted Position Radar INTerpolation
SSMIS	Special Sensor Microwave Imager/Sounder
V05B	GPM Version 5B products
V06A	GPM Version 6A products
WFF	Wallops Flight Facility
WPRF	Wallops Precipitation Research Facility
WSR-88D	"Weather Service Radar-1988 Doppler"
XPOL	The University of Iowa X-band polarimetric radar
ZDR	Differential reflectivity
ZH	Horizontal reflectivity
dp_products	Python toolkit to retrieve various products from radar data
ΦDP	Differential phase
σ(ΦDP)	Variable of differential phase

References

[1] Silberstein D. S., Wolff D. B., Marks D. A., Atlas D., and Pippitt, J. L. 'Ground clutter as a monitor of radar stability at Kwajalein, RMI'. *J. Atmos. Ocean. Technol.* 2018, vol. 25, pp. 2037–2045.

[2] Wolff D. B., Marks D. A., and Petersen W.A. 'General application of the relative calibration adjustment (RCA) technique for monitoring and correcting radar reflectivity calibration'. *J. Atmos. Ocean. Technol.* 2015, vol. 32, pp. 496–506.

[3] Gorgucci, E., Scarchilli G., and Chandrasekar V. 'Calibration of radars using polarimetric techniques'. *IEEE Trans. Geosci. Remote Sens*. 1992, vol. 30, pp. 853–858.

[4] Gorgucci E., Scarchilli G., and Chandrasekar V. 'A procedure to calibrate multiparameter weather radar using properties of the rain medium'. *IEEE Trans. Geosci. Remote Sens*. 1999, vol. 34, pp. 269–276.

[5] Scarchilli G., Gorgucci E., Chandrasekar V., and Dobaie A. 'Self-consistency of polarization diversity measurement of rainfall'. *IEEE Trans. Geosci. Remote Sens*. 1996, vol. 34, pp. 22–26.

[6] Vivekanandan J., Zhang G., Ellis S., Rajopadhyaya D., and Avery S. 'Radar reflectivity calibration using differential propagation phase measurement.' *Radio Sci*. 2003, vol. 38, p. 8049. doi:10.1029/2002RS002676.

[7] Ryzhkov A.V., Giangrande, S. E., Melnikov V. M., and Schuur T. J. 'Calibration issues of dual-polarization radar measurements'. *J. Atmos. Ocean. Technol.* 2005, vol. 22, pp. 1138–1154.

[8] Tokay A., D'Adderio L. P., Wolff, D. B, and Petersen, W. A. 'Development and evaluation of the raindrop size distribution parameters for the NASA Global Precipitation Measurement Mission ground validation program'. *J. Atmos. Ocean. Technol.* 2020, vol. 37, pp. 115–128. https://doi.org/10.1175/JTECH-D-18-0071.1.

[9] Andsager K., Beard K. V., and Laird N. F. 'Laboratory measurements of axis ratios for large raindrops'. *J. Atmos. Sci.* 1999, vol. 56, pp. 2673–2683.

[10] Beard, K. V., and C. Chuang. 'A new model for the equilibrium shape of raindrops'. *J. Atmos. Sci.* 1987, vol. 44, pp. 1509–1524.

[11] Beard K. V. 'Terminal velocity and shape of cloud and precipitation drops aloft'. *J. Atmos. Sci.* 1976, vol. 33, pp. 851–864.

[12] Pippitt J. L., Marks D. A., and Wolff D. B. 'Dual-polarimetric quality control for NASAs global precipitation measurement (GPM) mission ground validation program'. *36th Conf. Radar Meteorol.*, Amer. Meteorol. Soc. Breckenridge, CO, USA, 2013. https://ams.confex.com/ams/36Radar/webprogram/Handout/Paper228522/36radarposter.pdf.

[13] Rinehart R. E. 'On the use of ground return targets for radar reflectivity calibration checks'. *J. Appl. Meteorol.* 1978, vol. 17, pp. 1342–1350.

[14] Helmus J. J., and Collis, S. M. 'The python ARM radar toolkit (Py-ART), a library for working with weather radar data in the python programming language'. *J. Open Res. Softw.* 2016, vol. 4(1), p. e25. DOI: http://doi.org/10.5334/jors.119.

[15] Lang T., Dolan B., Guy N., Gerlach C., and Hardin J. 'CSU-Radarmet/CSU_RadarTools: CSU_RadarTools v1.3 (v1.3)'. *Zenodo*. 2019. https://doi.org/10.5281/zenodo.2562063.

[16] Ryzhkov A. V., and Zrnić D. S. 'Polarimetric rainfall estimation in the presence of anomalous propagation'. *J. Atmos. Ocean. Technol.* 1998, vol. 15, pp. 1320–1330. https://doi.org/10.1175/1520-0426(1998)015<1320:PREITP>2.0.CO;2.

[17] Pippitt J. L.,Wolff D. B., Marks D. A., Pabla C. S., and Gardner B. M. 'A python-based radar data processing system for the NASA GPM ground validation program'. *AGU Fall Meeting, Virtual*, 13–17, Dec 2021.

[18] Zrnic D. S., and Ryzhkov A. V. 'Polarimetry for weather surveillance radars'. *Bull. Amer. Meteorol. Soc.* 1999, vol. 80, pp. 389–406.

[19] Cifelli R., Petersen W. A., Carey L. D., Rutledge S. A., and Da Silva Dias, M. A. 'Radar observations of the kinematic, microphysical, and precipitation characteristics of two MCSs in TRMM LBA'. *J. Geophys. Res. Atmos.* 2002, vol. 107(D20), LBA-44.

[20] Marks D. A., Wolff D. B., Carey L. D., and Tokay A., 'Quality control and calibration of the dual-polarization radar at Kwajalein, RMI'. *J. Atmos. Ocean. Technol.* 2011, vol. 28, pp. 181–196.

[21] Hubbert J., and Bringi V. N. 'An iterative filtering technique for the analysis of copolar differential phase and dual-frequency radar measurements'. *J. Atmos. Ocean. Technol.* 1995, vol. 12, pp. 643–648.

[22] James C. N., and Houze Jr. R. A. 'A real-time four-dimensional Doppler dealising scheme'. *J. Atmos. Ocean. Technol.* 2001, vol. 18, pp. 1674–1683.

[23] Cifelli R., Chandrasekar V., Lim S., Kennedy P. C., Wang Y., and Rutledge, S. A. 'A new dual-polarization radar rainfall algorithm: Application in Colorado precipitation events'. *J. Atmos. Ocean. Technol.* 2011, vol. 28(3), pp. 352–364.

[24] Dolan B., Rutledge S. A., Lim S., Chandrasekar V., and Thurai M. 'A robust C-band hydrometeor identification algorithm and application to a long-term polarimetric radar dataset'. *J. Appl. Meteorol. Climatol.* 2013, vol. 52(9), pp. 2162–2186.

[25] Tokay A., D'Adderio,L. P., Marks D. A., Pippitt J. L., Wolff D. B., and Petersen W. A. 'Comparison of raindrop size distribution between NASA's S-band polarimetric radar and two-dimensional video disdrometers'. *J. Appl. Meteorol. Climatol.* 2020, vol. 59, pp. 517–533.

[26] Seto S., Iguchi T., Meneghini T. R., *et al.* 'The precipitation rate retrieval algorithms for the GPM dual-frequency precipitation radar'. *J. Meteorol. Soc. Japan*. 2021, *Ser. II*, vol. 99, pp. 205–237.

[27] Liao L., and Meneghini R. 'GPM DPR retrievals: Algorithm, evaluation, and validation'. *Remote Sens*. 2022, vol. 14, p. 843. https://doi.org/10.3390/rs14040843.

[28] Skofronick-Jackson G., Petersen W. A., Berg W., *et al.* 'The global precipitation measurement (GPM) mission for science and society'. *Bull. Amer. Meteorol. Soc.* 2017, vol. 98, pp. 1679–1695. https://doi.org/10.1175/BAMS-D-15-00306.1.

[29] Wingo S. M., Petersen W. A., Gatlin P. N., Pabla C. S., Marks D. A., and Wolff D. B. 'The system for integrating multiplatform data to build the atmospheric column (SIMBA) precipitation observation fusion framework'. *J. Atmos. Ocean. Technol.* 2018, vol. 35, pp. 1353–1374.

[30] Pabla C. S., Wolff D. B., Marks D. A., Wingo S. M., and Pippitt J. L. 'GPM ground validation at NASA wallops precipitation research facility'. *J. Atmos. Ocean. Technol.* 2022, vol. 39, pp. 1199–1215. https://doi.org/10.1175/JTECH-D-21-0122.

[31] Hou A. Y., Kakar R. K., Neek S., *et al.* 'The global precipitation measurement mission'. *Bull. Amer. Meteorol. Soc.* 2014, vol. 95, pp. 701–722. https://doi.org/10.1175/BAMS-D-13-00164.1.

[32] Simpson J., Alder R. F., and North G. 'A proposed tropical rainfall measuring mission (TRMM)'. *Bull. Amer. Meteorol. Soc.* 1998, vol. 69, pp. 278–295.

[33] Thurai M., Bringi V., Wolff D., Marks D., and Pabla C. 'Testing the drop-size distribution-based separation of stratiform and convective rain using radar and disdrometer data from a mid-latitude coastal region'. *Atmosphere*. 2021, vol. 12(3), p. 392. https://doi.org/10.3390/atmos12030392.

[34] Gatlin P. N., Thurai M.,Williams C., and Adirosi E. 'Measurement and modeling of the precipitation particle size distribution'. *Atmosphere*. 2021, vol. 12(7),p. 819. https://doi.org/10.3390/atmos12070819.

[35] Jensen M. P., Petersen W. A., Bansemer A., Bharadwaj N., Carey L. D., and Cecil D. J. 'The midlatitude continental convective clouds experiment

(MC3E)'. *Bull. Amer. Meteorol. Soc.* 2016, vol. 97, pp. 1667–1686. https://doi.org/10.1175/BAMS-D-14-00228.1.

[36] Skow A., Poellot M., Delene D., Askelson M., North K., and Oue M. 'A multi-platform in situ kinematic and microphysical analysis of a hybrid parallel–trailing stratiform mesoscale convective system'. *Mon. Weather Rev.* 2022, vol. 150, pp. 927–948. https://doi.org/10.1175/MWR-D-21-0009.1.

[37] Bringi V. N., Tolstoy L., Thurai M., and Petersen W. A. 'Estimation of spatial correlation of drop size distribution parameters and rain rate using NASA's S-band polarimetric radar and 2D video disdrometer network: Two case studies from MC3E'. *J. Hydrometeorol.* 2015, vol. 16, pp. 1207–1221. https://doi.org/10.1175/JHM-D-14-0204.1.

[38] Mishra K. V., Krajewski W. F., Goska R. G., *et al.* 'Deployment and performance analyses of high-resolution Iowa XPOL radar system during the NASA IFloodS campaign'. *J. Hydrometeorol.* 2016, vol. 17, pp. 455–479. https://doi.org/10.1175/JHM-D-15-0029.1.

[39] Beauchamp R. M., Chandrasekar V., Chen H., and Vega M. 'Overview of the D3R observations during the IFloodS field experiment with emphasis on rainfall mapping and microphysics'. *J. Hydrometeorol.* 2015, vol. 16, pp. 2118–2132. https://doi.org/10.1175/JHM-D-15-0023.1.

[40] Adirosi E., Baldini L., and Tokay A. 'Rainfall and DSD parameters comparison between micro rain radar, two-dimensional video and Parsivel2 disdrometers, and S-band dual-polarization radar'. *J. Atmos. Ocean. Technol.* 2020, vol. 37, pp. 621–640. https://doi.org/10.1175/JTECH-D-19-0085.1.

[41] Seo B.-C., Krajewski W. F., Quintero F., *et al.* 'Comprehensive evaluation of the IFloodS radar rainfall products for hydrologic applications'. *J. Hydrometeorol.* 2018, vol. 19, pp. 1793–1813. https://doi.org/10.1175/JHM-D-18-0080.1.

[42] Chen S., Gourley J. J., Hong Y., *et al.* 'Using citizen science reports to evaluate estimates of surface precipitation type'. *Bull. Amer. Meteorol. Soc.* 2016, vol. 97(2), pp. 187–193. https://doi.org/10.1175/bams-d-13-00247.1.

[43] Barros A. P. NASA GPM-ground validation: Integrated precipitation and hydrology experiment *NASA Tech. Rep. [online]*. 2014. Available from https://doi.org/10.7924/G8CC0XMR.

[44] Miller D. K., Hotz D., Winton J., and Stewart L. 'Investigation of atmospheric rivers impacting the pigeon river basin of the Southern Appalachian Mountains'. *Weather Forecast.* 2018, vol. 33, pp. 283–299.

[45] Chandrasekar V., Baldini L., Bharadwaj N., and Smith P. L. 'Calibration procedures for global precipitation-measurement ground-validation radars'. *URSI Radio Sci. Bull.* 2015, vol. 355, p. 4573. https://doi.org/10.23919/URSIRSB.2015.7909473.

[46] Chen H., and Chandrasekar V. 'Characterization and estimation of light rainfall using NASA D3R observations during GPM IFloodS and IPHEx field campaigns'. *IEEE Int. Geosci. Remote Sens. Symp.: Proc., IEEE*, 2015, 26262629, https://doi.org/10.1109/IGARSS.2015.7326351.

[47] Houze R. A., McMurdie L. A, Petersen W. A., *et al.* 'The Olympic Mountains Experiment (OLYMPEX)'. *Bull. Amer. Meteorol. Soc.* 2017, vol. 98, pp. 2167–2188. https://doi.org/10.1175/BAMS-D-16-0182.

[48] Dolan B., Rutledge S. A, and Rasmussen K. L. 'Multiscale interactions contributing to enhanced orographic precipitation in landfalling frontal systems over the olympic peninsula'. *Mon. Weather Rev.* 2022, vol. 150, pp. 1207–1231. https://doi.org/10.1175/MWR-D-21-0160.1.

[49] DeLaFrance A., McMurdie L., and Rowe A. 'Orographically modified ice-phase precipitation processes during the Olympic Mountains Experiment (OLYMPEX)'. *J. Atmos. Sci.* 2021, vol. 78, pp. 3815–3833. https://doi.org/10.1175/JAS-D-21-0091.1.

[50] Zagrodnik J. P., McMurdie L. A., and Houze R. A. 'Stratiform precipitation processes in cyclones passing over a coastal mountain range'. *J. Atmos. Sci.* 2018, vol. 75, pp. 983–1004. https://doi.org/10.1175/JAS-D-17-0168.1.

[51] Kidd C., and Levizzani V. 'Status of satellite precipitation retrievals'. *Hydrol. Earth Syst. Sci.* 2011, vol. 15, pp. 1109–1116.

[52] Iguchi, Seto T. S., Meneghini R., Yoshida N., *et al. GPM/DPR Level-2 Algorithm Theoretical Basis Document[online]*. NASA Goddard Space Flight Center, 2017, p. 81 Available from https://arthurhou.pps.eosdis.nasa.gov/Documents/ATBD_DPR_201708_whole_1.pdf.

[53] Grecu M., Olson W. S., Munchak S. J., *et al.* 'The GPM combined algorithm'. *J. Atmos. Ocean. Technol.* 2016, vol. 33, pp. 2225–2245.

[54] Schwaller M. R., and Morris K. R. 'A ground validation network for the global precipitation measurement mission'. *J. Ocean. Atmos. Tech.* 2011, vol. 28, pp. 301–319.

[55] Williams C. R., Bringi V. N., Carey L. D., *et al.* 'Describing the shape of raindrop size distributions using uncorrelated raindrop mass spectrum parameters'. *J. Appl. Meteorol. Climatol.* 2014, vol. 53, pp. 1282–1296.

[56] Seto S., Shimozuma T., Iguchi T., and Kozu T. 'Spatial and temporal variations of mass-weighted mean diameter estimated by GPM/DPR'. *IEEE Int. Geosci. Remote Sens. Symp.*, Beijing, China, IEEE, 2016, 39383940. https://doi.org/10.1109/IGARSS.2016.7730023.

[57] Chase R. J., Nesbitt S. W, and McFarquhar G. M. 'Evaluation of the microphysical assumptions within GPM-DPR using ground-based observations of rain and snow'. *Atmosphere.* 2020, vol. 11, p. 619. https://doi.org/10.3390/atmos11060619.

[58] Maahn M., Burgard C., Crewell S., *et al.* 'How does the spaceborne radar blind zone affect derived surface snowfall in polar regions?'. *J. Geophys. Res. Atmos.* 2014, vol. 119, pp. 13,604–13,620.

[59] Skofronick-Jackson G., Kulie M., Milani S., Munchak S.J., Wood N.B, and Levizzani V. 'Satellite estimation of falling snow: A global precipitation measurement (GPM) core observatory perspective'. *J. Appl. Meteorol. Climatol.* 2019, vol. 58, pp. 1429–1448.

[60] Valdivia J. M., Gatlin P. N., Kumar S., Scipión D., Silva Y., and Petersen W. A. 'The GPM-DPR blind zone effect on satellite-based radar estimation of

precipitation over the Andes from a ground-based Ka-band profiler perspective'. *J. Appl. Meteorol. Climatol.* 2022, vol. 61, pp. 441–456.

[61] Kirstetter P., Hong Y., Gourley J. J., *et al.* 'Toward a framework for systematic error modeling of spaceborne precipitation radar with NOAA/NSSL ground radar based national mosaic QPE'. *J. Hydrometeorol.* 2022, vol. 13 (4), pp. 1285–1300. https://doi.org/10.1175/jhm-d-11-0139.1.

[62] Huffman G. J., Bolvin D. T., Braithwaite D., *et al. NASA Global Precipitation Measurement (GPM) Integrated Multi-satellitE Retrievals for GPM (IMERG). Algorithm Theoretical Basis Document (ATBD)* Version 6 [online]. 2020. Available from https://gpm.nasa.gov/sites/default/files/2020-05/IMERG_ATBD_V06.3.Pdf [Accessed 28 July 2022].

[63] Wang J., Petersen W. A., and Wolff D. B. 'Validation of satellite-based precipitation products from TRMM to GPM'. *Remote Sens.* 2021, vol. 13, p. 1745. https://doi.org/10.3390/rs13091745.

[64] Wilks, D. S. *Statistical Methods in the Atmospheric Sciences, 3rd ed. International Geophysics Series.* Academic Press: Waltham, MA, USA, 2011; vol. 100, p. 704.

[65] Tian Y., Peters-Lidard C. D., Eylander J. B., *et al.* 'Component analysis of errors in satellite-based precipitation estimates'. *J. Geophys. Res.* 2009, vol. 114, p. D24101. DOI: 10.1029/2009JD011949.

[66] Habib E., Larson B. F., and Graschel J. 'Validation of NEXRAD multisensor precipitation estimates using an experimental dense rain gauge network in south Louisiana'. *J. Hydrol.* 2009, vol. 373, pp. 463–478.

[67] Kidd C., Tan J., Kirstetter P.-E., and Petersen W. A. 'Validation of the version 05 level 2 precipitation products from the GPM Core Observatory and constellation satellite sensors'. *Q. J. R. Meteorol. Soc.* 2018, vol. 144, pp. 313–328. https://doi.org/10.1002/qj.3175.

[68] Wang J., Wolff D. B., Marks D. A., Pippitt J. L., Tan J., and Huffman G. J. 'Validation of IMERG oceanic precipitation over Kwajalein'. *Remote Sens.* 2022, vol. 14, p. 3753. https://doi.org/10.3390/rs14153753.

[69] Bringi V., Seifert A., Wu W., Thurai M., Huang G.-J., and Siewert C. 'Hurricane Dorian outer rainbBand observations and 1D particle model simulations: A case study'. *Atmosphere.* 2020, vol. 11(8), p. 879. https://doi.org/10.3390/atmos11080879.

[70] Thurai M., Wolff D., Marks D., Pabla C., and Bringi V. 'Separation of stratiform and convective rain types using data from an S-band polarimetric radar: A case study comparing two different methods'. *Environ. Sci. Proc.* 2021, vol. 8(1), p. 1. https://doi.org/10.3390/ecas2021-10358.

[71] Thurai M., Bringi V., Wolff D., Marks D. A., Gatlin P. N., and Wingo M. T. 'Retrieving rain drop size distribution moments from GPM dual-frequency precipitation radar'. *Remote Sens.* 2021, vol. 13(22), p. 4690. https://doi.org/10.3390/rs13224690.

Chapter 7
NASA high altitude airborne weather radars

Gerald M. Heymsfield[1], Lihua Li[2], Matthew L. Walker McLinden[2], Liang Liao[3], Charles N. Helms[1,4] and Stephen Guimond[1,5]

7.1 Introduction

7.1.1 Motivation for high altitude radars

Airborne weather radar has played an important role in studying extreme weather systems (e.g., hurricanes, organized convective systems, fronts, and snow storms) that has led to significant advancements in the understanding and modeling of these phenomena. A few instruments are as capable of providing the coverage, spatial and temporal resolution, and geophysical quantities as airborne radars. For example, a single airborne Doppler radar can provide the full three-dimensional wind field, latent heating, precipitation, and other variables over the hurricane's inner-core at convective scale resolution [1–3]. This type of information cannot be currently obtained by space and ground-based systems that are limited by their sampling domain without the ability to measure ocean-based phenomena.

Airborne radars have also been important in the early development of precipitation retrievals from space beginning in the 1980s leading up to the Tropical Rainfall Measurement Mission (TRMM) [3–6], and with later missions such as Cloud-Sat [7]. There have been numerous airborne radar systems beginning in the 1980s [1,8–11] that have flown at various altitudes less than 12–15 km (40,000–50,000 ft), but a few aircraft are capable of high altitudes >18 km (60,000 ft) that would be necessary to fly over deep weather systems such as convection and hurricanes. These deeper weather systems are often not safe to fly through with instrumented low- to mid-altitude aircraft. This provides a unique role for high-altitude aircraft that can fly over these weather systems with significant science interests. NASA has conducted radar measurements from high-altitude

[1]NASA, Goddard Space Flight Center, Mesoscale Atmospheric Processes Lab, USA
[2]NASA, Goddard Space Flight Center, Microwave Instruments Technology Branch, USA
[3]GESTAR-II, Morgan State University and NASA Goddard Space Flight Center, USA
[4]Earth System Science Interdisciplinary Center, University of Maryland, College Park, USA
[5]Department of Atmospheric and Planetary Sciences, Hampton University, Hampton, USA

aircraft since the early 1990s targeting deep weather systems to address various science questions and to provide simulators for satellite missions such as CloudSat and Global Precipitation Measurement (GPM) (e.g., [12]). Several airborne radars have been developed at the NASA Goddard Space Flight Center (GSFC) for high-altitude operation at frequencies ranging from X-band to W-band, and with both fixed nadir beams and scanning beams. These radars include the X-band (9.6 GHz) **E**R-2 **Dop**pler Radar (EDOP), the W-band (94 GHz) **C**loud **R**adar **S**ystem (CRS), the Ku (13 GHz)/Ka-band (35 GHz) **Hi**gh-altitude **W**ind and **R**ain **A**irborne **P**rofiler (HIWRAP), and the X-band (9.6 GHz) **E**R-2 **X**-band **Rad**ar (EXRAD). The primary NASA high-altitude aircraft have been the ER-2, WB-57, and Global Hawk Unmanned Aerial System (UAS) (Figure 7.1). These aircraft are all capable of altitudes >18 km and have flight durations of 8, 6, and 26 h, respectively (Figure 7.2).

Airborne radars at high altitudes present several hardware challenges compared with their lower-altitude counterparts. Typical environmental conditions at 18–20 km altitude are 50–70 hPa and <−60 °C making the environment very hostile for aircraft instruments. These conditions cause significant challenges for radar subsystems design such as transmitters with high-voltage power supply, high RF power, thermal management, and data storage. Although some instrument payload areas of these high-altitude aircraft are pressurized at about 300 hPa

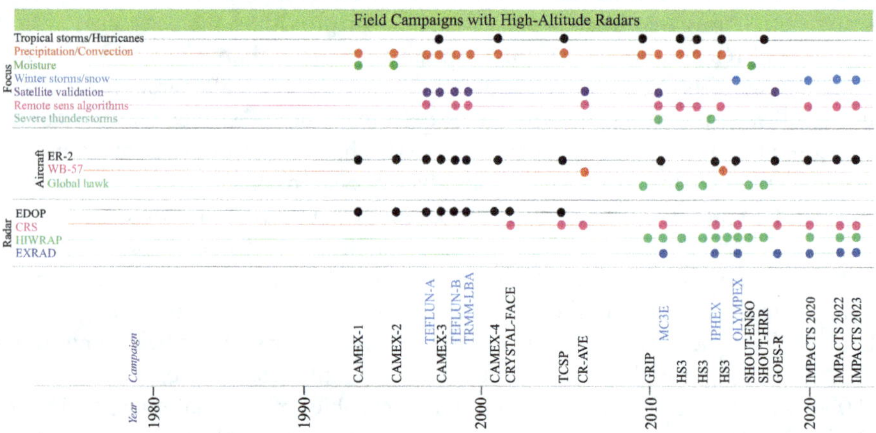

Figure 7.1 Field campaigns were conducted with the NASA high-altitude radars. These campaigns used the three aircraft in Figure 7.2.

Figure 7.2 NASA aircraft used for high-altitude radar measurements: ER-2 (left), WB-57 (center), and Global Hawk (right)

(30 kft), pressure vessel and conduction cooling are necessary for high-power radar transmitters. Additionally, all of the NASA high-altitude aircraft require turn-key instrument operation since there is single pilot on the ER-2, a pilot and a "back-seater" that operates instruments on the WB-57, while the unmanned Global Hawk, instruments are controlled remotely using a satellite link. The instruments are turned on, off, or reset by the pilot or the sensor operator on ER-2 or WB57, or remotely by the ground crew for the Global Hawk. Along with the expansion of satellite link service, all of the high-altitude-piloted aircraft currently have the option to monitor and control the radars remotely from the ground using a communication link to the aircraft.

In the following, we will discuss various aspects of the NASA high-altitude airborne radars including their design considerations, hardware characteristics, measurement capabilities, calibration, data processing algorithms, and application of the data for science retrieval, such as cloud and precipitation microphysical characteristics and storm wind fields.

7.1.2 ER-2 Doppler radar (EDOP)

The X-band (9.6 GHz) EDOP was developed for studying severe deep convection, but its emphasis rapidly shifted to tropical precipitation after the launch of the TRMM satellite [12]. The nominal EDOP specifications are shown in Table 7.1; Figure 7.4 depicts a schematic view of the EDOP configuration in the ER-2 nose. EDOP was built with two fixed pointing beams (nadir and 33° forward of nadir; Figure 7.3) to provide horizontal (along-track) and vertical winds in precipitation and clouds within a curtain below the aircraft. The design criteria for EDOP required that the radar took advantage of the existing side-looking radar (SLR) nose built for the US Air Force U-2 program in the 1970s. This nose, used by EDOP since 1992, is ~0.9 m longer than the standard ER-2 nose. Being the first weather Doppler radar built for use on a high-altitude aircraft created a number of additional design challenges. The use of a 25-kW-peak-power traveling wave tube (TWT)-based high-voltage transmitter and a high-speed digital receiver/processor introduced potential concerns with high voltage/RF arching, thermal dissipation, high data storage requirements, weight limitations, and unattended operation.

The first EDOP flights occurred in 1993 and the instrument continued to fly until 2007, when it was retired due to reliability issues resulting from aging components. During its 14 years of operation, EDOP primarily took part in field experiments studying tropical convection and hurricanes (e.g., [13,14] and Figure 7.1). In addition to the successful collection of science data during numerous field campaigns, the instrument provided valuable experience that was leveraged during the development of CRS, HIWRAP, and EXRAD.

An example of EDOP data is shown in Figure 7.5. This curtain of radar data was collected from Hurricane Georges (1998) as the eyewall was passing over the mountains in Hispaniola. The unusually strong convective updrafts (bottom panel, blue colors) were enhanced by the near-surface hurricane flow over the mountains, as described in [15]. A radar bright band is evident at ~4.5 km above the surface (top panel), while updrafts in the convection had magnitudes exceeding 20 ms^{-1}.

Table 7.1 Specifications for high-altitude radars

	EDOP	EXRAD	HIWRAP	CRS
Frequency (GHz)	9.60 (X-band)	9.596 (scan) (X-band) 9.624 (nadir)	13.47, 13.91 (Ku-band) 33.72, 35.56 (Ka-band)	94.00 (W-band)
Transmitter	TWT	TWT	SSPA	SSPA
RF peak power (W)	25,000	9,000	80 (Ku)/40 (Ka)	30
Antenna size (m)	0.76 (reflector)	0.66 (slotted waveguide flat plate)	0.51 (reflector)	0.51 (reflectarray)
Antenna beam width (degree)	2.9	3.4	3.0 (Ku)/1.2 (Ka)	0.46
Antenna gain (dB)	36.0	34.5	35.4 (Ku)/42.2 (Ka)	52.0
PRF (Hz)	4,400	5.000/4,000	5.000/4,000	5.000/4,000
Range resolution (m)	37.5–150	75–150	37.5–150	37.5–150
Footprint at surface (m)	1,010	1,190	1,050 (Ku), 420 (Ka)	160
Beam pointing	Nadir and forward	Nadir & conical scan	Conical scan (GH, WB57) nadir (ER-2)	Nadir
Measurement products	Reflectivity, Doppler, LDR	Reflectivity, Doppler	Reflectivity, Doppler, LDR	Reflectivity, Doppler, LDR
Minimum detectable reflectivity (dBZe @10 km range)	−5	−15 (nadir)/−8 (scan)	−10 (Ku)/ −12 (Ka)	−28
Doppler Nyquist range (m/s)	34.4	156.2	107.8 (Ku)/42.2 (Ka)	15.9
Aircraft has flown	ER-2	ER-2, Global Hawk	ER-2, Global Hawk, WB57	ER-2, WB57

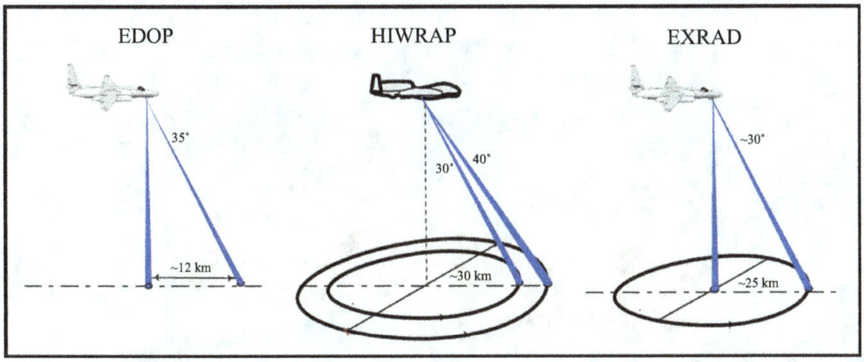

Figure 7.3 Schematic depictions of the multi-beam or scanning configurations for EDOP (left), HIWRAP (center) in its scanning configuration, and EXRAD (right). Beam tilt angles are approximate and the rotation direction can vary between data sets.

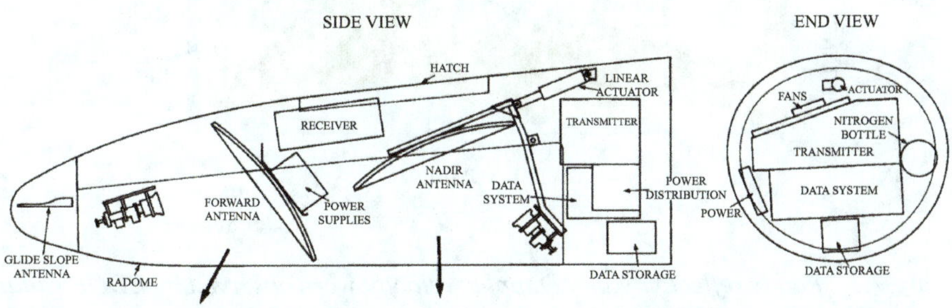

Figure 7.4 EDOP two fixed antenna configuration in ER-2 nose

7.1.3 Cloud radar system (CRS)

The W-band (94 GHz) CRS is a nadir pointing radar that was originally built in the 1990s and primarily flies in the ER-2 superpod tail cone, although it has also flown on the WB-57 aircraft during the Costa-Rica Aura Validation Experiment (CR-AVE) mission [16]. The original CRS system utilized a high-peak power extended interaction klystron (EIK) transmitter with a nominal transmit power of 1,600 W. The radar electronics, including the RF front-end, were housed in a canister which was pressurized at one atmosphere and installed in the ER-2 tail cone (Figure 7.6). A 0.28 m × 0.35 m reflectarray antenna was used to produce a nadir beam through a window opening at the bottom of wing pod tail cone. This system flew in many field campaigns, both for scientific data collection and for calibration and validation of spaceborne instruments.

After approximately one decade of service, CRS underwent an extensive electrical redesign [17]. The aging EIK-based transceiver was replaced with a 30-W solid-state transceiver that supports pulse compression. By transmitting a

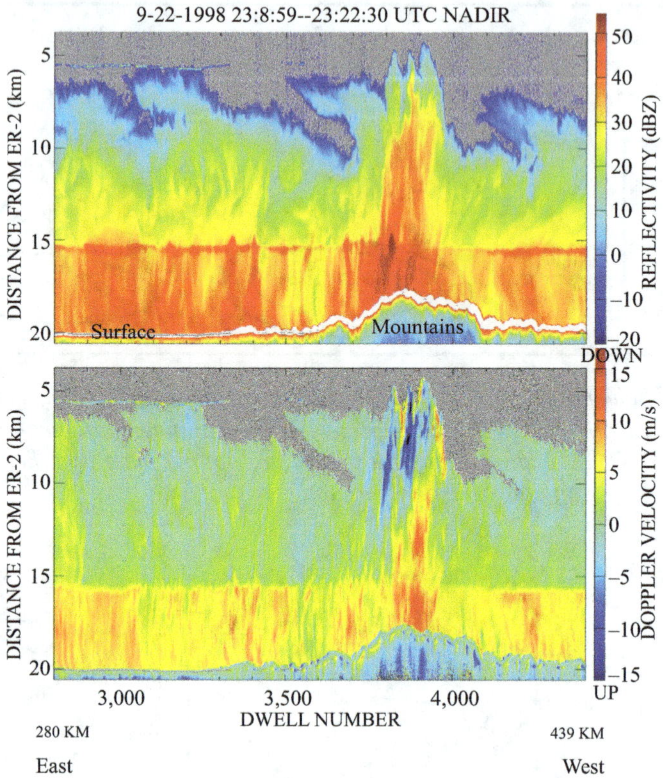

Figure 7.5 Radar reflectivity and Doppler velocity were observed by EDOP radar during CAMEX-3 field campaign when the ER-2 passed over Hurricane Georges (1998). The cross-section passes east to west across Hispaniola. The Doppler velocities shown are not corrected for hydrometeor fallspeed and they indicate strong updrafts (dark blue) and downdrafts (red). Further details are in [15]. © American Meteorological Society. Used with permission.

frequency- or phase-modulated signal, pulse compression allows a radar to utilize a long pulse with relatively lower power to achieve system sensitivity equivalent to a high power tube-based radar. The antenna was also upgraded with a 0.51 m × 0.51 m reflectarray such that, with pulse compression, the sensitivity of the radar remained largely unchanged despite the lower transmit peak power. The traditional tube-based radar transmitter requires a high-voltage power supply and modulator to drive a high-peak-power RF amplifier. Therefore, the radar usually requires pressurization of its transmitter and front-end subsystem for operation in the low-pressure environment at high altitudes. This usually results in a radar system with large size, heavy weight, and poor system reliability. Recent advancements in the communication industry have led to higher power solid-state

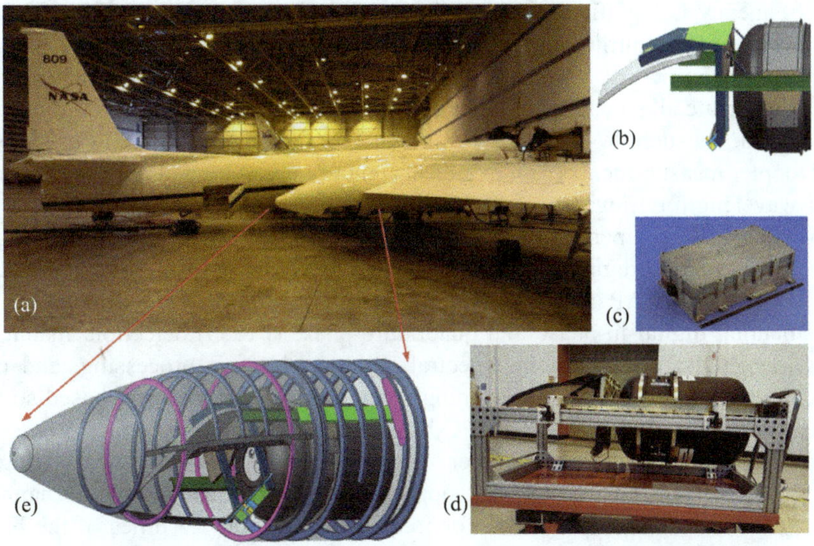

Figure 7.6 CRS radar hardware and installation in ER-2 wingpod tailcone: (a) ER-2 aircraft; (b) 0.51 m reflectarray antenna; (c) 30 W W-band solid state power amplifier; (d) CRS radar transceiver and antenna assembly on a lab cart prior to installation in ER-2; (e) 3D model on CRS transceiver and antenna assembly in ER-2 tail cone

amplifiers with average output powers comparable to that of tube-based power amplifiers. Compared to tube-based transmitters, solid-state power amplifiers (SSPA) can be built in much more compact sizes with lower phase noise and better output power stability. In addition, SSPAs can be operated at a much higher duty cycle (up to 100%). This enables the use of versatile waveforms, such as frequency diversity pulses for increasing the number of independent samples or long frequency-modulated chirps for implementing pulse compression. One challenge of using pulse compression for a nadir-looking airborne radar is that pulse compression range sidelobes couple the strong reflection from the surface into the adjacent range cells and in turn contaminate the backscattering signals from these cells. Many factors, including system phase noise, affect the pulse compression range sidelobe level. Phase noise is the random fluctuations in the phase of a radar signal. Phase noise in the frequency-modulated chirp signal will result in the degradation of the pulse compression range sidelobe [17,18]. In addition, phase noise has adverse effects on Doppler measurement accuracy. Therefore, it is desirable to utilize hardware with low-phase noise, such as an SSPA transmitter and low-phase noise local oscillators, in radar development.

CRS is highly sensitive (−28 dBZ at 10 km range) due to its combination of high frequency (94 GHz), high-gain antenna (51 dBi), and reasonably high average transmit power (5 W; Table 7.1). The 94 GHz frequency in particular provides for strong backscatter from clouds and small hydrometeors. The drawback of such a high frequency is significant non-Rayleigh scattering for heavy clouds and even light precipitation. Furthermore, attenuation due to water vapor is always present, even in the absence of clouds and rain. Moderate and heavy precipitation can rapidly attenuate all signals such that even the strongest targets (such as the surface) are masked. As such, W-band radars like CRS are ideal for providing reflectivity and Doppler measurements from clouds or light precipitation and are very useful in multi-wavelength retrievals (e.g., [19]).

CRS uses a commercial high-speed Xilinx Virtex-5 field programmable gate array (FPGA)-based digital receiver and signal processor subsystem, which is also used by the HIWRAP and EXRAD radars. This subsystem is responsible for implementing digital in-phase and quadrature-phase (I & Q) detection, match filtering (i.e., pulse compression), spectral and/or pulse-pair processing, and data reduction through data profile averaging. This subsystem is network based so that the data can be easily ingested by a PC-based data system where further processing algorithms may be applied. Another novel feature of this FPGA-based digital receiver is that each IF channel will support up to eight independent sub-channels, with more than 90 dB of rejection for frequencies beyond 1.5 times of the bandwidth from the center frequency of the given sub-channel. By using a programmable digital waveform generator and a FPGA-based digital receiver for data acquisition and data processing, CRS is able to transmit a pulse sequence that consists of a number of pulses with slightly different center frequencies. The echo signals from these pulses are separated by the numerically controlled oscillators (NCOs) and digital filters implemented within the digital receiver.

7.1.4 High-altitude wind and rain airborne profiler (HIWRAP)

HIWRAP was originally designed for operation aboard the Global Hawk as a conically scanning radar with two off-nadir beams (30° and 40° incidence) to measure tropospheric winds by collecting Doppler profiles from cloud and precipitation volume backscatter measurements ([20] and Figure 7.3). It has also operated on the ER-2 aircraft in a nadir pointing mode with a dual-frequency single-lens antenna. The selection of Ku- and Ka-band frequencies was based on the tradeoff among hardware size, spatial resolution, backscatter efficiency, atmospheric attenuation, and cost. The Ku-band provides reasonable storm penetration capability with less attenuation, but its backscatter efficiency and spatial resolution (with the same antenna aperture) are lower than that of Ka-band. On the other hand, the Ka-band signal experiences more significant attenuation caused by atmospheric gas (water vapor and oxygen), by ice cloud and mixing phase particles in and above the melting layer, and by precipitation between the melting layer and the surface. However, with the use of Ka-band, HIWRAP is more sensitive to cloud particles and small droplets than lower frequency radars, enabling it to provide three-dimensional winds in lower reflectivity regions of storms. Because the operation frequencies are similar to those used by the NASA GPM Dual-frequency Precipitation Radar (DPR), HIWRAP also provides airborne validation data for the GPM mission. In addition to volumetric measurements from atmospheric targets, ocean surface backscatter measurements using the dual-frequency, dual conical scanning beams allow the retrieval of ocean surface vector winds similar to scatterometers [21] as well as atmospheric attenuation [22]. As illustrated in Figure 7.1, for most volume cells within the swath (except those located along the flight track or at the edge of the swath), HIWRAP will view a given cell at both Ku and Ka-band from two different incidence angles and four different azimuth angles (forward and backward looking while the aircraft flies over the cell), from which the three components of the wind can be derived (see Section 7.5 for details). Analysis of different incidence angle combinations showed that operating at 30° and 40° incidence angles maximized the number of retrieval cells meeting the wind vector uncertainty requirements [20].

The design of HIWRAP involved trades to address the scientific requirements and hardware limitations imposed by the Global Hawk payload and flight environment. Table 7.1 provides performance specifications for HIWRAP and Figure 7.7. The radar hardware including the intermediate frequency and local oscillator (IF/LO) subsystem, the RF transceivers, and the digital receiver and processor are not pressurized and are mounted on a rotating structure that typically spins at 16 rpm (Figure 7.8).

The HIWRAP scanning antenna subsystem for Global Hawk operation consists of a flight frame that also functions as the mounting structure of the radar transceiver hardware, a 0.5 m parabolic offset reflector, and two dual-frequency feeds to form the 30° and 40° beams. Each of the two feeds supports simultaneous operation at Ka and Ku bands with the 30° beam for a given band having a slightly different frequency than that of the 40° beam. The frequency separation between these two feeds allows HIWRAP's receivers to separate the return signals from each beam and allows simultaneous transmission on one beam while receiving from the other [23]. The 30° incidence beams are horizontally polarized while the 40° beams are vertically polarized. The use of orthogonal polarizations reduces wind direction

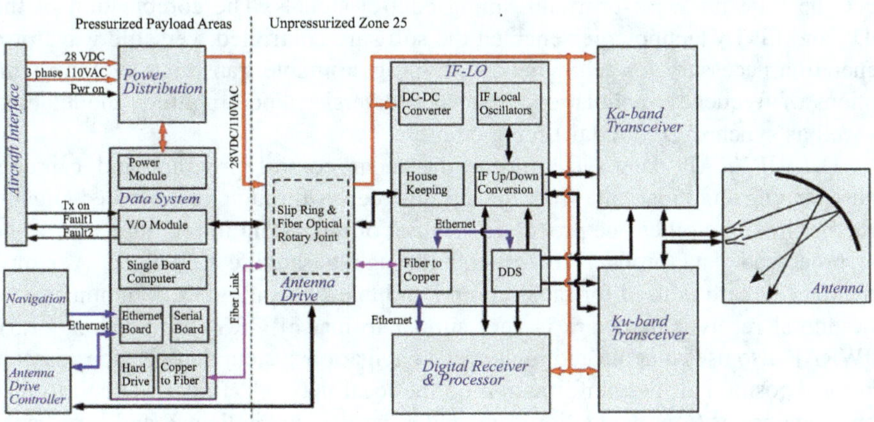

Figure 7.7 HIWRAP Ku- and Ka-band block diagram. Further details are in [20].

Figure 7.8 HIWRAP scanner and other hardware in Global Hawk

ambiguity associated with ocean scatterometry [24]. The HIWRAP beam configuration on Global Hawk is consistent with the QuikScat inner and outer beam polarization arrangement [25]. For operation on ER-2, a 0.51-m diameter dual-frequency dual-polarization lens antenna is used to produce a fixed nadir pointing beam. This antenna is installed in the unpressurized ER-2 wing pod middle body.

HIWRAP uses commercial solid-state power amplifiers as its transmitters. A direct-digital-synthesizer-(DDS)-based custom PCB board was developed to produce the transmit waveform and timing control signals. The combination of the DDS and FPGA technologies enabled the software controlled, versatile waveform generation necessary for achieving a fully-programmable transmit waveform with options of frequency modulation, frequency diversity, and amplitude modulation, as well as synchronized radar timing control.

The HIWRAP data system runs the radar control program and captures housekeeping and radar data from the digital receiver through a high-speed Gigabit Ethernet link, the other one processes a subset of radar data in real time and sends the processed data products to ground through a satellite downlink. A multi-function I/O card is used for radar control and housekeeping data communication. The digital receiver hardware is very similar to that of CRS (see Section 7.1.3). HIWRAP also uses a dedicated high-speed navigation system that provides precise platform position/altitude information on the local network. Radar status information, scanner position, navigation data, and radar data are collected and saved on a solid-state disk array by the host computer.

7.1.5 ER-2 X-band radar (EXRAD)

Following the successful development of a compact, lightweight successor to EDOP was initiated in 2004 for studies of tropical storm formation using the Global Hawk Unmanned Aerial System (UAS); this radar would become ER-2 X-band radar (EXRAD). EXRAD operation has combined precipitation radar and scatterometer objectives for measuring both the 3D cloud/precipitation structure and surface winds with fixed nadir and conical/cross-track scanning beams (Figure 7.9). Although the cross-track scanning capability is useful for coordinating with cross-track radiometers on the ER-2 payload, the cross-track scanning configuration has only seen limited use. The first version of EXRAD was flown between 2014 and 2022 with a 9-kW peak power TWT transmitter that was more compact and lighter than its predecessor. EXRAD shares a number of key subsystem technologies with HIWRAP and CRS, including the power distribution unit, digital receiver, and data system.

In contrast to the pair of differently polarized conical scanning beams of HIWRAP, EXRAD uses a 0.66-m diameter slotted waveguide antenna with a single linear polarization for a conical/cross-track scanning beam plus a second 0.66-m linear polarized slotted waveguide antenna for a nadir beam. Despite this difference, the wind retrieval algorithms (Section 7.5) are similar between EXRAD and HIWRAP, although the presence of a nadir beam on EXRAD enables direct measurement of vertical Doppler motions. The EXRAD antenna setup is also an improvement over the pair of fixed-angle beams employed by EDOP, which were only capable of providing the vertical Doppler motion and the along-track horizontal wind component (Section 7.1.2).

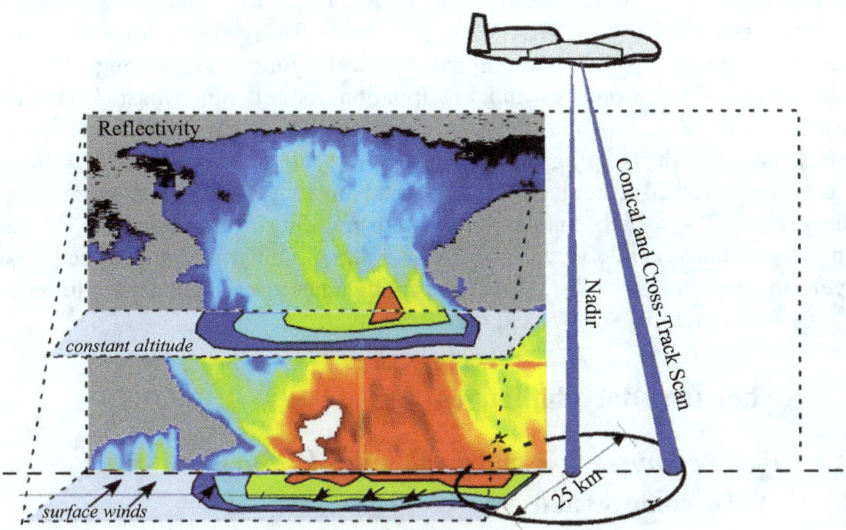

Figure 7.9 EXRAD measurement concept showing two beam geometry. The white shades in the image are reflectivities exceeding 50 dBZ.

Figure 7.10 EXRAD configuration in ER-2 SLR nose. Two flat plate slotted waveguide antennas are used for a fixed nadir pointing and scanning beams. From [12]. Published 1996 by the American Meteorological Society.

The development of EDOP, CRS, and HIWRAP provided considerable experience with autonomously operated high-altitude radars. The design approach for EXRAD focused on maximizing the use of off-the-shelf components and shared technologies, developed through the other NASA high-altitude radars, to reduce development and maintenance costs without compromising the performance of the system. This was aided by advances in the commercial availability of a number of the subsystems that had previously been custom built for previous high-altitude radars. Furthermore, EXRAD was designed to use the same ER-2 nose section that had been used by EDOP. Figure 7.10 shows the configuration for the EXRAD antennas in the ER-2 nose.

Since its completion, EXRAD has been successfully flown for a number of field campaigns on ER-2 and Global Hawk (Figure 7.3). During the 2022 Investigation of Microphysics and Precipitation for Atlantic Coast Threatening Snowstorms (IMPACTS) field deployment, the EXRAD TWT transmitter started to show issues with its aging high-voltage power supply and the TWT tube. An upgrade was initiated to replace the TWT transmitter with a 900-W compact and lightweight SSPA, which enables higher transmit-duty-cycle operation for pulse compression and versatile waveforms. A new digital unit with a high-speed digital waveform generator and a digital receiver module has also been adapted to replace the previous Virtex-5 FPGA-based digital receiver.

7.2 Airborne radar challenges

7.2.1 Pulse compression and surface backscatter

7.2.1.1 Pulse compression

Pulse compression radars offer a number of advantages over the traditional tube-based high-peak-power short-pulsed radars, such as having no requirements for high-power RF circuitry, no need for high-voltage electronics, having a more compact size, being lighter weight, having an improved range resolution, and having improved reliability. Modern digital waveform generation, digital receiver, and SSPA technologies enable versatile radar waveforms and programmable pulse compression. Linear frequency-modulated (LFM) or non-linear frequency-modulated (NLFM) chirps with desirable bandwidths and pulse widths can be generated using high-speed FPGA and digital-to-analog (DAC) devices. Pulse compression processes, including FFT and a matched filter are implemented in FPGA-based digital receivers.

Pulse compression, however, introduces range-side lobes that pose a particular challenge for airborne or spaceborne near-nadir-pointing cloud and precipitation radars as the side-lobe surface returns may mask the weaker returns from nearby cloud or precipitation particles. To minimize the pulse compression range-side lobe artifacts, amplitude, phase, or frequency modulation are applied on the radar waveforms. In addition, a sample of the transmit pulse, obtained using an internal calibration loop, is used for the construction of matched filter for better pulse compression performance.

7.2.1.2 Waveforms and frequency diversity

Digital waveform generation technology allows versatile programmable radar waveforms and frequency diversity. Within the same transmit and receive cycle, multiple pulses, including LFM or NLFM chirps and single-tone pulses are transmitted. These chirps and pulses are programmed with different time delays and different frequency offsets. Therefore, the digital receiver is capable of separating and receiving data from each subchannel simultaneously.

7.2.2 Doppler algorithms

Achieving precise and accurate Doppler retrievals from fast-moving airborne platforms such as the NASA high-altitude aircraft poses unique challenges. The platform is unstable, with pitch and roll changing throughout the flight. The high speed of the platform can cause strong apparent Doppler signatures that must be estimated and removed by tracking the precise pointing of the radar antenna. Further, non-uniform reflectivity within the radar beam (generally referred to as non-uniform beam filling or NUBF) can cause apparent Doppler signatures by altering the effective along-track weighting of the Doppler moments and causing some of the forward motion of the aircraft to be present in the Doppler estimate even after corrections based on aircraft attitude [26,27]. Any off-nadir pointing of the radar beam provides an avenue for strong horizontal winds to impact the Doppler signature, providing non-trivial apparent vertical motion. Finally, the fast motion of the aircraft combined with the relatively high frequencies of the radar (Ku through W-bands) mean that measured Doppler velocities will routinely exceed the Doppler Nyquist velocity (e.g., [20]) often folding multiple times. Precise unfolding algorithms are required to produce reliable data sets.

7.2.2.1 Staggered PRF

For the HIWRAP and EXRAD scanning radars, the ground speeds range from 170 to 205 m s^{-1} and the radars measure strong Doppler velocities due to the platform motion and horizontal winds. For the nadir-pointing radars, strong vertical motions such as those present in updrafts and downdrafts also cause large vertical velocities that exceed the Nyquist interval and become folded. Different techniques, such as dual pulse repetition frequency (Dual-PRF) and frequency diversity Doppler processing (e.g., [28] and Section 7.4), could be used for unfolding Doppler velocities. When these methods do not provide a large enough velocity interval and there is still folding, an environmental wind profile is assumed for unfolding radial velocity. The maximum Doppler velocity that a conventional pulsed radar can measure without ambiguity is

$$v_{\max} = \frac{\lambda}{4T_s}, \tag{7.1}$$

where λ is the wavelength and T_s is the pulse repetition time. High-altitude airborne radars have a minimum pulse repetition time that is set by the altitude of the aircraft and the pulse length so as to avoid range ambiguities for all targets between the aircraft and the surface, a range of ~20 km. The resulting pulse repetition time for NASA's high-altitude radars is typically 200–250 μs corresponding with a pulse repetition frequency (PRF) of 4,000–5,000 Hz. The resulting unambiguous velocity is as little as 3.5 m s^{-1} for the highest-frequency NASA high-altitude radar, the W-band CRS.

To enable accurate Doppler measurements even in the presence of motion well beyond the Doppler Nyquist limit, NASA's high-altitude radars use a staggered PRF (alternating between two pulse repetition times) to extend the unambiguous velocity. By staggering between two PRFs with a 4/5 ratio, the unambiguous velocity is extended by a factor of 4, dramatically simplifying the process of estimating accurate Doppler velocities.

The significant drawback of a direct estimate of Doppler velocity using staggered PRF data is the decrease in precision as uncertainties of Doppler phase from two pulse pairs with different PRFs are compounded [28]. To overcome this challenge, single-PRF and dual-PRF data are used together, as detailed in [17], to provide a Doppler velocity estimate that has the extended unambiguous velocity of the dual-PRF estimate combined with the high precision of single-PRF estimates.

Conceptually, the single-PRF Doppler data provides a precise but ambiguous velocity estimate that has been folded an unknown number of times about the Nyquist interval. The dual-PRF Doppler data provides a more noisy but unfolded velocity estimate. The estimated Doppler velocity produced from NASA's high-altitude radar data is the single-PRF Doppler folded value nearest to the dual-PRF Doppler estimate. In this way, the single-PRF data is unfolded using the dual-PRF estimates to provide a precise and accurate result.

The challenge with using the single- and staggered-PRF data together is that an error in the unfolding of the single-PRF measurement (generally caused by noisiness in the initial staggered-PRF estimate) may cause an increase in error. In these cases, unfolding errors would be off by twice the single-PRF Nyquist velocity. This algorithm minimizes uncertainty by avoiding directly estimating N directly. In doing so, it avoids break-points associated with velocities falling near the folding intervals and is resistant to uncertainties less than the whole single-PRF Nyquist interval. If additional precision is required, the single-PRF data is then used to unfold the Doppler estimate of "second-lag" (skipped-PRF) data.

7.2.2.2 Non-uniform beam filling (NUBF)

Lower-frequency airborne radars have the possibility of significant Doppler velocity errors due to the effects of NUBF. Assuming uniform reflectivity within a radar beam, the Doppler signature caused by platform motion (including forward velocity) can be directly estimated using aircraft navigation data. In the case that radar reflectivity varies in the along-track direction of the aircraft, the effective weighting of the Doppler estimate will be shifted in the direction of the primary aircraft motion. Since the aircraft may travel at 150–200 m s^{-1}, even a small change in the direction of the Doppler velocity weighting function can cause significant Doppler offsets. This is also a significant concern for spaceborne Doppler radar, where velocities are on the order of 7,200 ms^{-1}. This has been studied extensively for both airborne and spaceborne radars by [26,27,29].

From [26,27], the apparent velocity caused by an along-track gradient in the dBZe field within the antenna beam is

$$\Delta v_{\mathrm{NUBF}} = \alpha \nabla_y(\mathrm{dBZe}) \tag{7.2}$$

where $\nabla_y(\mathrm{dBZe})$ is the along-track gradient of the equivalent reflectivity. The coefficient α is

$$\alpha = v_{\mathrm{platform}} \frac{\ln(10)}{160 \ln(2)} \beta^2 R, \tag{7.3}$$

where β is the beamwidth in radians and R is the range.

This result shows that the sensitivity of a radar to NUBF velocity errors due to platform motion, range from the radar, and beamwidth squared. Table 7.2 shows α for the NASA high-altitude radars as well as for some spaceborne radars. With real-world reflectivity gradients, only the EXRAD (X-band) and HIWRAP Ku-band

Table 7.2 Theoretical NUBF correction coefficients based on approximate radar specifications

Radar	Speed (m s^{-1})	Beamwidth (degrees)	Range km	Coefficient m s^{-1} (db km^{-1})$^{-1}$
CRS	200	0.4	15	0.003
HIWRAP (Ka)	200	1.2	15	0.03
HIWRAP (Ku)	200	2.9	15	0.16
EarthCARE	7,200	0.10	400	0.18
EXRAD	200	3.3	15	0.21
CloudSat	7,200	0.12	710	0.46
RainCube or GPM	7,200	1.1	400	22

channels have a significant impact on NUBF on Doppler estimates. The susceptibility of those channels to NUBF is comparable to the EarthCare W-band radar, and the effects must be compensated for in processing.

7.2.2.3 Horizontal wind contamination

While the NASA high-altitude radars are typically in a near-nadir configuration, they do not have steerable antennas to maintain exact nadir-pointing throughout flights. As such, it is not uncommon to have the realized beam pointing be as much as 1° or 2° off-Nadir. Such off-nadir pointing causes the horizontal motion to appear in the Doppler measurement. As vertical motion in the atmosphere is often very small, even a small intrusion of horizontal movement into the near-nadir Doppler measurement can create challenges in interpreting the data in terms of vertical hydrometeor motions.

Such horizontal motion is most prominently caused by the aircraft itself, moving at ~200 m s^{-1}. This is compensated for using aircraft navigation and GPS data to directly estimate mitigate any offsets. More challenging to correct is the Doppler signature from horizontal motion of hydrometeors in the atmosphere caused by strong environmental winds. With a 1.5° off-nadir pointing, a 1 m s^{-1} Doppler signature can be caused by a 40 m s^{-1} horizontal wind. Such strong winds are common in the upper troposphere.

Starting in 2020, NASA high-altitude radar data has been corrected for the estimated impacts of horizontal winds using a numerical forecast model. Wind speeds from the NOAA High-Resolution Rapid Refresh (HRRR) analysis or NASA Modern-Era Retrospective analysis for Research and Applications (MERRA) are interpolated to the location and time of the radar data. With the estimated horizontal winds from models and a precise record of radar antenna pointing, the impact of these horizontal winds can be estimated and subtracted from data. While the winds from HRRR or MERRA are not completely accurate, they serve to significantly reduce the magnitude of errors caused by horizontal winds. Building on the 40 m s^{-1} example, even if the model analysis has a 10 m s^{-1} error in wind magnitude, removing the erroneous model analysis wind from the Doppler velocity observations would still reduce the 1 m s^{-1} Doppler velocity error due to the horizontal winds to a 0.25 m s^{-1} error. Figure 7.11 shows an example of the correction scheme applied to CRS and EXRAD nadir Doppler velocity measurements.

Ideally, the horizontal wind correction would be made using coincident observations of horizontal wind. As will be discussed in Section 7.5, the EXRAD scanning beam can be used to retrieve horizontal wind measurements via one of three retrieval methods: the coplanar technique, the VAD technique, and the three-dimensional variational (3DVAR) technique. These horizontal wind retrievals are, however, limited by the lower sensitivity of EXRAD in areas of cloud or light precipitation. As such, EXRAD horizontal wind retrievals are typically unavailable near cloud top where the higher frequency radars are still able to produce reliable Doppler velocity measurements. Depending on the desired application and meteorological conditions, the loss of information near cloud top may or may not be problematic.

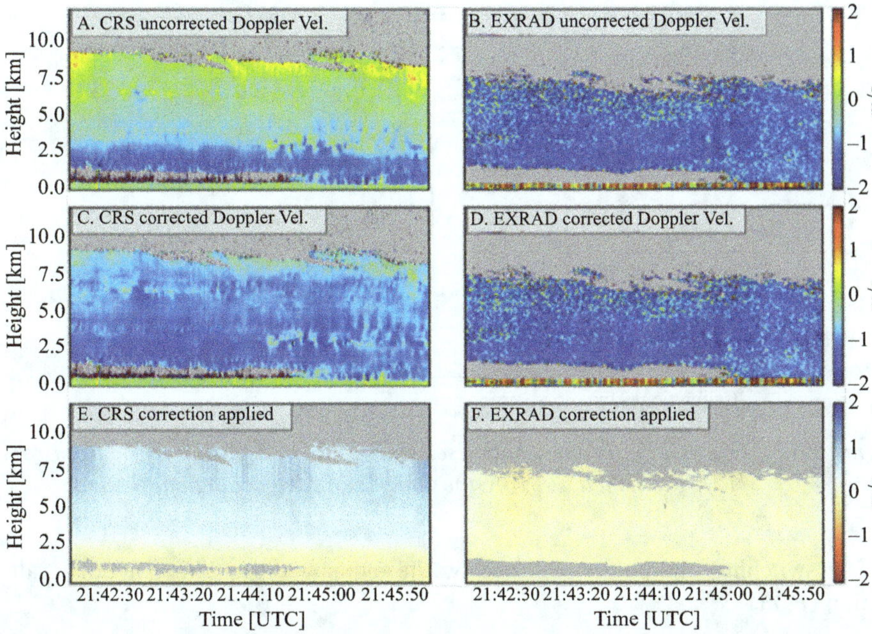

Figure 7.11 Doppler horizontal wind correction for nadir CRS and EXRAD measurements. Shown are (A) and (B) uncorrected CRS velocities, (C) and (D) corrected Doppler velocity using HRRRv3 model, and (E) and (F) difference uncorrected Doppler–corrected Doppler. Corrected CRS and EXRAD are much closer in magnitude to the correction algorithm. CRS antenna is pointing more off-nadir, hence, the Doppler correction is larger.

7.3 Calibration

The NASA high-altitude airborne radars are typically calibrated using two approaches: relative calibration using an internal calibration and external absolute calibration using the ocean surface as a reference (e.g., [16,17]). This is preferable to calibration based on individual system components due to the large amount of bookkeeping that would be required and the numerous opportunities for error. Further, as most NASA high-altitude radars utilize pulse compression, accurately estimating the range weighting function becomes an added challenge. As instrument performance will drift over time, relative calibration is maintained by directing a small portion of the transmitted power to the receiver through a stable path, tracking changes in the product of the transmitted power, receiver gain, and range weighting function.

7.3.1 Internal calibration

For internal calibration, a small portion of the transmitted power is coupled into the receiver prior to the low-noise amplifier, as illustrated in Figure 7.12. Assuming

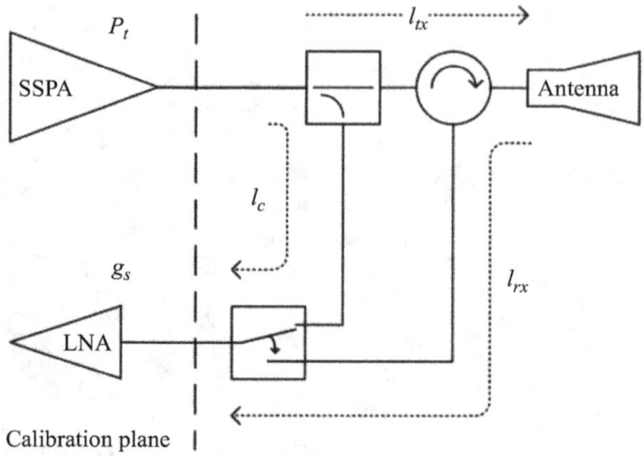

Figure 7.12 Internal calibration path used in HIWRAP, EXRAD, and CRS. From [17]. Published 2021 by the American Meteorological Society.

perfect switching, the power recorded by the radar receiver during internal calibration (P_c) is

$$P_c = \frac{P_t g_s}{l_c}, \tag{7.4}$$

where P_t is the transmitted power, g_s is the receiver gain including any matched filtering, and l_c is the total loss of the calibration loop-back path. Integrating the recorded calibration power over the full transmitted waveform provides an estimate of the product of the transmitted power, receiver gain, and the integrated range weighting function, as

$$l_c \sum_{\text{tx}} P_c[r'] \Delta r = P_t g_s \int |w_s|^2 dr, \tag{7.5}$$

where $P_c[r']$ is the recorded power during transmission at apparent range-gate r' and Δr is the spacing of neighboring range gates. The integrated range weighting function is $\int |w_s|^2 dr$ (for a pulsed radar the integrated range weighting function is $c\tau/(2l_r)$, where c is the speed of light, τ is the pulse length, and l_r is the finite bandwidth loss). Thus, the internal calibration allows the overall performance of the major active RF components of the radar to be tracked over time.

The idealized internal calibration is complicated by the presence of alternative paths for transmitted energy to reach the RF receiver, either through radiation or limited receiver protection switch isolation. The signal loss through these alternate paths is shown as l_{rx} in Figure 7.12. The desired calibration signal (through l_c) and the undesired leakage signal (through l_{rx}) will combine coherently, with the overall combined power depending on the relative phase between these two paths. Because this relative phase cannot be assumed to be stable, there is a resulting potential calibration uncertainty based on the ratio of the two path losses, as shown in Figure 7.13. As a general

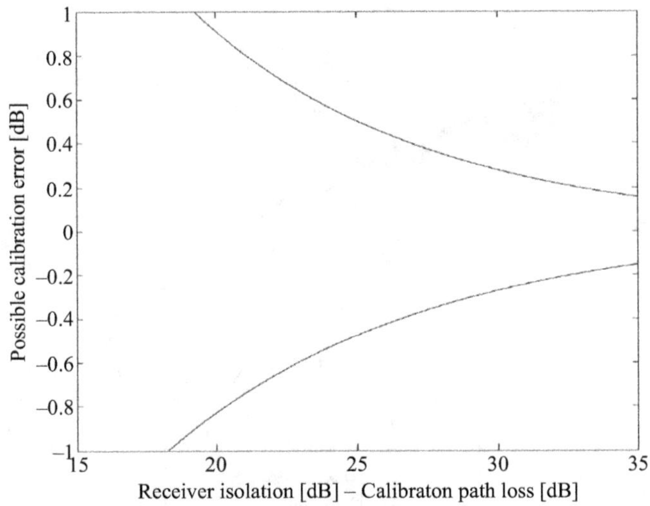

Figure 7.13 Internal calibration error. From [17]. Published 2021 by the American Meteorological Society.

design goal, the receiver protection loss is set to 30 dB greater than the calibration path loss, resulting in an overall internal calibration uncertainty on the order of 0.3 dB.

7.3.2 Ocean calibration

The ocean surface has been widely used as a calibration target for airborne and spaceborne microwave radars and radiometers. Numerous studies of ocean surface scattering have been performed at microwave and millimeter-wave frequencies. In general, the ocean surface normalized radar cross-section (NRCS) is a function of radar wavelength, radar beam incidence angle, polarization, ocean surface wind speed, and wind direction. For incidence angles smaller than 15°, NRCS is dominated by large-scale surface waves, and, at microwave frequencies, the quasi-specular scattering theory has been shown to work well in this region [30–32]. When the incidence angle is larger than 15°, Bragg scattering produced by small-scale waves becomes more significant, and, therefore, two-scale scatter models have been used since they take into account both quasi-specular scattering and Bragg scattering [33,34].

Figure 7.14 shows CRS radar ocean surface backscatter measurements versus incidence angle from the Cirrus Regional Study of Tropical Anvils and Cirrus Layers Florida Area Cirrus Experiment (CRYSTAL-FACE) mission. The solid, dashed, and dotted lines are theoretical models under different surface wind conditions. Both the measurements and model results indicate that the radar NRCS is largely insensitive of surface wind speed near 8–10° degrees incidence angle. Therefore, the modeled NRCS values in this incidence angle have been used as radar external calibration references [35–37]. A summary of airborne radar calibration at Ka- and W-band is found in [38].

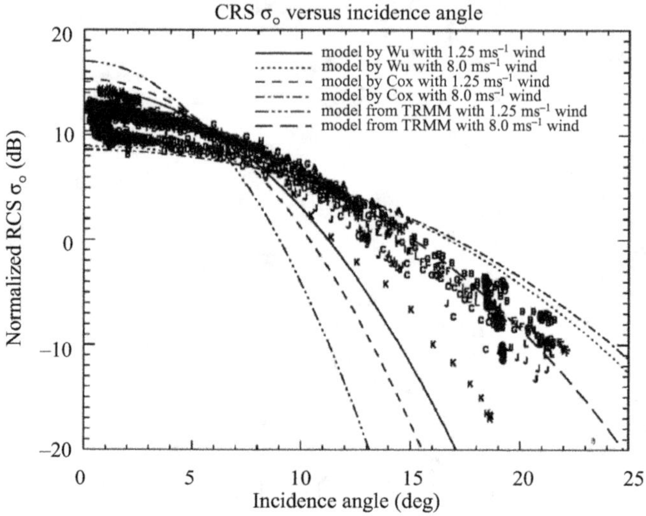

Figure 7.14 The measured σ_0 by 94 GHz CRS radar versus incidence angle from different flights during the CRYSTAL-FACE mission. The solid, dashed, and dotted lines are theoretical models under different surface wind conditions. More details are in [35]. © American Meteorological Society. Used with permission.

To obtain the near 8–10° incidence angle measurements for a nadir-pointing radar, the aircraft performs a series of calibration maneuvers. Each calibration maneuver is composed of a series of several sharp banks of the aircraft in alternating directions. For instance, the ER-2 has a maximum bank angle of 22°, giving a nominal incidence angle between 0° and 22°. Depending on where the radar is mounted (wing versus fuselage), there may be some deviation from this nominal incidence angle due to flexing in the aircraft wing. To mitigate this deviation during calibration and aircraft motion removal, the attitude angles are typically measured at the instrument using a high-precision navigation system, although some attitude bias correction will still be needed to remove any attitude difference between the attitude sensor and the instrument antenna.

Weather radar is designed to measure volume backscatter from cloud and precipitation particles using a range gating technique, while ocean surface backscatter is the radar signal echo from a two-dimensional flat surface. There is a direct relationship derived in [17] between the normalized radar cross-section of the surface and the integrated apparent volume reflectivity from the surface reflection,

$$\sigma^\circ = \sum_{\text{surf}} \eta[r]\Delta r \phi_0 \tag{7.6}$$

where σ_0 is the normalized radar cross-section of the surface, ϕ_0 is the off-nadir pointing angle of the radar, $\eta[r]$ is the apparent volume reflectivity at range r, and Δr is the spacing in the range between radar reflectivity records.

Although calibration occurs under clear-air conditions, atmospheric attenuation is non-negligible for calibration purposes, especially at higher frequencies. The two-way atmospheric attenuation for the ocean surface return is computed based on the local thermodynamic profile, such as temperature, humidity, and pressure, measured by a nearby radiosonde, dropsonde, or buoy sensors [35].

Due to aircraft-maximum-roll-angle limitations, the scanning beam reflectivity cannot be directly calibrated with the 8–10°-incidence-angle approach described above. Depending on the availability of a co-located nadir-pointing radar, the ocean calibration will be applied either via cross-calibration with a nadir-pointing radar or via the application of a geophysical model function (GMF) to estimate the expected NRCS for higher incidence angles under given surface conditions. These GMF models have been used for airborne or spaceborne scatterometers, such as the spaceborne QuikSCAT [39] and the Imaging Wind and Rain Airborne Profiler (IWRAP) [40].

When a nadir-pointing radar is flown alongside the scanning radar, cross-calibration against the nadir beam is used to calibrate the scanning beam. The cross-calibration is performed under non-clear-sky conditions. Specifically, the nadir and scanning beam NRCS are compared in weak-reflectivity regions near the top of cloud or precipitation features (depending on the frequency) where any attenuation due to cloud and/or precipitation will be minimal. The scanning-beam NRCS is then adjusted to match the nadir-beam NRCS. Alternatively, the GMF has been used for HIWRAP calibration as in [41]. This latter paper has performed a more precise calibration required for ocean scatterometry.

7.4 Applications: precipitation physics

7.4.1 Differential measurements

Airborne radar observations of cloud and precipitation at two or more frequencies can provide important information on the microphysical properties of hydrometeors. One of the important radar measurements is the dual-frequency ratio (DFR), which is defined as the difference of the radar reflectivity factors, expressed in decibels, between two frequencies. Additionally, the dual-frequency Doppler velocity (*DDV*), defined as differences in the mean Doppler velocities between two frequencies, has been used for inferring hydrometeor sizes and air motion [19,42,43].

Early ER-2 dual-frequency measurements used EDOP (X-band) and CRS W-band described in Sections 7.1.2 and 7.1.3. Figure 7.15 illustrates a typical example of the measured radar reflectivity (left panels) and mean Doppler velocities (right panels) from EDOP and CRS on July 7, 2002, from 20:15 to 20:25 UTC, during the CRYSTAL-FACE field campaign in South Florida in 2002. The vertical profiles are plotted at selected locations along the flight line. With a range resolution of 37.5 m, the signatures of the bright band, produced by melting snow, are clearly detected by both radars at an altitude of around 4 km throughout the flight line. As a result of the presence of the bright band, the storm is classified as stratiform rain in

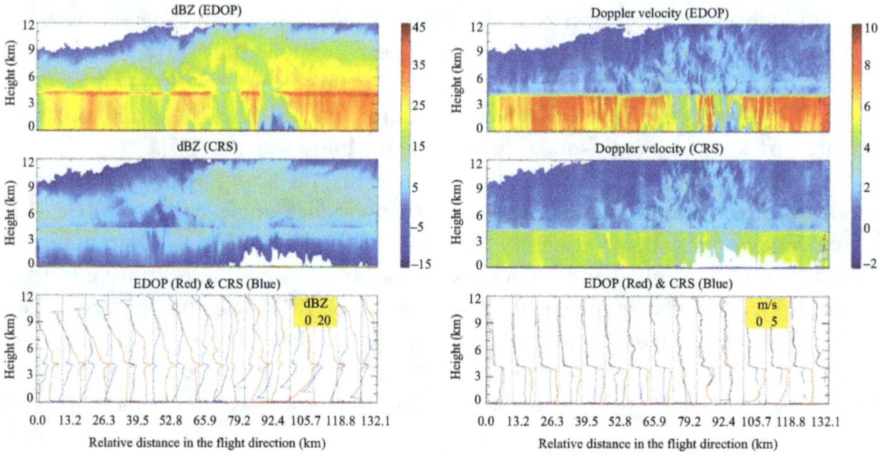

Figure 7.15 Measured radar reflectivity factors (top and middle panels of left column) and measured mean Doppler velocities (top and middle panels of right column) from the EDOP (X-band) and CRS (W-band) nadir-looking airborne radars over a 130-km flight line over stratiform rain. The radar reflectivity and mean Doppler velocity profiles in the selected locations, indicated by the dashed lines, are shown in the bottom panels, where the red and blue curves represent the EDOP and CRS profiles, respectively. Adapted from [43]. © American Meteorological Society. Used with permission.

which the snow, melting snow, and rain appear in three distinct regions. Above the bright band snow is exclusively present, which then becomes rain after the snow falls through the melting region. The melting layer, which separates snow and rain, is normally about 0.2–0.5 km in thickness [44].

As shown in Figure 7.15, the X- and W-band radar reflectivities are comparable in magnitude near the storm top as snow particle size distribution (PSD) is primarily composed of small particles and the radar returns are dominated by the Rayleigh scattering in which the radar reflectivity factor is approximately independent of frequency and equal to the sixth moment of the drop size distribution [45]. They tend to diverge as the signals progress deeper into the storm, however. The differences of the radar reflectivities measured from EDOP and CRS in snow are the result of two factors, i.e., differential attenuation and differential deviations from Rayleigh scattering in the radar backscattering cross-sections between X- and W-band. In dry snow, the attenuation at X-band is negligibly small, whereas the attenuation at W-band becomes noticeable only at higher mass densities (like graupel) and large number concentrations. Cloud water attenuates the radar signal at W-band, but it can usually be neglected at X-band.

At the W-band, the attenuation by rain is so severe that the radar returns often fall below the noise level in moderate and heavy rain, as shown in the x-axis segment from 72 to 118 km, where the EDOP detects rain down to the surface and the

CRS experiences a loss of signal before reaching the surface. In rain, the difference of the measured radar reflectivity factors or DFR between EDOP and CRS is the combined result of path attenuation and non-Rayleigh backscattering effects. The attenuation corrections to the radar returns in rain, particularly for the CRS, are the key to accurately extract the DFR.

The mean Doppler velocities at X- and W-bands are significantly smaller in snow than in rain. Moreover, the differences in mean Doppler velocities, i.e., *DDV*, at X- and W-bands are small in snow but can be substantial in rain. Weak *DDV* signal in snow largely limits its capability in inferring snow size parameters while the large dynamic range of *DDV* in rain provides a useful means to estimate rain size parameters.

7.4.2 Microphysical retrieval method

The measured or apparent radar reflectivity factor Z_m (mm^6 m^{-3}) of the hydrometeors at frequency f (Hz) and at a range of r (m) can be described as [45]

$$Z_m(f) = Z_e(f)e^{-2\int_0^\infty k(f)dD}, \tag{7.7}$$

where Z_m (mm^6 m^{-3}) and k (dB m^{-1}) are the effective radar reflectivity factor and the specific attenuation, respectively, which are related to the particle size distribution $N(D)$ (mm^{-1} m^{-3}), where D (mm) is the liquid-equivalent particle diameter. They are given as

$$Z(f) = \frac{10^{12}c^4}{\pi^5 f^4 |K_W|^2}\int_0^\infty N(D)\sigma_b(D,f)dD, \tag{7.8}$$

$$k(f) = 10^{-6}\int_0^\infty N(D)\sigma_e(D,f)dD, \tag{7.9}$$

where K_w (unitless) is used to designate the dielectric factor and c (m s^{-1}) is the light speed. σ_b (mm^2) and σ_e (mm^2) are the backscattering and extinction cross-sections of the hydrometeors. The measured Doppler velocity V_m (m s^{-1}), which is the sum of the reflectivity-weighted mean of the particle fall velocity V (m s^{-1}) and the air motion vector $\vec{w}$ (m s^{-1}) along the range direction $\hat{r}$ (defined as a unit vector pointing away from radar), is given by

$$V_m(f) = [-V(f)\hat{z} + \vec{w}] \cdot \hat{r}, \tag{7.10}$$

where

$$V(f) = \frac{\int_0^\infty N(D)\sigma_b(D,f)V_t(D)dD}{\int_0^\infty N(D)\sigma_b(D,f)dD}, \tag{7.11}$$

and $\hat{z}$ is a unit vector along the zenith direction. $V_t(D)$ (m s^{-1}) is the terminal fall velocity of a particle of diameter D. The DFR in decibels is given by

$$DFR(f_1,f_2) = 10log_{10}\left(\frac{Z(f_1)}{Z(f_2)}\right), \tag{7.12}$$

where $Z(f_1)$ and $Z(f_2)$ are the radar reflectivity factors at frequencies f_1 and f_2. Similarly, the *DDV* (m s^{-1}) is written as

$$DDV(f_1,f_2) = V_m(f_1) - V_m(f_2) = -[V(f_1) - V(f_2)]\hat{z} \cdot \hat{r}. \tag{7.13}$$

If *N*(*D*) is described by the Gamma distribution, expressed as [46–48]

$$N(D) = N_w f(\mu)(\frac{D}{D_m})^{\mu}\exp(\Lambda D), \tag{7.14}$$

where

$$f(\mu) = \frac{6(4+\mu)^{\mu+1}}{4^4\Gamma(\mu+4)}, \tag{7.15}$$

$$\Lambda = (4+\mu)/D_m, \tag{7.16}$$

where Γ is the Gamma function, N_w (mm^{-1} m^{-3}) the normalized intercept, and μ the shape factor of the Gamma distribution. D_m (mm) is the mass-weighted diameter, defined by

$$D_m = \frac{\int_0^{\infty} D^4 N(D)dD}{\int_0^{\infty} D^3 N(D)dD}, \tag{7.17}$$

Both *DFR* and *DDV* are exclusively a function of D_m for a given μ. The independence of *DFR* and *DDV* on N_w is a useful feature that allows the estimation of D_m directly from either *DFR* or *DDV*. A large number of applications of *DFR* and *DDV* for the estimation of hydrometeor size parameters can be found in the studies [4,5,19,42,49–63].

As shown in (7.13), the *DDV* is unaffected by air motion. This invariance is useful not only from the perspective of deriving the particle size distribution [19,42,64,65] but also because of its potential application in identifying the hydrometeor phase in a fast-changing convective system [66,67]. One of the important characteristics of the *DDV* is that unlike DFR, it is immune to hydrometeor attenuation. This is useful for the retrieval of hydrometeor profiles comprised of liquid and mixed-phase particles as it circumvents the need to correct for the attenuation in rain and mixed-phase hydrometeors. Once D_m is obtained, N_w can be derived from radar equation or (7.8) at either frequency. With assumed μ and estimated D_m and N_w, the specific attenuation *k* can be computed from *N*(*D*) as in (7.9), and the precipitation rate *R* (mm h^{-1}) is derived by

$$R = 6\pi \times 10^{-4} \int_0^{\infty} N_w f(\mu)(\frac{D}{D_m})^{\mu}\exp(-\Lambda D)D^3 V(D)dD. \tag{7.18}$$

7.4.2.1 Rain retrieval

Figure 7.16 shows the relationships between $DFR(Ku, Ka)/DDV(Ku, Ka)$ and D_m for the case of rain when the raindrop size distribution is assumed to follow a Gamma distribution with a fixed μ of 3. Under the Gamma distribution assumption

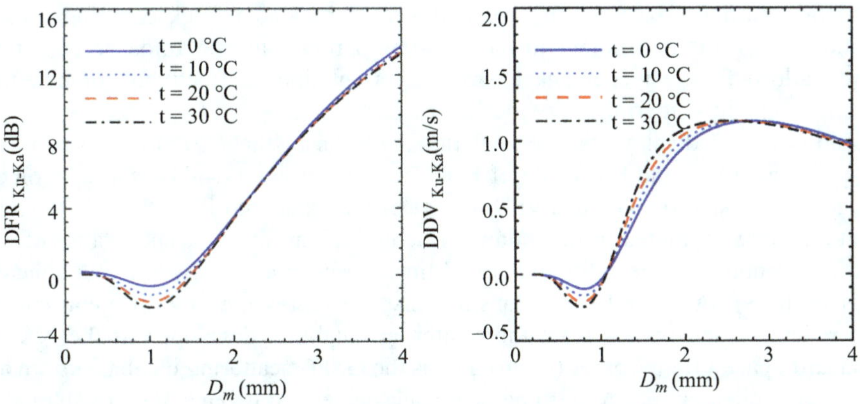

Figure 7.16 *DFR(Ku, Ka) versus D_m (left) and DDV(Ku, Ka) versus D_m (mm) (right) assuming for various raindrop temperatures t (°C). For both plots, a gamma raindrop distribution with μ = 3 is assumed.*

with fixed μ, D_m can be uniquely determined for a given *DFR* if the *DFR* values are greater than 0. There exist two solutions of D_m for a negative *DFR*, however, and an incorrect selection of D_m will lead to an error which, in turn, will affect the estimates of N_w, *k*, and *R*. Similarly, *DDV* and D_m have one-to-one relation when *DDV* is greater than 0 m s^{-1} and D_m is less than ~ 2 mm. A negative *DDV* results in two solutions of D_m. With values of D_m greater than ~ 2 mm, the *DDV* saturates and becomes insensitive to D_m. For the pair of Ku- and Ka-band frequencies, D_m is accurately derived from DFR for the values of D_m greater than ~ 1.5 mm, whereas the *DDV* can be used to accurately estimate D_m in the range from ~ 1 to 2 mm. The temperature has little impact on the $DFR - D_m$ relations for the positive *DFR* but does have an influence for negative *DFR*. There are slight variations of the $DDV - D_m$ relations with respect to temperature.

The results shown in Figure 7.16 are for the case of the Ku- and Ka-band dual-frequency radar, where the raindrops are prescribed as oblate spheroids with axial ratios that follow the shape–size relations reported by [68]. The T-matrix method is used to compute the scattering properties of single particles [69]. For other radar frequency pairs, such as X/Ku-W and Ka-W bands, the features of the $DFR - D_m$ and $DDV - D_m$ relations resemble those obtained for the Ku–Ka combination, i.e., double values of D_m for negative *DFR* or *DDV*, but with different dynamic ranges in *DFR* and *DDV*. For example, larger dynamic ranges of *DFR* and *DDV* are found for the Ku-W frequency pair than for the Ka-W frequency pair.

7.4.2.2 Snow retrieval

Snow particles exhibit a wide variety of shapes and structures in nature – from single crystals to aggregates and graupel [70,71]. Modeling snow particles for the computation of their scattering properties is challenging. There are several scattering databases available, most of which, however, are limited to small and moderate particle sizes. In the calculations of the radar backscattering cross-sections, it is often sufficient to model snow particles as spheroids because this model allows for changes in shape, size, and orientation to effectively compute the electromagnetic scattering properties of the ensemble.

Figure 7.17 displays the computational results of the *DFR* for two pairs of frequencies of Ku- and Ka-band (left) and Ku- and W-band (right) as a function of D_m for μ=0 so that the Gamma PSD becomes an exponential PSD. The $DFR - D_m$ relations were computed from randomly oriented spheroids with aspect ratios of 0.5 and 0.7, denoted by the solid and dotted lines, respectively, as well as for spheres (dashed lines) for several snow densities that are represented by different colors. For reference, the results from the scattering database, developed at the NASA Goddard Space Flight Center (referred to as the GSFC scattering database) from an enormous collection of crystals and aggregates generated using a 3D growth model [72], are included and given by the thick green lines. The D_m in the plots is the liquid equivalent mass-weighted diameter.

It is important to note that a fixed mass density (ρ_s in g cm^{-3}) is used in Figure 7.17 for computations of scattering properties from the simple (spheroid/sphere) models, i.e., the density is constant and independent of particle sizes. This is called the fixed-density model, which differs from the spheroid/sphere scattering models used in much of the literature, e.g., [72–75], which assumes that the mass density of individual snow particles varies with particle sizes, i.e., the density is given by a density-size ($\rho_s - D$) relation. In contrast to the fixed-density snow scattering model, this size-dependent density is referred to as the variable density scattering model. Computations show that there exist large differences in the

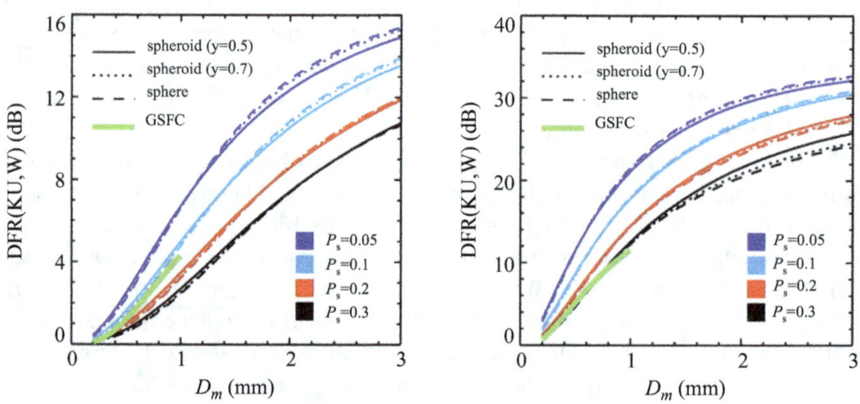

Figure 7.17 DFR(Ku, Ka) (left) and DFR(Ku, W) (right) versus D_m in snow for four snow densities ρ_s assuming a fixed μ = 0 gamma PSD. Randomly oriented spheroids with aspect ratio γ of 0.5 (solid lines) and 0.7 (dotted lines) and sphere (dashed lines) are used for computations of radar reflectivity factors. The results from the GSFC scattering database are given by the heavy green solid lines.

scattering results of snow particles, when their masses are fixed from those obtained from size-dependent density particle models [76]. The former coincides reasonably well with the scattering results computed numerically from the complex snowflake models [72,77] while the latter deviates significantly from the numerical results.

As clearly shown in Figure 7.17, the $DFR - D_m$ relations are not sensitive to eccentricities of the simple models for a given density at both Ku–Ka and Ku–W frequency pairs but vary largely with snow density. The GSFC scattering database is in better agreement with the results of the simple models with the density of 0.1 g cm^{-3} for the Ku- and Ka-band and the density of 0.3 g cm^{-3} for the Ku- and W-band. The maximum diameters in the GSFC scattering database are around 3 mm. To avoid truncation error caused by limited maximum sizes of the database (~3 mm in D), the largest D_m of the GSFC is calculated up to 1 mm. It is worth noting that the snow *DDV* at either frequency pairs are close to the measurement noise level, illustrated by both theoretical computations (not shown) and measurements as shown in Figure 7.15. As such, the application of the *DDV* for estimates of snow size parameters is practically prohibited unless a large number of the samples are averaged [2,42]. Unlike the situation for rain, the *DFR* of snow has one-to-one relation with D_m, implying unique solution of D_m.

7.4.3 Attenuation correction

Shown in Figures 7.16 and 7.17 are the results of attenuation-corrected *DFR*. Accurate attenuation correction to the radar measurements is important for the estimation of precipitation profiles. The Hitschfeld–Bordan (HB) [78] equation provides the attenuation-corrected radar reflectivity factor along the radar column in terms of the measured radar reflectivity factor for a single frequency. The solution to the HB equation tends to become unstable when the path integral attenuation (PIA) along the range is relatively large. An independent estimate of PIA can be used to constrain the solution and improve its convergence [79]. For down-looking dual-frequency airborne radar, forward and backward recursive procedures are often used for attenuation correction. In the forward approach, the retrieval starts from the storm top and then proceeds into the storm gate by gate until reaching the surface or final clutter-free gate. Conversely, the backward approach begins at the surface with a known PIA that is usually estimated by the surface reference technique (SRT) or other independent measurements (such as radiometers), with the correction moving upward until reaching the storm top. Benefits and demerits of the forward and backward approaches depend on the types of error sources involved in the algorithms. The SRT, widely used for air/space-borne radar, provides an effective means to estimate PIA. It is based on the differences in the normalized radar cross-section (NRCS or σ°) of the surface between rain-free and rainy areas. The details of how the SRT a implemented for down looking radar have been well documented in the literature [80–84].

A new dual-frequency (Ku- and Ka-band) attenuation correction technique for HIWRAP (Section 7.1.4) was recently developed to achieve the goal of improving

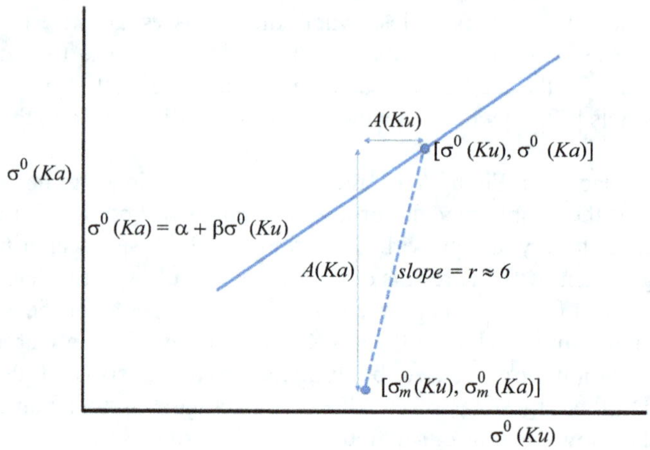

Figure 7.18 Schematic of the procedure to estimate the attenuation-corrected NRCS values $[\sigma^\circ(Ku), \sigma^\circ(Ka)]$ and corresponding path attenuations, A(Ku) and A(Ka), given the apparent or measured values $[\sigma^\circ_m(Ku), \sigma^\circ_m(Ka)]$. From [22]. Published 2019 by the American Meteorological Society.

wind retrievals in the presence of rain [21,22]. This method is based on the relative insensitivity of the differential-frequency surface cross-section, $\delta\sigma^\circ = \sigma^\circ(f_1) - \sigma^\circ(f_2)$, to wind speed and direction, where f_1 and f_2 denote two radar frequencies. If we consider measurements of the NRCS plotted on the $\sigma^\circ(Ku)$–$\sigma^\circ(Ka)$ plane, as shown in Figure 7.18, the location of a point can be thought of as the result of two processes: the first, a traversal along a rain-free regression line (solid line) that depends only on the wind vector; and second, a downward trajectory (dashed line) from this line, with slope $r = A(Ka)/A(Ku)$ (where $A(Ku)$ and $A(Ka)$ are the PIA at Ku- and Ka-band, respectively), that depends only on the ratio of atmospheric path attenuations at the two frequencies. To reverse the attenuation effect, the measured dual-frequency NRCS is translated upward along a slope determined by the ratio $A(Ka)/A(Ku)$ until it intersects the rain-free regression line. From the location of this translated point, relative to its original location, the path attenuations, and corrected values of the NRCS can be read directly.

Note that $[\sigma^\circ_m(Ku), \sigma^\circ_m(Ka)]$ are the measured or apparent surface cross-sections, in the presence of rain, that are translated along a slope r to the rain-free regression line. As noted above, the point of intersection yields the path attenuations $A(Ku)$ and $A(Ka)$ as well as the attenuation-corrected surface cross-sections. The equations that describe the method are given below:

$$\sigma^\circ(Ka) = \frac{r\alpha - \beta\gamma}{r - \beta}, \tag{7.19}$$

$$\sigma^\circ(Ku) = \frac{\alpha - \gamma}{r - \beta}, \tag{7.20}$$

$$A(Ka) = \sigma^\circ(Ka) - \sigma^\circ_m(Ka) = \frac{r\alpha - \beta\gamma}{r - \beta} - \sigma^\circ_m(Ka), \tag{7.21}$$

$$A(Ku) = \sigma^\circ(Ku) - \sigma^\circ_m(Ku) = \frac{\alpha - \gamma}{r - \beta} - \sigma^\circ_m(Ku), \tag{7.22}$$

$$\delta A_\sigma = A(Ka) - A(Ku), \tag{7.23}$$

where

$$\gamma = \sigma^\circ_m(Ka) - r\sigma^\circ_m(Ku), \tag{7.24}$$

where α and β are the regression coefficients through the rain-free NRCS data ($\sigma^\circ(Ka) = \alpha + \beta\sigma^\circ(Ku)$) and r is the slope of the regression line through the NRCS data observed in rain.

The procedures in estimating Ku- and Ka-band path attenuations can be briefly summarized as following basic steps: (1) obtaining α and β from linear regression using nearby rain-free Ku- and Ka-band NRCS data; (2) computing slope (r) on the $\sigma^0(Ku)$–$\sigma^0(Ka)$ plane from measured Ku- and Ka-band NRCS in the presence of rain; (3) plugging in known parameters (α, β, and r) into (7.19–7.22) and using measured $\sigma^0_m(Ku)/\sigma^0_m(Ka)$ to obtain rain attenuation-corrected $\sigma^0(Ku)/\sigma^0(Ka)$ as well as $A(Ku)$ and $A(Ka)$.

Shown in Figure 7.19 are HIWRAP NRCS data plotted in the $\sigma^\circ(Ku)$–$\sigma^\circ(Ka)$ plane. These results were obtained by scanning through the data files choosing regions containing rain data as well as segments of nearby rain-free data. The scatter plots show data from the inner conical swath of 30° under rain-free (left) and rain

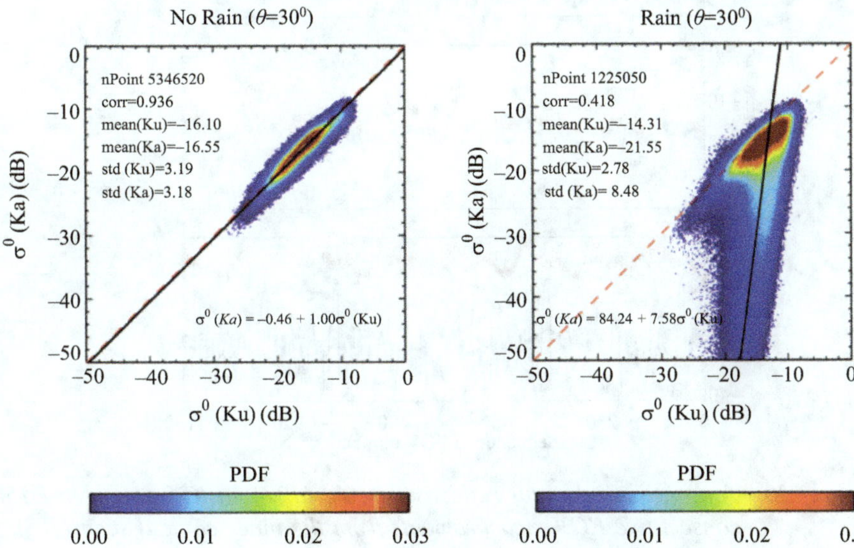

Figure 7.19 Scatter plots of $[\sigma^\circ_m(Ku), \sigma^\circ_m(Ka)]$ or the inner swath (30°) in the absence of rain (left) and the presence of rain (right)

(right) conditions. Figure 7.20 depicts a time sequence of radar reflectivity measurements from a single field of view (FOV) at Ku- and Ka-band in the top two plots. As a full conical scan is made every 3.75 s, the sequence of 1,150 FOVs, representing a time period of about 30 s, roughly has 8 conical scans during which the aircraft flew a distance of about 6 km. For these data, the rain is primarily stratiform with light to moderate rain rates and with a clearly defined bright-band or melting layer at about 4.5 km from the surface. Measured and attenuation-corrected (using the $\delta\sigma^\circ$ approach described above) NRCS data are shown in the lower two panels where the results in blue give the measured or apparent σ° while the results in red show the attenuation-corrected value of σ°. The results indicate that for this stratiform case, the attenuation correction at Ku-band is modest whereas the Ka-band attenuation is much more substantial, reaching values of nearly 30 dB.

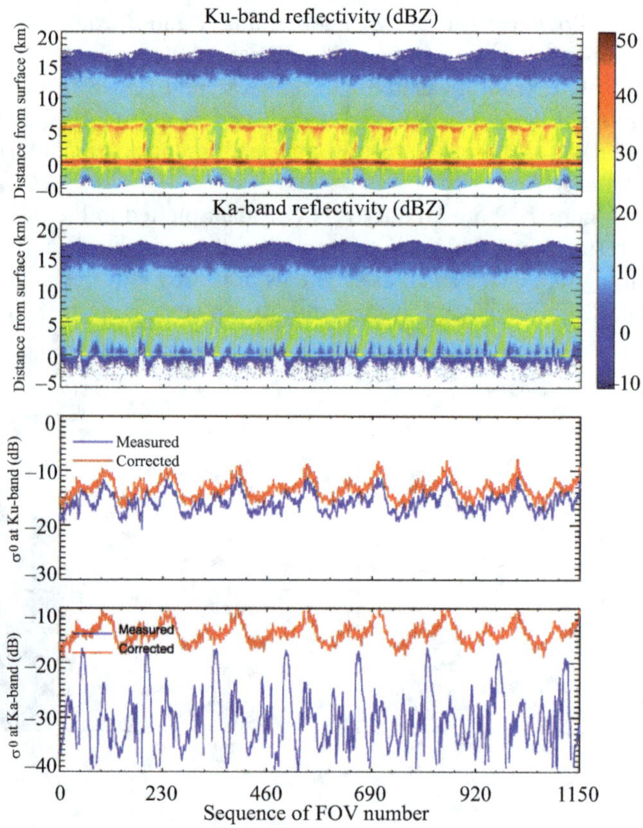

Figure 7.20 Top: measured radar reflectivities, Z_m (dBZ), at Ku band over 1,150 consecutive FOV measurements from the inner conical scan (30°); second from top: the same but for Ka-band; third from top: measured (blue) and attenuation-corrected (red) $\sigma^\circ(Ku)$. Bottom: the same but for $\sigma^\circ(Ka)$.

7.4.4 Four frequency nadir measurements

The above illustrates how two widely spaced frequencies, X- and W-band have the capability to infer microphysical properties. With CRS, HIWRAP, and EXRAD flying together, we have four frequency measurements that provide a wealth of information on the microphysics in weather systems. Figure 7.21 is a single pass of the ER-2 from a wintertime snow storm. Rain is occurring at the surface before ~21:06 UTC as evidenced by the higher reflectivities >30 dBZ at low levels and corresponding low Doppler velocities associated with the higher fall velocities of rain. The different DFR pairs show a region of enhanced DFR in the snow region centered at ~15:39:00 UTC. Recent results have shown that ice particles are on average larger and less dense when the DFR is larger and supports the process by which crystals aggregate to form larger particles [86]. The LDR plots in Figure 7.21 show that high LDR values uniquely define the melting layer/bright band region at all frequencies. In addition, LDR at W- and Ka-bands show enhanced values similar but more well-defined than DFR.

Four frequency radar measurements present stronger constraints than typical two-frequency radar in solving the radar equations and, therefore, provide an excellent opportunity to improve accuracy for retrieval of microphysical properties of hydrometeors. With the combined use of *DFR*s and *DDV*s obtained from different pairs of frequencies, the ambiguous retrieval regions, where the double-solution occurs for a negative *DFR* or *DDV* as shown in Figure 7.16, will be mitigated to a great extent. As such, better DSD estimates will be achieved. Additionally, simple simulation study shows four frequency measurements offering a potential means to identify hydrometeor phase states. In short, retrievals described above and more recently in [86] can be applied to four frequency data sets to provide a more consistent understanding of the microphysics.

7.5 Applications: horizontal wind retrievals

A key benefit of mounting a radar antenna in a non-nadir configuration (at a fixed forward or rearward angle or on a conically scanning mount) is that it enables horizontal wind retrievals. There are four techniques used to retrieve horizontal winds from the Doppler observations collected using NASA high-altitude radars: the dual-Doppler technique, the coplane technique, the **V**elocity-**A**zimuth **D**isplay (VAD) technique, and the three-dimensional variational technique. Table 7.3 provides a list of which radars can be used for each of these techniques. Unless otherwise noted, we will assume that the following corrections have been made to the Doppler velocities before proceeding to the wind retrieval techniques: (1) aircraft motion removal, (2) velocity ambiguity correction (see Sections 7.2.2.2 and 7.2.2.3), and (3) instrument attitude correction.

7.5.1 Dual-Doppler technique

The dual-Doppler wind retrieval technique [12,87,88] requires Doppler velocity measurements of a volume collected from two viewing angles (i.e., from two

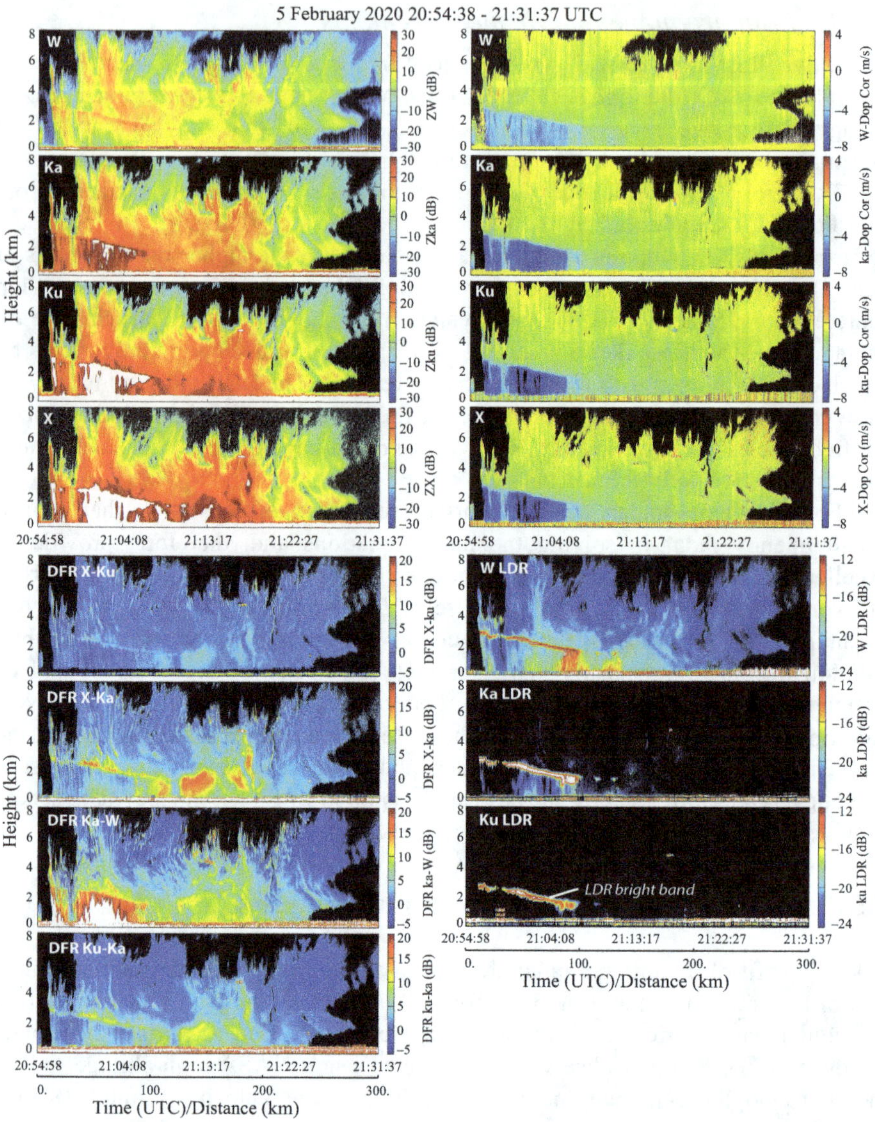

Figure 7.21 Multifrequency radar measurements from a winter storm in New York State during IMPACTS Campaign [85,86]. Left upper panels: radar reflectivity; right upper panels: Doppler velocity corrected for NUBF and horizontal winds; lower left panels: DFR for different frequency pairs; lower right panels: LDR for Ku, Ka, and W-bands.

Table 7.3 List of available horizontal wind retrieval techniques for each radar. Nadir-only radar setups are excluded as no horizontal wind retrieval is possible with a nadir-only configuration.

Radar	Dual-Doppler	Coplane	VAD	3DVAR
EDOP	•			
HIWRAP (scanning)	•	•	•	•
EXRAD	•	•	•	•

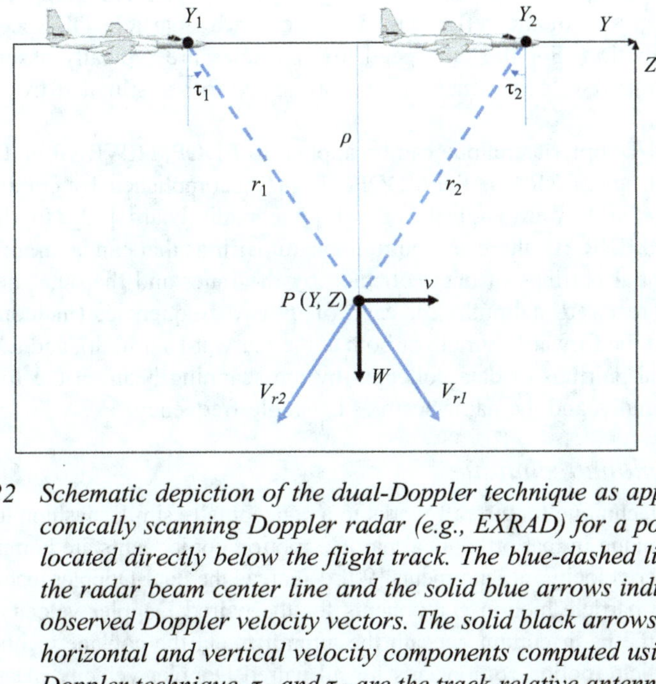

Figure 7.22 Schematic depiction of the dual-Doppler technique as applied to a conically scanning Doppler radar (e.g., EXRAD) for a point, P, located directly below the flight track. The blue-dashed line indicates the radar beam center line and the solid blue arrows indicate the observed Doppler velocity vectors. The solid black arrows indicate the horizontal and vertical velocity components computed using the dual-Doppler technique. τ_1 and τ_2 are the track-relative antenna tilt angles.

separate locations). As such, this is the only wind retrieval technique used with the NASA high-altitude radar data that does not require a conically scanning radar; however, the dual-Doppler technique, as applied to airborne radar data, cannot retrieve the across-track component of the horizontal wind. An example of this setup is schematically depicted in Figure 7.22. These two Doppler velocity measurements can then be used to compute a pair of orthogonal wind vectors in the plane of the observations; for Figure 7.22, this plane is defined by points P(Y,Z), Y_1, and Y_2. While the horizontal motion can be computed from any two sufficiently different view angles (see [89] for a discussion of what constitutes "sufficiently different"), the dual-Doppler technique can only retrieve vertical motion if the point, $P(Y,Z)$, is directly below the flight track.

Referencing the configuration depicted in Figure 7.22, the horizontal motion, v, and vertical motion, W, are calculated, per [90], as

$$v = \frac{r_1 V_{r1} - r_2 V_{r2}}{Y_2 - Y_1} \text{ and} \tag{7.25}$$

$$W = \frac{-r_1(Y - Y_2)V_{r1} + r_2(Y - Y_1)V_{r2}}{\rho(Y_2 - Y_1)}, \tag{7.26}$$

where $r_1 = \sqrt{\rho^2 + (Y - Y_1)^2}$ and $r_2 = \sqrt{\rho^2 + (Y - Y_2)^2}$. The computed vertical motion is the sum of the reflectivity-weighted mean particle fall speed and the vertical air motion. For the fall speed, the particles are typically assumed to be falling at terminal velocity, enabling the fall speed to be estimated from the radar observations.

The dual-Doppler technique can be applied to EDOP, HIWRAP in its scanning configuration, and EXRAD. For EDOP, this is accomplished by combining data collected by the forward-angle beam with the nadir beam [12]. In the case of HIWRAP and EXRAD, there are multiple beam pairings that can be used: HIWRAP has six potential pairings of data collected by the inner and the outer beam at the forward and rearward azimuths for each of its two frequencies (including pairing both beams at the forward azimuth or both at the rearward azimuth), and EXRAD has three potential pairings of data collected by the scanning beam at the forward and rearward azimuths and the nadir beam at its single frequency.

7.5.2 Coplane technique

The coplane technique for aircraft works in a conceptually similar fashion to the dual-Doppler technique in that orthogonal particle motion components are computed from paired Doppler velocity measurements [90]. Whereas the dual-Doppler technique only computes the particle motion components for the paired Doppler velocity measurements along the nadir curtain beneath the aircraft track, the coplane technique computes the particle motion components for all paired Doppler velocity measurements. For the off-nadir measurement pairs, the particle motion components lie on the coplane defined by the aircraft positions at the time of the two Doppler velocity measurements and the location of the paired measurements (see Figure 7.23).

To transform the coplanar particle motion components to horizontal wind and vertical particle motion, the particle motion normal to the coplane must first be determined. The coplane-normal particle motion is computed by integrating outward from the nadir curtain, where a boundary condition is applied. This nadir curtain boundary condition is computed using data located at small off-nadir angles on either side of the nadir curtain. For details on how the boundary condition calculations are performed, see [90]. An additional surface impermeability boundary condition (i.e., zero vertical motion at the surface) must also be assumed. A key limitation of the coplane technique is that due to the reliance on integration to determine the coplane-normal component, the technique requires valid data at all

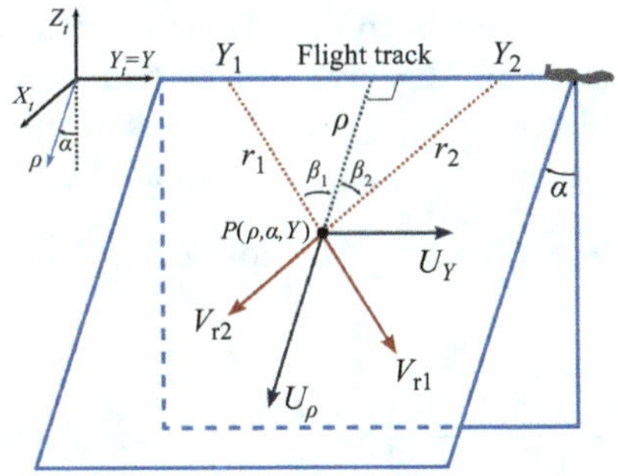

Figure 7.23 Cylindrical coordinate system used for the coplane dual-Doppler retrieval. P is the position of the target in the cylindrical coordinate system defined by the radius (ρ), coplane angle (α), and flight track distance (Y). The red-dashed lines and red arrows represent line segments and vectors, respectively, along the radar beam. The black-dashed line and black arrows within the coplane represent a line segment and vectors in the cylindrical coordinate system. The symbols r1 and r_2 are the ranges of the target from the radar positions at Y_1 and Y_2. β_1 and β_2 are the positive angles between the ranges and the radius ρ. V_{r1} and V_{r2} are the radial Doppler velocities at point P, and U_ρ and U_Y are the corresponding cylindrical coordinate velocity components in the coplane. The axes X_t, Y, and Z_t define the corresponding track-following Cartesian coordinate frame where Y_t is coincident with Y [90].

points between the nadir curtain and the point of interest; any errors in the intervening data will be propagated outwards with the integration. Integrating outward from the nadir curtain, however, also means that any errors that are introduced only affect wind retrievals along that cross-track line of integration in contrast to the VAD technique (Section 7.5.3), where the error can impact multiple retrievals in time, or the three-dimensional variational technique (described in Section 7.5.4), where the error is propagated throughout the entire domain.

The coplanar technique requires Doppler velocity observations from multiple viewing angles. As such, the only NASA high-altitude radars it is applicable to are EXRAD and HIWRAP. While only one set of wind retrievals can be made using the EXRAD data, the dual-beam/dual-wavelength setup of HIWRAP means that potentially four estimates of three-dimensional winds can be retrieved within the swath of the inner beam. Given the need for full data coverage to perform the necessary integration, however, attenuation may prevent retrievals from being made using the Ka-band Doppler velocities in regions of heavy liquid precipitation.

7.5.3 Velocity-azimuth display (VAD) technique

The VAD technique is typically used to retrieve vertical profiles of horizontal winds using conically scanning ground-based radars [91,92]. The technique works by decomposing the wave pattern seen on a Doppler-velocity-versus-azimuth plot (the namesake velocity–azimuth display) into its harmonics. On the ground, this can be accomplished by determining the coefficients of the following equation for the Doppler velocity, V_r, as a function of azimuth angle, θ:

$$V_r = C_0 + C_1 \cos\theta + C_2 \sin\theta + D_1 \cos(2\theta) + D_2 \sin(2\theta), \tag{7.27}$$

where

$$C_0 = -W_0 \sin\Phi + \left(\frac{\partial u}{\partial x} + \frac{\partial v}{\partial y}\right)\frac{r\cos^2\Phi}{2}, \quad C_1 = v_0 \cos\Phi, \quad C_2 = u_0 \cos\Phi,$$

$$D_1 = \frac{r\cos^2\Phi}{2}\left(\frac{\partial v}{\partial y} - \frac{\partial u}{\partial x}\right), \quad \text{and} \quad D_2 = \frac{r\cos^2\Phi}{2}\left(\frac{\partial u}{\partial y} + \frac{\partial v}{\partial x}\right).$$

For these VAD coefficients, W is the vertical particle velocity; Φ is the antenna elevation angle, assumed to be constant; y and x are the distances from the antenna in the direction of the 0° and 90° azimuths, respectively; v and u are the winds in the direction of the 0° and 90° azimuths, respectively; r is the along-beam range from the antenna; and the zero subscripts indicate the value is an average over the ingested data points. In this way, the VAD technique can retrieve the horizontal winds and the horizontal deformations; the vertical particle motion or the divergence can also be retrieved by making certain assumptions or by providing values for the vertical particle motion or divergence taken from other sources. To arrive at (7.27) from the u, v, and W, two critical assumptions must be made: first, that the horizontal wind varies only linearly in space across the ingested data points; and, second, that the vertical particle motion is constant across the ingested data points.

Some adjustments are required when adapting the VAD technique to the high-altitude conically scanning airborne NASA radars [93,94]. First, the 0° azimuthal (i.e., positive y) direction is aligned to the positive along-track direction, typically by adding an offset to the azimuth angle. Second, the along-track motion of the moving antenna platform (i.e., the aircraft) must be accounted for within the y coordinate in the VAD equation (7.27). While a single range gate of a conically scanning ground-based radar traces out a circle, a conically scanning airborne radar will trace out a spiral pattern, as demonstrated in Figure 7.24. This second adjustment adds two additional terms to the VAD equation to account for this difference, resulting in the airborne VAD equation:

$$\begin{aligned} V_r = {} & C_0 + C_1 \cos\theta + C_2 \sin\theta + D_1 \cos(2\theta) \\ & + D_2 \sin(2\theta) + E_1(\theta \cos\theta) + E_2(\theta \sin\theta), \end{aligned} \tag{7.28}$$

where

$$E_1 = \frac{U_a\tau}{2\pi}\cos\Phi\frac{\partial v}{\partial y} \quad \text{and} \quad E_2 = \frac{U_a\tau}{2\pi}\cos\Phi\frac{\partial u}{\partial y}.$$

For these additional VAD coefficients, U_a is the aircraft ground speed and τ is the time required to complete one full rotation of the antenna. For HIWRAP and

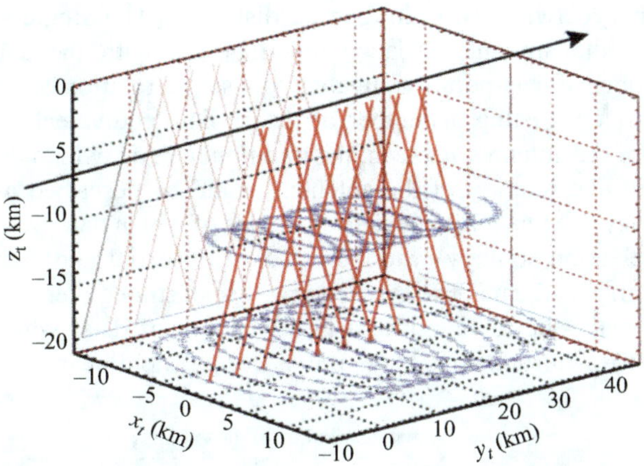

Figure 7.24 Beam trajectory (blue) at two different heights for HIWRAP at an incidence angle of 30°. The red line shows the fore and aft beams in the vertical cross-section under the flight track. From Figure 1(a) in [93]. © American Meteorological Society. Used with permission.

EXRAD products, the VAD coefficients are retrieved via a multiple linear regression [93,94]. Because θ cos θ and θ sin θ suffer from colinearity issues with cos θ and sin θ, the E_1 and E_2 terms cannot be retrieved without contaminating the retrievals of the horizontal winds. As a result, an additional adjustment must be made to the ground-based VAD technique: it must be assumed that the antenna platform is stationary, ideally at the temporal midpoint of the ingested data [94]. Making this additional assumption, the C_0, C_1, C_2, D_1, and D_2 coefficients can then be retrieved (the E_1 and E_2 coefficients and their independent variables, θ cos θ and θ sin θ are simply excluded from the multiple linear regression) and used to compute the horizontal wind and deformation components. Because the wind and deformation retrievals are oriented relative to the along-track direction, the stretching deformation, computed from D_1, has an axis of dilatation oriented in the across-track direction and the shearing deformation, computed from D_2 has an axis of dilatation oriented 45° counterclockwise of the along-track direction when viewed from above. Furthermore, the retrievals are technically in the plane orthogonal to the antenna axis of rotation, which can deviate slightly from true horizontal even during nominally level flight. The effects of the deviations from true horizontal retrievals are usually negligible unless the aircraft is banking.

While the moving antenna platform introduces a source of error, it also creates the possibility of altering the footprint of the VAD retrieval to increase the robustness of the retrievals to errors. Several strategies for data selection that would alter the VAD footprint are discussed in [94]. These data selection strategies are defined by two pairs of traits: sequential/synthetics and single scan/multiscan. The sequential/synthetic trait determines which data points are ingested into the VAD with the sequential trait ingesting all data between two times and the synthetic trait ingesting all data between two along-track distances. The single-scan/multiscan trait determines the amount of data being ingested with the single-scan trait ingesting a number of data points equivalent to a single rotation of the antenna and the multiscan trait ingesting a number of data points equivalent to more than a single rotation of the antenna. As such, the traditional VAD, which uses all the data collected during one rotation of the antenna, would be described as a sequential single-scan retrieval. Figure 7.25 depicts examples of which data would be ingested by each data selection strategy in terms of both the spatial and the temporal distribution of the ingested data. The synthetic multiscan strategy tends to produce the best wind retrievals due to ingesting a large amount of data, which reduces the influence of error-introducing data points (e.g., nonlinear winds or large

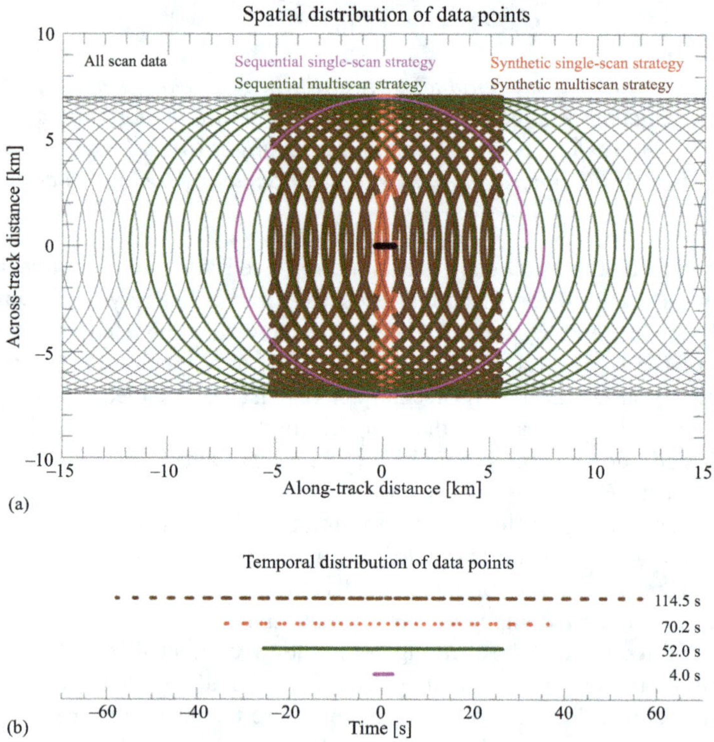

Figure 7.25 Examples of the (a) spatial and (b) temporal distributions of ingested data points for the four data selection strategies resulting from the sequential/synthetic and the single-scan/multiscan trait pairings. In this example, the two single-scan strategies ingest exactly the same number of points as the other; this is also the case between the two multiscan strategies. The black points indicate the aircraft location during the sequential single-scan strategy data collection period. From Figure 1 in [94] © American Meteorological Society. Used with permission.

fluctuations in vertical particle motion), while having a relatively small footprint for the amount of ingested data, which reduces the number of adjacent wind retrievals contaminated by a single error-introducing data point [94].

This VAD technique for downward-pointing conically scanning airborne Doppler radars has been applied to both EXRAD and HIWRAP data. As HIWRAP has four data streams of conically scanning Doppler velocities (Ku and Ka frequencies for the inner and outer beams), four retrievals of the horizontal wind can be performed. The choice between Ku and Ka retrievals will largely come down to how much attenuation is expected versus how important the motions of the smaller particles are to getting a complete view of the winds. In terms of the inner beam versus the outer beam, the VAD retrievals using the inner beam will have a smaller footprint, making the linear wind field and constant vertical velocity assumptions easier to satisfy, but any variations in the vertical particle motion that do occur will have a greater impact on the Doppler velocities and, therefore, on the VAD retrievals. As EXRAD only has a single conical scanning beam, there will only be a single set of VAD retrievals.

7.5.4 Three-dimensional variational technique

The three-dimensional variational (3DVAR) technique provides solutions for the three-dimensional (3D) Cartesian wind vector over the full scanning volume of the radar at horizontal and vertical grid spacings consistent with the along-track and vertical sampling of the instrument, respectively. This type of information is crucial for understanding the dynamics of atmospheric processes, especially those on the mesoscale and microscale, where 3D motions are prevalent. The description of the 3DVAR method follows that described in [95] and full details can be found in this paper.

The calculation of the 3D wind vector requires at least one conical scanning beam to provide sufficient azimuth diversity to solve for the wind field. Figure 7.26 shows the scan pattern and grid structure methodology for the HIWRAP radar. The radar swath is divided up into a series of control volumes with horizontal lengths of 1 km (slightly larger than the along-track sampling) and the along-beam gate spacing of HIWRAP when it was in its scanning configuration. The solutions for the wind field are valid at the centers of each control volume and cover the entire conical scan.

The 3DVAR approach defines a function that includes terms that describe the error between observations (Doppler radial velocities) and the observation operator plus constraints that control noise and nudge the solution towards dynamical laws, such as the conservation of mass:

$$J(\vec{u}) = \|\mathbf{R}^{-1}\mathbf{W}(\mathbf{f} - \mathbf{E}\mathbf{g})\|^2 + \alpha_m\|\nabla \cdot \rho\vec{u}\|^2 + \alpha_s\|\nabla^2\vec{u}\|^2. \tag{7.29}$$

In the first term on the right-hand side of (7.29), $\mathbf{R}^{-1}$ is the inverse of the covariance matrix for the observations, $\mathbf{W}$ is the distance weight from the observations to the control volume center, $\mathbf{f}$ is a column vector of m observations, $\mathbf{E}$ is a rotation matrix that maps the radar spherical coordinates to Cartesian space and

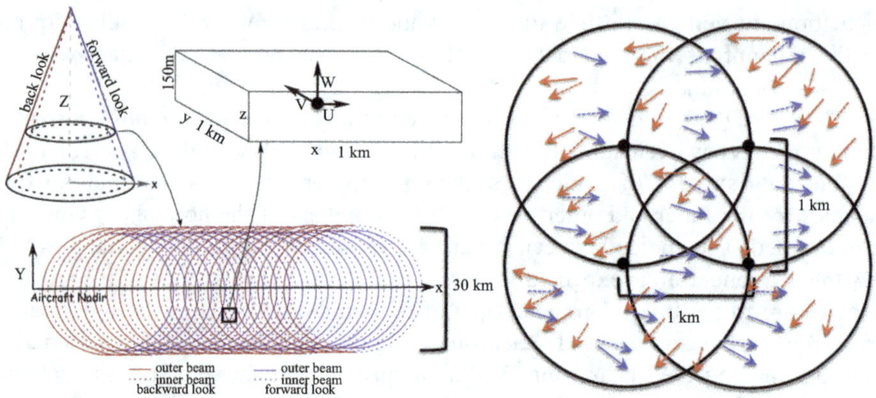

Figure 7.26 Scan pattern and grid structure methodology for HIWRAP. The forward and backward portions of the scan are labeled in blue and red, respectively. The inner beam (30°) and outer beam (40°) are shown in dashed and solid lines, respectively. The influence radii shown are only 1 km for illustration but are larger for calculations. The arrows inside the influence radii represent the HIWRAP radial velocities with the color and line representations denoted in the figure. From Figure 2 in [95]. © American Meteorological Society. Used with permission.

$\mathbf{g} = [uvw]^T$ is a column vector of the three unknown Cartesian wind components. The $\|\cdot\|$ operation denotes the Euclidean or L^2 norm. The second term describes the conservation of mass per unit volume with the anelastic approximation, $\rho = \rho(z)$, and $\vec{u}$ is the three-dimensional velocity vector. The last term is a Laplacian diffusion operation that smooths the wind field to control noise. The coefficients α_m and α_s are weights that control the contribution of the anelastic mass continuity equation and smoothing terms, respectively, in the total function. Readers can refer to [95] for details on how these coefficients are determined.

The **W** in (7.29) is an $m \times m$ diagonal matrix of Gaussian weights given by:

$$W_i = exp\left[-\left(\frac{r_o^i}{\gamma\delta}\right)^2\right], \tag{7.30}$$

where r_o^i is the ith observation from the cell center point, γ is a shape parameter that determines the width of the weighting function, and δ is the influence radius expressed as

$$\delta = s\beta\left(1 - \frac{L_k}{H}\right) + s, \tag{7.31}$$

where s is the along-track sampling of the radar, β is a chosen smoothing factor, L_k is the kth vertical level of the analysis grid, and H is the mean height of the radar. More information on the free parameters (β and γ) can be found in [95].

The rotation matrix **E** is an $m \times 3$ matrix:

$$\mathbf{E} = \begin{pmatrix} x_1 r_1^{-1} & y_1 r_1^{-1} & z_1 r_1^{-1} \\ \vdots & \vdots & \vdots \\ x_m r_m^{-1} & y_m r_m^{-1} & z_m r_m^{-1} \end{pmatrix}, \tag{7.32}$$

where r_m is the range for the mth observation and the Earth-relative coordinates centered on the radar are given by (subscript m is dropped here)

$$\begin{pmatrix} x \\ y \\ z \end{pmatrix} = r \begin{bmatrix} \cos H(a) + \sin H \sin \tau(b) + \sin H(c) \\ -\sin H(a) + \cos H \sin \tau(b) + \cos H(c) \\ \sin \tau(\sin P \cos \theta - \cos P \sin R \sin \theta) - \cos P \cos R \cos \tau \end{bmatrix}, \tag{7.33}$$

where

$$\begin{pmatrix} a \\ b \\ c \end{pmatrix} = \begin{bmatrix} \cos R \sin \theta \sin \tau - \sin R \cos \tau \\ \cos P \cos \theta + \sin P \sin R \sin \theta \\ \sin P \cos R \cos \tau \end{bmatrix} \tag{7.34}$$

and P, R, θ, H, and τ are the pitch, roll, azimuth, heading, and tilt angles, respectively. The azimuth and tilt angles of the antenna are defined here as aircraft-relative variables following the convention in [96].

To compute the "best-fit" wind field solutions, the global minimum of (7.29) is found using numerical methods. Specifically, the vector of three unknown Cartesian wind components ($\vec{u}$) is determined using an iterative, nonlinear conjugate gradient algorithm that requires evaluations of the function in (7.29) and gradient of J with respect to $\vec{u}$

$$\frac{\partial J}{\partial \vec{u}} = 0 \tag{7.35}$$

for each iterative search for the minimum. Eq. (7.35) is evaluated analytically and the resulting partial differential equations, as well as those from (7.29), are discretized to the second-order accuracy. Boundary conditions are imposed on the vertical velocity at the surface through the impermeability condition (w=0 at z=0).

Figure 7.27 shows a composite image of EXRAD reflectivity in dBZ overlaid with horizontal wind vectors calculated with the 3DVAR method for several legs of data collected during the NASA IMPACTS field campaign on February 1, 2020 between 12:04 UTC and 15:23 UTC. The retrieval data is placed onto a storm-following grid for each leg with the origin at the approximate circulation center of a developing extratropical cyclone. The retrievals show a clear cyclonic circulation at low levels (z = 1.25 km) to the northeast of the low-pressure center with several convectively active regions shown by patches of reflectivity near 40 dBZ. These regions represent samples of precipitation bands that are being studied to understand their formation and evolution mechanisms. The 3DVAR winds provide dense sampling of the flow structures in precipitating features that can be used to address the science objectives of field campaigns that target mesoscale processes.

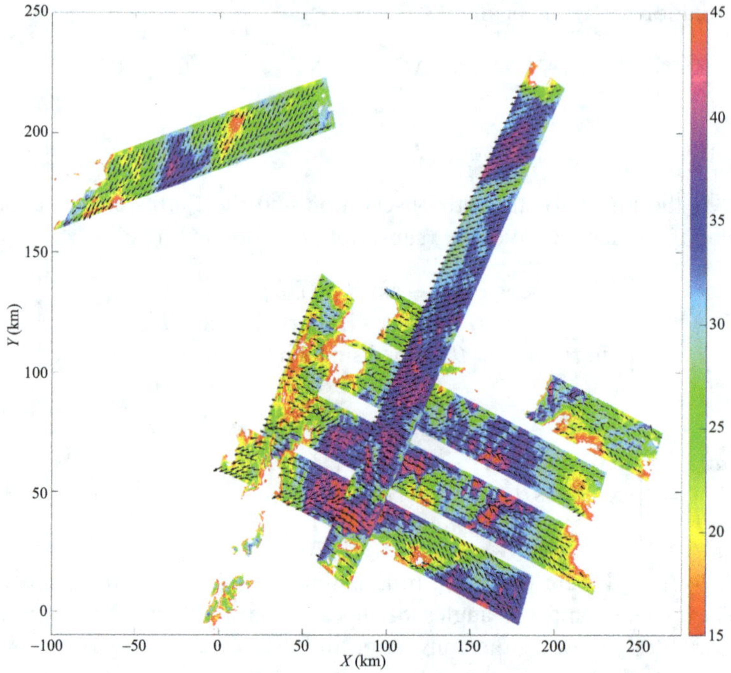

Figure 7.27 Composite image of EXRAD reflectivity (dBZ) overlaid with horizontal wind vectors obtained from the 3DVAR method for several flight legs during the NASA IMPACTS field campaign on February 1, 2020 between 12:04 UTC and 15:23 UTC. The data in this image are placed onto a storm-following grid with the origin at the approximate circulation center and height at 1.25 km. Note that the wind vectors are only intended to highlight the overall flow structure as it can be difficult to identify the wind magnitude over a large region.

Quantification of uncertainty is an important element of any data source. The 3DVAR method described here has been extensively analyzed for uncertainty quantification utilizing theory, simulations, and validation with independent, in situ observations. The random errors in the retrievals have been evaluated using an error propagation analysis with least squares theory and a HIWRAP simulator using a numerical model of a hurricane [95]. These analyses showed that the along-track and vertical winds are most accurate at nadir with a linear decay in accuracy towards the swath edges. This structure occurs because the solutions approximately collapse to an analytical method ("dual-Doppler") at nadir where the across-track wind component is minimized. As the antenna scans off-nadir, more across-track wind is sampled by the beams introducing more error into the along-track and vertical winds. The across-track wind errors have a more complicated structure that depends on the azimuth diversity, influence radius, and magnitude of the wind. Simulations of typical rotated figure-four flight patterns through hurricanes show that the zonal and meridional wind speed errors are ~1.5–2.0 m s^{-1}. Validation with in situ flight level data shows relative errors of ~5–10% in the horizontal winds and larger percentages for the vertical wind with absolute errors of ~1–2 m s^{-1} [95,97,98]. The vertical winds have larger errors due to fall speed uncertainties and smaller magnitudes. The correlation coefficients of all wind components are high with values of 0.80–0.99 for the horizontal and 0.60–0.80 for the vertical.

7.6 Future developments

The earlier generation high-power weather radars were built using tube-based transmitters, such as Magnetron, Klystron or TWT, which require high-voltage power supplies and modulators, and hermetic enclosures for operation in high-altitude environments. This results in heavy hardware and reliability issues. In recent decades, solid-state power amplifier technology has advanced significantly, especially the high-power highly efficient GaN device. SSPAs are currently available with an output power up to kilowatt level at X- and Ku-band from the commercial market. At Ka- and W-band, the state-of-the-art SSPAs are available approaching the hundred watts power level. Compared to the high peak power, low-dutycycle tube-based transmitters, SSPA supports high-dutycycle operation, in most cases up to continuous wave (CW) mode. This enables versatile radar waveform design and pulse compression, in turn, to achieve comparable system sensitivity as the tube-based radars. The additional benefits of an SSPA transmitter include wide bandwidth and lower phase noise compared to tube-based counterparts, see [99].

Pulse compression techniques allow for the transmission of low-to-medium power and long-duration RF pulses and attain fine-range resolution and improved system sensitivity. Traditionally pulse compression has been implemented using surface acoustic wave (SAW) filters or phase-coded pulse. Along with the advance of digital technologies, high-speed FPGA and DAC devices enable the generation of versatile programmable radar waveforms. With a FPGA-based digital receiver, frequency diversity pulses, including frequency-modulated chirps, can be transmitted during a transmit/receive cycle and the backscattering signals from clouds and precipitation can be received simultaneously. Meanwhile, SSPA enables high dutycycle operation and pulse compression implementation. Although pulse compression provides benefits to modern radar development, range sidelobe associated with pulse compression has limited its application for weather radars, especially for spaceborne and airborne cloud and precipitation radars since range sidelobe may mask weak echos from cloud or precipitation particles near the surface or intense cells. For example, at 35 GHz (Ka-band), to achieve a −10 dBZe minimum detectable reflectivity near the surface, range sidelobes need to be about −70 dB down from the main lobe. Different techniques, including linear and non-linear frequency modulation of the chirp pulse, have been studied to range sidelobe reduction. Currently, the best range sidelobe levels that have been achieved by airborne and spaceborne remote sensing radars are around −55 to −60 dB (e.g., [17,18]). Further reduction of the range sidelobe level is not trivial. Airborne CRS and spaceborne RainCube data have indicated that the system phase noise might have played a limiting role in pulse compression sidelobe performance. Future efforts need to investigate system phase noise improvement using new local oscillator technologies, such as the photonic-RF signal generation technology.

In addition to radar hardware advancements, new remote sensing techniques are being developed such as atmospheric pressure sounding using differential absorption radar (DAR). The microwave barometric and sounder (MBARS) is a new airborne instrument under development by NASA/Goddard Space Flight Center and NASA/Langley Research Center with funding from the NASA Earth Science Technology Office (ESTO). This instrument, planned to fly on the NASA ER-2 aircraft, will enable the estimation of the atmospheric column dry air mass using differential absorption measurements from the ocean backscatter on the upper-side of the V-band oxygen absorption feature (65–70 GHz). This DAR measurement, combined with passive retrievals of atmospheric temperature and water vapor, will enable estimation of the full atmospheric column mass (and thus surface air pressure) [100]. To date, no remote sensing technology has demonstrated an ability to provide this critical atmospheric state variable with the precision approaching that is required for numerical weather prediction or research.

List of acronyms

CR-AVE	Costa-Rica Aura validation experiment
CRS	cloud radar system
CRYSTAL-FACE	cirrus regional study of tropical anvils and cirrus layers Florida area
CW	continuous wave
DAC	digital-to-analog
DAR	differential absorption radar
DDS	direct-digital-synthesizer
DPR	dual-frequency precipitation radar
DFR	dual-frequency ratio
DDV	dual-frequency Doppler velocity
Dual-PRF	dual pulse repetition frequency
EDOP	ER-2 Doppler radar
ESTO	NASA Earth Science Technology Office
EXRAD	ER-2 X-band radar
FPGA	field programmable gate array
GMF	geophysical model function
GPM	global precipitation mission

GSFC	Goddard Space Flight Center
LFM	linear frequency modulated
HB	Hitschfeld–Bordan
HIWRAP	high-altitude wind and rain airborne profiler
HRRR	high-resolution rapid refresh
IF/LO	intermediate frequency and local oscillator
IMPACTS	Investigation of Microphysics and Precipitation for Atlantic Coast-Threatening Snowstorms
IWRAP	imaging wind and rain airborne profiler
MBARS	microwave barometric and sounder
MERRA	modern-era retrospective analysis for research and applications
NCO	numerically controlled oscillators
NLFM	non-linear frequency modulated
NRCS	normalized radar cross-section
NUBF	non-uniform beam filling
PRF	pulse repetition frequency
SLR	side-looking radar
SSPA	solid-state power amplifiers
SAW	surface acoustic wave
TRMM	Tropical Rainfall Measurement Mission
TWT	traveling wave tube
UAS	unmanned aerial system
VAD	velocity–Azimuth display
3D	three-dimensional
3DVAR	three-dimensional variational

References

[1] Marks F and Houze R. Inner core structure of Hurricane Alicia from airborne Doppler Radar observations. *Journal of the Atmospheric Sciences*. 1987;44:1296–1317.

[2] Guimond SR, Bourassa MA, and Reasor PD. A latent heat retrieval and its effects on the intensity and structure change of Hurricane Guillermo (1997). Part I: the algorithm and observations. *Journal of the Atmospheric Sciences*. 2011;68:1549–1567.

[3] Weinman JA, Meneghini R, and Nakamura K. Retrieval of precipitation profiles from airborne radar and passive radiometer measurements: comparison with dual-frequency radar measurements. *Journal of Applied Meteorology and Climatology*. 1990;29(10):981–993.

[4] Meneghini R, Kozu T, Kumagai H, *et al.* A study of rain estimation methods from space using dual-wavelength radar measurements at near-nadir incidence over ocean. *Journal of Atmospheric and Oceanic Technology*. 1992;9(4):364–382.

[5] Meneghini R and Kumagai H. Characteristics of the vertical profiles of dual-frequency, dual-polarization radar data in stratiform rain. *Journal of Atmospheric and Oceanic Technology*. 1994;11(3):701–711.

[6] Liao L and Meneghini R. On modeling air/spaceborne radar returns in the melting layer. *IEEE Transactions on Geoscience and Remote Sensing*. 2005;43(12):2799–2808.

[7] Stephens GL, Vane DG, Tanelli S, *et al.* CloudSat mission: performance and early science after the first year of operation. *Journal of Geophysical Research: Atmospheres*. 2008;113(D8):717.

[8] Okamoto K, Yoshikado S, Masuko H, *et al.* Airborne microwave rain-scatterometer/radiometer. *International Journal of Remote Sensing*. 1982;3 (3):277–294.

[9] Sadowy GA, Berkun AC, Chun W, *et al.* Development of an advanced airborne precipitation radar. *Microwave Journal*. 2003;46(1):84–98.

[10] Pelon J, Vali G, Ancellet G, *et al.* Radar and lidar observations. In: Wendisch, Brenguier (eds.). *Airborne Measurements for Environmental Research: Methods and Instruments*. Wiley Press; 2013.

[11] Hagen M, Delanoé J, Ellis S, *et al.* Chapter 39: Airborne radar. In: Foken T (eds.). *Springer Handbook of Atmospheric Measurements*. Springer; 2022.

[12] Heymsfield GM, S Bidwell IJC, Ameen S, *et al.* The EDOP radar system on the high-altitude NASA ER-2 aircraft. *Journal of Atmospheric and Oceanic Technology*. 1996;13:795–809.

[13] Heymsfield GM, Halverson JB, Simpson J, *et al.* ER-2 Doppler radar investigations of the eyewall of Hurricane Bonnie during the Convection and Moisture Experiment-3. *Journal of Applied Meteorology*. 2001;40: 1310–1330.

[14] Heymsfield GM, Tian L, Heymsfield AJ, *et al.* Characteristics of deep and subtropical convection from nadir viewing high-altitude airborne radar. *Journals of the Atmospheric Sciences*. 2010;67:285–308.

[15] Geerts B, Heymsfield GM, Tian L, *et al.* Hurricane Georges's Landfall in the Dominican Republic: detailed airborne Doppler radar imagery. *Bulletin of the American Meteorological Society*. 2000;81:999–1018.

[16] Li L, Heymsfield GM, Racette PE, *et al.* A 94-GHz cloud radar system on a NASA high-altitude ER-2 aircraft. *Journal of Atmospheric and Oceanic Technology*. 2004;21(19):1378–1388.

[17] McLinden MLW, Li L, Heymsfield GM, *et al.* The NASA GSFC 94-GHz airborne solid-state Cloud Radar System (CRS). *Journal of Atmospheric and Oceanic Technology*. 2021;38:1001–1017.

[18] Beauchamp RM, Taneli S, and Syl OO. Observations and design considerations for spaceborne pulse compression weather radar. *IEEE Transactions on Geoscience and Remote Sensing*. 2021;59:4535–4546.

[19] Tian L, Heymsfield GM, Li L, *et al.* Properties of light stratiform rain derived from 10 and 94 GHz airborne Doppler radars. *Journal of Geophysical Research: Atmospheres*. 2007;112(D11):12.

[20] Li L, Heymsfield G, Carswell J, *et al.* The NASA high-altitude imaging wind and rain airborne profiler. *IEEE Transactions on Geoscience and Remote Sensing*. 2016;54:298–310.

[21] Meneghini R, Liao L, and Heymsfield GM. Relationship between horizontal wind velocity and normalized surface cross section using data from the HIWRAP dual-frequency airborne radar. *Journal of Atmospheric and Oceanic Technology*. 2021;38(3):423–439.

[22] Meneghini R, Liao L, and Heymsfield GM. Attenuation correction over ocean for the HIWRAP dual-frequency airborne scatterometer. *Journal of Atmospheric and Oceanic Technology*. 2019;36(10):2015–2030.

[23] Creticos J and Schaubert DH. A compact Ku/Ka band feed. In: *2007 IEEE Antennas and Propagation Society International Symposium*. 2007; pp. 3237–3240.

[24] Yueh SH, Wilson WJ. Polarimetric radar remote sensing of ocean surface wind. *IEEE Transactions on Geoscience and Remote Sensing*. 2002;40:793–800.

[25] Wu C, Graf JE, Freilich M, *et al.* The SeaWinds scatterometer instrument. In: *Proceedings of IGARSS'94 – 1994 IEEE International Geoscience and Remote Sensing Symposium*, vol. 3. 1994. pp. 1511–1515.

[26] Kollias P, Tanelli S, Battaglia A, *et al.* Evaluation of EarthCARE cloud profiling radar Doppler velocity measurements in particle sedimentation regimes. *Journal of Atmospheric and Oceanic Technology*. 1990;31(2):366–388.

[27] McLinden MLW, Loftus AM, Li L, *et al.* Application of nonuniform beam filling (NUBF) Doppler velocity error correction on airborne radar measurements. *IEEE Geoscience and Remote Sensing Letters*. 2022;19:5.

[28] Doviak RJ and Zrnić DS. *Doppler Radar and Weather Observations* (Second Edition). San Diego, CA: Academic Press; 1993.

[29] Sy OO, Tanelli S, Takahashi N, *et al.* Simulation of EarthCARE spaceborne Doppler radar products using ground-based and airborne data: effects of aliasing and nonuniform beam-filling. *IEEE Transactions on Geoscience and Remote Sensing*. 2014;52:1463–1479.

[30] Valenzuela GR. Theories for the interaction of electromagnetic and oceanic waves—a review. *Boundary-Layer Meteorology*. 1978;13:61–85.

[31] Barrick DE. Wind dependence of quasi-specular microwave sea scatter. *IEEE Transactions on Antennas and Propagation*. 1974;22:135–136.

[32] Brown G. In: W Plant and G Geernart (eds.). *Quasi-specular Scattering from the Air–Sea Interface Surface Waves and Fluxes*. Kluwer Academic. 1978;2:1–40.

[33] Brown G. Backscattering from a Gaussian-distributed perfectly conducting rough surface. *IEEE Transactions on Antennas and Propagation*. 1978;26:472–482.

[34] Chan HL and Fung AK. A theory of sea scatter at large incident angles. *Journal of Geophysical Rresearch*. 1977;82:3439–3444.

[35] Li L, Heymsfield GM, Tian L, *et al.* Measurements of ocean surface backscattering using an airborne 94-GHz cloud radar—implication for calibration of airborne and spaceborne W-band radars. *Journal of Atmospheric and Oceanic Technology*. 2005;22:1033–1045.

[36] Battaglia A, Wolde M, D'Adderio LP, *et al.* Characterization of surface radar cross sections at W-band at moderate incidence angles. *IEEE Transactions on Geoscience and Remote Sensing*. 2017;55:3846–3859.

[37] Tanelli S, Durden SL, and Im E. Simultaneous measurements of Ku- and Ka-band sea surface cross sections by an airborne radar. *IEEE Geoscience and Remote Sensing Letters*. 2006;3(3):359–363.

[38] Ewald F, Groß S, Hagen M, *et al.* Calibration of a 35 GHz airborne cloud radar: lessons learned and intercomparisons with 94 GHz cloud radars. *Atmospheric Measurement Techniques*. 2019;12(3):1815–1839. Available from: https://amt.copernicus.org/articles/12/1815/2019/.

[39] Yueh SH, Stiles BW, Tsai WY, *et al.* QuikSCAT geophysical model function for tropical cyclones and application to Hurricane Floyd. *IEEE Transactions on Geoscience and Remote Sensing*. 2001;39:2601–2612.

[40] Fernandez DE, Kerr EM, Castells A, *et al.* IWRAP: the Imaging Wind and Rain Airborne Profiler for remote sensing of the ocean and the atmospheric boundary layer within tropical cyclones. *IEEE Transactions on Geoscience and Remote Sensing*. 2005;43:1775–1787.

[41] Coto J, Jones WL, and Heymsfield GM. Validation of the high-altitude wind and rain airborne profiler during the Tampa Bay Rain Experiment. *Climate*. 2021;9(6). Available from: https://www.mdpi.com/2225-1154/9/6/89.

[42] Liao L, Meneghini R, Tian L, *et al.* Retrieval of snow and rain from combined X- and W-band airborne radar measurements. *IEEE Transactions on Geoscience and Remote Sensing*. 2008;46(5):1514–1524.

[43] Liao L, Meneghini R, Tian L, *et al.* Measurements and simulations of nadir-viewing radar returns from the melting layer at X and W bands. *Journal of Applied Meteorology and Climatology*. 2009;48(11):2215–2226.

[44] Fabry F and Zawadzki I. Long-term radar observations of the melting layer of precipitation and their interpretation. *Journal of Atmospheric Sciences*. 1995;52(7):838–851.

[45] Battan LJ. *Radar Observation of the Atmosphere*. Chicago IL: University Chicago; 1973.

[46] Bringi VN and Chandrasekar V. *Polarimetric Doppler Weather Radar—Principles and Applications*. Cambridge University Press; 2001.

[47] Ulbrich CW. Natural variations in the analytical form of the raindrop size distribution. *Journal of Climatology & Applied Meteorology*. 1983;22 (10):1764–1775.

[48] Testud J, Oury S, Black RA, *et al.* The concept of 'normalized' distribution to describe raindrop spectra: a tool for cloud physics and cloud remote sensing. *Journal of Applied Meteorology*. 2001;40(6):1118–1140.

[49] Matrosov SY. Radar reflectivity in snowfall. *IEEE Transactions on Geoscience and Remote Sensing*. 1992;30(3):454–461.

[50] Meneghini R, Kumagai H, Wang JR, *et al.* Microphysical retrievals over stratiform rain using measurements from an airborne dual-wavelength radar-radiometer. *IEEE Transactions on Geoscience and Remote Sensing*. 1997;35(3):487–506.

[51] Mardiana R, Iguchi T, and Takahashi N. A dual-frequency rain profiling method without the use of a surface reference technique. *IEEE Transactions on Geoscience and Remote Sensing*. 2004;42(10):2214–2225.

[52] Liao L and Meneghini R. A study of air/space-borne dual-wavelength radar for estimates of rain profiles. *Advances in Atmospheric Sciences*. 2005;22:841–851.

[53] Rose CR and Chandrasekar V. Extension of GPM dual-frequency iterative retrieval method with DSD-profile constraint. *IEEE Transactions on Geoscience and Remote Sensing*. 2006;44(2):328–335.

[54] Rose CR and Chandrasekar V. A GPM dual-frequency retrieval algorithm: DSD profile-optimization method. *Journal of Atmospheric and Oceanic Technology*. 2006;23(10):1372–1383.

[55] Grecu M, Tian L, Olson WS, *et al.* A robust dual-frequency radar profiling algorithm. *Journal of Applied Meteorology and Climatology*. 2011;50 (7):1543–1557.

[56] Seto S and Iguchi T. Applicability of the iterative backward retrieval method for the GPM dual-frequency precipitation radar. *IEEE Transactions on Geoscience and Remote Sensing*. 2011;49(6):1827–1838.

[57] Seto S, Iguchi T, and Oki T. The basic performance of a precipitation retrieval algorithm for the global precipitation measurement mission's single/dual-frequency radar measurements. *IEEE Transactions on Geoscience and Remote Sensing*. 2013;51(12):5239–5251.

[58] Liao L, Meneghini R, and Tokay A. Uncertainties of GPM DPR rain estimates caused by DSD parameterizations. *Journal of Applied Meteorology and Climatology*. 2014;53(11):2524–2537.

[59] Liao L and Meneghini R. A dual-wavelength radar technique to detect hydrometeor phases. *IEEE Transactions on Geoscience and Remote Sensing*. 2016;8:1–7.

[60] Gorgucci E and Baldini L. A self-consistent numerical method for microphysical retrieval in rain using GPM dual-wavelength radar. *Journal of Atmospheric and Oceanic Technology*. 2016;33(10):2205–2223.

[61] Liao L and Meneghini R. A modified dual-wavelength technique for Ku- and Ka-band radar rain retrieval. *Journal of Applied Meteorology and Climatology*. 2019;58(1):3–18.

[62] Liao L, Meneghini R, Tokay A, *et al.* Assessment of Ku- and Ka-band dual-frequency radar for snow retrieval. *Journal of the Meteorological Society of Japan Ser II*. 2020;98(6):1129–1146.

[63] Seto S, Iguchi T, Meneghini R, *et al.* The precipitation rate retrieval algorithms for the GPM dual-frequency precipitation radar. *Journal of the Meteorological Society of Japan Ser II*. 2021;99(2):205–237.

[64] Meneghini R, Liao L, Bidwell SW, *et al.* On the feasibility of a Doppler weather radar for estimates of drop size distribution using two closely spaced frequencies. *IEEE Transactions on Geoscience and Remote Sensing*. 2001;39(10):2203–2216.

[65] Meneghini R, Bidwell SW, Rincon R, *et al.* Differential-frequency Doppler weather radar: theory and experiment. *Radio Science*. 2003;38(3):5-1–5-5.

[66] Liao L and Meneghini R. A study on the feasibility of dual-wavelength radar for identification of hydrometeor phases. *Journal of Applied Meteorology and Climatology*. 2011;50(2):449–456.

[67] Liao L, Meneghini R, Tokay A, *et al.* Retrieval of snow properties for Ku- and Ka-band dual-frequency radar. *Journal of Applied Meteorology and Climatology*. 2016;55(9):1845–1858.

[68] Thurai M, Huang GJ, Bringi VN, *et al.* Drop shapes, model comparisons, and calculations of polarimetric radar parameters in rain. *Journal of Atmospheric and Oceanic Technology*. 2007;24(6):1019–1032.

[69] Mishchenko MI and Travis LD. Capabilities and limitations of a current FORTRAN implementation of the T-matrix method for randomly oriented, rotationally symmetric scatterers. *Journal of Quantitative Spectroscopy and Radiative Transfer*. 1998;60(3):309–324.

[70] Heymsfield AJ. Properties of tropical and midlatitude ice cloud particle ensembles. Part II: applications for mesoscale and climate models. *Journal of the Atmospheric Sciences*. 2003;60(21):2592–2611.

[71] Heymsfield AJ and Miloshevich LM. Parameterizations for the cross-sectional area and extinction of cirrus and stratiform ice cloud particles. *Journal of the Atmospheric Sciences*. 2003;60(7):936–956.

[72] Kuo KS, Olson WS, Johnson BT, *et al.* The microwave radiative properties of falling snow derived from nonspherical ice particle models. Part I: an extensive database of simulated pristine crystals and aggregate particles, and their scattering properties. *Journal of Applied Meteorology and Climatology*. 2016;55(3):691–708.

[73] Westbrook CD, Ball RC, and Field PR. Radar scattering by aggregate snowflakes. *Quarterly Journal of the Royal Meteorological Society*. 2006;132(616):897–914.

[74] Botta G, Aydin K, and Verlinde J. Modeling of microwave scattering from cloud ice crystal aggregates and melting aggregates: a new approach. *IEEE Geoscience and Remote Sensing Letters*. 2010;7(3):572–576.

[75] Petty GW and Huang W. Microwave backscatter and extinction by soft ice spheres and complex snow aggregates. *Journal of the Atmospheric Sciences*. 2010;67(3):769–787.

[76] Liao L, Meneghini R, Nowell HK, *et al.* Scattering computations of snow aggregates from simple geometrical particle models. *IEEE Journal of Selected Topics in Applied Earth Observations and Remote Sensing*. 2013;6(3):1409–1417.

[77] Nowell H, Liu G, and Honeyager R. Modeling the microwave single-scattering properties of aggregate snowflakes. *Journal of Geophysical Research: Atmospheres*. 2013;118(14):7873–7885.

[78] Hitschfeld W and Bordan J. Errors inherent in the radar measurement of rainfall at attenuating wavelengths. *Journal of Atmospheric Sciences*. 1954;11(1):58–67.

[79] Iguchi T and Meneghini R. Intercomparison of single-frequency methods for retrieving a vertical rain profile from airborne or spaceborne radar data. *Journal of Atmospheric and Oceanic Technology*. 1994;11(6):1507–1516.

[80] Meneghini R, Iguchi T, Kozu T, *et al.* Use of the surface reference technique for path attenuation estimates from the TRMM precipitation radar. *Journal of Applied Meteorology*. 2000;39(12):2053–2070.

[81] Meneghini R, Jones JA, Iguchi T, *et al.* A hybrid surface reference technique and its application to the TRMM precipitation radar. *Journal of Atmospheric and Oceanic Technology*. 2004;21(11):1645–1658.

[82] Meneghini R, Liao L, Tanelli S, *et al.* Assessment of the performance of a dual-frequency surface reference technique over ocean. *IEEE Transactions on Geoscience and Remote Sensing*. 2012;50(8):2968–2977.

[83] Meneghini R, Kim H, Liao L, *et al.* An initial assessment of the surface reference technique applied to data from the dual-frequency precipitation radar (DPR) on the GPM satellite. *Journal of Atmospheric and Oceanic Technology*. 2015;32(12):2281–2296.

[84] Meneghini R, Kim H, Liao L, *et al.* Path attenuation estimates for the GPM dual-frequency precipitation radar (DPR). *Journal of the Meteorological Society of Japan Ser II*. 2021;advpub:2021–010.

[85] McMurdie LA, Heymsfield GM, Yorks JE, *et al.* Chasing snowstorms: the investigation of microphysics and precipitation for Atlantic coast-threatening snowstorms (IMPACTS) campaign. *Bulletin of the American Meteorological Society*. 2022;103(5):E1243–E1269.

[86] Finlon JA, McMurdie LA, and Chase RJ. Investigation of microphysical properties within regions of enhanced dual-frequency ratio during the IMPACTS field campaign. *Journals of the Atmospheric Sciences*. 2022;79 (10):2773–2795.

[87] Armijo L. A theory for the determination of wind and precipitation velocities with Doppler radars. *Journal of Atmospheric Sciences*. 1969;26: 570–573.

[88] Miller LJ and Strauch RG. A dual Doppler radar method for the determination of wind velocities within precipitating weather systems. *Remote Sensing of Environment*. 1974;3:219–235.

[89] Klimowski BA and Marwitz JD. The synthetic dual-Doppler analysis technique. *Journal of Atmospheric and Oceanic Technology*. 1992;9: 728–745.

[90] Didlake AC, Heymsfield GM, Tian L, *et al.* The coplane analysis technique for three-dimensional wind retrieval using the HIWRAP airborne Doppler radar. *Journal of Applied Meteorology and Climatology*. 2015;54:605–623.

[91] Lhermitte RM and Atlas D. Precipitation motion by pulse Doppler radar. In: *Proceedings of the Ninth Weather Radar Conference*, Kansas City, MO: Amer. Meteor. Soc.; 1961. pp. 218–223.

[92] Browning KA and Wexler R. The determination of kinematic properties of a wind field using Doppler radar. *Journal of Applied Meteorology and Climatology*. 1968;7:105–113.

[93] Tian L, Heymsfield GM, Didlake AC, *et al.* Velocity-azimuth display analysis of Doppler velocity data for HIWRAP. *Journal of Applied Meteorology and Climatology*. 2015;54:1792–1808.

[94] Helms CN, Walker McLinden ML, Heymsfield GM, *et al.* Reducing errors in Velocity–Azimuth Display (VAD) wind and deformation retrievals from airborne Doppler radars in convective environments. *Journal of Atmospheric and Oceanic Technology*. 2020;37:2251–2266.

[95] Guimond SR, Tian L, Heymsfield GM, *et al.* Wind retrieval algorithms for the IWRAP and HIWRAP airborne Doppler radars with applications to Hurricanes. *Journal of Atmospheric and Oceanic Technology*. 2014;31: 1189–1215.

[96] Lee WC, Dodge P, Marks FD, *et al.* Mapping of airborne Doppler radar data. *Journal of Atmospheric and Oceanic Technology*. 1994;11:572–578.

[97] Guimond SR, Heymsfield GM, Reasor PD, *et al.* The rapid intensification of Hurricane Karl (2010): new remote sensing observations of convective bursts from the global Hawk platform. *Journal of the Atmospheric Sciences*. 2016;73:3617–3639.

[98] Guimond SR, Zhang J, Sapp J, *et al.* Coherent turbulence in the boundary layer of Hurricane Rita (2005) during an eyewall replacement cycle. *Journal of the Atmospheric Sciences*. 2018;75:3071–3093.

[99] Haque SM, Hoppe DJ, and Epp LW. Solid state power amplifier and travelling wave tube amplifier additive phase noise characterization at Ka-band operation. In: *IEEE Radar Conference*, 4–8 May, 2009, Pasadena, CA, USA, DOI: 101109/RADAR20094977053. 2009.

[100] Lin B and Hu Y. Numerical simulations of radar surface air pressure measurements of O_2 bands. *IEEE Geoscience and Remote Sensing Letters*. 2005;2(3):324–328.

Chapter 8

Ocean-going weather and profiling radar for clouds and precipitation

P.T. May[1], B. Dolan[2], M. Katsumata[3], P.A. Kucera[4], V. Louf[5], A. Protat[5] and C.R. Williams[6]

Shipborne weather and profiling radars have been crucial for our understanding of precipitating weather systems over the past 40 years and there are now several new platforms with state-of-the-art dual-polarization weather radar including in the United States, Japan, and Australia with new capabilities planned for Taiwan and on an Australian icebreaker. This chapter will review some key weather radar developments and experiments over the last 40 years (Section 8.2) and the new dual-polarization systems (Section 8.3). This is followed by a more technical discussion on the new-generation dual-polarization radars, challenges of ship operations, quality control processing, and science opportunities (Section 8.4). We then discuss profiling radars including some historical context as well as the science capabilities (Section 8.5). The chapter concludes with some discussion on new opportunities.

8.1 Introduction

Over two-thirds of the Earth is covered by oceans and many crucial components of the weather and climate system are predominantly focused over ocean waters. For these reasons, understanding key processes across the global oceans has been a science priority for decades. Of course, this brings unique challenges because of the remoteness of many of these areas. Even where there are suitable islands to base equipment, the islands themselves can influence the local meteorology. This challenge alone has led to large investments in ocean-going technology to support meteorological research and the deployment and use of radars of various kinds.

[1]School of Earth, Atmosphere and Environment, Monash University, Australia
[2]Department of Atmospheric Sciences, Colorado State University, USA
[3]Japan Agency for Marine-Earth Science and Technology, Yokosuka, Japan
[4]COMET, University Corporation for Atmospheric Research, Boulder, Colorado, USA
[5]Science and Innovation/Research, Bureau of Meteorology, Australia
[6]Ann and H.J. Smead Aerospace Engineering Sciences Department, Center for Astrodynamics Research, University of Colorado, Boulder, USA

Some of the earliest achievements involved the deployment of weather radar, but over the last 20 years profiling radars (and lidars) are also being increasingly used for comprehensive observations of precipitating and non-precipitating clouds.

The applications of these ocean-going radars have been extremely broad. As will be discussed, the application of these radars has led to an improved understanding of fundamental processes with concepts applied globally (e.g., [1]) and provided critical ground validation data and data for algorithm development for satellite programs over remote oceans (e.g., [2,3]). Radar observations have also been used to better understand a wide variety of other phenomena such as ocean waves, but that is beyond the scope of this chapter. This chapter also is limited to shipborne radar while noting the enormous contributions of airborne radars over the ocean. While a large number of experiments will be mentioned, this is not meant to be an exhaustive list.

This chapter is organized as follows. It will begin with a historical perspective (Section 8.2) followed by a description of some current state-of-the-art weather radars (Section 8.3) and a discussion of the technical challenges specific to shipborne dual-polarization weather radar applications and the required radar processing (Section 8.4). The technology and science arising from profiling radars are discussed in Section 8.5. The chapter concludes with a discussion regarding future directions and opportunities (Section 8.6).

8.2 Some historical perspectives

This section will examine some of the pivotal periods and experiments in terms of the deployment of new shipborne radar technologies and the profound impact they had on our science from the 1960s through to the present. Radar has been crucial for many major programs and innovations that will be discussed next.

8.2.1 Shipborne radars by the Japan Meteorological Agency (JMA)

In the 1960s, the Japan Meteorological Agency (JMA) launched two research vessels, *Ryofu-maru* and *Keifu-maru*, with weather radar. These C-band conventional radars were stabilized by actively controlling their pedestals. Early data were taken by photographing and sketching scope, and/or using a printer [4]. The *Keifu-maru* radar was upgraded in 1987 to enable recording data digitally, as well as stabilizing the radar beam by controlling azimuth and elevation of the antenna to cancel the ship motion [5] – an approach still used today. These radars were in operation for a wide variety of field campaigns in the north-western Pacific to observe phenomena such as typhoons [6], the Baiu-front [7], and winter snow clouds [8], until the retirement of the *Keifu-maru* in 2000.

8.2.2 GATE (Global Atmospheric Research Program (GARP) Atlantic Tropical Experiment – 1974)

GATE was a visionary experiment of unprecedented scale that transformed our understanding of tropical meteorology. The legacy of GATE continues to this day from the seminal papers describing tropical convective storms, the influence of the environment on convection, and how that convection impacted the environment. Even today, the well-documented GATE convection provides the basis for evaluating numerical model simulations. GATE was also notable for the impact it had on the careers of young scientists around the world who went on to be global science leaders. The vision of GATE was to combine observations from radar, satellite, aircraft, soundings, moorings, etc. to understand oceanic convective systems and their two-way interaction with the larger environment. The experiment itself was focused on the equatorial Atlantic and coordinated from an operations center in Senegal, West Africa. This itself was a major undertaking as it involved 40 research ships and 12 research aircraft from 20 countries – a scale of research activity unlikely to be repeated. Four of these research vessels carried a weather radar, representing the first large-scale deployment of weather radar at sea (Table 8.1). These radars recorded the returned power and hence the reflectivity (Z) that was pivotal for the GATE science program.

The radars on all four ships from the United States and Canada were operating at the C-band with beamwidths between 1° and 2°. The range resolution of 1–2 km for three of the radars seems modest by today's standards but was good at the time and allowed detailed observations of the reflectivity structure of deep convection in the experimental area. The four ships were spaced in a rhomboid with sides of approximately 200 km (Figure 8.1). All the radars operated on stabilized platforms. Furthermore, this is one of the earliest experiments where the radar data was digitally stored [9]. The data processing stream and examples of both photographs and digital data products were discussed by [9]. All forms of these data were extensively exploited, for example, in detailed studies of a squall line [1] and convective organization (e.g., [10,11]).

The scientific outcomes combining the variety of datasets provided key insights into the structure and organization of oceanic deep tropical convection as well as its impacts on the environment. Parts of this used the radar data for the calculation of large-scale heat and moisture (the so-called Q1 and Q2) budgets associated with these systems using area-averaged rainfall and soundings (e.g., [12]). The analysis of these unprecedented GATE radar observations enabled decades of research in tropical meteorology.

Table 8.1 Outlining the characteristics of the 4 shipborne radars during GATE (after [1])

Radar characteristics	RV Gilliss	RV oceanographer	RV researcher	RV quadra
Wavelength (cm)	5.3	5.3	5.35	5.35
Peak power (kW)	250	215	225	1,000
Beamwidth (deg)	1.45	1.5	2	1
Range bin size (km)	0.25–1	2	2	1
Azimuth recording increment (deg)	1	2	2	1

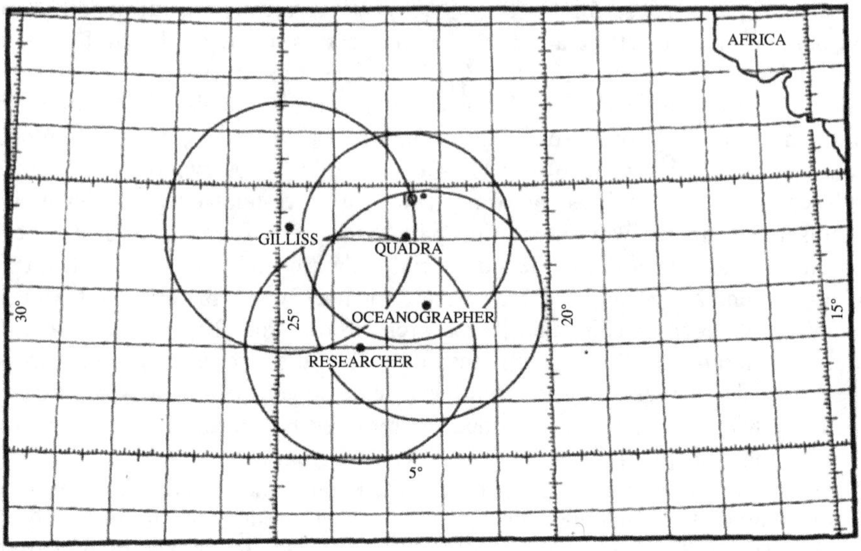

Figure 8.1 Position of the ships carrying weather radars during Phase 3 of GATE (30 August–10 September 1974) (after [1]) published by the American Meteorological Society

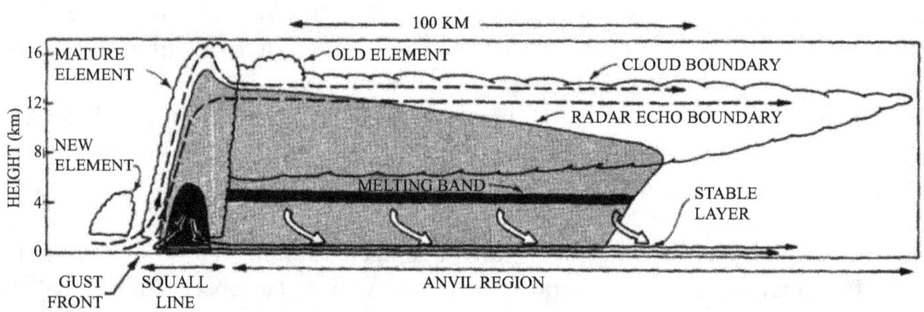

Figure 8.2 Schematic cross-section through a squall line system. Streamlines show the relative flow, dashed lines the updraft circulations, and the solid lines the downdraft circulation associated with a mature squall line (after [1]). Published by the American Meteorological Society.

The GATE squall line conceptual model (Figure 8.2) from [1] is still relevant and recognizable in current conceptual descriptions of both oceanic and land-based squall lines, as well as still being a test case for convection resolving models. The evolution of squall line structure and interactions between cells was analyzed and the resulting idealized models of storm evolution [13] had an enormous impact. For the first time, the apparent propagation of convective systems associated with the generation of new cells on storm outflows and the decay of old cells into the trailing stratiform areas of squall lines were documented. The impact of these datasets was neatly summarized by a quote from Joanne Simpson, collected by Anneliese Sherer on 23 July 1999 (https://www.ametsoc.org/sloan/gate/index.html): "I believe that the GATE dataset is probably the most useful dataset in tropical meteorology."

8.2.3 TOGA-COARE: Doppler radars and profilers

The TOGA-COARE field campaign was conducted in the equatorial western Pacific warm pool during the period of November 1992–February 1993. The four main goals of the experiment were to describe and understand (1) air–sea coupling in the tropical warm pool region, (2) processes that organized the convection, (3) ocean responses, and (4) multi-scale interactions [14]. Like GATE, the TOGA-COARE program was a seminal experiment advancing our knowledge of precipitation systems, air sea exchange, and mechanisms behind El Nino and the Madden-Julian Oscillation.

The array of atmospheric and oceanic instrumentation including shipborne Doppler and conventional radars, in situ and airborne radar observations, surface buoys, soundings, disdrometers, and vertical profilers provided an opportunity to examine tropical convection and the supporting large-scale atmospheric-ocean characteristics in the western Pacific warm pool [14–18]. The TOGA COARE observations area was centered at 2°S, 156°E within an Intensive Flux Array (IFA) of enhanced instrumentation (see Figure 8.3).

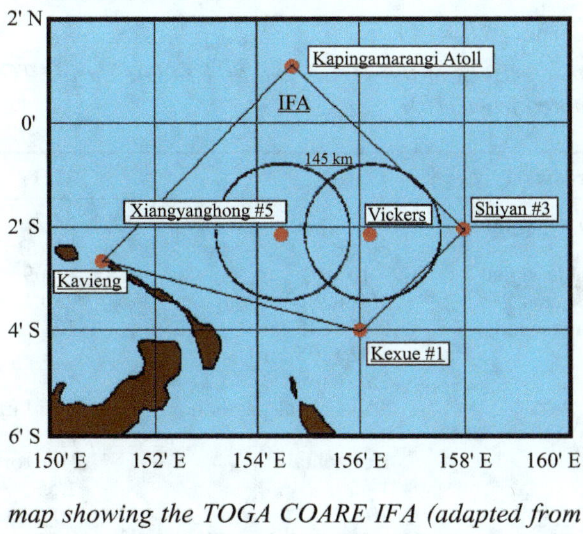

Figure 8.3 A map showing the TOGA COARE IFA (adapted from [16]). © American Meteorological Society. Used with permission.

The TOGA Doppler radar (operated by the National Aeronautical and Space Administration (NASA)) was deployed on the R/V *Xiang Yang Hong #5* from the People's Republic of China (PRC#5). The Massachusetts Institute of Technology (MIT) Doppler radar (operated by Colorado State University) was installed on the R/V *John V. Vickers* (supported by the United States National Oceanic and Atmospheric Administration) and the Japanese C-band weather radar was installed on the *R/V Keifu-Maru* [19]. A comparative analysis of radar reflectivity observations show consistent precipitation characteristics for the three radars [20]. The characteristics of the TOGA and MIT radar are shown in Table 8.2. The R/V *PRC #5* and the R/V *Vickers* were nominally deployed about 150 km W–E from each other when both ships were on station (see Figure 8.3 for ship locations). During two periods of the field campaign, the TOGA and MIT radars were brought to within 40–50 km of each other to obtain the first-ever ocean-going dual-Doppler observations [21].

The advancement since GATE in radar antenna stabilization hardware and software made it possible to have geo-referenced, fully stabilized radar observations from ships. TOGA COARE also was the first ship experiment to obtain Doppler velocity observations since the GATE radars were reflectivity only. The TOGA COARE stabilization approach was developed through a collaboration between SIGMET, Inc., the National Center for Atmospheric Research (NCAR), NASA, and Colorado State University. This centered on integrating antenna stabilization software that was developed to ingest navigation data (heading, position, roll, pitch, heave, and timing) from an inertial navigation system (INS) and output commands to the antenna servo system to drive the antenna to compensate for the ship motion in real time. The INS system was also integrated with a GPS satellite receiver that provided updates on the geodetic position. During that time, GPS position uncertainty was on the order of 100 m [22] that could result in antenna instability [23]. Therefore, the radar antenna INU was manually updated at least

Table 8.2 Characteristics of the TOGA and MIT radars during TOGA COARE (adapted from [16])

Radar characteristics	TOGA	MIT
Wavelength (cm)	5.35	5.37
Dynamic range	80 dB	80 dB
Minimum detectable signal	−105	−109
Peak power	250 KW	250 KW
Pulse width (microseconds)	0.48 (145 km) 1.82 (295 km) 1.0 (Cruise 3)	1.0 (short/long range)
Antenna	Solid metal parabolic	Solid metal parabolic
Reflector size (m)	2.4	2.4
Polarisation	Horizontal	Horizontal
Beam width (deg)	1.55	1.55
Antenna gain (dB)	40	40
Side lobe level (dB)	−22 to 10 deg off-axis −30 > 10 deg off-axis	−22 to 10 deg off-axis −30 > 10 deg off-axis

every 15 min during TOGA COARE, which resulted in a high-quality shipborne radar dataset [16] that has been used for comprehensive studies to further understand tropical convection in the equatorial Pacific, following the initial legacy of GATE.

With the integration of the ship motion into the radar scanning software, detailed intercomparisons could be conducted with the geo-referenced radar observations. One of the innovative analyses that was developed was the relative intercomparison of the radar calibration between the TOGA and MIT radars. A technique was developed by Kucera [15] to match each radar bin in three-dimensional space for the overlap region between the two radars (see the overlap region in the map shown in Figure 8.3). The radar reflectivity distributions were compared for observed convection in the overlap region. The results from the method were used to determine the post-experiment calibration adjustments. Based on this analysis, the MIT radar had a +2.4 dB adjustment applied to the radar reflectivity [16,24] and a variety of bias adjustments were applied to the TOGA radar reflectivity that accounted for issues with a malfunctioning amplifier; errors in noise sampling; hardware and software adjustments and empirical adjustments to match the MIT radar reflectivity [16,25].

These well-calibrated shipborne radar datasets provided a framework for investigative studies of tropical convection. The Short *et al.*'s [16] study provided a detailed overview of the development of a comprehensive TOGA COARE merged radar rainfall and convective/stratiform-gridded dataset using the TOGA and MIT radars (an example rainfall map is shown in Figure 8.4). This dataset was used to examine the variability in tropical precipitation observed during TOGA COARE. Short *et al.* [16] used the dataset to confirm the different scales of precipitation

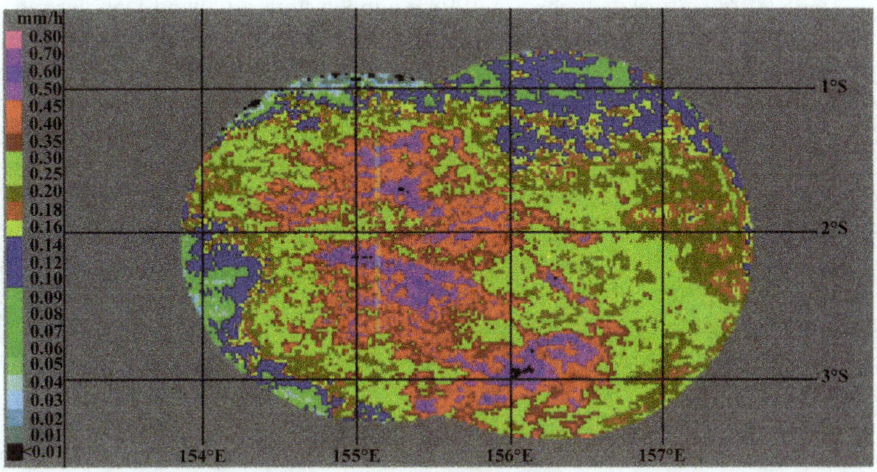

Figure 8.4 Merged radar rainfall estimates using the TOGA and MIT radars for the observation period of 21 December 1992–10 January 1993 [16]. © American Meteorological Society. Used with permission).

observed in the Western Equatorial Pacific. The different patterns observed included a diurnal maximum that peaks during the night, intense convection associated with westerly wind bursts (WWB), and large-scale precipitation events associated with the Madden-Julian Oscillations (MJO).

The TOGA and MIT radar datasets were used for a variety of investigations that explored the vertical structure of tropical convection [16,26,27], the horizontal scales, lifecycle, and small-scale variability of tropical convection [24,28], the nature of electrification in oceanic convection [29], the coupling between surface fluxes and mesoscale convection [30], and the kinematics of observed tropical convection inferred from dual-Doppler observations [21]. For example, Petersen *et al.* [21] investigated several Dual-Doppler case studies (see Figure 8.5) using the TOGA and MIT radar datasets that were only made possible with the antenna stabilization technology that was developed and implemented during TOGA COARE.

8.2.4 Doppler weather radar applications post-TOGA COARE

Shipborne Doppler radar became the norm after TOGA-COARE. New radar capability was deployed on the NOAA R/V *Ronald H. Brown (R/V Brown)*, and on the JAMSTEC R/V *Mirai*.

The C-band Doppler radar on the R/V *Brown* had an enhanced antenna stabilization system installed that provided position and ship motion information into the radar software to track and compensate for ship motion in the operational scanning. This radar was deployed for many experiments including TEPPS (1997), KWAJEX (1999), JASMINE (1999), EPIC (2001), and VOCALS (2008)* [31–33] with the aim of studying tropical convection, its structure, and variability on various time scales [32,34–36], wave structure [37], and monsoon circulations [38].

The R/V *Mirai* was launched in 1998 equipped with a C-band Doppler radar (M-DOP). As with the TOGA-COARE radars, the antenna direction was stabilized by controlling the antenna azimuth and elevation to account for the ship's motions detected by an inertial navigation system. The radar was in operation in major cruises studying the atmosphere–ocean interaction mainly in the western Pacific, e.g., for ITCZ convection [39], boreal summer intraseasonal variations [40], and cyclone structure in the Arctic Ocean [41].

Combinations of ship and island sites for dual-Doppler applications also became important. The South China Sea Monsoon Experiment (SCSMEX) took place in May and June 1998 in the northern portion of the South China Sea (SCS). The Intensive Flux Array of instrumentation included a dense network of surface meteorological observations, atmospheric soundings, aerosondes, oceanic observations, and two scanning radars [42–45]. The TOGA radar was deployed on the Chinese Research R/V *Shiyan* 3 and took advantage of the same antenna stabilization system as that used during TOGA COARE. The Bureau of Meteorology C-Band polarimetric Doppler

*TEPPS, Tropical Eastern Pacific Process Study; KWAJEX, Kwajalein Experiment, JASMINE, Joint Air–Sea Monsoon Interaction Experiment; EPIC, East Pacific Investigation of Climate Processes in the Coupled Ocean-Atmosphere Systems; VOCALS, VAMOS Ocean-Cloud-Atmosphere-Land Study.

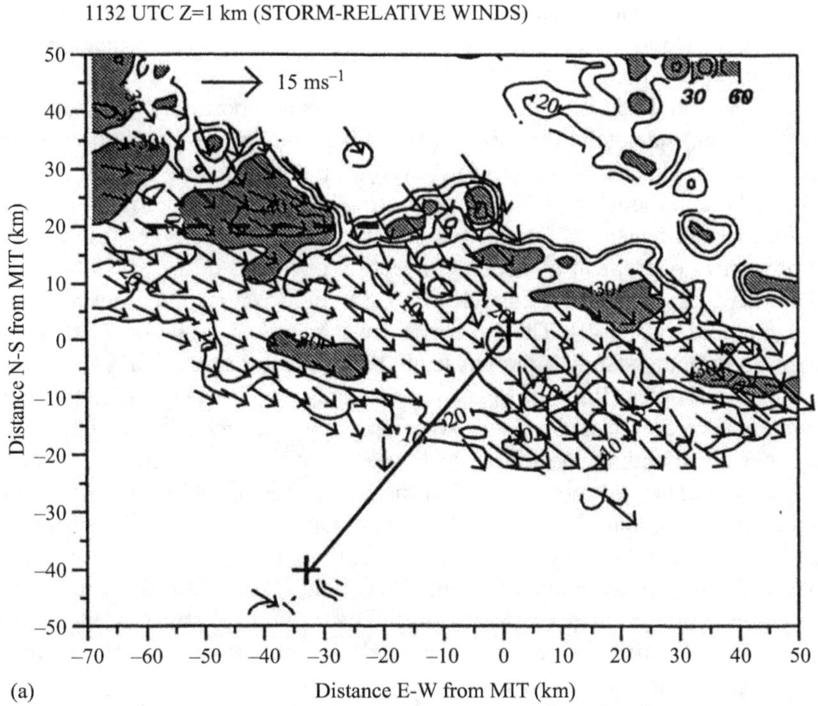

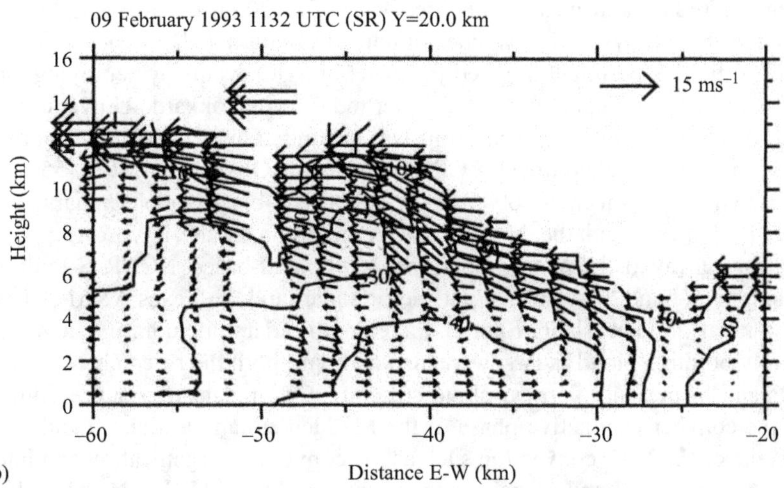

Figure 8.5 Example of dual-Doppler storm-relative wind flow retrievals from TOGA COARE [21]. © American Meteorological Society. Used with permission.

radar (CPOL) was deployed on Dongsha Island (20°N, 116°E) near the TOGA radar to create a dual-Doppler network that was used for a variety of investigative studies of Monsoon Convection.

The TOGA and CPOL radar datasets provided an opportunity to conduct radar studies to better understand the rainfall variability and spatial distribution of drop-size distributions in the SCS [42,46]. Several studies used the TOGA shipborne Doppler radar and the island-based polarimetric CPOL radar data to investigate the lifecycle, three-dimensional convective structures, cloud microphysics, and kinematic properties of tropical monsoon convection observed in the SCS [47–49].

The Kwajalein Experiment (KWAJEX) field campaign took place in July–September 1999 [50]. The experiment was centered on the Kwajalein Atoll and focused on understanding the physical characteristics of tropical convection in support of the Tropical Rainfall Measurement Mission (TRMM) satellite program. Observations were collected using a wide array of instrumentation including aircraft, upper-air sounding, profilers, disdrometers, rain gauges, and two scanning Doppler radars. The two radars included the S-Band dual-polarized Doppler radar located on the Kwajalein Atoll and R/V *Brown* Doppler radar. The R/V *Brown* was nominally located 40 km away for the Kwajalein radar to provide a high-resolution Doppler radar network. Spooner [51] produced and analyzed dual-Doppler retrievals to characterize the lifecycle of storm intensity, vertical structure, divergence, and vertical velocities to better understand the characteristics of convection observed over the tropical Pacific Ocean.

The Madden-Julian Oscillation (MJO) became a key international science focus area in the Indian and western Pacific Oceans. In 2006, the Mirai Indian Ocean cruise for the Study of the Madden-Julian Oscillation (MJO)-convection Onset (MISMO) field experiment targeted the onset of MJO events by deploying the R/V *Mirai* with M-DOP and an X-band Doppler radar from Hokkaido University in Gan Island [52–54]. The international effort was expanded in 2011 with the coordinated international field experiments CINDY/DYNAMO/AMIE/LASP[†] [55]. These coordinated field projects involved two shipborne C-band Doppler radars (M-DOP on the R/V *Mirai* and the NASA TOGA radar on board R/V *Roger Revelle*), forming an array of radars together with the island-based NCAR S-Polka dual-wavelength (S-band /Ka-band), polarimetric radar, and the Texas A&M SMART-R C-band radar on Gan Island. Shorter wavelength radars, including W-band (3mm) vertically pointing cloud radars, were also on board both the research vessels and at Gan Island to measure vertical cloud structure. The radar array successfully captured the convectively active phase of the Madden-Julian Oscillation and revealed the details of the MJO convection such as the convective organization and intensity [56,57], meridional and zonal connective variability [58], the diurnal cycle (Figure 8.6, from [59] and detailed air-sea interaction, e.g., [60]). Radar

[†]CINDY: Cooperative Indian Ocean Experiment on Intraseasonal Variability in Year 2011.
Dynamo: Dynamics of the MJO.
AMIE: ARM Madden-Julian Oscillation Investigation Experiment.
LASP: Littoral Air–Sea Process.

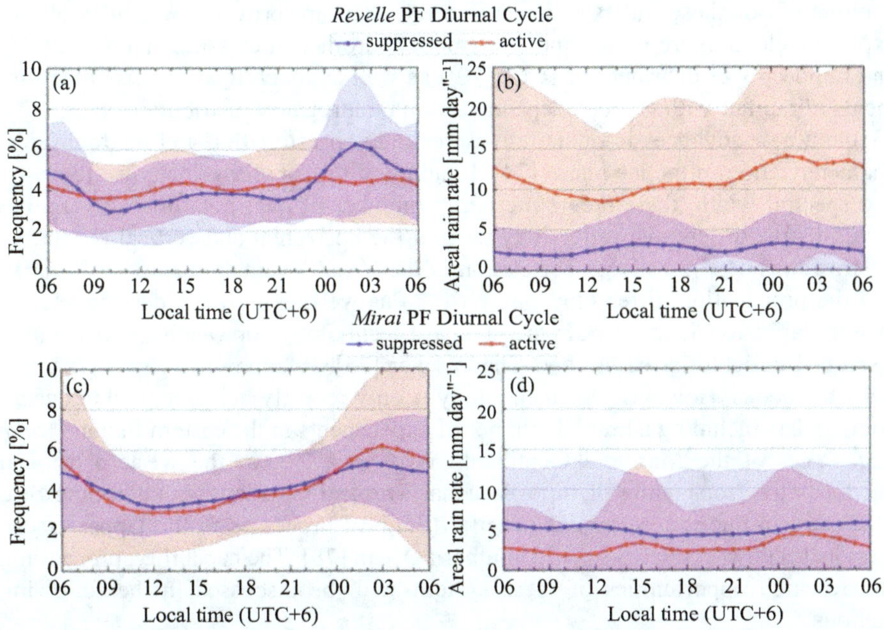

Figure 8.6 Diurnal cycle of rain frequency (a, c) and rain rate (b, d) from the Revelle (top) and Mirai (bottom) precipitation features (PFs) sampled during DYNAMO. [59]. © American Meteorological Society. Used with permission

observations were also used to understand the convective variability in the context of surface drop size distributions (DSD) observed during DYNAMO [61].

The innovative antenna stabilization technology provided an opportunity to evaluate satellite radar retrievals from the TRMM satellite. The radar rainfall estimated dataset during SCSMEX provided an early opportunity to validate satellite precipitation retrievals early in the TRMM satellite mission [44,62,63]. The radar data collected during KWAJEX supported the study by Anagnostou *et al.* [64] to evaluate the performance of the TRMM satellite precipitation radar (PR). The evaluation methodology developed by [64] was also applied to other globally located surface-based radar systems to understand the sampling characteristics of the TRMM PR but also provided a reference platform to examine the relative calibration of adjacent radars.

Recently marine navigation radar data has also been demonstrated to have the potential to provide useful data [65]. These operational radars have also been exploited to provide wave observations.

8.2.5 The coming of shipborne dual-polarization radars

The last decade marked the implementation of dual-polarization weather radar on research vessels in Australia, the United States, and Japan [66–68]. While the exploitation of these radars is in its infancy, they are proven powerful tools for exploring cloud microphysics including bulk hydrometeor classification (e.g., [69]) and raindrop size distributions (e.g., [70]) as well as much more precise measurements of rainfall with reduced dependence on raindrop size distributions (e.g., [71–73]) and new advances in quantitative precipitation estimation and microphysical characterization in the ice phase [74]. In addition to Doppler velocity, reflectivity, and spectral width, these radars measure a number of dual polarimetric variables including the differential reflectivity ratio of the horizontal and vertical (H and V) polarizations (Z_{DR}), the cross-correlation of the H and V signals at zero lag ($\rho_{HV}(0)$) and the propagation differential phase (Φ_{DP}) as well as its range derivative K_{DP}. Dual-polarization radar is now the standard technology for weather radar and is discussed in detail by Bringi and Chadrasekar [75], Ryzkhov and Zrnic [76], and other textbooks. However, this technology is only recently being applied to ocean-going radars including a limited number of experiments in the eastern Indian Ocean (e.g., Year of the Maritime Continent, YMC [67,77]), in the western tropical Pacific (e.g., Propagation of Intra-Seasonal Tropical OscillatioNs Philippine Sea, PISTON), in the eastern tropical Pacific (Salinity Processes in the Upper Ocean Regional Study – SPURS-2), and Southern Ocean [78]. The capability, processing, and research opportunities of these systems will be discussed in the following sections.

8.3 Three current shipborne C-band dual-polarization radars

The deployment of weather radar on a ship is a major investment but one with significant payoffs. There are currently three polarimetric radars available for ship-based research. Two of these radars are permanently affixed to ships, those being the OceanPol radar on the Australian R/V *Investigator* operated by Australia's Commonwealth Scientific and Industrial Research Organisation (CSIRO), and the C-band radar on the Japanese R/V *Mirai* (M-POL). The third radar is the Sea Going Polarimetric radar (SEA-POL), a C-band polarimetric radar maintained and operated by Colorado State University. Rather than be affixed permanently to one ship, SEA-POL was designed in a way to allowed it to operate from several ships in the US fleet. SEA-POL was deployed on the R/V *Roger Revelle*, R/V *Thomas G. Thompson*, and R/V *Sally Ride* between 2017 and 2019. A number of research vessels capable of carrying dual-polarization C- and X-band radars are being planned or are already being commissioned. This section will provide a very brief description of the radars before discussing the technical challenges and approaches to mitigate these issues and the radar performance in Section 8.4.

The Ocean-Pol radar (Figure 8.7) was manufactured by Enterprise Electronics Corporation with a 250-kW peak power and range resolution from 60 m to 300 m and employs the simultaneous H/V transmission mode. The radar is routinely operated on all cruises by the CSIRO Marine National Facility and data is publicly available. Most operations use 0.8 microsec pulse for a typical range resolution of

Figure 8.7 The R/V Investigator with the OCEAN-POL radar mounted at the high point of the ship. Image courtesy of CSIRO.

125 m. The 1.3° beamwidth antenna is mounted at the top of the vessel to minimize any blockage and avoid safety issues so that it can run continuously and does not have any beam-blocked sectors or requirements for sector blanking. The transmitter is located 15 m below the antenna while the receiver is located in the radome to minimize losses. The receiver system uses simultaneous reception via two completely separate receiver chains. The RF signal is down converted to 60 MHz IF, digitized, and then sent to the process computer in the transmitter room. The minimum detectable signal for routine operation at 150 m resolution is $\sim$ 10 dBZ at 100 km.

Similar to the stabilization on other research vessels, the system ingests the movement/attitude of the ship via an inertial measurement unit located in the pedestal, which is fed directly into the radar servo controllers and used to "offset" the antenna pointing so that it is relative to real-world horizon.

The SEA-POL radar was motivated by the donation of the former Ronald Brown C-band radar (NOAA) to Colorado State University. NOAA gave this radar to CSU with the understanding that the radar would provide the basis for a new ocean-going C-band polarimetric radar. SEA-POL evolved from this process, using the transmitter and pedestal from the NOAA C-band radar with the remainder of the radar built in-house or acquired from outside vendors. The design of SEA-POL provides a highly portable radar, which facilitates shipping and installation on multiple ships or on land. The new design with dual-polarization capability is containerized with the dedicated radome mount above the container to allow rapid deployment on a variety of ships and safe operation (Figure 8.8 [66]). The radar

Figure 8.8 Photo of SEA-POL mounted on the foredeck of the R/V Roger Revelle in San Diego before departing on the SPURS-2 cruise. SEA-POL is the large radome near the bow of the ship. The smaller radome mounted above the bridge houses the ship's satellite communications antenna. [Credit: Steven Rutledge; this figure is reprinted from [66] under the terms of the Creative Commons Attribution 4.0 International License (https://creativecommons.org/licenses/by/4.0/).

specifications are similar to the OceanPOL radar on the R/V *Investigator* in many respects with a 250-kW peak power magnetron transmitter but SEA-POL has a 1° beamwidth antenna and a sensitivity of −7 dBZ at 100 km. SEA-POL is installed on the forward deck of its host ship and, therefore, is subject to beam blockage and a banking sector when the radar looks in the aft direction toward the wheelhouse. Depending on the ship, the blocked sector ranges from 90° to 120°.

The R/V *Mirai* (Figure 8.9) had a major radar upgrade in 2014. The fundamental characteristics of the existing radar were retained such as the wavelength (C-band) and antenna stabilization, but with major upgrades enabling dual-polarization measurements, a solid-state transmitter, and a new antenna for better spatial resolution (1° beamwidth) mounted at a higher location to reduce obstructions by ship structures [79]. Though the peak power is 6 kW each (H and V), a pulse-compression technique is utilized to keep the sensitivity for typical volume scans with a 32-μs pulse and 1.0 MHz bandwidth providing 150 m range resolution to detect at least 10 dBZ reflectivity at 100 km range. Taking full advantage of the

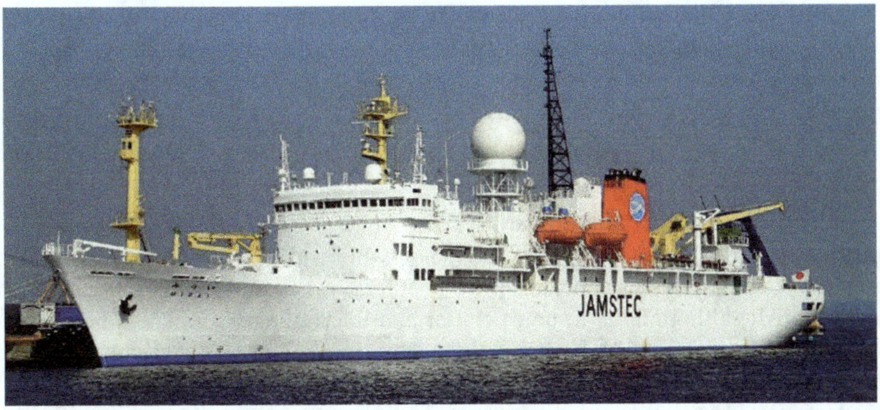

Figure 8.9 Photo of R/V Mirai with the upgraded M-POL radar mounted at the middle of the vessel

solid-state electronics, the new radar has been continuously operated during most of the research cruises ranging from the tropical ocean to the polar regions including both the Arctic Ocean and the Southern Ocean.

8.4 Technical challenges and approaches for shipborne dual-polarization weather radar

The operating environment of a weather-capable radar is extremely challenging on a ship. By definition, the radar is on a moving platform with relatively high vibration levels on top of the pitch, roll, and heave of the vessel itself. Furthermore, the ship superstructure and equipment itself are major sources of clutter, beam blockage, and interference depending on the radar installation configuration. The radio frequency environment is relatively noisy with other radars and communication systems operating at frequencies near the weather radar's transmit frequency. These can have a major effect on data quality and all the current operating weather radars have filters in place to minimize such interference. However, their receiver noise floor can still be adversely affected by the presence of interference which affects data quality and radar sensitivity. In addition, there is the issue of calibration of the radars. Ground-based radars often utilize fixed clutter and networks of disdrometers and/or rain gauges to facilitate calibration and the conversion of weather radar data to rainfall estimation, which is difficult for shipborne systems. There is a further challenge with dual-polarization on the moving platform and the need for calibration of the differential reflectivity. All of these factors mean that the shipborne radars require extensive quality control of their data and the overall performance may be lower than similar ground-based systems.

Dual-polarization radars have only recently been installed on research vessels. As a result, the development of tailored post-processing techniques to optimize radar performance for science outcomes is in flux. In the remainder of this section, we will discuss the solutions to minimize the impact of ship movement, clutter impacts, and current approaches for quality control and provide some examples of the radar performance. It will conclude with a summary of areas where work is still underway and further analysis is required.

8.4.1 Blockage

Mounting a radar on a ship requires attention to both beam blockage and the radiation hazard posed by the radar itself. This can be relatively straightforward for vertically pointing instruments such as profilers, cloud radars, and Micro Rain Radars (MRR) although the issue of clutter and interference from sidelobes means that the radars need careful positioning. This is a much greater concern with scanning radars. Limited space on the vessel may force the antenna to be mounted on a lower level such that ship structures block the radar beam. In addition, for the shipborne radars, antenna stabilization may force the elevation angle to go negative relative to the ship body. The blocked sectors need to be blanked to avoid radiation hazards and data from the missing sectors need to be set as the missing values to avoid biasing climatologies. For example, Geng and Katsumata identified blocked sections by analyzing reflectivity data [79].

This can be minimized where the radar is a permanent installation mounted high on the vessel such as the OceanPol radar on R/V *Investigato*r. There is an engineering issue of having heavy instruments mounted high on the vessel and implications for ship stability. For example, the R/V *Investigator* is a multi-science platform and has very heavy equipment mounted deep inside the hull as a counterweight. The alternative is to install the radar lower on the ship to manage the blockage and operate the radar focusing on sectors. This gives more flexibility in terms of operation on various ships as has been done with SEA-POL.

8.4.2 Antenna stabilization

For all weather radars, it is crucial to understand where the radar beam is pointing and avoid radiating towards the sea surface. The early deployment of shipborne weather radars in the 1960s and 1970s utilized mechanically stabilized platforms to level the pedestal (Sections 8.2.1 and 8.2.2). The alternative to stabilized platforms that must deal with significant weight and forces associated with moving a large antenna in the case of a weather radar is to use the antenna control in azimuth and elevation to adjust the beam direction in real time using the ship attitude data. This approach began to be implemented in the 1980s on the radars in TOGA COARE (Section 8.2.3) as well as the updated R/V *Keifu-Maru* radar and has become the standard that is utilized on the current dual-polarization radars.

One drawback of using the antenna pedestal to control the antenna pointing direction is that as the ship pitches and rolls, the horizontal and vertical polarizations are systematically rotated from the ground relative to vertical (horizontal). This has been modeled for C-band radars in rain using T-matrix calculations where the impact of the inclination of the polarization plane is cast as an apparent mean canting angle of the raindrop [80]. This analysis showed that for modest ship motions, the impact of the roll on Z_{DR}, etc. was relatively small and in the worst case could potentially be corrected noting the potential for canting angle errors to contribute to significant Z_{DR} errors, particularly if circular polarized waves are transmitted [81,82].

8.4.3 Sea clutter

While many near-shore-based radars observe sea clutter, it is the norm for ocean-going radar. Radar scatter from ocean waves is from surface waves of half the radar wavelength, i.e. ~ 2.5 cm for the current dual polarimetric weather radars. These are short wavelength gravity waves induced by wind stress and where the restoring force is gravity. Note that X-band radars will observe capillary waves where the restoring force is the ocean surface tension as the cut-off between these modes is about 1.7 cm capillary waves and is a function of wind stress. Radar scatter from ocean waves is discussed in detail in a review by Raynal and Doerry [83]. Wave sensing and ocean current sensing radars use longer wavelengths so that they observe the Bragg scatter from the waves. The situation is further complicated by the presence of wave breaking that can dominate the returns [83].

Sea clutter has the potential to bias rainfall estimates but needs to be treated separately from techniques used for ground clutter as sea clutter has non-zero Doppler shift [84]. In fact, the Doppler signals align reasonably well with the radial component of the near-surface wind. Since the signal is from scatter on a surface rather than volume scatter, the signal has a different range dependence – with the power returned decreasing as the range cubed (R^3). Fortunately, there are clear dual-polarization signatures associated with sea clutter. An example of sea clutter seen by the Ocean-POL radar on the R/V *Investigator* is shown in Figure 8.10. The decrease in returned power with range follows an R^{-3} dependence quite closely. It is rare to see sea clutter returns extending far beyond about 20 km. In this example, the corresponding Z_{DR} is biased to large negative values with the largest values in the sector propagating away from the ship. This negative Z_{DR} signal is common, but there are occasions where the distribution is much closer to zero, so simple Z_{DR} thresholds are not always reliable for sea clutter detection. However, the relatively low values of $\rho_{HV}(0)$ can be used as an additional filter.

There may still be issues when echoes from sea clutter and weather echoes are overlaid [84]. The effective $1/R$ dependence in sea clutter radar reflectivity (with the returned power being adjusted with the R^2 correction for reflectivity) ensures that the impact will be limited in range but may still be significant near the ship. Generally, the Z_H will be overestimated because of the multiple sources, although the Z_V will often have a greater enhancement than Z_H as per the Z_{DR} in Figure 8.10. At a range of ~ 10 km the sea clutter in this example has an effective Z of about 8 and 15 dBZ for the H and V polarizations. This is sufficient to be an issue for light rain and ice precipitation measurements. In addition to biasing the reflectivity, the large negative Z_{DR}'s of the clutter will bias the Z_{DR} low and also suppress the value of the $\rho_{HV}(0)$. These biases in turn will affect the accuracy of retrievals of raindrop size distributions, quantitative precipitation estimation (QPE), and microphysical retrievals. There are promising methods to minimize such biases using a fuzzy logic approach [84].

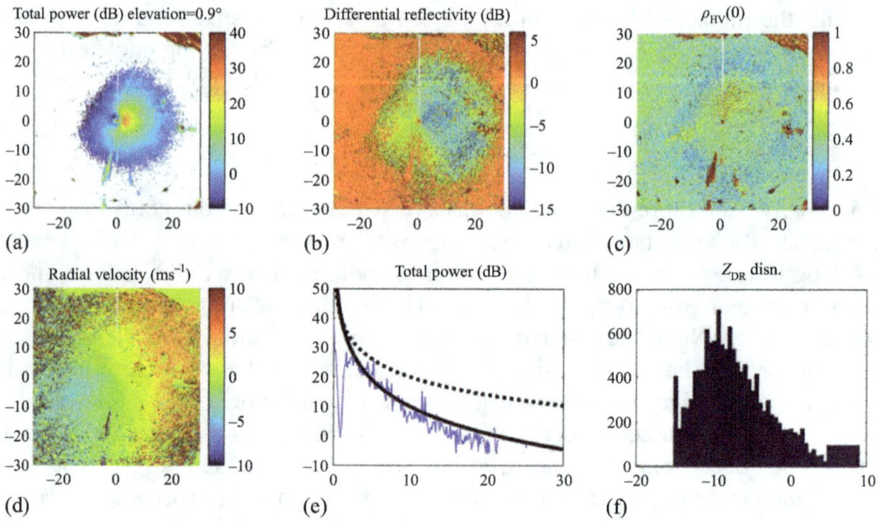

Figure 8.10 Panels showing an example of sea clutter observed with the OceanPOL radar on 19 November 2019 at an elevation of 0.9° near the Darwin coast. The panels are (a) the returned power without any range correction, (b) the corresponding Z_{DR} field, (c) the $\rho_{HV}(0)$ field, (d) radial velocity field, (e) the profile of the returned power as a function of range at an azimuth of 90° along with a $1/R^3$ (solid) and $1/R^2$ (dotted) line for reference, and (f) a histogram of the Z_{DR} values in dB within a range of 15 km.

8.4.4 Quality control: calibration

8.4.4.1 Reflectivity calibration

By virtue of being at sea, many of the calibration measures traditionally undertaken with ground-based radars cannot usually be used including comparisons with disdrometers and rain gauge networks and using fixed clutter to monitor potential changes in the calibration of the radar (e.g., [85]). Observations on the ship itself are of limited use as by definition they are in the radar cone of silence. Further flow distortion around the ship superstructure remains an issue for onboard rain observations, although there have been successful deployments of disdrometers on several research vessels including the R/V *Investigator* [86–88]. Other methods, such as solar calibration, are limited to periods when the radar is at shore because of ship movement and this only tests the receiver chain of the radars. Other methods are possible at shore, e.g., using tethered spheres, but this has not yet been demonstrated.

All is not lost for the use of reasonably standard methods. There is the potential for direct comparisons of the ocean-going radars with other well-calibrated radars. For example, as illustrated in Figure 8.11, routine comparisons with shore-based radars have been demonstrated with the R/V *Investigator* [68] as well as using the

Figure 8.11 The concept of shipborne (OceanPOL on the R/V Investigator) and ground-based inter-comparisons for calibration using the GPM Ku-band spaceborne radar. The 150 km radius is shown by a yellow circle and the ship track by a white line. © 2021 Google Earth Map Data: SIO, NOAA, US Navy, NGA,GEBCO; Map image Landsat/ Copernicus (after [68]).

satellite radar (TRMM/PR and GPM/DPR) as a standard for calibration [68,85, 89,90] although the number of overpasses during a given cruise is limited. Where the radar can be removed, it is also possible to relocate and compare. For example, the SEA-POL radar is returned to the CSU CHILL radar site [91] between deployments and can be intercompared and calibrated against CHILL, as was done prior to its first deployment in 2016.

There are a number of error sources in cross-calibration with other radars [90,92], including temporal and spatial mismatches, errors in gridding, incomplete overlap of volumes, potential non-Rayleigh impacts, and attenuation corrections noting that the GPM radar is at Ku-band and the current ocean radars are at C-band. The processing steps used to calibrate the OceanPOL radar against satellite measurements follow Warren *et al.* [90]. These require at least 10 satellite profiles for processing a volume. The melting layer is detected using GPM products and excluded because of the wavelength differences of the systems. Matched volumes are calculated and a 95% beam filling and 35 dBZ maximum reflectivity threshold criteria are applied (Figure 8.12). Where there is sufficient data, an absolute calibration uncertainty of a volume is $\sim$ 2 dB [68,90], but this error is appreciably reduced when using a higher number of overpasses. This is easily extended to comparisons with shore-based radar (Figure 8.13).

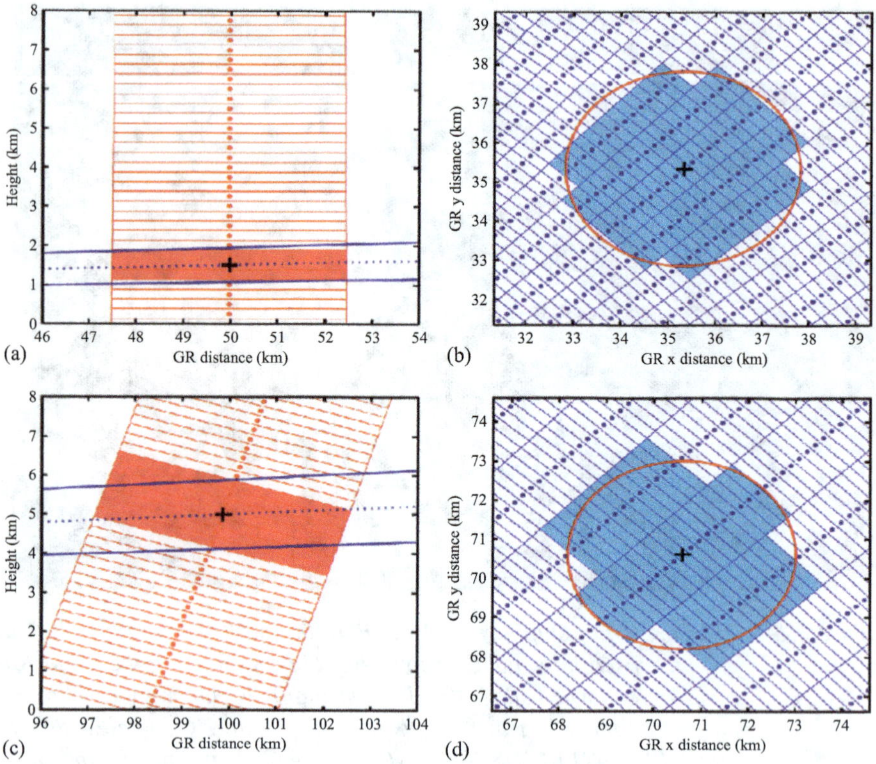

Figure 8.12 Matching sampling volumes [90]. © American Meteorological Society. Used with permission.

8.4.4.2 Calibration of Z_{DR}

Calibration of the Z_{DR} is a particular issue for all dual-polarization radars but is particularly difficult for shipborne radars where noise, sea clutter, and blockage are additional issues. Furthermore, there is no disdrometer data under the radar coverage for comparisons.

Some radars, such as the M-POL and SEA-POL radar, use near vertical (70–89°) bird bath scans where the antenna is rotated and the resulting signals from light precipitation are examined for both off-set from zero and any azimuthal dependence. The Z_{DR} offset for SEA-POL has been fairly constant at about 0.3 dB using this method.

An alternative is applying the technique described by Katsumata and Geng (2023) [93] to look at weak echoes by liquid hydrometers which theoretically should be spherical to result in $Z_{DR}{\sim}0$. An example of the results of both the birdbath and weak echo calibrations from M-POL is shown in Figure 8.14. These two methods show excellent agreement with stable results over a 3-week period. The bias of about 1.05 dB early in the record is clear and can be corrected. These

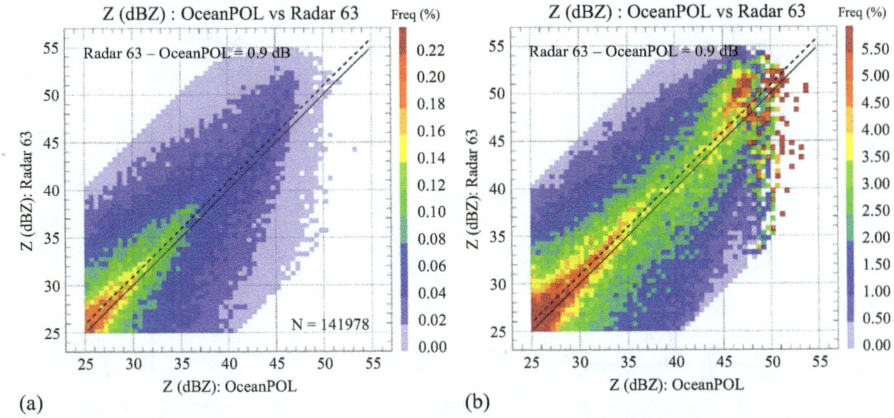

Figure 8.13 Illustration of 2D joint frequency histograms of reflectivity used to compare quantitatively the OceanPOL radar (x-axis) and any of the ground-based radars (y-axis), here for the Berrimah radar (63) for 1 day (21 November 2019) of the YMC experiment. For each plot, the 1:1 line is drawn as a solid line, and the calibration difference estimate is written and shown as a dashed line. The colors show the frequency of points falling in each reflectivity pixel 0.5 dB in the resolution of the 2D joint histograms, either expressed as the % of the total number of points (a) or as a % of the sum of points for each value of OceanPOL reflectivity (i.e., the sum of all points along the y-axis at each constant value of the x-axis) (b). The number of samples N for this case is 141,978 (see panel a) (after [68]). Used with permission.

analyses also highlight and enable real-time monitoring of radar technical issues. For example, the radar had a fault developing on the 9 December 2015 with a loss of power on the H channel.

Another alternative is to examine the distribution of Z_{DR} from weather echoes. In particular, choosing low-intensity rain echoes or light dry snow where the signal-to-noise ratio is still high (at a relatively close range) can be used to monitor stability and test for offsets noting that the Z_{DR} for light rain ($Z\sim$ 20 dBZ) is non-zero and dependent on the details of the drop size distribution (e.g., [94]). Therefore, aggregates and rimed particles are better candidates because of their low intrinsic Z_{DR} values (e.g., [95,96]) but low reflectivity echoes from heights at riming temperatures can still have strong asymmetries and substantial Z_{DR} [97]. An example of the Z_{DR} distribution as a function of elevation angle for a temperature range of a Southern Ocean case with a reflectivity between 20 and 28 dBZ is shown in Figure 8.15. First, a clear offset in Z_{DR} is seen with median values for the -15 to -5°C data having a peak around 0.4 dB. Furthermore, the position of this peak does not change much with the elevation angle whereas if it was a true estimate of the Z_{DR}, the value should decrease as the cosine of the elevation angle.

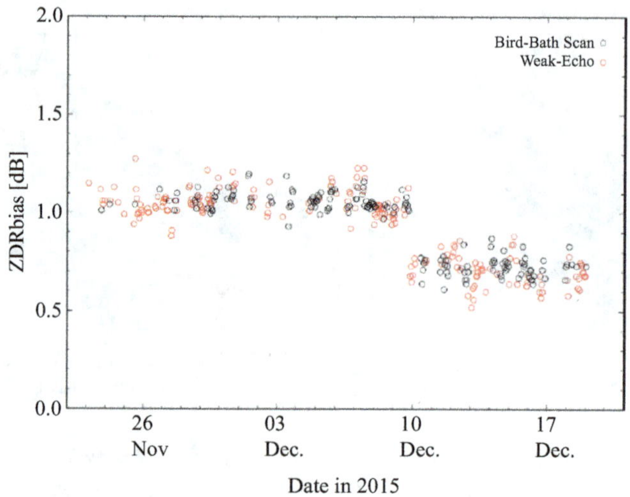

Figure 8.14 Superposition of the calibration of the M-POL using the birdbath (black) and weak echo (red) methods illustrating excellent agreement for a period in November–December 2015. The jump in mid-record was associated with a radar technical fault illustrating the capability of these diagnostics to monitor radar performance (from [93] under a Creative Commons Attribution 4.0 International [CC BY 4.0]).

8.4.5 Quality control: dual-polarization weather radar post-processing

While there are variations in the detail of QC processing applied to dual-polarization weather radar data, all of the radar groups leading the analysis of current shipborne dual-polarization radar follow similar methodologies. There is inevitably some dependence on the results depending on how harsh the QC control is versus the tolerance for spurious data. The basic approaches are:

Step 1. Correct for calibrations of Z and Z_{DR} using values obtained from the approaches outlined above. SEA-POL data is also examined for Z_{DR} azimuthal asymmetries using the birdbath scans.

Step 2. Apply noise corrections to the $\rho_{\mathrm{HV}}(0)$ and Z_{DR} for example using (6.55) and (6.58) in [76] or (5) and (6) in [98].

Step 3. Apply basic QC steps for data outside certain thresholds to remove non-meteorological echoes. Options used are:

- NCP > 0.3; $\rho_{\mathrm{HV}}(0) > 0.6$; $-2.0 < Z_{\mathrm{DR}} < 7.0$; $-20 < Z_{\mathrm{H}} < 80$ (OceanPOL)
- NCP$>$ 0.35 + SNR > 5.7; $\rho_{\mathrm{HV}}(0) > 0.8$; s.d. $(\Phi_{\mathrm{DP}}) < 13$ $+Z > 10$ dBZ (SEA-POL)
- NCP$>$ 0.3; $\rho_{\mathrm{HV}}(0) > 0.65$; s.d. $(\Phi_{\mathrm{DP}} < 40°$ (M-POL)

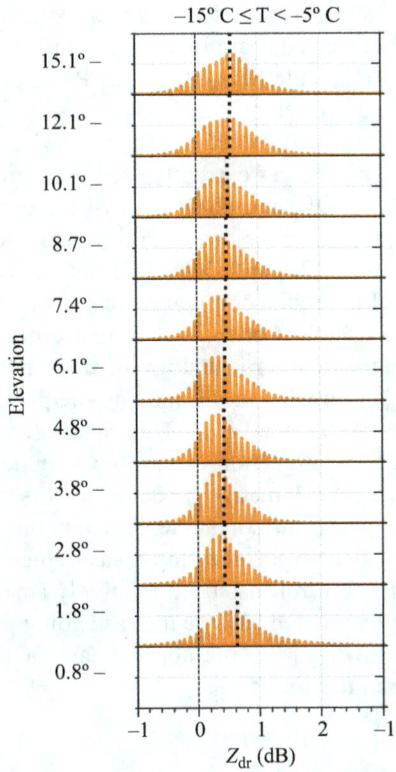

Figure 8.15 Histograms of the measured Z_{DR} as a function of elevation angle for the temperature range −15 to −5°C where Z_{DR} values are expected to be ~ 0.1 dB or less. The offset here is used to calibrate the Z_{DR}.

Where NCP is the normalized coherent power. As discussed above, the $\rho_{HV}(0)$ thresholds effectively remove sea clutter unless it is overlaid by weather echoes. Standard SEA-POL processing then applies a four-gate de-speckling filter and second trip echoes are removed with a smoothed five-gate Gaussian filter and NCP <=0.12 threshold.

Step 4. With this clean data, the Φ_{DP} data can have any receiver offsets subtracted and be de-aliased. The M-POL processing chain has an additional step by first filtering the radar Φ_{DP} data to remove spurious jumps in the phase of backscatter phase jumps that will affect the estimation of the K_{DP} using the method of Hubbert and Bringi [99] prior to calculating the K_{DP}. Options used for the K_{DP} include Wang and Chandrasekar [100] on SEA-POL, a linear K_{DP} processing method [99] on OceanPOL, and a linear fit to the Φ_{DP} data over an interval of 1.5 km (M-POL).

Step 5. Additional QC is also needed for the Doppler measurements from the radar to remove aliasing and, in some cases, use dual-PRF unfolding [101].

Dual-PRF methods also have a risk of incorrectly propagating unfolding errors and themselves should be corrected carefully (e.g., [102]). Ocean-Pol uses the UNRAVEL algorithm [103] while SEA-POL uses Py-ART‡ [104] phase unwrapping and M-POL uses the HMP algorithm [105].

Step 6. Calculate attenuation corrections for Z and Z_{DR}. Again, there are many options in the literature. The choices currently used for Z are described in [100] for SEA-POL, [106] for OceanPOL, and [107] for M-POL exploiting the differential phase information (Φ_{DP}). Corrections due to differential attenuation are also made for Z_{DR} with SEA-POL using [75] and OceanPOL using [108].

Step 7. Subtraction of the radial component of the ship steaming motion to the observed radial velocities which is often done in real time.

While the radar beam direction is stabilized, there is still a potential further step associated with the ship pitch and roll moving the plane of polarization of the radar from the true vertical/horizontal [80]. This can be represented as an artificial effective canting angle of the projection of the ship attitude along the radar beam. They [80] demonstrated using T-matrix modeling of disdrometer observed drop size distributions that this was a surmountable issue that only had a major impact at quite large angles. In practice so far, the ship measurements have not applied the potential correction as the ship roll has been relatively small§ for the cases examined so far, although that may not always be the case going forward as research ship roll can exceed 10°. Following [80], the approximate fit for C-band radars is to apply a correction given in the form:

$$\begin{aligned} p(0) &= [c_1(\beta)p(\beta)] + c_2(\beta) \text{ with} \\ c_1(\beta) &= c_{10} + [c_{11}\beta] + [c_{12}\beta^2] + [c_{13}\beta^3] \\ c_2(\beta) &= c_{20} + [c_{21}\beta] + [c_{22}\beta^2] + [c_{23}\beta^3] \end{aligned} \tag{8.1}$$

where $p(\beta)$ represents the measured Z_{DR}, K_{DP}, or $\rho_{HV}(0)$ when the apparent canting angle due to the ship orientation is β (°), and $p(0)$ represents the corresponding parameter when $\beta = 0$.

The values of the fitted coefficients are given in Table 8.3.

As for all weather radars, for many applications, it is useful to interpolate the data from the measured (r, θ) coordinates to a Cartesian grid CAPPI format. This of course entails dealing with the beams rising above the curving Earth, data gaps in the vertical, and increasing beam diameter with range. Furthermore, rigorous additional QC is required to remove any bad values as even single bad pixels will potentially be spread horizontally and vertically and impact further analyses. Propagation of these errors varies with elevation and this can cause issues for both the immediate estimates and the derivation of products such as microphysical classifications. Various methods for interpolation are used (e.g., [109] and many still use bi-linear interpolation (e.g., [110,111]).

‡Available at https://github.com/ARM-DOE/pyart

§For example, Mirai observations show the effective roll is less than ±3° for cruises so far and mostly < ±1°.

Table 8.3 Coefficients from least squares fits of Eq. (1) from scattering simulations performed by Thurai et al. [80]

Parameter	Z_{DR}	K_{DP}	$\rho_{HV}(0)$
c_{10}	1	1	1
c_{11}	1.1756×10^{-2}	1.1958×10^{-2}	1.60080×10^{-2}
c_{12}	-8.6695×10^{-4}	-8.6856×10^{-4}	-7.4594×10^{-4}
c_{13}	5.5340×10^{-5}	5.5903×10^{-5}	7.6936×10^{-5}
c_{20}			
c_{21}			-1.6086×10^{-2}
c_{22}			7.4674×10^{-4}
c_{33}			-7.6963×10^{-5}

Next-level products include the estimation of the microphysical habit of the scatterers (e.g., [69,112–114] and others) as well as properties such as quantitative estimates of the rainfall (e.g., [71–73]) and raindrop size distribution retrievals (e.g., [70]).

8.4.6 Dual-polarization weather radar performance

This section will discuss the data quality following the quality-controlled procedures outlined in the previous section, noting that the radar sensitivity has already been discussed in Section 8.3. There are multiple ways to assess data quality. First, we will examine the quality of the data using scatter plots of key variables as a function of the reflectivity. This will then be followed with some objective metrics consistent with those defined by Marks *et al.* [115]. The section will then show some examples of data illustrating the real-world capabilities of these shipborne radars.

Figure 8.16, adapted from [66], shows scatterplots between Z, Z_{DR}, and K_{DP} in convective rain over the tropical eastern Pacific. First, the systematic structure of the data in these plots illustrates the high quality of data that gives confidence that it is suitable for microphysical interpretation. For example, we see the rapid increase in K_{DP} with high Z typical of these plots, but also a population of high Z, low K_{DP}, and large Z_{DR} indicating the presence of clouds with small numbers of large drops. This is also clear in panel 2 focusing on the K_{DP}–Z_{DR} scatter. Likewise, a large range of Z–K_{DP} values are seen for $Z_{DR} \sim 1$ dB illustrating the presence of a wide range of rain rates and, therefore, water concentration over a narrow range of drop-size distribution shapes corresponding to this Z_{DR}.

Now that the ability of the radars to see systematic and realistic features in their inter-relationships is demonstrated, the question of quantitative data quality is next. A quantitative assessment methodology was proposed by Marks *et al.* [115] where they assessed a number of research radars. To minimize the impact of variations in microphysics as a function of rain rate and potential contamination of hail, they selected data from regions with relatively light rain between 20 and 28 dBZ and calculated several metrics. These include the medians and standard deviations of the estimates of $\rho_{HV}(0)$ and K_{DP} where the expected values are close to 1 and 0, respectively. The measurement of Z_{DR} is critical for improved QPE and microphysical estimates and is assessed using an average absolute deviation (AAD) parameter defined as:

$$AAD[Z_{DR}] = \sum_{i=1}^{N} |Z_{DR}(i) - \bar{Z_{DR}}|/N$$

where $\bar{Z_{DR}}$ is the mean value of Z_{DR} for the sample.

The quality and the stability of the radar were also checked using the standard deviation (s.d.) of the Φ_{DP} calculated over a 2.2-km window centered on each pixel. Only values within the 20–28 dBZ reflectivity range are used. Table 8.4 shows the values for the M-POL, SEA-POL, and OceanPOL radars along with different environments with the CSU-CHILL radar from [115] as a reference.

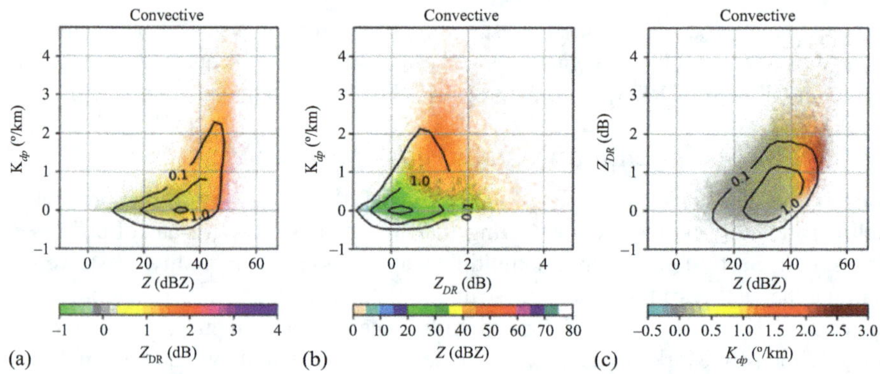

Figure 8.16 Scatter plot of Z, Z_{DR}, and K_{DP} with points colored for the third variable for rain data collected by SEA-POL during the SPURS2 cruise (adapted from [66]) © American Meteorological Society. Used with permission.

Table 8.4 Performance metrics for the three current ocean-going C-band dual-polarization radars along with the S-band CSU-CHILL radar as a reference following the method of Marks et al. [115] as described in the text

Radar	**Median *Z* (dBZ)**	**Median $\rho_{HV}(0)$**	**Median K_{DP} (°/km)**	**AAD (dB)**	**s.d.(Φ_{DP}) (°)**
CSU-CHILL	25.5	0.98 (0.01)	0.0 (0.0)	0.2	1.2 (0.3)
SEA-POL tropical	24.1	0.994 (0.01)	0.024 (0.21)	0.38	2.02 (1.1)
M-POL tropical	24.0	0.995 (0.01)	0.018 (0.23)	0.39	0.4 (0.2)
M-POL mid-latitude	24.5	0.997(0.01)	0.016 (0.21)	0.16	0.4 (0.2)
OceanPOL tropical	23.4	0.990 (0.02)	0.091 (0.34)	0.51	2.0 (0.9)
OceanPOL high latitude	24.3	0.992(0.03)	0.027 (0.18)	0.39	1.6 (1.1)

The values in brackets are the standard deviations of the metrics. The CSU data is from [115].

From these values, it can be seen that the performance of all these radars is comparable but there are variations from cruise to cruise for individual radars illustrating location-dependent performance statistics as well as differences between radars. SEA-POL and OceanPOL have slightly larger values of the AAD compared with the M-POL and CSU-CHILL, but are comparable to other radars considered by [115]. One clear feature is the s.d. (Φ_{DP}) for M-POL is substantially lower (i.e., much better) than the other radars. This radar applies the Hubbert and Bringi phase filtering [99] which may contribute to this improved estimate of performance.

8.4.7 Some examples of dual-polarization data and science applications

The last part of Section 8.4 will use some examples from recent field programs to illustrate the capability for science using these new platforms. Examples from all three current dual-polarization radars will be shown.

8.4.7.1 Tropical thunderstorm structure

Figure 8.17 shows an example of SEA-POL data collected in the West Pacific east of Luzon, Philippines during the 2019 deployment for PISTON on the R/V *Sally Ride*. During this time, a tropical depression was in the vicinity of the ship, yielding a large MCS to the west-northwest of SEA-POL. The MCS produced an extensive ice anvil cloud evident above 6 km in a vertical (range height indicator) scan cross-section (Figure 8.17). The winds were strongly sheared throughout the column, with strong upper-level easterlies evident by the inbound radial velocities nearing 20 ms^{-1} in the anvil cloud and strong low-level westerlies, manifested as 10–20 ms^{-1} outbound velocities in the convection below (Figure 8.17(b)). Isolated convective cells were continually developed at the surface but were rarely able to grow deep enough to penetrate the mixed- and ice-phase clouds due to the strong shear.

SEA-POL data show mammatus undulations on the underside of the anvil cloud that was characterized by near-zero Z_{DR} values (Figure 8.17(c)). This could indicate the presence of smaller crystals that are preferentially sorted to the bottom of the cloud by the strong shear and convective motions, while larger oriented ice crystals comprise the main anvil cloud as shown by the non-zero Z_{DR} (Figure 8.17(c)). In the convection below the anvil, the convective cores are characterized by Z_{DR} values larger than 2 dB (Figure 8.17(c)), suggesting large, flattened raindrops which grow in the warm phase through collisions and coalescence. The prevalence of high Z_{DR} cores in tropical isolated warm rain convection has been noted from SEA-POL observations [116]. K_{DP} values of more than 4°/km demonstrate that these cells are efficient rain producers with high liquid water content.

These detailed observations from SEA-POL reveal the complex interactions between the large scale (the tropical depression and anvil cloud) and the small scale (the small but ubiquitous convection) and provide insights into the microphysics, kinematics, radiative balance, and surface freshwater flux in data-sparse regions such as the tropical oceans.

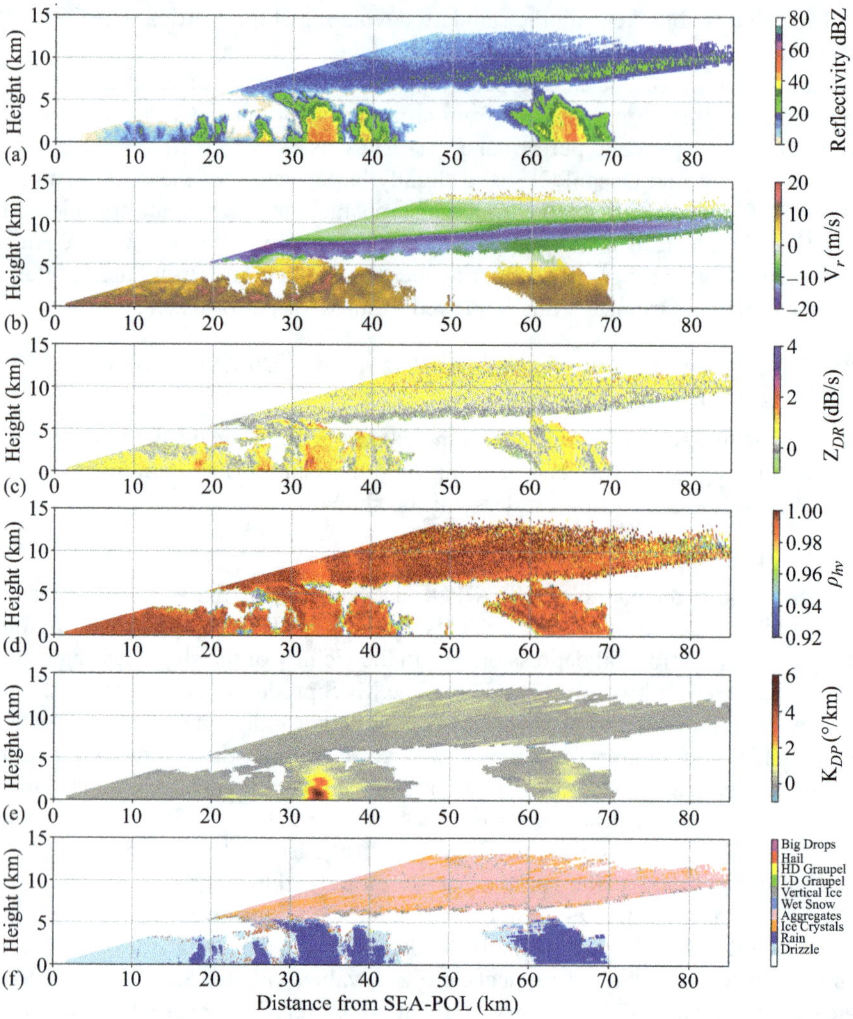

Figure 8.17 SEA-POL observations of convection in the West Pacific from 17 September 2019 0359 UTC. (a) Reflectivity, (b) radial velocity, (c) differential reflectivity, (d) correlation coefficient, (e) specific differential phase, and (f) hydrometeor classification.

8.4.7.2 Rainfall features in the Southern Ocean

Clouds and precipitation over the Southern Ocean have been a major international focus over the past decade with a number of large field programs in the remote ocean because of major weather and climate model biases associated with cloud processes [78]. The paucity of observations means that the precipitation structure of the cloud

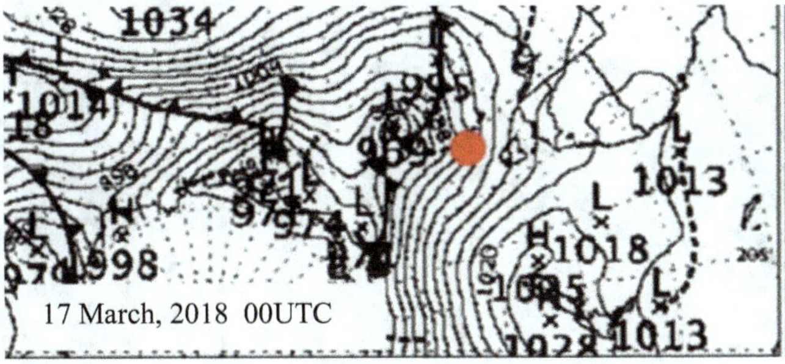

Figure 8.18 A section of the Bureau of Meteorology synoptic map showing the deep cyclone and associated front to the SW of Tasmania. The ship location (46.9°S, 141.95°E) is marked by a red dot.

systems is relatively poorly documented and understood. The R/V *Investigator* has been a key platform towards this goal and OceanPOL data is now being exploited. The vessel also visits the Southern Ocean many times a year for non-meteorological experiments, but the radar continues operation. This region is characterized by frequent deep low-pressure systems and associated fronts that have never been analyzed using weather radar. The following case study from 17 March 2018 is from one of these non-meteorological experiments. The vessel was located around 45°S approximately 600 km SW of Tasmania as a well-defined frontal system associated with a 965 hPa low to the SW approached and passed over the ship (Figure 8.18).

Figure 8.19 shows a snapshot of the dual-polarization data as the front approached the radar with high reflectivity and deep echoes to the west. Note the high values of $\rho_{HV}(0)$ throughout the profiles. Larger values of Z_{DR} coincide with the convective high reflectivity elements and echoes extend to the tropopause. As the front approached the vessel, it was apparent that the highest reflectivities characterized by convective elements embedded in the rainband as defined using the Steiner *et al.* [117] definitions were to the west on the domain. Approximately 8% of the echo area is classified as convective and by 1850 UTC 45% of the 300 km diameter radar domain is covered by precipitation. This area decreased afterwards as the precipitation weakened and the convective fraction decreased apart from a small burst around 1900 UTC (Figure 8.20). Drop size distributions (DSD) estimated from the Z_{DR}, using a method similar to Bringi *et al.* [70], show that the rainfall had a relatively narrow distribution of median volume diameter (D_0) values centered around 1.5 mm. An increasing number of clouds with smaller mean drop sizes were observed as the precipitation weakened. Cross-sections show the highest values of Z and Z_{DR} below the freezing level.

8.4.7.3 Microphysical retrievals

Variations in precipitation microphysical characteristics are a key issue for ground validation of spaceborne radar and have been a major activity for the TRMM and GPM satellite science programs. Dual-polarization radars offer the capability to estimate microphysical information such as raindrop size distribution (e.g., [70]) and hydrometer identification (e.g., [118]). As an illustration of this capability, the

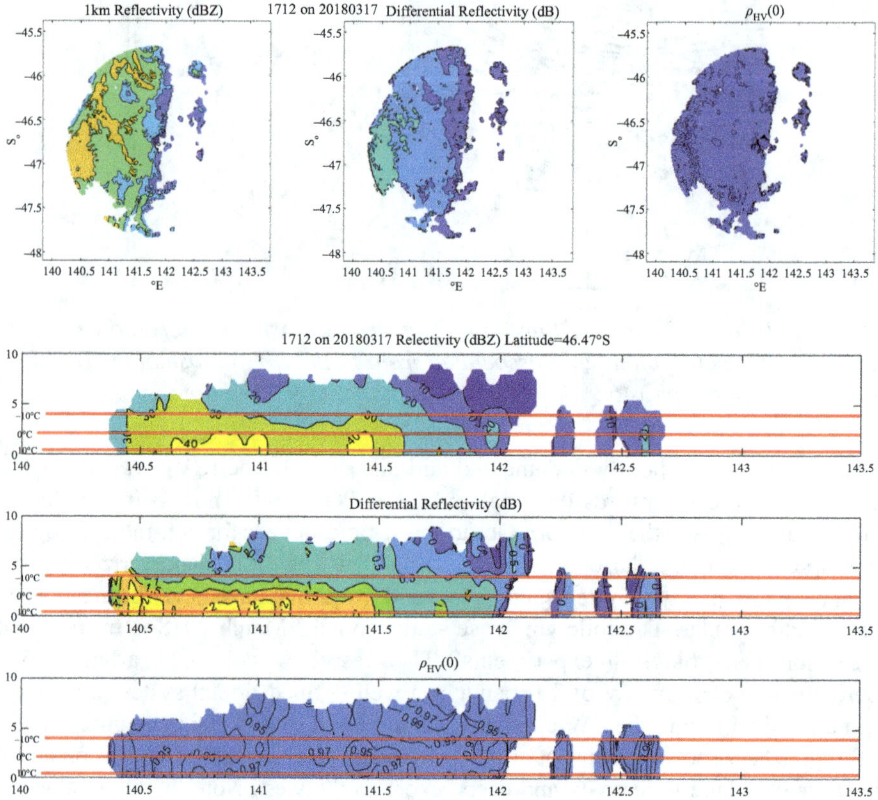

Figure 8.19 A constant altitude scan of the frontal rainband showing the reflectivity, Z_{DR}, and $\rho_{HV}(0)$ along with an east-west cross-section through this line

DSD retrieval method of Bringi *et al.* [70] is applied to data from the M-POL C-band radar and is compared to those from GPM retrievals for a summertime mid-latitude front over the Pacific Ocean northeast of Japan (Figure 8.21). For rainfall greater than 0.5 mm h^{-1}, there is an excellent agreement in the reflectivity although the D_m values from the ship were systematically lower than the GPM estimates.

Applying microphysical retrievals to shipborne radar continues to be an active area of research. The quality of Z_{DR}, which is the key parameter to retrieve the shape of hydrometeors, must be a priority including calibration (Section 8.4.4.2) and attenuation correction. The hydrometeor characteristics (e.g., particle size distribution) over the ocean can differ from those over land which have been well studied and retrieval algorithms need to be optimized. Over-ocean validation using other data sources at the surface (e.g., shipborne observation of drop size distribution as in [88]) and aloft (e.g., particle imaging by aircraft or by balloon-attached instruments [119]) are also required for DSD retrievals as well as hydrometer identification as shown in the results from PISTON (Figure 8.17(f)). Improved microphysical retrievals in remote areas are also important for model development (e.g., [120]).

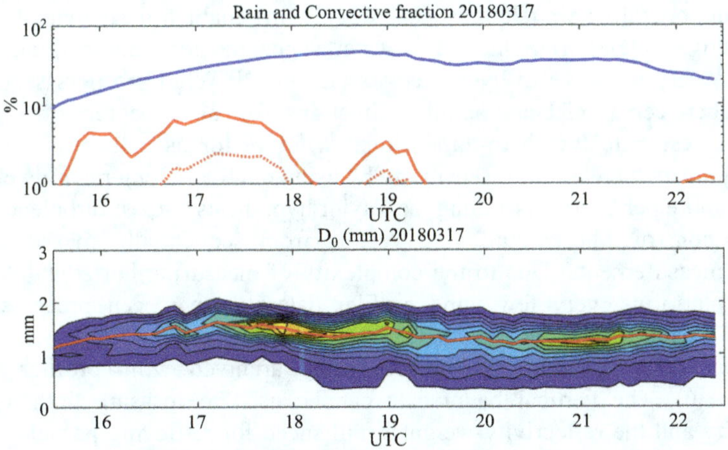

Figure 8.20 Upper. A time series of the total fraction of the radar domain covered by 20 dBZ at an altitude of 1.5 km indicating rain area (blue) and the convective fraction normalized against the rain area (solid red line) and against the total area (dotted red line). The lower panel shows the time series of the distribution of the radar estimate of the median volume diameter D_0 as a function of time. Contours are drawn for every 1,000 km^2.

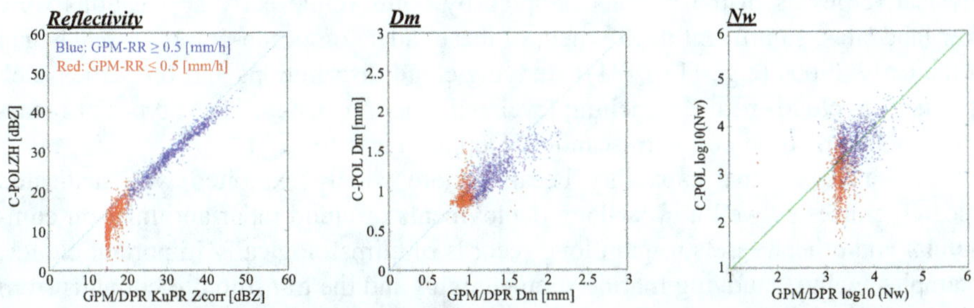

Figure 8.21 Comparisons of C-band M-POL reflectivity and derived drop size parameters against GPM dual-frequency estimates

8.5 Weather and cloud profiling radars on ships

8.5.1 History of cloud radar deployments and outcomes

As well as scanning radar, it is important to recognize the contributions of (near) vertically pointing profiling radars to our understanding of atmospheric structure and processes over the ocean. In the 1990s wind profiling radars with frequencies ~1 GHz were deployed both on stabilized platforms and using electronic beam steering (e.g. [121]) to detect "clear air" motion outside of clouds or precipitation.

These long operating wavelengths (i.e., 23–33 cm) enabled measurement of clear air refractive index variations due to water vapor and temperature gradients but were also very sensitive to hydrometeors [122–124]. Wind profilers sequentially alternate between 3 to 5 beam pointing directions from a few degrees off-vertical to vertical to estimate the three-dimensional wind vector as a function of height. Profiling radars increase their sensitivity by using relatively long radar dwell times, ~10 s or more per beam. Assuming the Taylor hypothesis (frozen turbulence), wind profilers construct horizontal wind vectors from sequentially measured radial velocity measurements. Due to the complexity of measuring horizontal winds on moving platforms, very few wind profiler datasets have been processed since TOGA COARE and in the Indian Ocean measuring boundary layer circulations [125] providing research opportunities using archived wind profiler datasets (Section 8.6). The vertical beam data can be used to measure both the radar reflectivity and the reflectivity-weighted fall speed for retrieving particle size distributions [122,126]. However, these radars do not have the sensitivity to observe non-precipitating hydrometeors, thus higher frequency cloud radars have been deployed on ships to study liquid and ice clouds.

The use of ground-based 35 and 95 GHz cloud radars became widespread in the 1990s for the study of cloud and precipitation properties for cloud heights ranging from the boundary layer through to cirrus including for both intense observing programs as well as for long-term deployments (e.g., [127–129]). These facilities often combined cloud radar with lidars to sample optically thin clouds below the radar's minimum detectable signal and provide additional constraints for cloud microphysical retrievals in the regions sampled by both. While early applications were over land sites, significant deployment of these radars on ocean-going vessels began in the early 2000s (e.g., [130,131]) studying cloud distributions and properties such as persistent clouds near the melting level and other processes. These types of radars have also been combined with scanning Doppler radar (e.g., [132]).

In the 2010s, this capability became more widely exploited with dedicated research cruises as well as year-long deployments of cloud radar and lidar on commercial and other vessels to gain long records of climatologically important clouds. Examples include studying marine stratocumulus and the transition to cumulus from drizzle (https://www.arm.gov/research/campaigns/amf2012magic; [133]), deployment in the Arctic Ice Sheet [134] and Southern Ocean multi-layer clouds including those with significant layers of supercooled liquid water which is a major problem for weather and climate models. For example, the CAPRICORN (Clouds, Aerosols, Precipitation, Radiation, and AtmospherIc Composition Over the SoutheRn OceanN) suite of campaigns in 2015, 2016, and 2018 [135–137] utilized the BASTA W-band cloud radar [138] and the Leosphere RMAN-510 lidar, while the MARCUS (Measurements of Aerosols, Radiation and Clouds over the Southern Ocean) campaign utilized the mobile DOE Atmospheric Radiation Measurement (ARM) facility including the stabilized Marine ARM Cloud Radar (M-WACR) 95 GHz radar and the MicroPulse Lidar on the Australian Icebreaker RSV Australis (e.g., [78,139]).

Some recent programs have augmented the cloud radars and lidar with radars dedicated to profiling precipitation at K-band (λ~2.2 cm) (e.g., [140,141]). This avoids the issues with attenuation and non-Rayleigh scatter limitations of mm-wave radars, although such radars cannot match the low minimum detectable signal of cloud radars. Thus, there is potentially a fully complementary set of profiling instruments capable of measuring clouds from optically thin cloud structures (with lidar) through to heavily precipitating clouds. These capabilities will be discussed in more detail in Section 8.5.3.

The above illustrates some examples, but this is far from a complete list of experiments and instruments. For example, the Physical Sciences Laboratory (PSL) of NOAA has been involved in more than 20 different experiments (https://psl.noaa.gov/data/cruises/).

8.5.2 Technical descriptions

Profiling radars were briefly introduced in Section 8.5.1. This section will discuss the associated technology as well as the capability and applications in more detail. Generally, these radars are vertically pointing and, therefore, require stabilization, although some early implementations integrated over a long enough period that the ship pitch and rolling average out (e.g., [142]) but this degrades both the time resolution and the quality of retrieval data. Some vertically pointing wind profilers also use phased array antennas that have the capability to electronically steer the radar beam, thereby maintaining the integrity of beam pointing angles (e.g., [121,143]). This is particularly important for wind profiling with relatively long radar dwell times and for re-constructing the horizontal wind vector using near vertical beams. However, most recent systems make use of the more accurate and robust ship data to mount the radars (and lidars) on stabilized platforms (e.g., Moran *et al.* [144]). Moran *et al.* designed a stabilized platform for a 94-GHz cloud radar and this design has been replicated for other systems including the 95 GHz BASTA radar that has operated on the R/V *Investigator* on several cruises [145]. Figure 8.22 shows a photo of a stabilized platform for wind profiling and vertically pointing antennas as well as photos of MRR and W-band radars mounted on stabilized platforms. These systems are readily mounted in containers. It has been recently demonstrated through an analysis of the platform data at 10 Hz that vertical stabilization was always better than 0.2° for sea states up to 6. In the pilot study presented in [146], a "profiler" mode has been demonstrated with the 95 GHz BASTA radar, consisting of a 120-s sequence with 15 s spent at the following eight pointing angles: vertical, +8° pitch; vertical, +8° roll; vertical, −8° pitch; vertical, −8° roll. This "profiler" mode enabled in-cloud horizontal and vertical wind estimates.

Profiling radars employ a far wider range of frequencies and system types compared with the weather radars discussed above. Various frequency millimeter wave radars have been used with a focus on cloud structure including non-precipitating cloud while profiling radars at centimeter wavelengths have been used to extend the capability to heavier precipitation and larger particle sizes to avoid the complications of Mie scatter. Some shipborne systems have been adapted from ground-based systems such as the commercial off the shelf Microwave Rain Radar (MRR: https://metek.de/product/mrr-pro/) and DOE ARM radars (KAZR and

Figure 8.22 An example of a stabilized platform with a radar and sitting below a hatch (from [144]). Other solutions are given in [121,141,145]. Used with permission.

M-WACR in Table 8.5) while the BASTA radar was developed as a lower cost solution for widespread deployments at observational sites, achieved by transmitting a small fraction of the power of pulsed radars such as the M-WACR but continuously (1.7 kW versus 0.5 W, see Table 8.5), therefore, reducing transmitter cost drastically [138]. Note that a number of cloud radars along with lidar have been deployed on the R/V *Mirai* over the past two decades as well [130,131,140,148]. By virtue of the short wavelengths, the cloud radar antennas are small and the radar beamwidths tend to be quite narrow. These radars often store the measured Doppler spectra to facilitate hydrometeor retrieval processes. FM-CW cloud radars use low-power solid-state transmitters (0.5 W) and estimate both reflectivity and Doppler velocity. The high PRF/FM-CW cycles and high radar frequencies enable 1 s time resolution observations with high Doppler measurement accuracy.

Table 8.5 Technical specifications from a sample of profiling instruments illustrating varying frequencies and methodologies

	24.4 GHz microwave rain radar	35 GHz KAZR [147]	95 GHz M-WACR [147]	NOAA Physical Sciences Laboratory W-band 94 GHz radar [144]	95 GHz BASTA [138]
Wavelength	2.2 cm	8 mm	3 mm	3 mm	3 mm
Antenna size	1.5°	1.8 m (0.3°)	0.3 m	0.3 m	0.6 m
Peak (mean) power	0.05W (FM-CW)	200 W (0.2 W)	1.7 KW (3W)	1.7 kW	0.5 W (FM-CW)
Range resolution	10 m	45 m pulse length spaced every 30 m	37.5 m	25 m	12.5–200 m
Max range (in low-resolution mode)	1.2 km	20 km	18 km	3 km	24 km
Record Doppler spectra	Yes	Yes	Yes	Yes	No (planned)

The MRR is a low-power vertically pointing Frequency Modulated – Continuous Wave (FM-CW) radar with sufficient sensitivity for precipitation studies (~–5 dBZ @ 1 km range) due to relatively long CW dwell times. These radars also store the radar Doppler spectra similar to the wind profilers, enabling post-cruise reprocessing [149,150]. The MRR range resolution can be as fine as 10 m in the vertical with 1 s dwell. They have been used over the last several years quite extensively for ground-based measurements but more recently have been applied to ocean field experiments, both in the Southern Ocean [139] and in the Atlantic [141].

8.5.3 Capabilities and science applications

Cloud radars provide outstanding capabilities to observe cloud and air motion characteristics, build in-depth climatologies as well as test satellite retrievals. Cloud radar, weather radar, lidar, and microwave radiometer combinations are particularly useful for describing cloud structure, turbulence, size distributions, and quantitative measurements of liquid water. There have been programs in many basins including for satellite validation [3,140] and cloud process studies (Figure 8.23 [141]).

Cloud radar capability is further illustrated by comprehensive observations over the Southern Ocean during the CAPRICORN experiment [78] which provided the first insights into the morphology, frequency, vertical distribution, water content, and phase of Southern Ocean clouds to complement existing satellite statistics [135–137] as well as enabling evaluation of Himawari-8 cloud products [2]. CAPRICORN observations have also informed the development and contributed to the evaluation of new and promising supercooled liquid water (SLW) and mixed-phase cloud detection algorithms for geostationary satellites [3]. First studies using CAPRICORN data have shown that statistical estimates of cloud phase from ground-based and satellite radar are disparate (e.g., [137,151]), owing to different

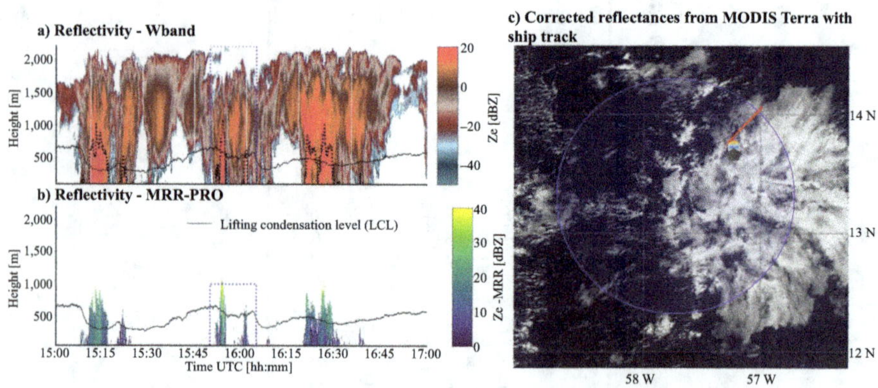

Figure 8.23 Time height cross-sections of 95 GHz radar data (top) and 2.2 cm (K band) from an MRR system (lower panel) with the corresponding MODIS image (right) (after [141]). Used with permission.

geometries of observations resulting in a different and problematic conditional sampling of cloud populations. Cloud radar observations of cloud systems close to the Antarctic coast made during MARCUS indicate the common occurrence of multiple SLW layers, along with the evidence of seeding of single-layer SLW by higher-altitude ice clouds present near the end of the life of cyclones (Figure 8.24, from Alexander *et al.* [139]).

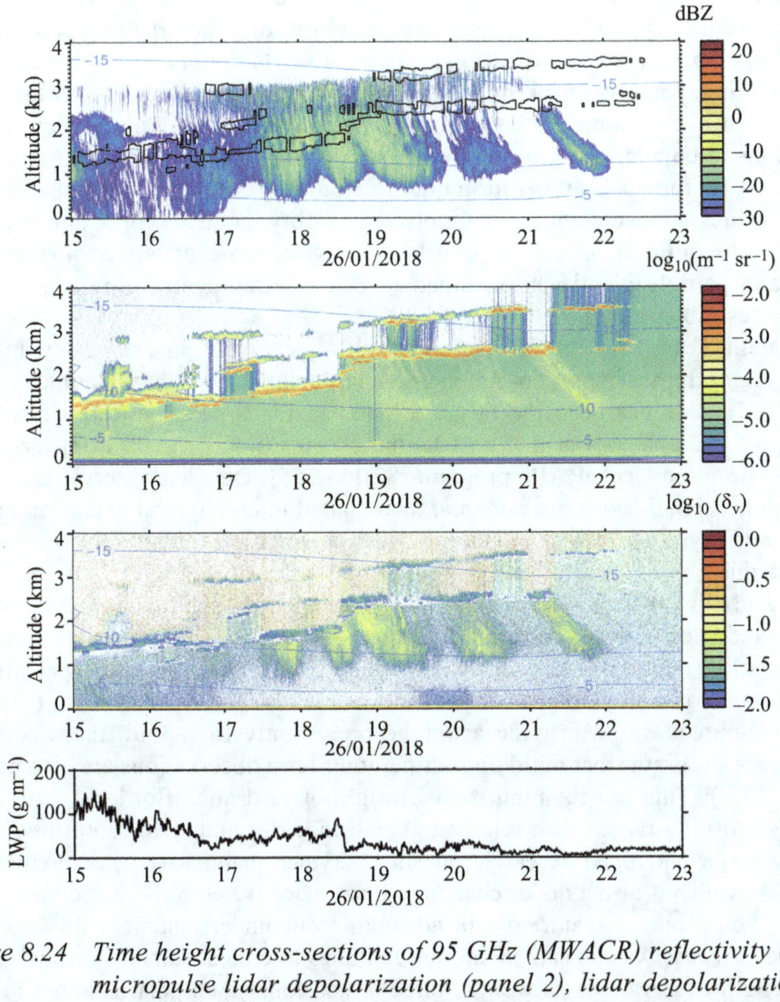

Figure 8.24 Time height cross-sections of 95 GHz (MWACR) reflectivity (top), micropulse lidar depolarization (panel 2), lidar depolarization ratio (panel 3), and derived liquid water path (panel 4). ERA5 temperatures are indicated on the first three panels (blue lines) and liquid water path derived from microwave radiometer, the MWACR, and micropulse lidar (from [139]). Used with permission.

8.6 Current and future capabilities and opportunities

In the previous sections, we have described some of the historical background and snapshots of major science achievements obtained with ocean-going precipitation and cloud measurements, as well as technical challenges and solutions with modern radars. In many respects, the field is in its infancy in terms of exploiting these new capabilities, especially multi-wavelength and dual-polarization capabilities. With high-quality, well-calibrated data, it is possible to derive the microphysical structure of the clouds including particle habit and size distributions from the characteristics of the dual-polarization radar fields and temperature. Potential additional information on both precipitating and non-precipitating clouds over the ship using combinations of MRR, cloud radar, lidar, and microwave radiometers has been demonstrated. This will allow new science questions to be addressed, but there are still gaps. For example, the methodologies used for the dual-polarization quality control are all quite dated and rely on fixed values in decision trees. Improved quality control, for example, with machine learning should be a priority. Likewise, more advanced methods for QPE and particle identification tuned for the environments being sampled are highly desirable (e.g., [74,114]).

Satellite-based rainfall estimation is still of relatively poor accuracy and different algorithms simply do not agree, particularly over high latitudes (e.g., [152,153]). This has been recently attributed to fundamental differences in precipitation microphysics at different latitudes using shipborne disdrometer observations from the OceanRAIN program [87,154,155]. Detailed observations of the three-dimensional cloud structure and the detailed microphysical characteristics of the cloud systems using shipborne dual-polarization radars are needed to resolve this.

Basic climate features such as the transition from shallow cumulus to congestus and deep convection in the trade regions need much better understanding and are still a major problem for climate models. Ship observations, especially the combination of cloud radar (to cover shallow clouds) and weather radar (to cover deeper clouds), would provide a unique opportunity to capture these complex processes. Programs that build on routine, long-term observations are also needed (e.g., [133]). This is becoming more straightforward and affordable with high-quality profiling radars with relatively small logistics and areal footprints.

A major untapped research question is the quantitative measurement of snowfall, which is also a major challenge from space (see [156] for a review). Ship radars should play a major role in advancing our understanding with improved methods because we can sample at high latitudes to collect suitable vertical multi-wavelength observations as well as three-dimensional dual-polarization on ships. This is important for key climate questions including the global water budget and its changes as well as addressing key science challenges such as the radiation biases over the Southern Ocean [78]. This is critical for improved weather and climate modeling as well as our understanding of climate processes, carbon uptake, and the evolution of high-latitude ecosystems.

Table 8.6 Some publicly available datasets and the associated web links for access

Dataset	Period	URL
R/V *Shiyan #3*	Nov 1992–Feb 1993	https://psl.noaa.gov/psd3/boundary/MstToga/index.html
R/V *Kexue #1*	Nov 1992–Feb 1993	https://psl.noaa.gov/psd3/boundary/MstToga/index.html
R/V *Moana Wave*	Nov 1992–Feb 1993	https://psl.noaa.gov/psd3/boundary/MstToga/index.html
R/V *Ka'imimoana*	June 1996–Nov 1996	https://psl.noaa.gov/psd3/boundary/MstToga/index.html
R/V *Ronald H. Brown*	Several months, 1996, 1997, and 1999	https://psl.noaa.gov/psd3/boundary/MstToga/index.html
Arctic Summer Cloud Ocean Study (ASCOS): Tjernström, M. *et al.* 2014. The Arctic Summer Cloud Ocean Study (ASCOS): overview and experimental design. *Atmospheric Chemistry and Physics*, 14, 2823–2869. https://doi.org/10.5194/acp-14-2823-2014	S-band vertically pointing data, Aug–Sep 2014	https://bolin.su.se/data/oden-ascos-2008-radar-1?n=ascos-radar
Surface heat budget of the Arctic Ocean Project (SHEBA)	S-band vertically pointing data, Nov 1997–Nov 1998.	https://data.eol.ucar.edu/dataset/13.312
EUREC4A datasets Acquistapace *et al.* (2022 a,b,c)	MRR-PRO radar dataset W band radar dataset W band radar dataset (V2)	https://doi.org/10.25326/233, 2021a https://doi.org/10.25326/156, 2021b https://doi.org/10.25326/235, 2021c
OceanPOL data	5 cm QC'd radar data from R/V *Investigator* from 2017.	https://www.openradar.io doi: 10.25914/5cb686a8d9450
MARCUS field campaign	Profiling radar data at high latitudes	https://adc.arm.gov/discovery/#/
Numerous profiling radar cruises	Links to data from numerous cruises	https://psl.noaa.gov/data/cruises/

Opportunities and needs for science and operations are boundless and growing with new vessels. These include exciting opportunities with additional C-band facilities planned for the Taiwanese R/V *New Ocean Researcher 1*, an X-band radar currently being commissioned on Australia's new icebreaker, RSV *Nuyina* and a planned radar on the new Japanese icebreaker to be launched in the late 2020s (https://www.nipr.ac.jp/arctic_info/e/columns/2021-02-16-1/). Just as for land-based weather radar, the default going forward will be dual-polarization.

Data is becoming increasingly available for both past and new experimental datasets as more institutions adopt variations of the FAIR (Findable, Accessible, Interoperable, Reusable) principles for data. Both raw and processed data are often available including sophisticated products such as gridded microphysical classifications routinely calculated from the new generation of ocean-going dual-polarization radar ([157], https://www.openradar.io doi: 10.25914/5cb686a8d9450). It should not be forgotten that there are many historical datasets from field campaigns that are fully available for further research and, in many cases, are under-exploited. For example, wind profiler data collected during TOGA-COARE have never been fully explored but can provide useful microphysical retrievals [158]. A sample of the available datasets is provided in Table 8.6.

These new and emerging facilities and capabilities together with increasing availability and discoverability of the data will enable science to tackle key weather and climate questions and facilitate much broader uptake and usage of this valuable data.

Acknowledgments

The authors would like to thank Merhala Thurai for her support and encouragement in writing this chapter as well as acknowledge assistance and discussions with Dusan Zrnic, Ben Evans, and Simon Alexander as well as the thoughtful review and discussions with Steve Rutledge, Elizabeth Thompson, Chris Fairall, and two anonymous reviewers. These have all greatly improved the chapter. Thanks also to Claudia Acquistapace for providing Figure 8.23.

The major national capability represented by modern radars on ocean-going vessels is only possible with strong national support and major teams. We would like to acknowledge the major contributions of Professors Rutledge and Chandrasekar from Colorado State University and the support of NSF for the SEAPOL radar, the contributions of Kunio Yoneyama, Biao Geng, and the technical staff of Nippon Marine Enterprise, Ltd., especially Mr Soichiro Sueyoshi, and Mr Kazuho Yoshida for the M-POL radar. OceanPOL is supported by the CSIRO, Australia's National Science Agency, and the Marine National Facility (MNF). We acknowledge the key contributions of Will Ponsonby as well as the support of the CSIRO leadership and the MNF Director, Toni Moate for their foresight and support in building this capability on the R/V *Investigator*.

Abbreviations/acronyms

AAD	average absolute deviation
AMIE	ARM Madden-Julian Oscillation Investigation Experiment
ARM	atmospheric radiation measurement program
ASCOS	Arctic Summer Cloud Ocean Study
BASTA	bistatic radar system for atmospheric studies

CAPRICORN	Clouds, Aerosols, Precipitation, Radiation, and AtmospherIc Composition Over the SoutheRn OceanN
C-band	5 cm wavelength
CHILL	Chicago-ILLinois radar (now operated by CSU)
CINDY	Cooperative Indian Ocean Experiment on Intraseasonal Variability in Year 2011
CPOL	C-band POLarimetric radar
CSIRO	Commonwealth Science and Industrial Research Organisation
CSU	Colorado State University
$\boldsymbol{D_0}$	median volume diameter of a drop size distribution
dB	Decibel
dBZ	unit of radar reflectivity
DYNAMO	dynamics of the MJO
EPIC	East Pacific Investigation of Climate Processes in the Coupled Ocean-Atmosphere Systems
EUREC4A	elucidating the role of clouds-circulation coupling in climate
FM-CW	frequency modulated – continuous wave
GATE	Global Atmospheric Research Program (GARP) Atlantic Tropical Experiment
GHz	Gigahertz
GPM	Global Precipitation Mission
GPM-DPR	Global Precipitation Mission Dual-frequency Precipitation Radar
GPS	global positioning system
IF	intermediate frequency
IFA	intensive flux array
INS	inertial navigation system
JAMSTEC	Japan Agency for Marine-Earth Science and Technology
JASMINE	Joint Air–Sea Monsoon Interaction Experiment
JMA	Japan Meteorological Agency
LASP	Littoral air–sea process
KAZR	Ka-band ARM Zenith Radar
kW	kilowatt
KWAJEX	Kwajalein Experiment
MARCUS	Measurements of Aerosols, Radiation, and Clouds over the Southern Ocean
MCS	Mesoscale Convective System
MISMO	Mirai Indian Ocean cruise for the study of the Madden-Julian Oscillation convection onset
MIT	Massachusetts Institute of Technology
MJO	Madden-Julian Oscillation

MODIS	moderate resolution imaging spectrometer
M-POL	Mirai POLarimetric radar
MRR	Micro Rain Radar
MWACR	Marine W-band ARM Cloud Radar
NCAR	National Center for Atmospheric Research
NCP	normalized coherent power
NOAA	National Oceanographic and Atmospheric Administration
OceanPOL	Ocean-going POLarimetric radar
PISTON	Propagation of Intra-Seasonal Tropical OscillatioNs Philippine Sea
PRF	pulse repetition frequency
pyART	Py-ART
QC	quality control
QPE	quantitative precipitation estimation
RF	radio frequency
R/V	research vessel
SLW	super-cooled liquid water
SCS	South China Sea
SCSMEX	South China Sea Monsoon Experiment
SEA-POL	Sea-going POLarimetric radar
SHEBA	Surface Heat Budget of the Arctic Ocean Project
SPURS-2	Salinity Processes in the Upper Ocean Regional Study
TEPPS	Tropical Eastern Pacific Process Study
TOGA	Tropical Ocean Global Atmosphere
TOGA-COARE	Tropical Ocean Global Atmosphere-Coupled Ocean Atmosphere Response Experiment
TRMM	Tropical Rainfall Measurement Mission
TRMM-PR	Tropical Rainfall Measurement Mission Precipitation Radar
VOCALS	VAMOS Ocean-Cloud-Atmosphere-Land Study
X-band	3 cm wavelength
WACR	W-band ARM Cloud Radar
W-band	3 mm wavelength
YMC	Years of the Maritime Continent
Z	radar reflectivity in dBZ
Z_{DR}	differential reflectivity
Φ_{DP}	propagation differential phase
K_{DP}	specific differential phase
$\rho_{HV}(0)$	cross-correlation of the H and V signals at zero lag
λ	wavelength

References

[1] Houze, R.A. Jr. 'Structure and dynamics of a tropical squall line system'. *Mon. Wea. Rev.* 1977; **105**, 1540–1567.

[2] Huang, Y., S. Siems, M. Manton, A. Protat, L. Majewski, and H. Nguyen. 'Evaluating Himawari-8 cloud products using shipborne and CALIPSO observations: cloud-top height and cloud-top temperature'. *J. Atmos. Oceanic Tech.* 2019; **36**, 2327–2347.

[3] Noh, Y.-J., S.D. Miller, A.K. Heidinger, G. Mace, A. Protat, and S. P. Alexander. 'Satellite-based detection of daytime supercooled liquid-topped mixed-phase clouds over the Southern Ocean using the Advanced Himawari Imager'. *J. Geophys. Res. Atmos.* 2019; 124. doi.org/10.1029/2018JD029524.

[4] Ono, G. 'Meteorological observation vessel "Keifu-maru"'. *Weather Service Bull.* 1970; **37**, 391–397 (in Japanese).

[5] Yokota, Y., T. Sakai, S. Nakamura, and K. Shutta. 'Maritime meteorological observation system of research vessel (I)—shipboard weather radar system and microwave-wavemeter'. *Weather Service Bull.,* 1990; **57**, 81–93 (in Japanese).

[6] Mori, K., S. Ishigaki, T. Maehira, M. Ohya, and H. Takeuchi. 'Structure and evolution of convection within typhoon Yancy (T9313) in the early developing stage observed by the Keifu Maru radar'. *J. Meteor. Soc. Jpn.* 1999; **77**, 459–482. https://doi.org/10.2151.jmsj1965.77.2_459.

[7] Ninomiya, K. 'Influence of the sea surface temperature on the stratification of airmass and the cumulus activity over the East China Sea in the Baiu season'. *Papers Meteorol. Geophys.* 1974; **25**, 159–175.

[8] Muramatsu, T., S. Ogura, and N. Kobayashi. 'The heavy snowfall arisen from small scale cyclone on the west coast of Hokkaido Island'. *Tenki* 1975; **22**, 369–379 (in Japanese). https://https://www.metsoc.jp/tenki/pdf/1975/1975_07_0369.pdf.

[9] Hudlow, M.D. 'Collection and handling of GATE shipborne radar data'. Preprints *16th Radar Meteorology Conference*, Houston, MA, Amer. Meteor. Soc., 1975. pp. 186–193.

[10] Houze, R.A., Jr. and C.-P. Cheng, 'Radar characteristics of tropical convection observed during GATE: mean properties and trends over the summer season'. *Mon. Wea. Rev.* 1977; **105**, 964–980.

[11] Cheng, C.-P. and R.A. Houze Jr. 'The distribution of convective and mesoscale precipitation in GATE radar echo patterns'. *Mon. Wea. Rev.* 1979; **107**, 1370–1381.

[12] Frank, W.M. and J.L. McBride. 'The vertical distribution of heating in AMEX and GATE cloud clusters', *J. Atmos. Sci.* 1989; **46**, 3464–3478.

[13] Leary, C.A. and R.A. Houze Jr. 'The structure and evolution of convection in a tropical cloud cluster'. *J. Atmos. Sci.* 1979; **36**, 437–457.

[14] Webster, P.J. and R. Lukas. 'TOGA COARE: the coupled ocean–atmosphere response experiment'. *Bull. Amer. Meteor. Soc.* 1992; **73**, 1377–1416. https://doi.org/10.1175/1520-0477(1992)073<1377:TCTCOR>2.0.CO;2.

[15] Kucera, P.A. 'Intercomparison of shipboard radars over the Pacific warm pool'. M.S. Thesis, Department of Geosciences, Texas Tech University, 1993. p. 190.

[16] Short, D.A., P.A. Kucera, B.S. Ferrier, J.C. Gerlach, S.A. Rutledge, and O. W. Thiele. 'Shipboard radar rainfall patterns within the TOGA COARE IFA'. *Bull. Amer. Meteor. Soc.* 1997; **78**, 2817–2836.

[17] TOGA COARE International Project Office, 'TOGA COARE operations plan'. University Corporation for Atmospheric Research, 1992. p. 500.

[18] TOGA COARE, 'TOGA COARE IOP operations summary'. TOGA COARE International Project Office, 1993. p. 493.

[19] Mori, K. 'Equatorial convection observed by the research vessel Keifu Maru during the TOGA COARE IOP, November 1992'. *J. Meteor. Soc. Jpn.* 1995; **73**, 491–508. https://doi.org/10.2151/jmsj1965.73.2B_491.

[20] Oki, P.P.A. Kucera, and D. A. Short. 'Triple shipborne observations of tropical mesoscale convective systems during TOGA COARE'. In: *Proceeding of the 22nd Conference on Hurricanes and Tropical Meteorology*, Fort Collins, CO, Amer. Meteor. Soc., 1997. pp. 323–324.

[21] Petersen, W.A., R.C. Cifelli, S.A. Rutledge, B.S. Ferrier, and B.F. Smull. 'Shipborne dual-Doppler operations during TOGA COARE: integrated observations of storm kinematics and electrification', *Bull. Amer. Meteor. Soc.* 1999; **80**, 81–97.

[22] Conley, R. 'GPS performance: what is normal?'. *Navigation: J. Inst. Navig.* 1993; **40**, 261–281.

[23] Masters, J. M. and J.A. Leise. 'Correction of inertial navigation with Loran C on NOAA's P-3 aircraft'. *J. Atmos. Oceanic Tech.* 1993; **10**, 145–154.

[24] Rickenbach, T.M. and S.A. Rutledge. 'Convection in TOGA COARE: horizontal scale, morphology and rainfall production'. *J. Atmos. Sci.* 1998; **55**, 2715–2729.

[25] Ferrier, B.S., J. Gerlach, P. Kucera, *et al.* 'Corrections and comparisons of TOGA COARE shipborne radar reflectivities'. Preprints, *27th Conference on Radar Meteorology*, Vail, CO, Amer. Meteor. Soc., 1995. pp. 675–677.

[26] DeMott, C.A. and S.A. Rutledge. 'The vertical structure of TOGA COARE convection. Part I: radar echo distributions'. *J. Atmos. Sci.* 1998; **55**, 2730–2747.

[27] DeMott, C.A. and S.A. Rutledge. 'The vertical structure of TOGA COARE convection. Part II: modulating influences and implications for diabatic heating'. *J. Atmos. Sci.* 1998; **55**, 2748–2762.

[28] Doggett, A. L., IV, 'Analysis of rainwater flux in the IFA during TOGA COARE using shipboard radar rainfall estimates'. Ph.D. Thesis, Texas Tech University, 1996. p. 174.

[29] Petersen, W.A., S.A. Rutledge, and R.E. Orville. 'Cloud-to-ground lightning observations in TOGA COARE: lightning location algorithms and selected results'. *Mon. Wea. Rev.* 1996; **124**, 602–620.

[30] Saxen, T.R. and S.A. Rutledge. 'Surface fluxes and boundary layer recovery in TOGA COARE: sensitivity to convective organization'. *J. Atmos. Sci.* 1998; **55**, 2763–2781.

[31] Cifelli, R., S.W. Nesbitt, S.A. Rutledge, W.A. Petersen, and S. Yuter. 'Radar characteristics of precipitation features in the EPIC and TEPPS regions of the East Pacific'. *Mon. Wea. Rev.* 2007; **135**, 1576–1595. https://doi.org/10.1175/2007JCLI2020.1.

[32] Cifelli, R., S.W. Nesbitt, S.A. Rutledge, W.A. Petersen, and S. Yuter. 'Diurnal characteristics of precipitation features over the tropical East Pacific: a comparison of the EPIC and TEPPS regions'. *J. Clim.* 2008; **21**, 4068–4086.

[33] Petersen, W.A., R. Cifelli, D.J. Boccippio, S.A. Rutledge, and C.W. Fairall. 'Convection and easterly wave structures observed in the Eastern Pacific warm pool during EPIC-2001'. *J. Atmos. Sci.* 2003; **60**, 1754–1773. https://doi.org/10.1175/1520-0469(2003)060<1754:CAEWSO>2.0.CO;2.

[34] Yuter, S.E. and R.A. Houze Jr. 'The 1997 Pan American Climate Studies Tropical Eastern Pacific Process Study, Part I: ITCZ region'. *Bull. Amer. Meteor. Soc.* 2000; **81**, 451–481. https://doi.org/10.1175/1520-0477(2000)081<0451:TPACST>2.3.CO;2.

[35] Yuter, S.E., Y.L. Serra, and R.A. Houze Jr. 'The 1997 Pan American Climate Studies Tropical Eastern Pacific Process Study, Part II: stratocumulus region'. *Bull. Amer. Meteor. Soc.* 2000; **81**, 483–490. https://doi.org/10.1175/1520-0477(2000)081<0483:TPACST>2.3.CO;2.

[36] Serra, Y.L. and R.A. Houze Jr. 'Observations of variability on synoptic timescales in the East Pacific ITCZ'. *J. Atmos. Sci.*, 2002; **59**, 1723–11743. https://doi.org/10.1175/1520-0469(2002)059<1723:OOVOST>2.0.CO;2.

[37] Straub, K.H. and G.N. Kiladis. 'Observations of a convectively coupled Kelvin wave in the Eastern Pacific ITCZ'. *J. Atmos. Sci.* 2002; **59**, 30–52. https://doi.org/10.1175/1520-0469(2002)059<0030:OOACCK>2.0.CO;2.

[38] Wood, R., C.R. Mechoso, R.A. Weller, *et al.* 'The VAMOS ocean-cloud-atmosphere-land study regional experiment (VOCALS-REx): goals, platforms, and field operations'. *Atmos. Chem. Phys.* 2011; **11**, 627–654. doi:10.5194/acp-11-627-2011.

[39] Katsumata, M. and K. Yoneyama. 'Internal structure of ITCZ mesoscale convective systems and related environmental factors in the tropical western Pacific: an observational case study'. *J. Meteor Soc. Jpn.* 2004; **82**, 1035–1056. https://doi.org/10.2151/jmsj.2004.1035.

[40] Katsumata, M., H. Yamada, Q. Moteki, H. Kubota, and R. Shirooka. 'Observed evolution of the northward-propagating intraseasonal variation over the western Pacific: a case study in boreal early summer'. *Mon. Wea. Rev.* 2013; **141**, 690–706. https://doi.org/10.1175/MWR-D-12-00011.1.

[41] Inoue, J., M. E. Hori, Y. Tachibana, and T. Kikuchi. 'A polar low embedded in a blocking high over the Pacific Arctic'. *Geophys. Res. Lett.* 2010; **37**, L14808. https://doi.org/10.1029/2010GL043946.

[42] Keenan, T. D., J. T. Wang, P. L. Lin, T. C. Chen, Y. C. Liou, and J. A. Kankiewicz. 'Mesoscale structure of convective weather systems observed during the South China Sea Monsoon Experiment (SCSMEX): a preliminary survey'. In: *Proceedings of the IX Pacific Science Inter-Congress,*

Asian Pacific Monsoon and Typhoon Meteorology Symposium, Taipei, Taiwan, 1998. pp. 65–66.

[43] Keenan, T.D., Z. Nan, F. Lei, V. Bringi, J. Nystuen, and M. Whimpey. 'A comparison of radar rainfall estimators during the South China Sea Monsoon Experiment (SCSMEX)'. In: *Proceedings of the 30th International Conference on Radar Meteorology,* 2001. 3 pp.

[44] Lau, K.M., Y. Ding, J. Wang, *et al.* 'A report of the field operations and early results of the South China Sea Monsoon Experiment (SCSMEX)'. *Bull. Amer. Meteor. Soc.* 2000; **81**, 1261–1270.

[45] Yihui, D., C. Qmgchen, Z. Yan, G. Zion, and Z. Xiuzhi. 'The South China Sea Monsoon Experiment (SCSMEX) implementation plan'. *Adv. Atmos. Sci.* 1997; **14**, 255–270.

[46] Liu, Y., V.N.Bringi, and T.D. Keenan 'Comparisons of raindrop size distributions from the TRMM precipitation radar and the C-POL ground radar'. Preprints, *31st International Conference on Radar Meteorology*, Seattle, WA, Amer. Meteor. Soc., P3A.9. 2003. Available online at https://ams.confex.com/ams/pdfpapers/63908.pdf.

[47] Wang, J.-J. and L. Carey. 'Structure and evolution of an oceanic squall line during South China Sea Monsoon Experiment'. *Mon. Wea. Rev.* 2005; **133**, 1544–1561.

[48] Wang, J.-J., X. F. Li, and L. Carey. 'Evolution, structure, cloud microphysical and surface rainfall processes of monsoon convection during the South China Sea Monsoon Experiment'. *J. Atmos. Sci.* 2007, **64**, 360–380.

[49] Johnson, R.H., S.L. Aves, P.E. Ciesielski, and T.D. Keenan. 'Organisation of oceanic convection during the onset of the 1988 East Asian monsoon'. *Mon. Wea. Rev.* 2005; **133**, 131–148. https://doi.org/10.1175/MWR-2843.1.

[50] Yuter, S.E., R.A. Houze, Jr., E.A. Smith, T.T. Wilheit, and E. Zipser. 'Physical characterization of tropical oceanic convection observed in KWAJEX'. *J. Appl. Met.* 2005; **4**, 385–415.

[51] Spooner, C.L. 'Dual Doppler analysis of an oceanic tropical mesoscale convective system during the Kwajalein Experiment (KWAJEX)'. M.S. Thesis, Texas Tech University. 2001. p. 112.

[52] Yoneyama, K., Y. Masumoto, Y. Kuroda, *et al.* 'MISMO field experiment in the equatorial Indian Ocean'. *Bull. Amer. Meteor. Soc.* 2008; **89**, 1889–1903. https://doi.org/10.1175/2008BAMS2519.1.

[53] Katsumata, M., R.H. Johnson, and P.E. Ciesielski. 'Observed synoptic-scale variability during the developing phase of an ISO over the Indian Ocean during MISMO'. *J. Atmos. Sci.* 2009; **66**, 3434–3448. https://doi.org/10.1175/2009JAS3003.1.

[54] Yamada, H., K. Yoneyama, M. Katsumata, and R. Shirooka. 'Observations of a super cloud cluster accompanied by synoptic-scale eastward-propagating precipitating systems over the Indian Ocean'. *J. Atmos. Sci.* 2010; **67**, 1456–1473. https://doi.org/10.1175/2009JAS3151.1.

[55] Yoneyama, K., C. Zhang, and C.N. Long. 'Tracking pulses of the Madden–Julian Oscillation'. *Bull. Amer. Meteor. Soc.* 2013; **94**, 1871–1891. https://doi.org/10.1175/BAMS-D-12-00157.1.

[56] Xu, W. and S.A. Rutledge. 'Convective characteristics of the Madden–Julian Oscillation over the Central Indian Ocean observed by shipborne radar during DYNAMO'. *J. Atmos. Sci.* 2014; **71**, 2859–2877. http://dx.doi.org/10.1175/JAS-D-13-0372.1.

[57] Xu, W. and S.A. Rutledge. 'Morphology, intensity, and rainfall production of MJO convection: observations from DYNAMO shipborne radar and TRMM'. *J. Atmos. Sci.* 2015; **72**, 623–640. http://dx.doi.org/10.1175/JAS-D-14-0130.1.

[58] Xu, W., S.A. Rutledge, C. Schumacher, and M. Katsumata. 'Evolution, properties, and spatial variability of MJO convection near and off the Equator during DYNAMO'. *J. Atmos. Sci.* 2015; **72**, 4126–4147. https://doi.org/10.1175/JAS-D-15-0032.1.

[59] Rocque, M.N. and S.A. Rutledge. 'Diurnal cycle of precipitation features observed during DYNAMO'. *J. Atmos. Sci.* 2021; **78**, 2287–2306. https://doi.org/10.1175/JAS-D-20-0215.1.

[60] Thompson, E.J., J.N. Moum, C.W. Fairall, and S.A. Rutledge. 'Wind limits on rain layers and diurnal warm layers'. *J. Geophys. Res. Oceans.* 2019; **124**, 897–924. https://doi.org/10.1029/2018JC014130.

[61] Thompson, E.J., S.A. Rutledge, B. Dolan, and M. Thurai. 'Drop size distributions and radar observations of convective and stratiform rain over the equatorial Indian and West Pacific Oceans'. *J. Atmos. Sci.* 2015; **72**(11), 4091–4125.

[62] Kummerow, C., J. Simpson, O. Thiele, *et al.* 'The status of the tropical rainfall measuring mission (TRMM) after two years in orbit'. *J. Appl. Meteor.* 2000; **39**, 1965–1982. doi:10.1175/1520-0450(2001)0402.0.CO;2.

[63] Simpson, J., R.F. Adler, and G.R. North. 'A proposed tropical rainfall measuring mission (TRMM) satellite'. *Bull. Amer. Meteor. Soc.* 1988; **69**, 278–295. doi:10.1175/1520-0477(1988)0692.0.CO;2.

[64] Anagnostou, E.N., C.A. Morales, and T. Dinku. 'The use of TRMM precipitation radar observations in determining ground radar calibration biases'. *J. Atmos. Oceanic Technol.* 2001; **18**, 616–628.

[65] Thompson, E.J., W.E. Asher, A.T. Jessup, and K. Drushka. 'High-resolution rain maps from an X-band marine radar and their use in understanding ocean freshening'. *Oceanography* 2019; **32**, 58–65. https://doi.org/10.5670/oceanog.2019.213.

[66] Rutledge, S.A., V. Chandrasekar, B. Fuchs, *et al.* 'SEA-POL goes to sea'. *Bull. Amer. Meteor. Soc.* 2019; **100**, 2285–2301. https://doi.org/10.1175/BAMS-D-18-0233.1.

[67] Geng, B. and M. Katsumata. 'Variation of radar-observed precipitation characteristics in relation to the simultaneous passages of a Madden–Julian Oscillation event and convectively coupled equatorial waves during the Years of the Maritime Continent Pilot Study'. *Mon. Wea. Rev.* 2021; 149, 3379–3399. https://doi.org/10.1175/MWR-D-20-0346.1.

[68] Protat, A., V. Louf, J. Soderholm, J. Brook, and W. Ponsonby. ‘Three-way calibration checks using ground-based, ship-based and spaceborne radars’. *Atmos. Meas. Tech.* 2022; **15**, 915–926. https://doi.org/10.5194/amt-15-915-2022.

[69] Vivekandanan, J., D.S. Zrnic, S.M. Ellis, R. Oye, A.V. Ryzhkov, and J. Straka. ‘Cloud microphysics retrieval using S-band dual-polarisation measurements’, *Bull. Amer. Meteor. Soc.* 1999; **80**, 383–388.

[70] Bringi, V.N., C.R. Williams, M. Thurai, and P.T. May. ‘Using dual-polarized radar and dual-frequency profiler for DSD characterization: a case study from Darwin, Australia’. *J. Atmos. Oceanic Tech.* 1999; **26**, 2107–2122.

[71] Pepler, A., P.T. May, and M. Thurai. ‘A robust error-based rain estimation method for polarimetric radar. Part I: development of a method’. *J. Appl. Meteor. Clim.* 2011; **50**, 2092–2103.

[72] Cifelli, R., V. Chandrasekar, S. Lim, P.C. Kennedy, Y. Wang, and S.A. Rutledge. ‘A new dual-polarisation radar rainfall algorithm: application in Colorado precipitation events’. *J. Atmos. Oceanic Tech.* 2011; **28**, 352–364.

[73] Thompson, E.J., S.A. Rutledge, B. Dolan, M. Thurai, and V. Chandrasekar. ‘Dual-polarisation radar rainfall estimation over tropical oceans’. *J. Appl. Meteor. Clim.* 2018; **57**, 755–775. doi:10.1175/JAMC-D-17-0160.1.

[74] Munchak, S.J., R.S. Schrom, C.N. Helms, and A. Tokay. ‘Snow microphysical retrieval from the NASA D3R radar during ICE-POP 2018’, *Atmos. Meas. Techniques Discussions.* 2021. doi.org/10.5194/amt-2021-264.

[75] Bringi, V.N. and Chandrasekar, V. ‘Polarimetric Doppler weather radar: principles and applications’. 2001. doi:10.1017/CBO9780511541094.

[76] Ryzhkov, A.V. and D.S. Zrnic. *Radar Polarimetry for Weather Observations*. 2019. Springer, p. 486. ISBN : 978-3-030-05092-4.

[77] Yoneyama, K. and C. Zhang. ‘Years of the maritime continent’. *Geophys. Res. Lett.* 2020; **47**, e2020GL087182. https://doi.org/10.1029/2020GL087182.

[78] McFarquhar, G., C.S. Bretherton, R. Marchand, *et al.* ‘Observations of clouds, aerosols, precipitation, and surface radiation over the Southern Ocean: an overview of CAPRICORN, MARCUS, MICRE, and SOCRATES’. *Bull. Amer. Meteor. Soc.* 2021; **102**, E894–E928. https://doi.org/10.1175/BAMS-D-20-0132.1.

[79] Geng, B. and M. Katsumata. ‘Beam blockage identification for weather radars on board the R/V *Mirai* using archived data’. *JAMSTEC Rep. Res. Dev.*, 2016; **22**, 1–12. https://doi.org/10.5918/jamstecr.22.1.

[80] Thurai, M., P.T. May, and A. Protat. ‘Shipborne polarimetric weather radar: impact of ship movement on polarimetric variables at C-band’, *J. Atmos. Oceanic Tech.* 2014; **31**, 1557–1563. doi:10.1175/JTECH-D-13-00242.1.

[81] Wang Y. and V. Chandrasekar. ‘Polarization isolation requirements for linear dualpolarization weather radar in simultaneous transmission mode of operation’. *IEEE Trans. Geosci. Remote Sensing.* 2006; **44**, 2019–2028.

[82] Hubbert, J.C., S.M. Ellis, M. Dixon, and G. Meymaris. ‘Modeling, error analysis, and evaluation of dual-polarization variables obtained from

simultaneous horizontal and vertical polarization transmit radar. Part II: experimental data'. *J. Atmos. Oceanic Tech.* 2010; **27**, 1599–1607. https://doi.org/10.1175/2010JTECHA1337.1.

[83] Raynal, A.M. and A.W. Doerry. 'Doppler characteristics of sea clutter', Sandia Report, SAND2010-3828, 2010. 39 pp.

[84] Dutta, A. and V. Chandrasekar. 'Detection, analysis and mitigation of sea clutter in polarimetric weather radar'. *United States National Committee of URSI National Radio Science Meeting (USNC-URSI NRSM).* 2019, pp. 1–2. doi: 10.23919/USNC-URSI-NRSM.2019.8712871.

[85] Louf, V., A. Protat, R.A. Warren, *et al.* 'An integrated approach to weather radar calibration and monitoring using ground clutter and satellite comparisons'. *J. Atmos. Oceanic Tech.* 2019; **36**, 17–39. doi: 10.1175/JTECH-D-18-0007.1.

[86] Klepp, C. 'The oceanic shipboard precipitation measurement network for surface validation – OceanRAIN'. *Atmos. Res., Special issue of the International Precipitation Working Group.* 2015; **163**, 74–90. doi:10.1016/j.atmosres.2014.12.014.

[87] Klepp, C., S. Michel, A. Protat, *et al.* 'Ocean rainfall and ice-phase precipitation measurement network – OceanRAIN. World Data Center for Climate (WDCC) at DKRZ'. 2017. https://doi.org/10.1594/WDCC/OceanRAIN-W; https://doi.org/10.1594/WDCC/OceanRAIN-M; https://doi.org/10.1594/WDCC/OceanRAIN-R.

[88] Klepp, C. 'OceanRAIN: a new in-situ shipboard global ocean surface-reference dataset of all water cycle components'. *Sci. Data.* 2018; **5**, 180122. https://doi.org/10.1038/sdata.2018.122.

[89] Katsumata, M., T. Ushiyama, K. Yoneyama, and Y. Fujiyoshi. 'Combined use of TRMM/PR and disdrometer data to correct reflectivity of ground-based radars'. *SOLA*, 2008; **4**, 101–104. https://doi.org/10.2151/sola.2008-026.

[90] Warren, R. A., A. Protat, S. T. Siems, *et al.* 'Calibrating ground-based radars against TRMM and GPM. *J. Atmos. Oceanic Tech.* 2018; **35**, 323–346. https://doi.org/10.1175/JTECH-D-17-0128.1.

[91] Brunkow, D., V.N. Bringi, P.C. Kennedy, *et al.* 'A description of the CSU–CHILL National Radar Facility'. *J Atmos. Oceanic Tech.* 2000; **17**, 1596–1608. https://doi.org/10.1175/1520-0426(2000)017<1596:ADOTCC>2.0.CO;2.

[92] Schwaller, M.R. and K.R. Morris. 'A ground validation network for the Global Precipitation Measurement mission', *J. Atmos. Oceanic Tech.* 2011; **28**, 301–319.

[93] Katsumata, M. and B. Geng. 'An attempt to estimate differential reflectivity bias using volume scan dataset obtained by C-band polarimetric radar on board R/V *Mirai*'. *SOLA*. 2023; **19**, 234–237. doi:10.2151/sola.2023-030.

[94] Bringi., V.N., G.-J. Huang, V. Chandrasekar, and E. Gorgucci. 'A methodology for estimating the parameters of a Gamma raindrop size distribution model from polarimetric radar data: application to a squall-line event from the TRMM/Brazil campaign'. *J. Atmos. Oceanic. Tech.* 2002; **19**, 633–645.

[95] Ryzhkov, A.V., S.E. Giangrande, V.M. Melnikov, and T.J. Schuur. 'Calibration issues of dual-polarization radar measurements'. *J. Atmos. Oceanic Tech.* 2005; **22**, 1138–1155. doi:10.1175/ JTECH1772.1.

[96] Kennedy, P., M. Thurai, C. Praz, V.N. Bringi, A. Berne, and B.N. Notaros. 'Variations in snow crystal riming and Z_{DR}: a case analysis', *J. Appl. Meteor. Clim.* 2018; **57**, 695–707. doi:10.1175/JAMC-D-17-0068.1.

[97] Williams, E.R. 'Measurements of differential reflectivity in snow storms and warm season stratiform systems'. *J. Appl. Meteor. Clim.* 2015; **54**, 573–595. doi:10.1175/JAMC-D-14-0020.1.

[98] Schuur, T., A. Ryzhkov, and P. Heinselman. 'Observations and classifications of echoes with the polarimetric WSR-88D radar'. *NOAA Tech Report*, 2003. 45 pp.

[99] Hubbert, J. and V.N. Bringi. 'An iterative filtering technique for the analysis of copolar differential phase and dual-frequency radar measurements'. *J. Atmos. Oceanic Tech.* 1995; **12**, 643–648.

[100] Wang, Y. and V. Chandrasekar. 'Algorithm for estimation of the specific differential phase'. *J. Atmos. Oceanic Tech.* 2009; **26**, 2565–2578. https://doi.org/10.1175/2009JTECHA1358.1.

[101] Geng, B. and M. Katsumata. 'An algorithm for detecting and removing the spurious differential phase observed by C-band polarimetric radar in the rain'. *J. Meteor. Soc. Jpn.* 2020; **98**, 585–613. https://doi.org/10.2151/jmsj.2020-031.

[102] Joe, P. and P.T. May. 'Correction of dual-PRF velocity errors for operational Doppler weather radars'. *J. Atmos. Oceanic Tech.*, 2003; **20**, 429–442.

[103] Louf, V., A. Protat, R.A. Jackson, and S.M. Collis. 'UNRAVEL: a modular velocity dealiasing algorithm for C-band Doppler radar'. *J. Atmos. Oceanic Tech.* 2020; **37**, 741–758. doi:10.1175/JTECH-D-19-0020.1.

[104] Helmus, J.J. and S.M. Collis. 'The Python ARM Radar Toolkit (Py-ART), a library for working with weather radar data in the Python programming language'. *J. Open Res. Software*. 2016; **4**, e25. https://doi.org/10.5334/jors.119.

[105] Yamauchi, H., O. Suzuki, and K. Akaeda. 'A hybrid multi-PRI method to dealias Doppler velocities'. *SOLA*. 2006; 2, 92–95.

[106] Gu, J-Y., A. Ryzhkov, P. Zhang, *et al.* 'Polarimetric attenuation correction in heavy rain at C band'. *J. Appl. Meteor. Clim.* 2011; **50**, 39–58.

[107] Carey, L.D., S.A. Rutledge, D.A. Ahijevych, and T.D. Keenan. 'Correcting propagation effects in C-band polarimetric radar observations of tropical convection using differential propagation phase'. *J. Appl. Meteor. Clim.* 2000; **39**, 1405–1433. https://doi.org/10.1175/1520-0450(2000)039<1405:CPEICB>2.0.CO;2.

[108] Bringi, V.N., T.D. Keenan, and V. Chandrasekar. 'Correcting C-band radar reflectivity and differential reflectivity data for rain attenuation: a self-consistent method with constraints'. *IEEE Trans. Geosci. Remote Sens.* 2001; **39**, 1906–1915. doi:10.1109/36.951081.

[109] Brook, J.P., A. Protat, J. Soderholm, R.A. Warren, and H. McGowan. 'A variational interpolation method for gridding weather radar data'. *J. Atmos. Oceanic Tech.* 2022; **39**, 1633–1654.

[110] Mohr, C.G., L.J. Miller, and R.L. Vaughan. 'An interactive software package for the rectification of radar data to three-dimensional Cartesian coordinates'. Preprints *20th Conference on Radar Meteorology*, Amer. Meteor. Soc., Boston, MA, 1981, pp. 690–695.

[111] Miller, L.J. 'SPRINT: Sorted Position Radar INTerpolation documentation', *NCAR*. 1999.

[112] Straka, J.M., D. Zrnic, and A. Ryzhkov. 'Bulk hydrometeor classification and quantification using polarimetric radar data: synthesis of relations'. *J. Appl. Meteor.* 2000; **39**, 1341–1372.

[113] Thompson, E.J., S.A. Rutledge, B. Dolan, V. Chandrasekar, and B.L. Cheong. 'A dual-polarization radar hydrometeor classification algorithm for winter precipitation'. *J. Atmos. Oceanic Tech.* 2014; **31**, 1457–1481. https://doi.org/10.1175/JTECH-D-13-00119.1.

[114] Wen, G., A. Protat, P.T. May, X. Wang, and W. Moran. 'A prototype-based method for hydrometeor classification using polarimetric variables. Part I: interpretation and analysis'. *J. Atmos. Oceanic Tech.* 2015; **32**, 1320–1340. http://dx.doi.org/10.1175/JTECH-D-13-00178.1.

[115] Marks, D.A., D.B. Wolff, L.D. Carey, and A. Tokay. 'Quality control and calibration of the dual-polarisation radar at Kwajalein, RMI'. *J. Atmos. Oceanic Tech.* 2011; **28**, 181–196. doi:10.1175/2010JTECHA1462.1.

[116] Chudler, K., S.A. Rutledge, and B. Dolan. 'Unique radar observations of large raindrops in tropical warm rain during PISTON'. *Mon. Wea. Rev.* 2022, **150**, 2719–2736. https://doi.org/10.1175/MWR-D-21-0298.1.

[117] Steiner, M., R.A. Houze Jr, and S.E. Yuter. 'Climatological characterization of three-dimensional storm structure from operational radar and rain gauge data'. *J. Appl. Meteor.* 1995; **34**, 1978–2007.

[118] Dolan, B., S.A. Rutledge, S. Lim, V. Chandrasekar, and M. Thurai. 'A robust C-band hydrometeor identification algorithm and application to a long term polarimetric radar data set'. *J. Appl. Meteor.* 2013; **52**, 2162–2186.

[119] Suzuki, K., K. Nakagawa, T. Kawano, *et al.* 'Videosonde-observed graupel in different rain systems during Pre-YMC project'. *SOLA*. 2018; **14**, 148–152. https://doi.org/10.2151/sola.2018-026

[120] Morrison, H., M. van Lier-Walqui, A.M. Fridlind, *et al.* 'Confronting the challenge of modeling cloud and precipitation microphysics'. *J. Adv. Modeling Earth Sys.* 2020; **12**, e2019MS001689. https://doi.org/10.1029/2019MS001689.

[121] Carter, D.A., W.L. Ecklund, K.S. Gage, *et al.* 'First test of a shipboard wind profiler'. *Bull. Amer. Meteor. Soc.* 1992; **73**, 1587–1592.

[122] Gossard, E.E. 'Measuring drop-size distributions in clouds with a clear-air-sensing Doppler radar. *J. Atmos. Oceanic Tech.* 1988; **5**, 640–649.

[123] Gossard, E.E., R.G. Strauch, and R.R. Rogers. 'Evolution of drop-size distribution in liquid precipitation observed by ground-based Doppler radar'. *J. Atmos. Oceanic Tech.* 1990; **7**, 815–828.

[124] Rogers, R.R., W.L. Ecklund, D.A. Carter, K.S. Gage, and S.A. Ethier. 'Research applications of a boundary-layer wind profiler'. *Bull. Amer. Meteor. Soc.* 1993; **74**, 567–580.

[125] Kawano, N., H. Hashiguchi, K. Yoneyama, and S. Fukao. 'Lower atmosphere observations over the equatorial Indian Ocean with a shipborne lower troposphere radar during MISMO field experiment', *Radio Sci.* 2009; **44**, RS6011. https://doi.org/10.1029/2008RS003885.

[126] Wakasugi, K., A. Mizutani, M. Matsuo, S. Fukao, and S. Kato. 'A direct method for deriving drop-size distribution and vertical air velocities from VHF Doppler radar spectra'. *J. Atmos. Oceanic Tech.* 1986; **3**, 623–629.

[127] Lhermitte, R. 'A 94 GHz Doppler radar for cloud observations'. *J. Atmos. Oceanic Tech.* 1987; **4**, *36–48.*.

[128] Illingworth, A.J., R.J. Hogan, E.J. O'Connor, *et al.* 'Cloudnet: continuous evaluation of cloud profiles in seven operational models using ground-based observations'. *Bull. Amer. Meteor. Soc.* 2007; **88**, 883–898. https://doi.org/10.1175/BAMS-88-6-883.

[129] Kollias, P., E.E. Clothiaux, T.P. Ackerman, *et al.* 'Development and applications of ARM millimeter-wavelength cloud radars. The Atmospheric Radiation Measurement (ARM) Program: the first 20 years', *Meteor. Monogr.*, No. 57, Amer. Meteor. Soc., 2016. https://doi.org/10.1175/AMSMONOGRAPHS-D-15-0037.1.

[130] Okamoto, H.T., T. Nishizawa, H. Takemura, *et al.* 'Vertical cloud structure observed from shipborne radar and lidar: midlatitude case study during the MR01/K02 cruise of the research vessel Mirai'. *J. Geophys. Res.* 2007; **112**, D08216.

[131] Yasunaga, K., K. Yoneyama, H. Kubota, *et al.* 'Melting-layer cloud observed during R/V *Mirai* cruise MR01-K05'. *J. Atmos. Sci.* 2006; **63**, 3020–3032.

[132] Webster, P.J., E.F. Bradley, C.W. Fairall, *et al.* 'The JASMINE pilot study'. *Bull. Amer. Meteor. Soc.* 2002; **83**, 1603–1630. doi: I0.II75/BAMS-83-11-1603.

[133] Brendecke, J., X. Dong, B. Xi, and P. Wu. 'Maritime cloud and drizzle microphysical properties retrieved from ship-based observations during MAGIC'. *Earth Space Sci.* 2021; **8**, e2020EA001588. https://doi.org/10.1029/2020EA001588.

[134] Schupe, M. 'Multidisciplinary drifting observatory for the study of Arctic Climate (MOSAiC) Field Campaign Report', *DOE/SC-ARM*-21-007. 2021.

[135] Protat, A., E. Schulz, L. Rikus, Z. Sun, and Y. Xiao. 'Shipborne observations of the radiative effect of Southern Ocean Clouds'. *J. Geophys. Res. Atmos.* 2017; **122**, 318–328.

[136] Mace, G. G. and A. Protat. 'Clouds over the Southern Ocean as observed from the RV investigator during CAPRICORN. Part 1: cloud occurrence and phase partitioning'. *J. Appl. Meteor. Clim.* 2018; **57**, 1783–1803.

[137] Mace, G.G. and A. Protat. 'Clouds over the Southern Ocean as observed from the RV investigator during CAPRICORN. Part 2: the properties of non-precipitating stratocumulus'. *J. Appl. Meteor. Clim.* 2018; **57**, 1805–1823.

[138] Delanoë, J., A. Protat, J.-P. Vinson, *et al.* 'BASTA: a 95-GHz FMCW Doppler radar for cloud and fog studies'. *J. Atmos. Oceanic Tech.* 2016; 33, 1013–1038. https://doi.org/10.1175/JTECH-D-15-0104.1.

[139] Alexander, S., G. McFarquhar, R. Marchand, *et al.* 'Mixed-phase clouds and precipitation in Southern Ocean cyclones and cloud systems observed poleward of 64°S by ship-based cloud radar and lidar'. *J. Geophys. Res: Atmos.* 2021; **126**, e2020JD033626. doi:10.1029/2020JD033626.

[140] JAMSTEC. 'The observational study of the heavy rainfall zone in the eastern Indian Ocean; R/V *Mirai* cruise report MR15-04'. 2016.

[141] Acquistapace, C., R. Coulter, S. Crewell, *et al.* 'EUREC4A's Maria S. Merian ship-based cloud and micro rain radar observations of clouds and precipitation', *Earth Syst. Sci. Data.* 2022; **14**, 33–55. https://doi.org/10.5194/essd-14-33-2022.

[142] Hashiguchi, H., M. Teshiba, and S. Fukao. 'Observations of the lower atmosphere over the equatorial Western-Pacific with a ship-borne lower troposphere radar'. In: *32nd International Conference on Radar Meteorology*, extended abstract, 2005.

[143] Law, D.C., S.A Mclaughlin, M.J. Post, *et al.* 'An electronically stabilized phased array system for shipborne atmospheric wind profiling'. *J. Atmos. Oceanic Tech.* 2002; **19**, 924–933.

[144] Moran, K., S. Pezoa, C. Fairall, *et al.* 'A motion-stabilized W-band radar for shipboard observations of marine boundary-layer clouds'. *Bound. Layer Met.* 2012; **143**, 3–24. https://doi.org/10.1007/s10546-011-9674-5.

[145] Filisetti, A., A. Marouchos, I. McRobert, B. Baldwinson, A. Protat, and B. Atkinson. 'Design of an instrument stabilizing system for in-situ measurements on a research vessel'. *OCEANS 2017 – Aberdeen*, 2017, pp. 1–7. doi:10.1109/OCEANSE.2017.8084695.

[146] Protat, A. and I. McRobert. 'Three-dimensional wind profiles using a stabilized shipborne cloud radar in wind profiler mode'. *Atmos. Meas. Tech.* 2020; **13**, 12. https://doi.org/10.5194/amt-13-3609-2020.

[147] Widener, K, N. Bharadwaj, and K. Johnson. 'Ka-Band ARM Zenith Radar (KAZR) Instrument Handbook'. United States. 2012. https://doi.org/10.2172/1035855.

[148] Suzuki, J., M. Fujiwara, A. Hamada, *et al.* 'Cloud-top height variability associated with equatorial Kelvin waves in the tropical tropopause layer during the Mirai Indian Ocean cruise for the study of the MJO-convection Onset (MISMO) campaign'. *SOLA*. 2010; **6**, 97–100. https://doi.org/10.2151/sola.2010-025.

[149] Maahn, M. and P. Kollias. 'Improved Micro Rain Radar snow measurements using Doppler spectra post-processing'. *Atmos. Meas. Tech. Discuss.* 2012; **5**, 4771–4808. http://doi.org/10.5194/amtd-5-4771-2012.

[150] Garcia-Benadi, A., S. Gonzalez, M. Udina, B. Codina, and J-F. Georgis. 'Precipitation type classification of micro rain radar data using an improved Doppler spectral processing methodology'. *Remote Sens*. 2020; **12**, 4113. https://doi.org/10.3390/rs12244113.

[151] Alexander, S.P. and A. Protat. 'Southern Ocean cloud properties as observed from the surface and satellite'. *J. Geophys. Res. Atmos.* 2018; **123**, 443–456. https://doi.org/10.1002/2017JD026552.

[152] Grecu, M., W.S. Olson, S.J. Munchak, *et al.* 'The GPM combined algorithm'. *J. Atmos. Oceanic Tech.* 2016; **33**, 2225–2245. https://doi.org/10.1175/JTECH-D-16-0019.1.

[153] Skofronick-Jackson, G., W.A. Petersen, W. Berg, C. Kidd, E.F. Stocker, and D.B. Kirschbaum. 'The Global Precipitation Measurement (GPM) mission for science and society'. *Bull. Amer. Meteor. Soc*. 2017; **98**, 1679–1695. https://doi.org/10.1175/BAMS-D-15-00306.1.

[154] Protat, A., C. Klepp, V. Louf, *et al.* 'The latitudinal variability of oceanic rainfall properties and its implication for satellite retrievals. Part 1: the latitudinal variability of drop size distribution properties'. *J. Geophys. Res. Atmos.* 2019; **124**, 13291–13311. https://doi.org/10.1029/2019JD031010.

[155] Protat, A., C. Klepp, V. Louf, *et al.* 'The latitudinal variability of oceanic rainfall properties and its implication for satellite retrievals. Part 2: the relationships between radar observables and drop size distribution parameters'. *J. Geophys. Res. Atmos*. 2019; **124**, 13312–13324. https://doi.org/10.1029/2019JD031011.

[156] Skofronick-Jackson, G., M. Kulie, L. Milani, S.J. Munchak, N.B. Wood, and V. Levizzani. 'Satellite estimation of falling snow: a global precipitation measurement (GPM) core observatory perspective'. *J. Appl. Meteor. Clim*. 2019; **58**, 1429–1448. https://doi.org/10.1175/jamc-d-18-0124.1.

[157] Louf, V. and A. Protat. 'OceanPOL Weather Radar Dataset. Electronic Dataset, National Computing Infrastructure'. doi: 10.25914/5fc4975c7dda8.

[158] Williams, C.R. and K.S. Gage. 'Raindrop size distribution variability estimated using ensemble statistics'. *Ann. Geophys*. 2009; **27**, 555–567. www.ann-geophys.net/27/555/2009/.

Chapter 9

A versatile stratosphere–troposphere radar at 205 MHz in the tropics

K. Mohanakumar[1], Titu K. Samson[2], P. Mohanan[1], K. Vasudevan[1], K.R. Santosh[1], V.K. Anandan[3], G. Viswanathan[4] and B.M. Reddy[5]

9.1 Stratosphere–troposphere (ST) radar

Atmospheric radar, also known as wind profiler radar (WPR), is a state-of-the-art system that continuously monitors horizontal and vertical winds in the troposphere and the stratosphere in all weather conditions with high resolution. The technique used in these radars is the coherent integration at very high frequency (VHF) that can detect signal returns arising from weak fluctuations in the atmospheric refractive index. These radars transmit pulses of electromagnetic radiation by steering the beams 15° off-zenith in different orthogonal directions to determine the three-dimensional wind vector. A small portion of energy transmitted in each direction is backscattered to the radar and is sampled in the receiver at discrete intervals [1,2]. The movement of the scattering media causes a Doppler shift in the backscatter returns and from these Doppler shifts, the wind velocity is determined.

Atmospheric radars can operate over a wide range of frequencies in the VHF (30–300 MHz) and the near ultra-high frequency (UHF) (300–1,400 MHz), depending on the nature of the scattering mechanism [2]. The mandatory height coverage and the resolution affect the choice of operating frequency. Lower frequencies radars are generally used for atmospheric studies, and higher frequencies radars are used for weather studies.

Three established ranges of frequencies are available for probing different heights of the atmosphere. The near-UHF radars are precisely designed for the boundary layer and lower-tropospheric studies [3]. They are in the range of 1,000 MHz frequency

[1]Advanced Centre for Atmospheric Radar Research, Cochin University of Science and Technology, India
[2]School of Earth and Space Exploration, Arizona State University, USA
[3]ISRO Telemetry, Tracking and Command Network (ISTRAC), India
[4]ISTRAC, Radar Development Area, Indian Space Research Organisation, CSIR – Bangalore, India
[5]National Geophysical Research Institute (NGRI), India

(~30 cm wavelength), usually known as the boundary layer radar, with a typical lowest range gate near 100 m. The maximum detectable signal varies with atmospheric conditions. Typically the observable altitude ranges from 1.5 to 5 km above ground level. The vertical sampling resolution mode varies from 60 to 100 m. The 1,000 MHz WPRs are smaller, less expensive, and transportable compared to other wind profilers at lower frequencies. For very short-term weather forecasting, air pollution monitoring, emergency response to chemical or nuclear incidents, and severe weather warnings for aviation and public protection, 1,000 MHz radar can provide vital wind information with very high temporal and spatial resolutions.

Atmospheric radars on the mesoscale system are used to operate around 400 MHz (~75 cm wavelength) to realize the needs concerning vertical range and resolution in the troposphere. They are used to measure and monitor the dynamics of the atmosphere up to 8–12 km. The 400 MHz radars have more powerful transmitters than 1,000 MHz radars. These WPRs are cost-effective and convenient instruments for meteorological analysis and weather forecasts. For the continuous monitoring of the three-dimensional wind vector profiles in the upper levels of the atmosphere, mesosphere–stratosphere–troposphere (MST) radars are generally used. Such radars operate at a frequency of around 50 MHz (~6 m wavelength) in the VHF range. These radars make continuous observations over the approximate altitude range of 2–25 km, covering most of the troposphere and the lower stratosphere. They can also monitor the sporadic observations over 60–90 km and the ionospheric E- and F-regions. MST radars provide valuable information on the dynamical and electrodynamical processes, including coupling the lower atmosphere to the ionosphere [1,4].

9.1.1 Concept of 200 MHz ST radar

Balsley and Gage [5] have described various factors influencing the choice of frequency for WPR as illustrated in Figure 9.1. The WPRs were developed around the conventional frequency bands, around 50, 400, and 1,000 MHz, and their merits and demerits are graphically shown in this figure. The dashed vertical line in the figure represents the ~200 MHz band (~1.46 m wavelength). The potential of radar around 200 MHz for wind profiling is not evident [6]. However, a feasibility study conducted by NASA for monitoring the wind for space shuttle launch suggested that frequencies of 50, 225, and 400 MHz are suitable for probing the stratospheric and tropospheric heights [7]. Earlier, no attempt was made to design and develop WPR in the frequency near 200 MHz because this frequency range is widely used for radio and TV broadcasting. However, recent broadcasting techniques changed from analog to digital, and the 180–220 MHz frequency ranges are also made available for meteorological observations. The advantage of this frequency band is that highly reliable wind measurement is possible in the entire troposphere and lower stratospheric levels.

Another major challenge was the height coverage of 200 MHz radars. The maximum observable height for the fixed value of power aperture product (PAP) as a function of frequency, as illustrated by Balsley and Gage [5], is shown in Figure 9.2. The maximum observable height in the VHF frequency range increases due to the rapid decrease of cosmic noise with frequency. For frequencies above 100 MHz, the cosmic noise contribution becomes small so that the increase in

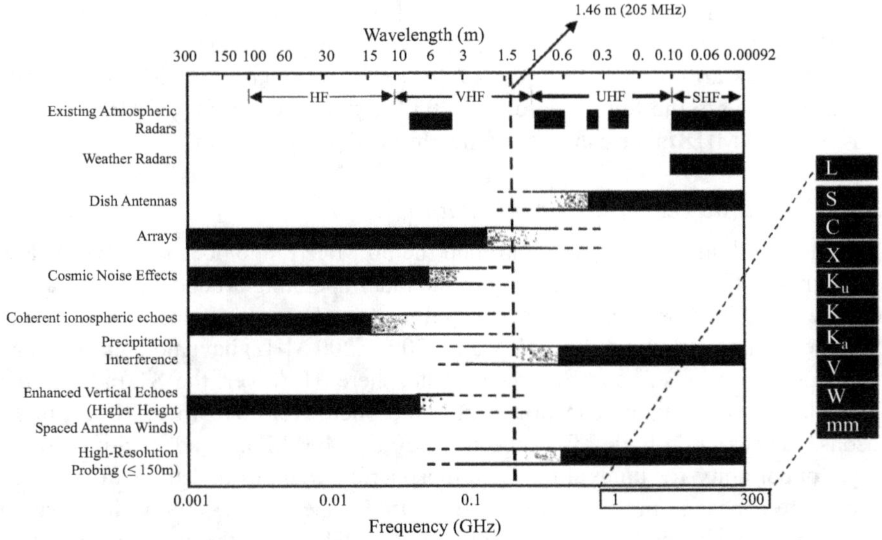

Figure 9.1 Graphical representation of the factors influencing the choice of frequency for WPR. (Balsley and Gage 1982) Courtesy: American Meteorological Society.

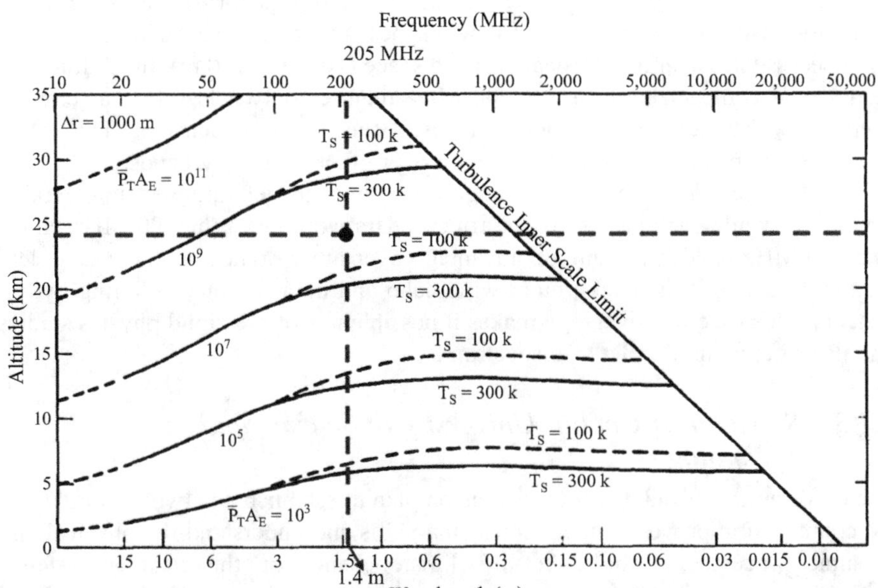

Figure 9.2 Maximum observational altitude as a function of frequency and PAP. (Balsley and Gage, 1982) Courtesy: American Meteorological Society.

observable altitude can be achieved by decreasing the system noise. By keeping the PAP 10^8 Wm^2, the 200 MHz radars can attain a height coverage of 20 km. The WPR operating around 50 MHz does not cover below 2 km altitude [6]. The vertical dashed line in the figure represents 205 MHz. By keeping the PAP in the 10^8 Wm^2, the 200 MHz radar can attain a height coverage of 20 km.

9.1.2 Advantages of 200 MHz radar

The wind profiling radar operating around the 50 MHz range does not cover below 2 km altitude. The large aperture array of the radar operating around 50 MHz cannot form a well-defined beam in the first few kilometers above the surface. On the other hand, the WPRs working in the UHF band (300–1,200 MHz) have no data coverage at the upper troposphere and the lower stratosphere. However, the ST radar in the 200 MHz frequency range can probe the atmosphere from 315 m to 20 km in all seasons. Hence, it is a trade-off between the 50 and 400 MHz bands and provides a unique opportunity for unraveling the atmospheric dynamics and the stratosphere–troposphere exchange processes in the upper troposphere lower stratosphere region [8]. The effective bandwidth of the ST radar is 3.3 MHz, while the allotted bandwidth for the MST radar is at or below 2 MHz. Hence, the vertical resolution of wind data is coarser (~150 m) for the 50 MHz bands compared to the 200 MHz range (45 m).

Another advantage of the 200 MHz band is that it is less affected by galactic or cosmic noise (~1,000 K), whereas the 50 MHz range is more vulnerable to this type of noise (~6,000 K) [9]. The galactic noise is less for 200 MHz, and the signal-to-noise ratio is improved for wind measurements. The 50 MHz antenna array necessitates a larger physical size and space (100 m × 100 m). In contrast, the modest and compact hardware of 200 MHz requires only a lesser area (circular aperture with ~27 m diameter) and is, hence, less expensive. This cost-effectiveness makes it an attractive option over other frequency bands.

WPRs operating in the UHF range are known to get saturated under rainy conditions, while such issues do not affect radars working in the 200 MHz range. The 200 MHz band has a unique advantage over other brands, as there is a clear separation between the background wind echo and the echo due to falling hydrometeors. This specific advantage makes it possible to explore cloud physics studies on fall velocity and droplet size spectrum.

9.1.3 ST radar at Cochin University of Science and Technology

The Government of India proposed a vision plan to establish a network of radars in the country to improve weather forecasting skills and understand the physical and dynamic processes of the effect of climate change on the weather systems. Accordingly, Cochin University of Science and Technology (CUSAT) proposed establishing a sophisticated VHF Stratosphere Troposphere (ST) radar in the 200 MHz range at Cochin to the Department of Science and Technology, Government of India, under its Intensification of Research in High Priority Area (IRHPA) scheme. The Wireless Planning and Co-ordination (WPC) of the Ministry

Figure 9.3 Aerial view of CUSAT ST radar

of Communications, Government of India, allocated 205 MHz frequency for the operation of the ST radar at CUSAT for meteorological purposes. An aerial view of CUSAT ST radar is shown in Figure 9.3.

The CUSAT ST radar operating at 205 MHz VHF range (wavelength: 1.4624 m) is the first WPR in the 200 MHz clubs, successfully functioning in the near-equatorial site to study the monsoon features. This radar measures zonal, meridional, and vertical winds from 315 m at lower altitudes up to a height of 20 km and above, at very high resolution in both time and vertical levels, continuously for 24 h a day. This radar can continuously monitor the atmosphere in all weather conditions. This unique frequency band has the upper hand over conventionally used frequencies in radars. Furthermore, it is highly cost-effective and requires lesser human interventions.

9.1.4 Geographical relevance

Kerala, located in southwest peninsular India, is the land of monsoons. This part of the country experiences both southwest summer monsoon and northeast monsoon rainfall. The first burst of the southwest monsoon over the Indian subcontinent takes over Kerala. Southwest monsoon arrives in this region in June and remains there till September. After the summer monsoon withdrawal, the northeast winter monsoon sets in for the October and November months. The timely onset, optimum duration, and reasonable strength of the southwest monsoon are vital to India's life and economy and the entire south Asian region.

Cochin, also known as Kochi, located at 10°N, 77°E, is a tropical coastal region in the central part of Kerala and is considered the Gateway of the Indian summer monsoon. Meteorologically, Cochin is a sensitive region bordered by the Arabian Sea on the one side and flanked by the Western Ghats on the other side. Rainfall over Cochin is influenced by the coastal and orographic effects due to its nearness to the sea and the hills. The highest mountain peak of Western Ghats, Anamudi, lies just 100 km east in the same latitude belt of Cochin. During the southwest monsoon season, the Arabian Sea behaves as a reservoir of water vapour, which gets pumped into the land region under strong westerly winds reaching a maximum speed of 15–25 ms^{-1}.

On the other hand, the presence of the Western Ghats, extending to an altitude of about 2 km, orographically lifts the strong wind. This vertical forcing of humid air mass enhances the cloud formation and precipitation on the windward side of the Western Ghats. The location map of ST radar at Cochin is presented in Figure 9.4.

Figure 9.4 Location map of the 205 MHz ST radar at Cochin, India

9.1.5 Refractive index structure constant (C_n^2) over Cochin

Variations in atmospheric refractive index structure constant, C_n^2, indicate the strength of turbulent motion in the atmosphere [10]. Vertical profiles of C_n^2 for the troposphere and the lower stratosphere were studied with the help of radiosonde and the ST radar over Cochin (Figure 9.5). Radiosonde data is used from the ground up to a height of about 30 km. It is observed that the values of C_n^2 decrease with altitude. A correlation analysis is performed to examine the contribution of temperature and relative humidity to the potential refractive index (M). It is seen that the contribution of moisture content is more dominant in the lower levels. The turbulent layers are flagged using the critical values of Reynolds's number and the Thorpe sorting method. Subsequently, the probability of turbulence is calculated in pre-monsoon and monsoon seasons. The probability of turbulence is maximum within the boundary layer and decreases gradually with height except above 8 km. The thickness of the turbulent layer shows the same trend as that of the probability distribution. A comparison between radar and radiosonde-derived C_n^2 values is made for selected dates, and a similar trend is observed.

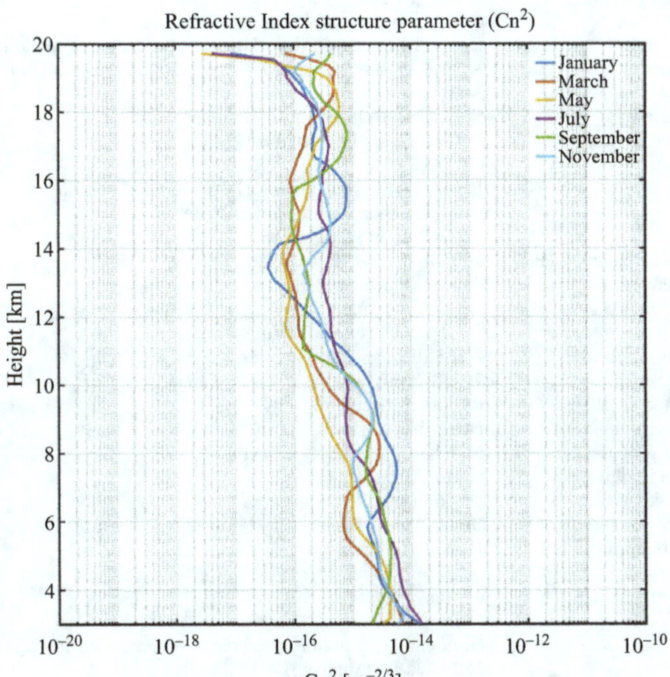

Figure 9.5 Vertical profiles of C_n^2 values over Cochin for different months

9.2 Technical specifications of CUSAT ST radar

Stratosphere–troposphere radar at Cochin is a pulse-Doppler radar capable of analyzing the backscattered signal to measure the Doppler shift from different heights. The system consists of 619 Yagi–Uda antennas arranged in a circular aperture with a diameter of 27 m. The PAP of the radar is designed in such a way that the weak backscattered signal coming from 20 km altitude should fall well above the sensitivity limit of the system. The system consists of two modes of operation: Doppler Beam Swinging (DBS) and Spaced Antenna Mode (SAM). Table 9.1 shows the specification of ST radar.

Table 9.1 Specification of 205 MHz WPR

Parameters	Specifications
Frequency	205 MHz
Bandwidth	5 MHz
Type of system	Active phased array with TRM
Antenna element	Three element Yagi–Uda antenna
Height coverage	315 m–20 km
Range gates	1,024 (programmable)
Modes of operation	Doppler beam swinging (DBS)/spaced antenna mode (SAM)
Height resolution	~45 m up to 4 km and 150 m or 300 m up to 20 km
Beam width	~3°
Off Zenith angle	Selectable from 0° to 30° in steps of 1°
Beam positions in Azimuth	0°–360° with 1° resolution
Pulse width	0.3–76.8 μs
Modulation	Binary phase shift keying (BPSK)-coded compression
Code	Complementary code/Barker code
Baud	0.3–4.8 μs in steps of 0.3
Pulse repetition frequency (PRF)	100 Hz to 16 kHz selectable
TR module transmit peak power	500 W (typical per element)
Duty ratio	Up to 15% (max)
Peak PAP	$\sim 1.3 \times 10^8$ (typical) Wm^2
Radar system sensitivity	−165 dBm (post-processing)
Dynamic range	70 dB (min)
Master reference oscillator	Rubidium oscillator
Type of receiver	Direct band with sampling
Type of signal processor	FFT-based frequency domain
Horizontal wind velocity	Up to 70 ms^{-1}
Vertical wind velocity	Up to 30 ms^{-1}
Radial velocity resolution	Better than 0.1 ms^{-1}
Accuracy in horizontal wind component	1 ms^{-1} (typical)
Accuracy in vertical wind component	0.1 ms^{-1} (typical)
Host PC	High-end workstation, preferably server based in Gigabit Ethernet
Input power	250 kW, 3ϕ (approx.)
Antenna array size	27 m × 27 m
Inter element spacing	$0.7 \times \lambda$ (~ 1.024 m)
Data displays	Power spectra (2D, 3D), UVW, speed and wind direction profiles, RTI, VTI plots

The radar has a beam-steering range of 0°–30° along the elevation and 0°–360° along the azimuth with a step of 1°. The resolution of 1° in both azimuth and off-zenith would enable a three-dimensional view of the atmosphere. The principal mechanism is the detection of backscattered electromagnetic waves caused by refractive index perturbations due to atmospheric turbulence. The Bragg scattering, which occurs due to the refractive index perturbations, gives rise to backscattered radar echoes modulated by the wind at that location.

The basic block diagram of ST radar is shown in Figure 9.6. The radar consists of five main subsystems: the antenna, the transmit/receive module (TRM), the distribution network, the radar computer, and the radar console.

Figure 9.7(a) shows the arrangement of the 619 Yagi antenna in a circular manner. Each dot represents a three-element Yagi antenna pointing toward zenith. The individual antenna has a gain of 7.5 dBi, and the whole array has a gain of 35 dBi. To avoid the grating lobes in the visible region, the inter-element spacing is kept at 0.7 λ. The entire array comprises a half-power beam width (HPBW) of ~3° and can tilt the beam 30° off-zenith without any grating lobes.

TRM acts as the interface between the feeder network and the antenna array. Based on the transmit/receive (TR) pulse and command from the radar processing computer, the module performs the phase shifting of the radio frequency (RF) signal for beam steering. It amplifies the RF signal to desired power level in both transmit and receive chains. The RF signal is routed to TRM through the feeder network from a coherent signal generator in which the power level is −10 dBm at 205 MHz. The block diagram of TRM, which mainly consisting of two paths, transmit and receive paths, is shown in Figure 9.7(b).

9.2.1 Radar-processing computer

The radar-processing computer (RPC) is considered the heart of the system. All the control signals, RF generation, and sampling modules are placed inside the RPC. It consists of a coherent signal generator, digital receiver, and control signal generator. The control signal generator produces the TR pulse, gain control, and phase programming of TR modules and monitors the health status of TRM. Coherent signal generators have the pulsed carrier signal at 205 MHz, amplified by the TR module and transmitted through the antenna array. The analog-to-digital converter (ADC) digitizes the backscattered signal with a sampling rate of 80 MHz. The signal from ADC is down-converted to 10 MHz using a direct digital converter (DDC). The RPC performs all the signal processing operations, and parameters like Doppler width, mean Doppler, and wind vectors are analyzed.

9.2.2 Coherent signal generator

The RF signal generator is field-programmable gate arrays (FPGA)-based dual-channel 14-bit digital-to-analog converter (DAC) card. It consists of two high-speed DAC. The external clock of 840 MHz is fed from the clock synthesizer unit. The DAC receives the baud, code length, and cord word from the radar controller

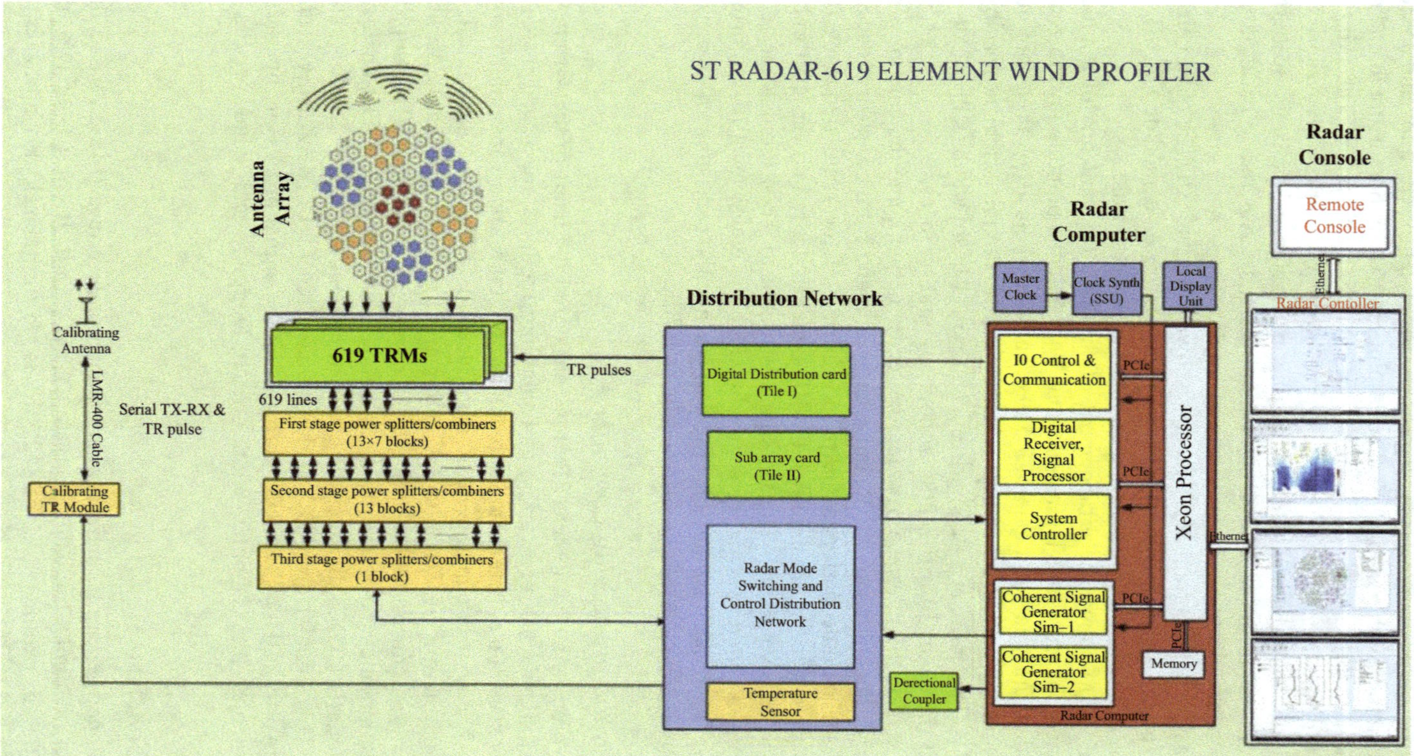

Figure 9.6 Block diagram of 205 MHz WPR

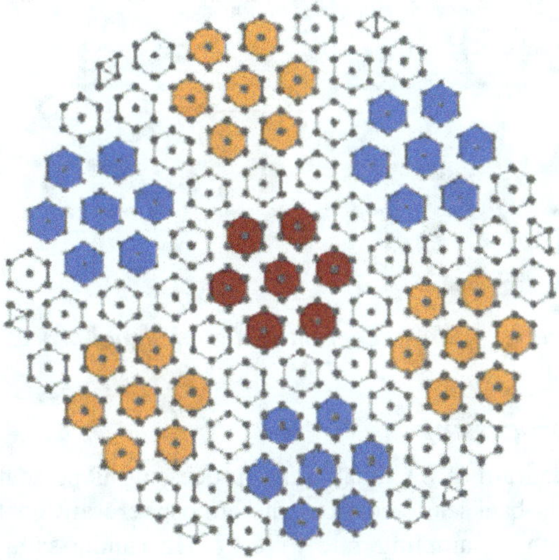

(a)

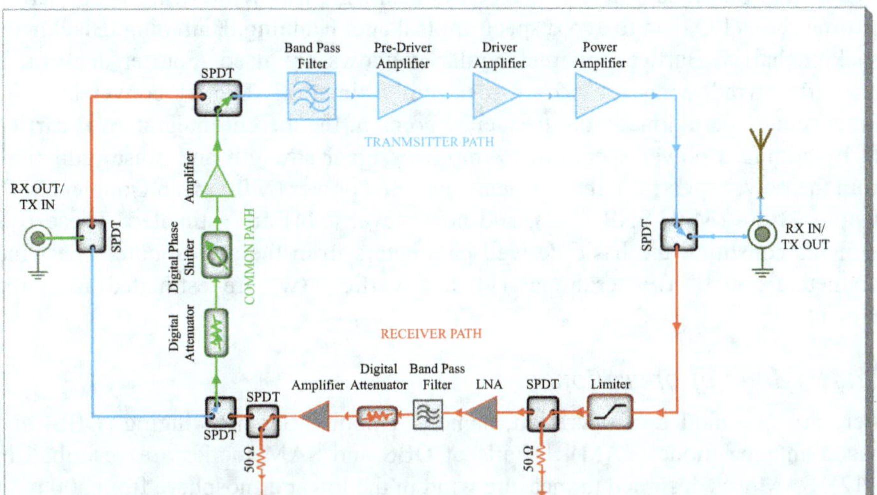

(b)

Figure 9.7 (a) Arrangement of 619 Yagi–Uda antenna. (b) Block diagram of transmit–receive module.

and generates the RF signal based on these parameters. The DAC is capable of generating complementary as well as barker-coded signals. The main features of DAC are (i) dual-channel DAC, (ii) 840 MHz sampling rate, (iii) 14-bit resolution, (iv) the output power level of −3 dBm, (v) spurious-free dynamic range (SFDR) −70 dBc, and (vi) 50-Ohm analog output [11].

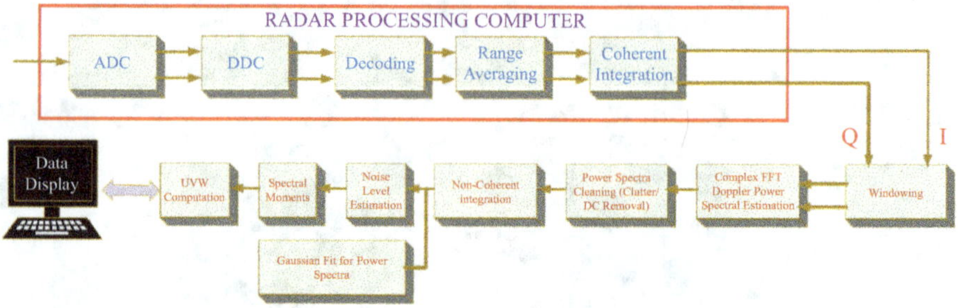

Figure 9.8 Block diagram of digital signal processing

9.2.3 Signal processing

The basic block diagram of ST radar signal processing is presented in Figure 9.8. RF signal, which is backscattered from the atmosphere at 205 MHz, is sampled using an ADC with a sampling rate of 80 MHz (under-sampling). The pulse decoding, range averaging, and coherent integration are done inside RPC using FPGA. RPC provides a time series of I/Q data samples. Windowing is applied in the time series I/Q data to avoid spectrum leakage. Hanning, Hamming, Blackman, Blackmanharris, Bartlet, and rectangular windows are used. Fourier analysis is done after windowing for power spectrum estimation. Digital conversion and clutter removal are done in the frequency domain. Incoherent integration is carried out by adding a power spectrum to improve signal strength and noise reduction. From the power spectrum, the moments reflected power (M0), mean Doppler (M1), Doppler width (M2), SNR (M3), and noise level (M4) are estimated. These five moments constitute the basic derived parameters from the atmosphere. The wind parameters zonal (u), meridional (v), and vertical (w) are estimated from the moments.

9.2.4 Mode of operation

There are two modes of operation, namely, Doppler Beam Swinging (DBS) and spaced antenna mode (SAM). Details of DBS and SAM modes are described in [5,12]. SAM was designed to measure wind in the lower atmosphere from 500 m to 4 km and DBS for higher heights. During the installation and validation time, it was observed that the DBS mode gives accurate measurements starting from 315 m to 20 km. So, the system setup for SAM is implemented but not used in the operational mode. The switching configuration for the SAM mode in which only seven clusters are active remaining all are idle, but in the case of DBS mode, all elements are active.

9.2.5 Doppler beam swinging

This mode is a monostatic radar, i.e., the antenna used for transmission and reception is the same. During the transmission time, signals from the coherent signal generator are distributed to 619 elements and transmitted [13]. The backscattered signal is received by the same antenna and power combined with the combiner/divider network in the reception time. For the analysis of wind in DBS mode, a 5-beam configuration is used, i.e., one beam toward the zenith and four oblique beams toward North, East, West, and South, respectively. Each beam will give the Doppler corresponding to radial wind movement.

Radial Doppler frequency is calculated using the equation below:

$$f_d = \frac{2V_r}{\lambda} \tag{9.1}$$

where f_d is the Doppler frequency and λ is the wavelength of the operation of radar.

For the wind vector estimation, it is assumed that the atmosphere observed by the radar is homogeneous. The zenith and four oblique beam configurations are most commonly used for wind vector estimation. The horizontal wind components are derived from the radial velocity (V_r) from the five oblique beams at a given height. Using the above equation, the wind vector can be calculated with the radial velocity toward the East (V_{re}), West (V_{rw}), North (V_{rn}), South (V_{rs}), and Zenith (V_{rz}).

The above wind components are calculated as:

$$East\,Wind, V_{re} = u \cos\,\theta + w \sin\,\theta \tag{9.2}$$

$$West\,Wind, V_{rw} = -u \cos\,\theta + w \sin\,\theta \tag{9.3}$$

$$North\,Wind, V_{rn} = v \cos\,\theta + w \sin\,\theta \tag{9.4}$$

$$South\,Wind, V_{rs} = -v \cos\,\theta + w \sin\,\theta \tag{9.5}$$

$$Vertical\,Wind, V_{rz} = w \tag{9.6}$$

where u is the zonal (east–west) wind, v is the meridional (north–south) wind, and w is the vertical (ascending–descending) wind components. The wind blowing from west to east is $+u$ (westerly wind), and the reverse is $-u$ (easterly wind). Similarly, the wind blowing from north to south is $-v$ negative (northerly wind), and south to north is $+v$ (southerly wind). In the case of wind moves upward wind, $+w$ (ascending wind), and the downward wind is denoted as $-w$ (descending wind). In normal atmospheric conditions, the vertical wind is two-order lower than the horizontal (zonal and meridional) winds.

9.2.6 Yagi–Uda antenna for ST radar

CUSAT ST radar is a phased array system using Yagi–Uda antenna. It can detect a backscattered signal from clear air up to an altitude of 20 km and above. The main parameter for designing an array is the half-power beamwidth (HPBW), maximum tilt angle, and PAP. The HPBW is approximately 3°, capable of tilting up to 30° off-zenith without any grating lobes. The PAP detects weak signals from 20 km. Yagi antennas are selected due to the better performance of high gain, stable and straightforward structure, wind resistance, and ease to mount. For the design of the Yagi–Uda antenna, the parameters, such as Gain, Front to Back lobe Ratio (F/B), voltage standing wave ratio (VSWR), bandwidth, and beamwidth were considered [14].

The physical parameters of the Yagi–Uda antenna had been optimized to achieve the maximum gain. Various design parameters for optimization are (i) the reflector length and spacing between the reflector and the driven element; (ii) the director length and spacing between the director and the driven element; and (iii) the element radius. Yagi–Uda antenna is designed and simulated for 205 MHz using computer simulation technology (CST).

9.2.7 *Total transmitted power*

The phased array system consists of 619 transmit/receive modules. Assume that each T/R module transmits a peak power of 400–500 W, for the calculation purpose, 400 W is considered. The transmitted power (P_t) is the product of the number of antenna arrays and the output power of one T/R module. The transmitted power of CUSAT ST radar is thus becoming 247,600 W, approximately 250 kW.

9.2.7.1 Effective aperture area

Effective aperture area (A_e) determines how effectively an antenna can transmit or receive electromagnetic radiation. It can be computed as

$$A_e = \frac{\lambda^2 G}{4\pi} \tag{9.7}$$

where G is the array gain and λ is the wavelength at operating frequency. In the case of 205 MHz radar, λ is 1.463 m. The effective aperture area thus becomes 538 m^2. The array gain (G) depends on the individual element gain and the total number of elements [15]. The simulated gain of the Yagi–Uda antenna is 7.5 dBi.

9.2.7.2 PAP

PAP is the product of total power (P_t) transmitted and effective aperture area (A_e) of the array antenna. The PAP is calculated as 1.33×10^8 Wm^2 for a T/R module with 400 W power output.

9.2.7.3 Received power

The backscattered signal power at the antenna from 20 km is calculated from the radar equation as:

$$P_r = \frac{\pi P_t A_e \Delta R \eta_r}{64R^2} \tag{9.8}$$

where

P_r is the received power (dBm)
P_t is the transmitted power (dBm)
A_e is the effective aperture area (m^2)
η_r = volume reflectivity
R is the range (m)
ΔR is the range resolution (m)

The volume reflectivity factor:

$$\eta_r = 0.38 C_n^2 \lambda^{-1/3} \tag{9.9}$$

where C_n^2 is the refractive index structure coefficient and λ is the wavelength in meters.

$$C_n^2 = a(\Delta n^2) L_0^{-2/3} \tag{9.10}$$

where Δn is the variation in refractive index and L_0 is the outer scale of turbulence.

The refractive index n is given by:

$$n = 1 + \frac{3.73 \times 10^{-1} e}{T^2} + \frac{77.6 \times 10^{-6} P}{T} - \frac{N_e}{2N_c} \tag{9.11}$$

Where

P is the atmospheric pressure (millibars)
e is the partial pressure of water vapor
T is the absolute temperature
N_e is the number density of electrons
N_c is the critical plasma density

Using the above values, the received power of 205 MHz radar is 8.1744×10^{-19} W, equivalent to -150.8754 dBm. The phased array system of the CUSAT radar can detect signals better than -150 dBm to cover an altitude of 20 km.

9.2.8 TRM

The main advantage of a phased array system is that the main lobe can be steered in any direction electronically without any mechanical tilting of the antenna. It will also reduce maintenance due to mechanical wear and tear. The beam scan time and beam-to-beam switching time can be reduced. All these are possible because of the development of individual TRM with programable digital phase shifters in the phased array system [13]. The TRM and feeder network form the interconnecting link between the antenna and the radar processing computer. The RF signal generated by the DAC is distributed to the 619 elements antenna array in transmission. In the receiving mode, all signals received by antennas are combined and connected to ADC.

TRM consists of a high-power transmit path, low power receives path, a 6-bit phase shifter common to both transmit/receive path, power supply cards, and digital control sections. It provides the required gain and phase for the transmitter and received RF signal along with the power amplifier.

9.2.8.1 Functions of a TRM

The TRM activates the transmit or receive path depending on the TR pulse from the radar processing computer. The TRM amplifies the -10 dBm input signal to 57 dBm output during the transmit cycle. The Rx path in the TRM gives a boost of 42 dB in the received signal. It provides an appropriate progressive phase shift for electronic beam steering using a 6-bit phase shifter during the Tx and Rx modes. The limiter in the Rx section protects the low-power Rx section from the high-power leakage from the Tx section during transmission. Table 9.2 shows the specification of TRM used in the 205 MHz ST radar wind profiler.

9.2.9 Radiation pattern measurement

The radiation pattern of the 205 MHz radar is measured using a radio star, emitting radio frequencies in a broad spectrum [16]. In this measurement, radar is configured in receive mode. The beam was tilted toward the North to track the movement of Virgo A. The observed received power is shown in Figure 9.9. The observation

Table 9.2 Specification of 500 W TRM

Parameters	**Specifications**
Frequency	205 MHz
Transmit input power	-10 ± 2 dBm
Output power	500 W (57 dBm)
Pulse width	0.3–76.8 μs
Duty ratio	Up to 15% (max)
Receiver chain gain	40 ± 2 dB
Receiver noise figure	<3.5 dB
Harmonic level at T_x	<−30 dBc
Rise time	<100 nS
Fall time	<100 nS
Digital attenuator	0.5–31.5 dB, with steps of 0.5 dB
Digital phase shifter	5.625°–354.375° with steps of 5.625°
Command interface	RS422 level digital interface
Supply voltage	48 V DC
Supply current	6 A

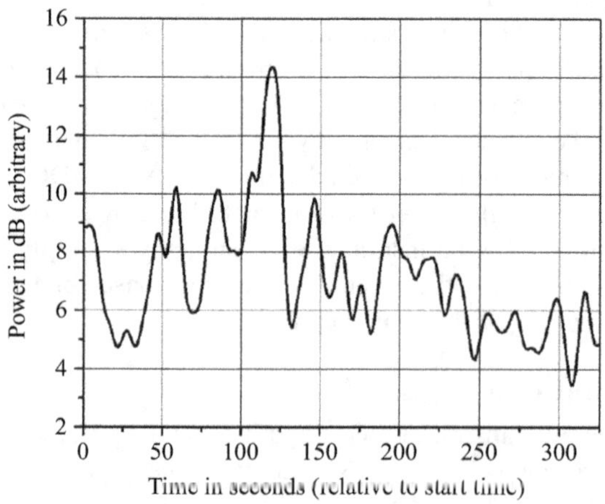

Figure 9.9 Received power from radio star Virgo-A

shows an increase in the received power when the radio star approaches the boresight direction of the array. The null-to-null beam width calculated from the figure is approximately 5.07°, and the HPBW is 3.36°, matching the simulation results [8].

9.2.9.1 Moon bouncing for beam-pointing testing

In this phased array system, the beam can be pointed toward 0°–360° in azimuth and 0°–30° off-zenith. Since this is a large array, testing the pointing accuracy is difficult. So, the Earth's natural satellite, the Moon, is used as the target. Moon can be approximated as a large sphere with a diameter of ~3,475 km, which can act as a perfect target, i.e., a high-power beam pointing toward the Moon will be reflected back. The radar beam is about ~3° beam-width, and the Moon is ~385,000 km away from Earth.

If we look into the trigonometry of the radar beam, it seems like a cone with a horizontal coverage of (2 × 385,000 × sin1.5°) ~20,000 km at the distance of the Moon. Figure 9.10 shows the Earth, radar beam, Moon, and moon track. The time taken by the Moon from rising at the horizon to setting at the horizon is ~12 h. At this time, the Moon will cover 180° (from east to west) with respect to Earth, i.e., in 1 min, it will move 0.25°.

The Moon takes 12 min to cover 20,000 km of the radar beam. The track of the Moon can be precisely predicted. The bounced-back signal can be captured by pointing the radar beam precisely to the Moon. This experiment was conducted on 5 April 2019. The Moon was precisely at 180° azimuth and 9.1° off-zenith at 12.44 pm. As per the calculation, the Moon is in the vicinity of the main beam approximately 4–6 min before and after 12.44 h local time. At this time, the radar beam is pointed toward it. The Moon is about 0.4 million kilometers from Earth, so the reflected signal is folded many times and some fall in the pulse period. The radar is configured to 180° azimuth and 9.1° off-zenith and started the experiment at 12:10:00 pm.

Figure 9.11 shows the series of Moon signatures in the spectrum. All the spectrum before 12:36:14 h and after 12:54:15 h have no moon Doppler signatures.

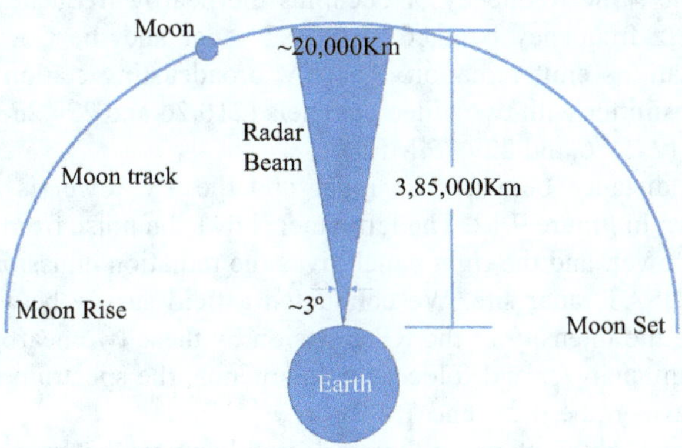

Figure 9.10 Moon track for point accuracy testing

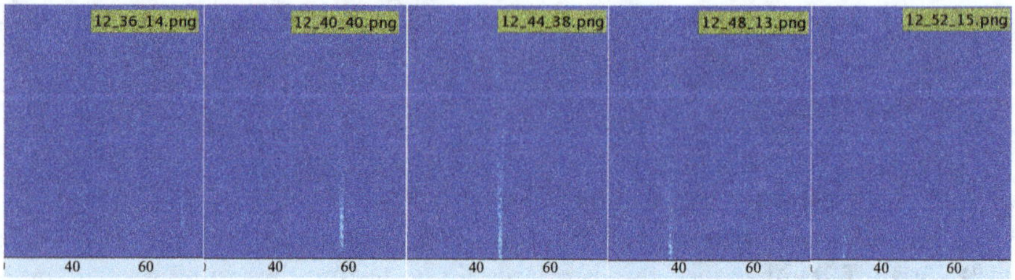

Figure 9.11 Reflected Doppler spectrum from the Moon

From the observation, it is clear that the Moon's signature was observed from 12:36:14 h to 12:52:15 h, i.e., ~16 min, which nearly matches the predicted time of watching the Moon. From this observation, beamwidth is calculated.

On that particular day, the rise and set timing of the Moon is 6:21:24 and 18:44:21 h, respectively. Approximately Moon takes 12 h to complete one-half circle, i.e., 180° of revolution around Earth, i.e., every 1 min, it covers 0.24°. In 16 min, it covered 3.89°, which matches the simulated HPBW of ~3.3°, neglecting the other errors in the experiment. This experiment was repeated for two consecutive days when the Moon was over-passing radar on 31 May 2019 and 1 June 2019. On 31 May 2019, radar was configured for off-Zenith 8°, Azimuth 117°, and off-Zenith 8°, Azimuth 245°. On 1 June 2019, the radar was pointed toward two-direction off-Zenith 2°, Azimuth 0° and off-Zenith 4°, Azimuth 290°; at the predicted time, the reflections from the Moon were observed. The observed reflection from the Moon shows that the beam pointing is precise.

9.2.10 Noise level survey between the radar site and a TV broadcasting station

One of the concerns of atmospheric radar is the emission of radio signals during its operation. The signals generated from the radar are noise for the outside world, which shares the same frequency or occupies the nearby frequency bands. From 202 to 240 MHz frequency band, CUSAT ST radar and the Cochin Television broadcasting stations emit radiation. The TV broadcasting station has a 10-kW high-power transmitter with two video channels (210.26 and 224.25 MHz) and two audio channels (215.76 and 229.75 MHz).

The aerial distance between the radar and the TV tower is approximately 3.6 km, as shown in Figure 9.12. The left panel shows the noise from CUSAT radar toward the TV tower, and the right panel gives the radiation emission from the TV tower to the CUSAT radar site. We conducted a field survey between these two sites to estimate the intensity of the RF emission by these two nearby transmitters. Using a spectrum analyzer and folded dipole antenna, the spectrum is measured at 34 locations between the radar and TV tower.

Spectrum peak observations at three selected locations between the TV broadcasting station and the CUSAT radar are illustrated in Figure 9.13. Three

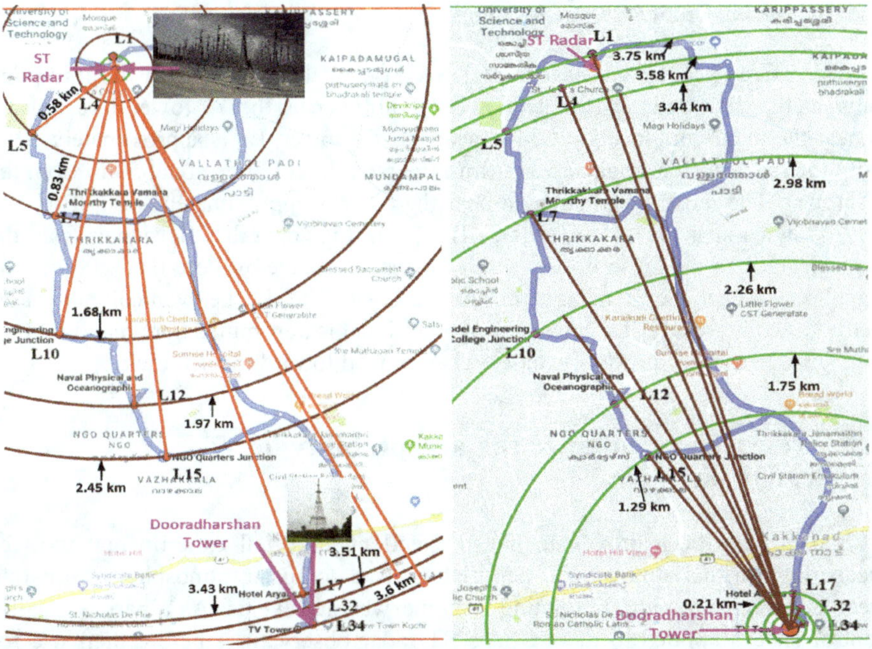

Figure 9.12 Location of CUSAT radar and Cochin Broadcasting Station. The concentric lines show selected 10 observational sites.

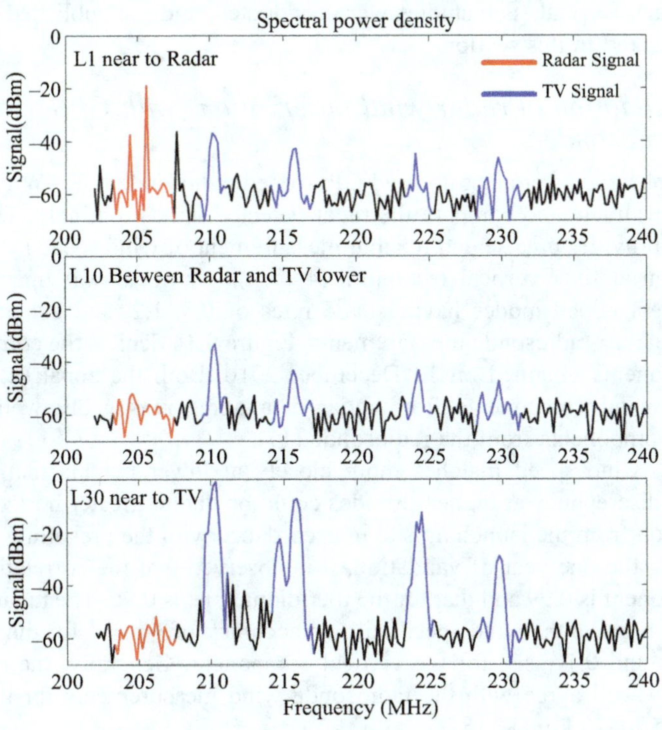

Figure 9.13 Noise signal power for three different locations

representative spectra, one near the CUSAT radar, the second is in the mid-point between the radar and the TV tower, and the third is near the TV tower. Location L1 is near the radar, and location L34 is near the TV tower. L1 is approximately 50 m from the radar, L10 is about 1.5 km from the radar, and L30 is about 50 m from the TV tower. The radar signal strength near the radar is approximately −20 dBm, and the TV signal near the TV tower is 0 dBm [11]. It is noted that the signals from the radar degrade to the noise floor when it moves a distance of more than 200 m. The signal power from the ST radar is transmitted toward the sky, i.e., vertically upwards. In the case of terrestrial TV broadcasting, the signal is transmitted horizontally. Thus, the signal power drastically increases when moved toward the TV tower.

9.3 Research to enhance the acceptability of ST radar data

ST radars have been proven to be an excellent and vital experimental tool for measuring various atmospheric parameters used for basic atmospheric research, weather forecasting, and predicting extreme weather events. As a part of quality control checks applied to the CUSAT ST radar observations before their use for scientific research in understanding the physical and dynamic process of the atmosphere, it is essential to filter out errors in the observed data sets. The ST radar data is validated with other accepted observational tools and assessed with standard available data. Several such studies were conducted, and the published results are briefly presented in this section.

9.3.1 Validation of radar wind observations with GPS radiosonde

The winds observed from a co-located GPS-based radiosonde (GRAW radiosonde) are used to validate and compare the radar-detected winds routinely. The GRAW radiosonde provides precise and reliable measurement of wind speed ($< 0.1\ \mathrm{ms}^{-1}$) and direction at finer vertical resolution (4–5 m). The radar was operated under three different coded modes having baud rates of 0.3, 1.2, and 2.4 μs and then compared against radiosonde measurements. Figure 9.14 depicts the comparison of wind components obtained on 13 December 2016. Both the zonal (u) and meridional (v) wind components from radar data agree reasonably well with the respective components from the radiosonde [17].

Further, wind speed matches more closely at lower heights than at higher levels. The discrepancy at higher altitudes could be due to greater horizontal drifts of the balloon from the launching site in accordance with the prevailing wind with time. During the one year of validation, it is observed that the correlation for the zonal component is 0.99 and that for the meridional one is 0.96. The radar system is designed to retrieve the wind speed with an accuracy of $1\ \mathrm{ms}^{-1}$ for the horizontal components and $0.1\ \mathrm{ms}^{-1}$ for the vertical component. Generally, the radar wind profiles very well agree with the radiosonde wind measurements for the altitude range of 315 m–20 km [17,18].

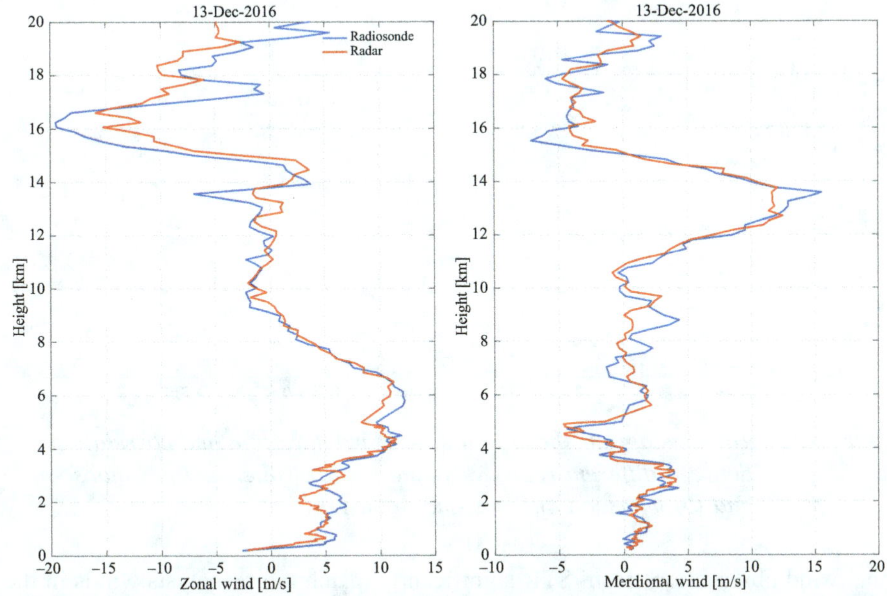

Figure 9.14 Comparison of derived wind (u and v components on the left and right, respectively) from radar with those from Radiosonde [18]

9.3.2 Evaluation of ST radar wind observations with reanalysis and model outputs

An evaluation study has been carried out with the horizontal and vertical winds obtained from atmospheric global reanalysis (ERA5, ERA-Interim, MERRA-2, and NCEP) data and the high-resolution regional model (WRF) output winds with the ST radar (STR) wind observations at Cochin. The contour plots of zonal wind for the four seasons (winter, pre-monsoon, monsoon, and post-monsoon) are illustrated by Sivan *et al.* [19]. The prominent features observed in the radar wind at tropospheric levels are also seen in reanalysis and model wind data.

The transition from the upper-tropospheric westerlies to the easterlies is evident in the pre-monsoon season. STR observations in May show large fluctuations due to thunderstorm activities, which are common over Cochin. Such localized variabilities are not able to be detected from reanalysis and model values. The low-level jet (LLJ) appearance in the lower troposphere during the monsoon onset period and variation in the intensity and vertical extension of tropical easterly jet-stream (TEJ) during the active and break phases of monsoon over Cochin are well captured by STR. Detailed evaluation of the observed zonal, meridional, and vertical winds from STR with the corresponding reanalysis and model output winds are reported [19].

Generally, the significant features observed in the horizontal winds in tropospheric levels by STR are reproduced in reanalysis data and model output. The

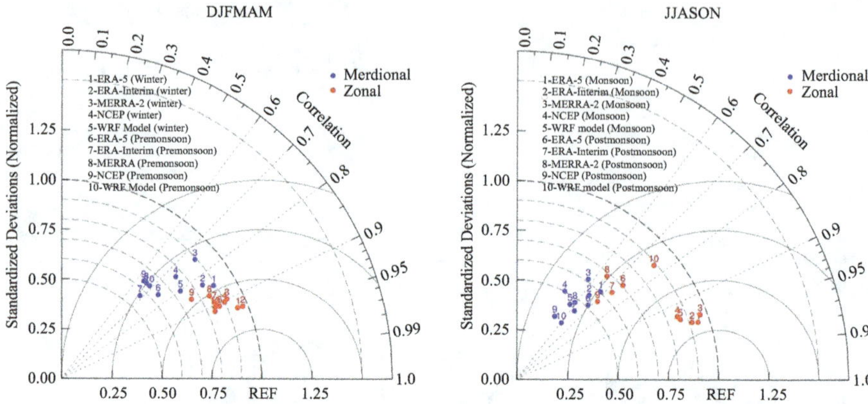

Figure 9.15 Taylor diagram showing the standard deviation and correlation between different reanalysis data for (a) winter and pre-monsoon and (b) monsoon and post-monsoon [19]

zonal wind observations from STR significantly match the reanalysis winds in the entire troposphere, especially during the winter and summer months (Figure 9.15). All reanalysis datasets show high-correlation coefficients (>0.90) during the winter and summer. The highest correlation coefficient (0.954) in ERA5 for zonal wind is seen in the summer monsoon. Due to intense convective activity in the post-monsoon season, the resemblance of the zonal wind pattern to STR is relatively poor. The small-scale fluctuations observed by STR are absent in reanalysis datasets and model outputs. Compared to the zonal winds, the meridional winds in reanalysis and WRF model outputs are not well represented. The deviation from the observed values is more during the summer monsoon season when the zonal winds become strong due to the appearance of LLJ and TEJ. The better similarity is seen in the relatively quiet winter season.

Vertical wind provided by the reanalysis and WRF model outputs does not match the observed winds by STR. The descending air motion in the lower troposphere and ascending flow in the upper troposphere noted during the summer monsoon warrants further investigation to understand the vertical structure of the monsoon and its dynamics. High-resolution and precise observations of winds in the tropospheric heights from a network of WPRs are needed to improve the reliability of reanalysis datasets used in scientific research and weather prediction models.

9.3.3 Radar observations under rainy conditions

The characteristics of the spectra during rainy conditions were studied by configuring the radar to observe the features of wind circulation patterns during the onset, active, and break phases of monsoon, rainfall intensity, etc. Figure 9.16 shows the Doppler signature of the early phase of typical monsoon rain in June for 2017, 2018, and 2019. In all three years, the signature of monsoon rain looks similar.

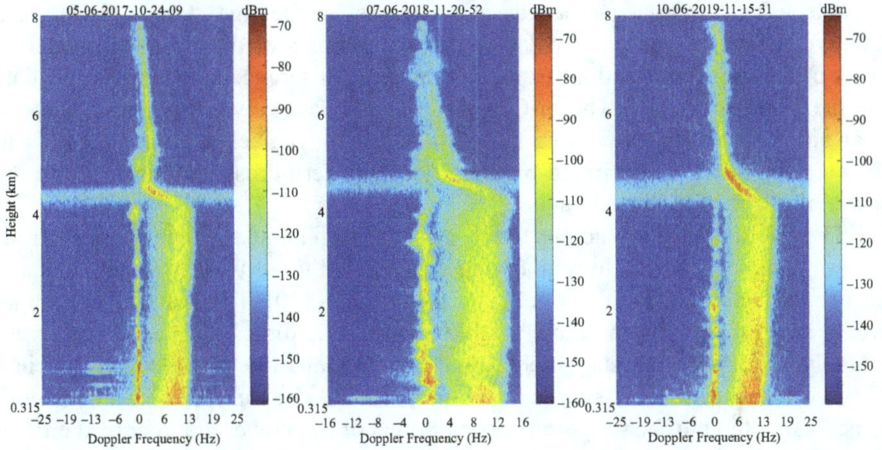

Figure 9.16 Doppler signature of the early phase of monsoon rain (June) in three consecutive years

There are two Doppler signatures seen below 4 km. The Doppler corresponds to the fall velocity of raindrops lying on the right part, and the Doppler of background wind keeps on the left side. It is a unique feature of WPR at 205 MHz, at the high VHF range in which the background wind Doppler is not contaminated by precipitation interference.

9.3.4 New technique for the identification of clear air and rain echoes

The unique feature of the 205 MHz frequency band is that the profiler can simultaneously detect echoes in clear air and rainy conditions. Due to the large spectral width and merging of clear and rain echoes during rainy periods, the conventional signal processing techniques become inept at distinguishing the two echoes. The Doppler spectrum of clear air from STR at Cochin is mixed with the rain echoes from precipitating hydrometeors or two well-separated echoes. Hence, identifying the turbulent background echo is essential for accurate computation of the wind components. A new technique was developed by Dhanya *et al.* [20] to isolate turbulent echoes from precipitation echoes using a hybrid approach.

A combination of an exponentially modified Gaussian model and a pure Gaussian model is used to adaptively fit the raw spectrum in the hybrid approach. The data-adaptive fitting procedure can identify two distinct peaks during the rainy period, and the appropriate process reduces to a single peak during non-rainy periods. The advantage of the hybrid approach is that the adaptive data fitting can identify the peaks without losing the raw I/Q data signal. The hybrid model is constructed with an exponentially decaying Gaussian model combined with a pure Gaussian model. Further, the peaks identified using this model are confirmed by checking the number of zero-crossing (NZC) points on the power gradient of the fitted data.

The hybrid approach proceeds in two steps. In the first step, a bi-Gaussian adaptive fitting is done on raw data using two variants of the Gaussian model. A series of iterations are used in fitting to reduce the difference between the raw data and the Gaussian model. The bi-Gaussian model adaptively fits a single peak for clear air and dual peaks for rain cases. In step two, the echoes corresponding to clear air and rain are confirmed by identifying the zero-crossing points [21] in the power gradient of the fitted data.

The hybrid algorithm adapts to the raw spectrum and can identify the clear air and rain echo peaks. The hybrid approach is applied to all weather conditions and has been effective in single and multi-peaked spectra [20]. The analysis shows that even a small trace of rain below the melting layer is distinctly picked. An exponentially modified Gaussian model and a pure Gaussian provided an efficient adaptive fitting without losing data on the tail end. The number of peaks is confirmed using the number of zero crossing points of the power gradient of the fitted data. The uniqueness of the new hybrid approach provides encouraging results and opens new avenues to ST radar signal processing.

9.3.5 *Validation of Aeolus satellite wind observations with the CUSAT ST radar*

The European Space Agency's Earth Explorer mission launched the Aeolus satellite on 22 August 2018. Aeolus is a polar-orbiting satellite revolving at an altitude of 320 km. It carries a Doppler wind lidar called ALADIN (Atmospheric LAser Doppler INstrument), the first of its kind in orbit. ALADIN delivers the line-of-sight (LOS) wind profiles perpendicular to satellite velocity from surface to 30 km altitude range. It operates in the ultraviolet spectral region (354.8 nm), emitting laser pulses of about 60 mJ at 50.5 Hz. The backscattered signal, which is Doppler shifted, is collected by a dual-channel receiver to measure signatures of both molecular (Rayleigh channel) and particular (Mie channel) scattering [18,22]. This study uses Aeolus L2B product, horizontal line-of-sight (HLOS) observations for baselines 10 (2B10) and 11 (2B11) over the period July 2019–September 2021.

Comprehensive validation of Aeolus horizontal line-of-sight wind over Cochin (10.04°N, 76.9°E), India using the state-of-the-art 205 MHz WPR has been carried out [23]. The quality of Aeolus level 2 wind baseline products 2B10 and 2B11 have been assessed for 26 months, from July 2019 to September 2021. The errors in the Aeolus wind profiles concerning radar between 1 and 18 km altitude are evaluated. The Aeolus wind profiles have been validated by comparing them with collocated radar measurements.

A one-to-one comparison between radar and Aeolus HLOS wind for an altitude range of 1–18 km from Rayleigh scattering for clear sky conditions ($Rayleigh_{clear}$) and Mie scattering for cloudy conditions (Mie_{cloudy}) is shown in Figure 9.17. The Aeolus wind indicates good agreement with radar with a Pearson correlation coefficient (R) of 0.93 for $Rayleigh_{clear}$ and 0.94 for Mie_{cloudy}. The linear regression slopes are pretty close to unity; these values are 1.017 and 0.99, respectively, for $Rayleigh_{clear}$ and Mie_{cloudy} winds. The bias, standard deviation,

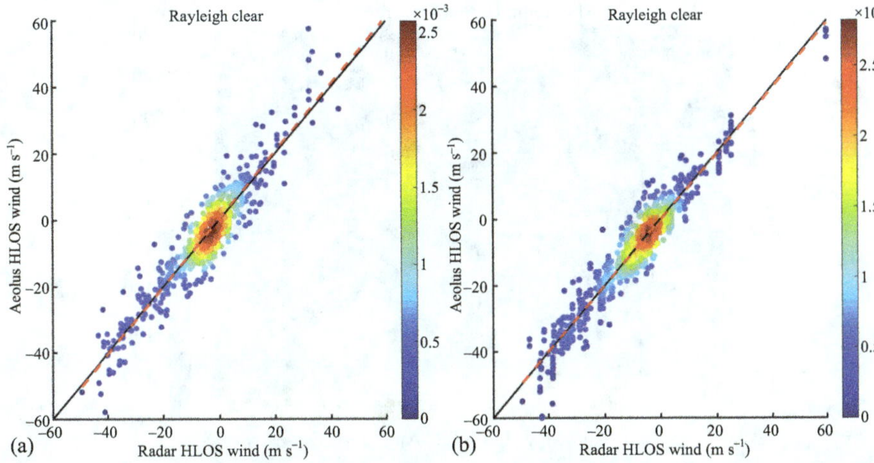

Figure 9.17 Scatter plot between radar and Aeolus wind for (a) Rayleigh clear and (b) Mie cloudy for 2B11. The black line is the diagonal line, and the dotted red line is the linear regression line. The color bar shows the probability density [23].

and scaled median absolute deviation (SMAD) observed in $Rayleigh_{clear}$ are -0.15 ms^{-1}, 5.49 ms^{-1}, and 3.30 ms^{-1}. In contrast, these values are lower for Mie_{cloudy} and are -0.12 ms^{-1}, 4.83 ms^{-1}, and 2.60 ms^{-1}, respectively.

Statistics have been worked out for the ascending and descending orbits of $Rayleigh_{clear}$ and Mie_{cloudy}, and the results are summarized. It can be seen that systematic and random errors do not differ much between the ascending and descending orbits and these values are close to that observed for pooled datasets for both Rayleigh and Mie winds. The $Rayleigh_{clear}$ and Mie_{cloudy} ascending orbits show a bias of -0.11 ms^{-1} and -0.12 ms^{-1}, respectively, while the corresponding values for SMAD are 3.18 ms^{-1} and 2.56 ms^{-1}. However, these values are higher for descending orbits and can be partially attributed to the fewer data points than the ascending orbits. The comparison results show that the systematic and random errors are lower in Mie_{cloudy} wind compared to $Rayleigh_{clear}$.

The Aeolus wind profiles match well with the observed wind speeds from the radar in all seasons [23]. A detailed analysis of Aeolus wind in different altitudes shows that the wind is more biased between 13 and 18 km compared to lower layers for both Rayleigh and Mie. Though Aeolus wind shows a significant bias in high wind speed ranges, the significance of the error estimate is less due to the limited number of collocated data sets. The systematic and random errors in Aeolus wind between ascending and descending orbits do not vary much for Rayleigh and Mie scatterings.

The study shows that Aeolus wind data is of good quality and reliable. Using the Aeolus data in the weather forecasting model could significantly improve the accuracy of weather forecasts over the tropics. Further, this study uses unique 205 MHz WPR data to validate satellite winds.

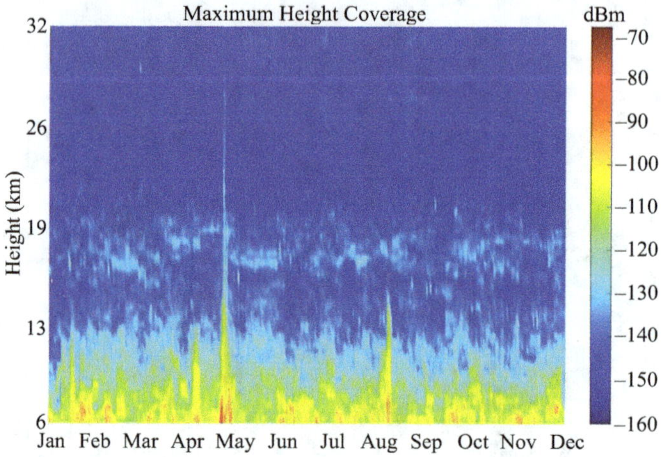

Figure 9.18 Higher height coverage of 205 MHz ST radar for the entire year 2018

9.3.6 Higher height coverage of CUSAT ST radar

WPRs over the tropical region with a height coverage beyond the tropopause poses challenges in the choice of frequency and optimization of performance parameters at the system level, like the PAP. The main reason is that tropical tropopause is around 16–18 km. The 205 MHz radar at the near-equatorial region, Cochin, is selected to study troposphere stratosphere exchange processes. This frequency band has a much higher PAP to meet the SNR needed for estimation accuracies, mainly because the scale size of irregularities giving rise to Bragg scatter is a direct function of height.

There are three different configurations in the regular and monsoon modes with height coverage of 0.3–8 km, 3.1–20 km, and 6–32 km. The radar observations from 6 to 32 km are plotted in time series to determine the seasonal variation of higher height coverage. Figure 9.18 shows the time series plot for 2018; it is evident that the height coverage was above 19 km in all the seasons in 2018.

9.4 Atmospheric studies using the 205 MHz ST radar

The MST/ST radars are now becoming accepted in the radar meteorology community as a valuable complement to the weather radar measurement techniques that make use of Rayleigh scattering from hydrometeors. These radars are used to detect echoes caused by Bragg scatter from refractive index structure due to variations in humidity and temperature in both clear and cloudy atmospheric conditions. In this section, some of the interesting studies are reported from the 205 MHz ST radar observations from a tropical coastal station in Cochin.

9.4.1 Vertical structure and evolution of Indian summer monsoon

The Indian summer monsoon (ISM) is a much-speculated annual phenomenon. The commencement of the monsoon brings about marked changes in atmospheric circulation. It is manifested by establishing two well-defined monsoon circulations, the low-level monsoon jet (LLJ) in the lower troposphere and the tropical easterly jet (TEJ) in the upper troposphere. The changes in the flow pattern of monsoon jet streams during the active/break phase can influence the progression as well as the distribution of monsoon rainfall over the Indian subcontinent [24,25]. To understand the monsoon processes in detail, monitoring the vertical structure of monsoon circulation is inevitable. Moreover, observations on monsoon circulation are vital for assessing the circulation features by model simulations.

The location of the CUSAT ST radar is unique, and it is situated at the entrance of the Indian summer monsoon. The radar provides an unprecedented amount of information on the evolution of monsoon circulation. Kottayil *et al.* [26] attempted to understand the vertical structure of the monsoon circulation by utilizing radar data from April to September for two consecutive years, 2017 and 2018. The study specifically looks into the possibility of defining predictors for the monsoon onset in an objective way.

The height–time variations in the zonal wind for 2017 and 2018 are shown in Figure 9.19. These two monsoon years of Kerala show zonal wind structure fluctuations within the height range of 315 m–20 km, beginning from April to

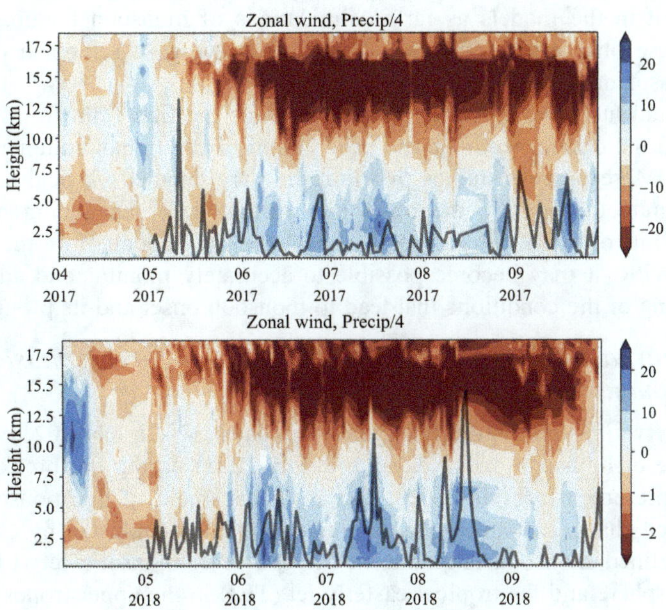

Figure 9.19 Distribution of zonal wind in the troposphere and lower stratosphere during the two consecutive monsoon years 2017 and 2018. The tick lines indicate the precipitation values [26].

September. By the middle of May, the monsoon low-level and the tropical easterly jet had started developing in the lower and the upper troposphere, gradually intensifying. There is a sharp increase in the strength of LLJ and TEJ when the monsoon onset over Kerala occurs [27]. The parameters representing LLJ and TEJ characteristics, such as westerly depth, LLJ core speed, core height, and TEJ core speed and height, are derived [25,28]. It is shown that these parameters can be used as predictors for monitoring the monsoon onset over Kerala (MOK). A preliminary analysis indicates that objective criteria for MOK can be developed from the 205-MHz wind profiler observations, which can verify the monsoon onset features.

The radar observations were used to understand the conditions that led to heavy rainfall in Kerala in 2018 [29]. In 2018, the core of LLJ was positioned below 2.5 km with high values of core speed compared to 2017. In general, radar observations point to an abnormal decrease in the core height of LLJ in 2018 compared to 2017. It implies the significant role of the mountain effect in the formation of heavy rainfall in 2018. The orographic lifting had been very intense and thus resulted in heavy rain over Kerala in 2018. We show that the WPR can ideally capture the transitions in the monsoon circulation before heavy rainfall events. There is a possibility of predicting extreme rainfall at least 2 days in advance through continuous monitoring of the low-level monsoon jet observed over Cochin [29].

Furthermore, the wind profiles from ST radar can assess the representation of monsoon circulation in the widely used analysis and reanalysis products. The initiation and progression of monsoon circulation observed from the radar have been used to evaluate models. This comparison reveals the areas that need improvement in the models as far as simulations of monsoon features are concerned. These observations from the radar can serve as a reference dataset for model assessments.

This study uses wind profiler information for the first time from Cochin to comprehend the commencement and progression of the Indian summer monsoon circulation. More process studies that look into the detailed vertical structure of monsoon circulation during the onset are warranted for advancing our understanding of one of the most challenging phenomena to model. With the advent of 205-MHz WPR, it may become possible to accurately monitor and advance our understanding of the conditions that lead to monsoon onset and its progression.

9.4.2 *Diurnal variations during the active and break phases of monsoon*

Special observations on the diurnal cycle for 5 days (17–21 July 2021) during the active phase of monsoon and 3 days (1–3 July 2021) during the break monsoon period are presented in Figure 9.20. The commencement of the monsoon brings about marked changes in atmospheric circulation. It is manifested by establishing two well-defined monsoon circulations, the low-level monsoon jet (LLJ) in the lower troposphere and the tropical easterly jet (TEJ) in the upper troposphere.

The core speed and heights of both LLJ and TEJ undergo significant diurnal variations, and LLJ exhibits a varying pattern of core height. The core speed of monsoon low-level jet is about 10–15 ms^{-1}, whereas tropical easterly Jetstream in

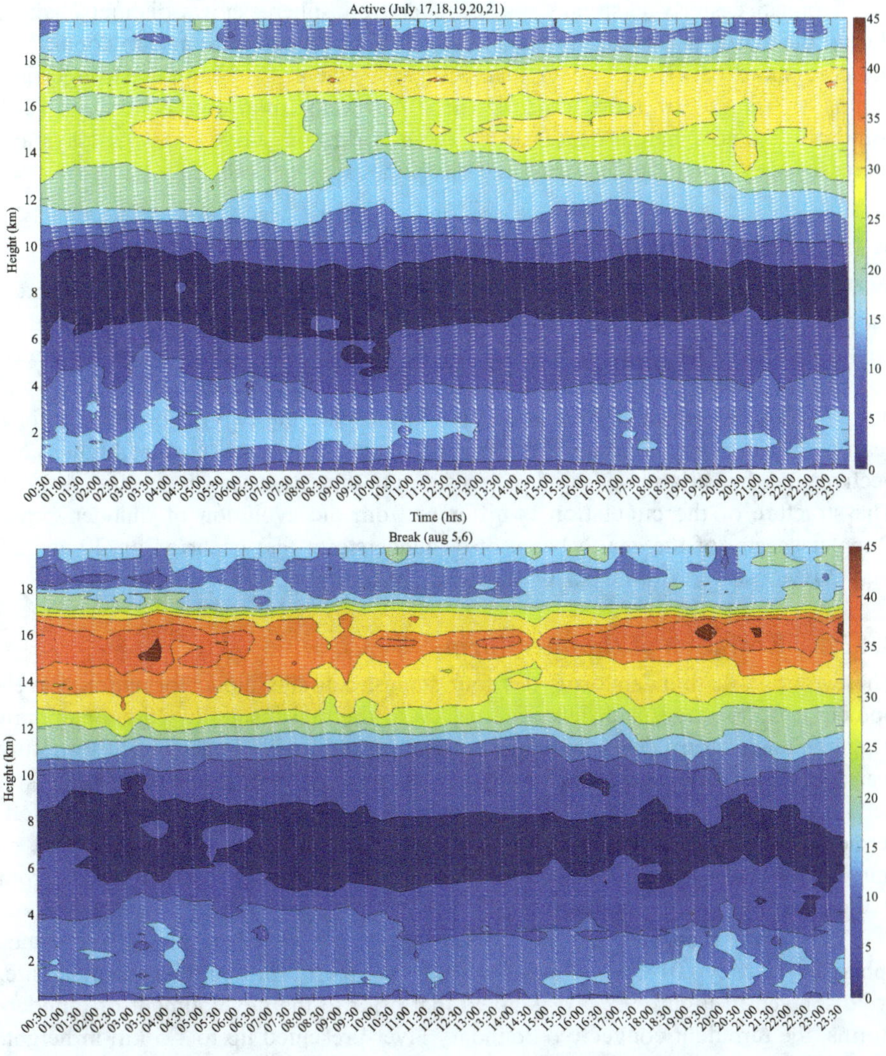

Figure 9.20 Diurnal variations of monsoon circulation during active and break phases

the upper troposphere attains a strength of about 40 ms^{-1}. In the early morning and late-night hours, LLJ core height is enhanced.

The diurnal variation differs from the monsoon's onset, active, break, and withdrawal phases. During the active phase of the monsoon, LLJ core height increases by more than 2 km, and the westerly domain also extends to an altitude of 10 km. In the break monsoon period, the LLJ core speed decreases, and the height of the core speed is around 1.5 km. The westerly domain in the lower atmosphere also shrinks during the break phase.

Tropical easterly jetstream in the upper troposphere shows distinct changes during the active and break monsoon periods. During active monsoon conditions, the core speed of TEJ is slower than 35 ms^{-1}, whereas it becomes stronger and attains a strength of more than 45 ms^{-1} in weak monsoon conditions [30]. Similarly, the core height decreases to about 15 km on active monsoon days and increases to 16 km altitude during the break monsoon period.

9.4.3 *Thunderstorms observations from ST radar*

WPR is an instrument that can continuously monitor the horizontal and vertical structure of wind information at high temporal and vertical resolution in a column within the convective region. It is the most efficient tool for diagnosing the development and evolution of thunderstorm activity. Pre-monsoon rain (April–May) and the post-monsoon period (October and November) are the two major thunderstorm seasons over Cochin. The thunderstorms during these two seasons are highly convective, and the variation of wind patterns during these two seasons is quite diverse. The structure of the circulation is different from the evolution of thunderstorms. Doppler spectra of the zenith beam of a thunderstorm that occurred on 20 and 25 May 2019 observed from CUSAT ST radar are illustrated in Figure 9.21.

9.4.3.1 Evolution of thunderstorm

The evolution of thunderstorms has four stages: cumulus cloud formation, development, mature stage, and dissipation. Due to thermal instability, the air updraft triggers thunderstorm development. As the air rises, it will condense and form tiny water droplets. If the updraft is strong, it will push the water droplets upward. At a certain point, the small water droplets combine to form a large droplet and fall, termed downdraft. An experiment is configured to observe the development of thunderstorms. In this experiment, the beam is configured to cover a height from 1.7 to 17 km.

The quiver plots of wind vectors in the entire troposphere and lower stratosphere on 10 May 2018, a day of a severe thunderstorm formed over the radar site, are displayed in Figure 9.22. It is noted that prior to the development of thunderstorms, the turbulent convective boundary layer presented up to 1.5 km in height. The horizontal winds in the upper troposphere are increasing their speed. The turbulent activity increases with height, even in the free troposphere, indicating that the atmosphere becomes even more unstable up to higher altitudes. Rain starts around 18:15 h LT and lasts up to around 21:00 h LT, reported $\sim$ 2 mm rainfall. The lower atmosphere returns to somewhat stable conditions during the post-rain event.

During the thunderstorm, the atmosphere is non-homogeneous and highly turbulent, and there is no symmetry between East–West and North–South beams. The Doppler signature of the monsoon rainfall from stratified clouds is entirely different from the precipitation during a thunderstorm, originating from highly convective clouds. The 205 MHz ST radar can detect the nature of rainfall which is unique to studying the precipitation pattern and its dynamics.

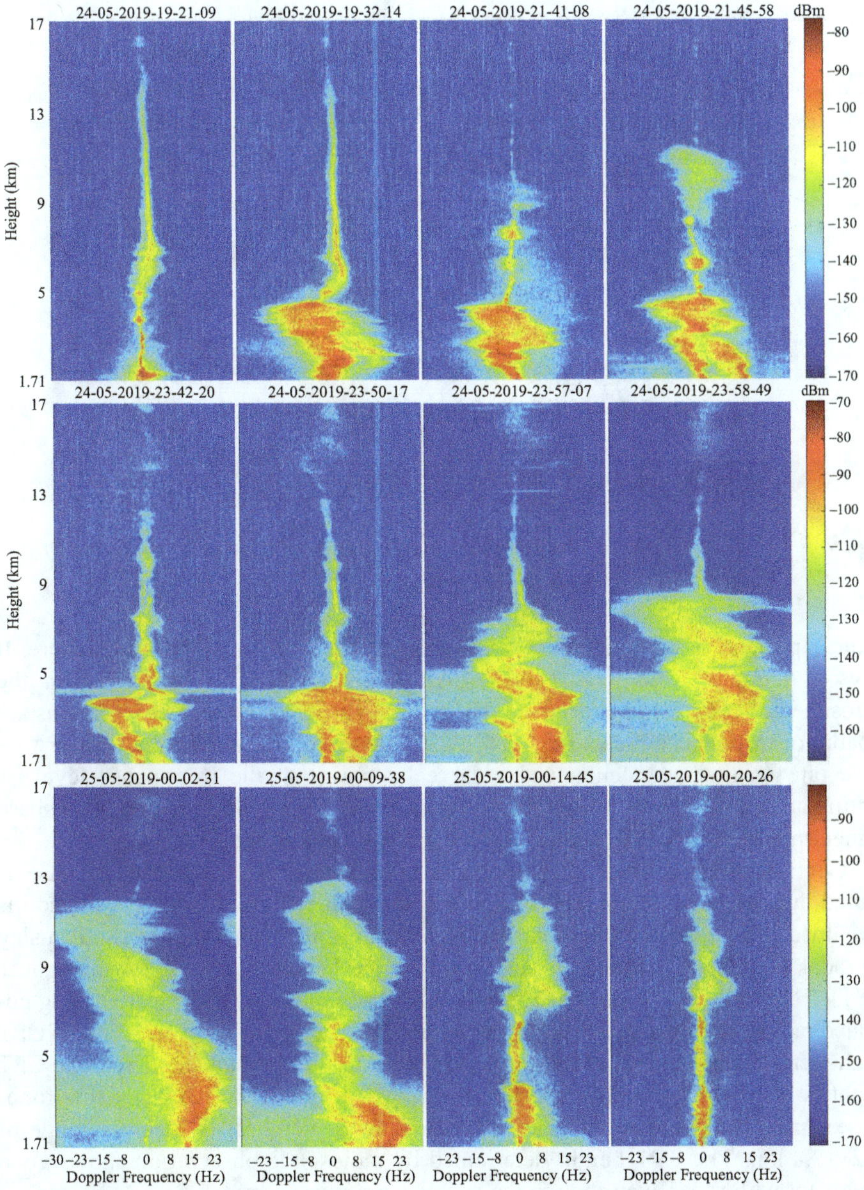

Figure 9.21 Doppler spectra of zenith beam of a thunderstorm occurred on 24 and 25 May 2019 observed from CUSAT ST radar

9.4.4 Determination of tropical tropopause using ST radar observations

The tropopause is the transition region that restricts the dynamic troposphere below and the stratified stratosphere above. This region is responsible for transporting

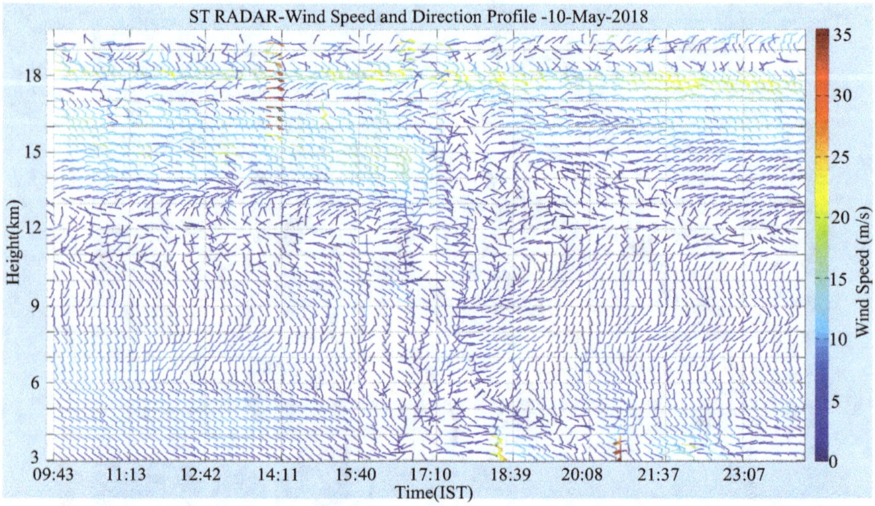

Figure 9.22 Quiver plot of horizontal winds during the day of occurrence of a severe thunderstorm

water vapor and other constituents in the upper troposphere to the stratosphere. It plays a dominant role in stratosphere–troposphere exchange processes in the atmosphere. Several studies showed that ground-based VHF radar observations are capable of tropopause measurements [31,32]. A sharp gradient in the temperature lapse rate results in an enhanced radar echo power at the tropopause level. A continuous radar reflectivity layer at the tropopause height is formed due to a sharp change in the refractive index around the tropopause altitude.

Continuous monitoring of the structure and variability of the tropopause is limited due to observational constraints. Using the 205 MHz ST radar, a method is developed to estimate the tropopause altitude from signal-to-noise ratio (SNR) using the radar [33]. SNR gradient is used to assess the height of cold point tropopause (CPT). The CPT altitude obtained from the STR observations is evaluated using co-located radiosonde measurements. Figure 9.23 shows a day-to-day variability of CPT height determined from radar and radiosonde over Cochin. A sudden increase in CPT height is associated with the monsoon onset over Kerala. The height of the tropopause during the summer monsoon season (June–September) is within a range of 16.2–17.8 km. The CPT height determined by radiosonde observation shows good agreement with radar even though the samples are minimum.

9.4.5 Tropopause variability and associated dynamics during monsoon

The variability of CPT height during the summer monsoon is studied, and the factors responsible for its variability are investigated. The presence of tropical easterly jet in the upper troposphere plays a role in modulating the CPT height. The TEJ core speed with the CPT height from the radar is evaluated. Figure 9.24

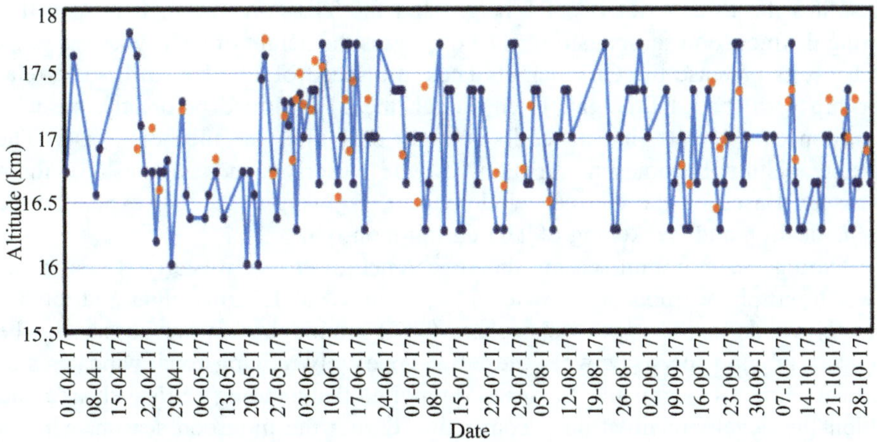

Figure 9.23 *Height of the tropopause was determined from radar (blue dots) and radiosonde (red dots) observations [33]*

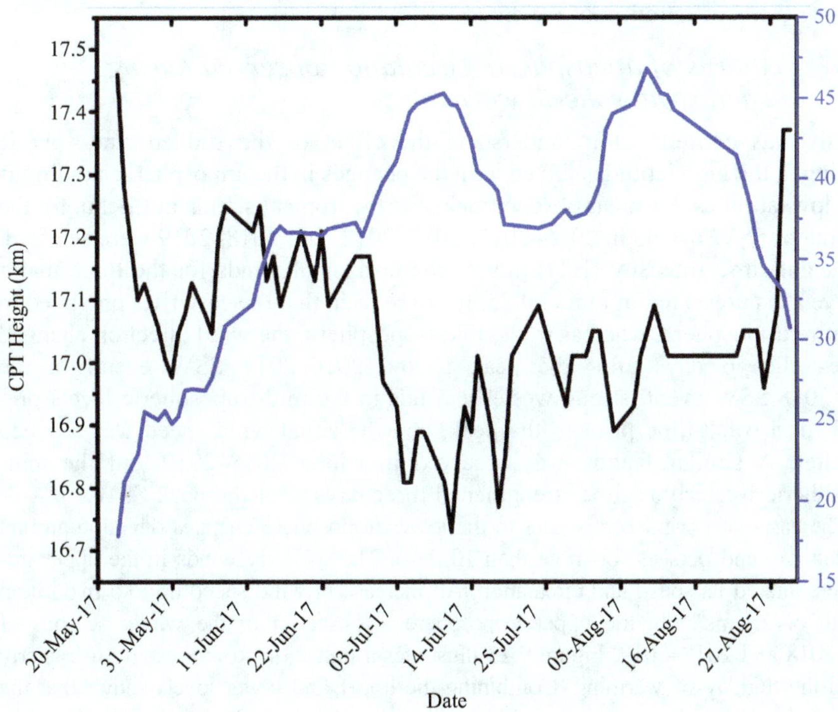

Figure 9.24 *Time series of CPT height (black line) and TEJ core speed (blue line) [33]*

illustrates the time series of CPT height and the TEJ core speed. It is seen that during the monsoon season, strong TEJ decreases the height of CPT, whereas weak TEJ results increase the CPT height. Thus, the speed of TEJ has an inverse relationship with the CPT height. Even though the TEJ formed before the onset of monsoon over Kerala, it reaches its strength only after the monsoon onset. The inverse relation becomes dominant when the easterly winds in the TEJ attain intensely. Based on the thermal wind concept, Fujiwara *et al.* [34] explained the intensification and weakening of TEJ during monsoon.

During the monsoon season, the appearance of strong westerly winds in the lower troposphere (monsoon low-level jet) centered at 1.5 km height and intense easterly winds in the upper troposphere (TEJ) around 14 km are observed. The intensity of wind speeds between these two oppositely moving zonal winds creates vertical shear. Goswami and Xavier [35] reported that a strong vertical shear could inhibit the development of deep convection during the monsoon season, while a weakened TEJ diminishes vertical shear strength. An assessment between TEJ core speed and ice cloud water path (ICWP) indicates that ICWP decreases as the strength of TEJ increases. The association between CPT height and ICWP is linear, which means that a higher CPT height coincides with a higher value of ICWP and vice versa. The elevated height of CPT for strong ICWP could be due to enhanced cloud latent heating. Thus, the study shows that TEJ affects tropopause height by changing the concentration of ice clouds.

9.4.6 Features of atmospheric circulation observed during sudden stratospheric warming

A study was carried out to understand the effect of the sudden stratospheric warming in the high latitudes linked with the changes in the atmospheric conditions in the low latitudes. ST radar observations over the tropical station at Cochin for the three major SSW events in 2016–2017, 2017–2018, and 2018–2019 were selected. The height–time–intensity (HTI) plot of the horizontal winds for the three major SSW years is presented in Figure 9.25. It can be seen that the easterlies prevailed in the entire troposphere, whereas in the mid-troposphere, the wind direction changed to westerlies 6 days after the peak in the 2016–2017 SSW event. In the 2017–2018 SSW event, strong westerly winds in the mid-tropospheric levels prevailed for a week-time prior to the peak, and the zonal wind speed was reduced thereafter. A similar feature was observed in winter 2018–2019, but the mid-tropospheric westerly regime strengthened three days after the peak SSW.

The east–west wind component in the lower troposphere changes direction around the peak day and persists for more than 10 days. The westerly winds in the upper troposphere altered its speed and circulation. An increase in wind speed up to a maximum strength of 25 ms^{-1} in the upper troposphere is observed in the winter seasons of 2017–2018 and 2018–2019. Figure 9.25 illustrates a transition from westerly to easterly around the midday of warming. Combining the upper and lower levels shows that the prevailing wind propagates downward but with reduced amplitude. It is evident from the wind profiler observations that wind reversal is not limited to the upper troposphere alone but extends to mid-tropospheric levels. It directly bears the lower troposphere and likely

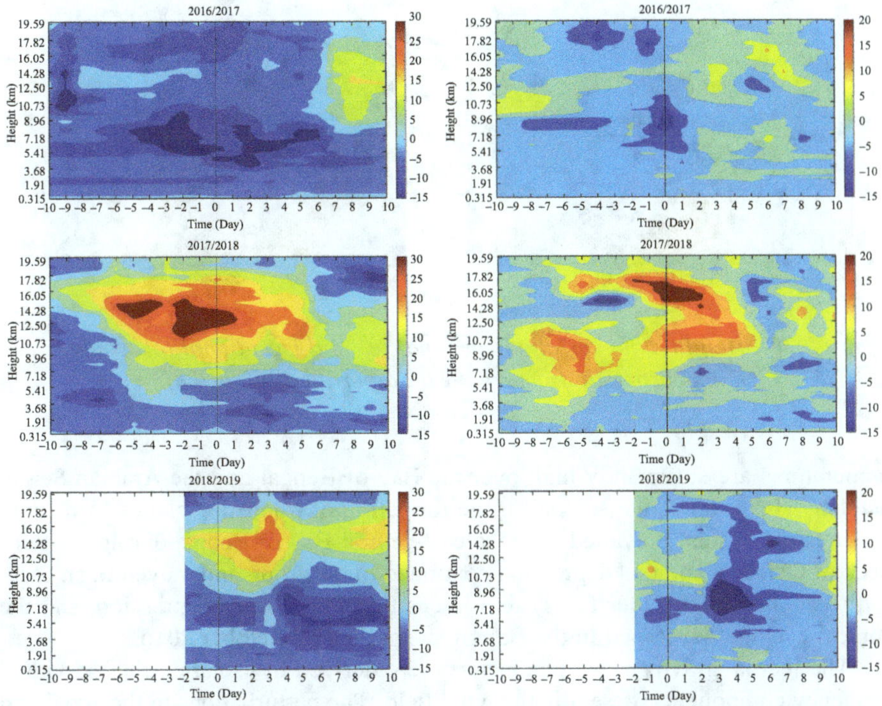

Figure 9.25 Height–time–intensity plot of zonal (Left panel) and meridional winds observed from ST radar observations at Cochin during sudden stratospheric events [36]

changes in surface weather parameters. Meridional winds associated with the sudden stratospheric warming in the higher latitudes are generally southerlies during SSW events. The observed changes in the meridional winds are minimal and not consistent [36].

9.4.7 Detection of inertia gravity waves

Cochin is a meteorologically sensitive region. The extensive Arabian Sea bounds this tropical coastal station in the east and the mountain ranges, known as the Western Ghats, in the east. This geographically unique site reveals the possibility of exploring and studying the waves generated by the orography and those originating through the development of tropical convective storms from the Arabian Sea or other nearby oceanic regions.

Kottayil *et al.* [37] presented the characteristics of inertia–gravity waves observed in the upper troposphere/lower stratosphere (UTLS) region (14–20 km) using the measurements from the 205 MHz WPR. The study identified the presence of inertia gravity waves (IGW) at 10°N. The WPR observations were made over 6 days during 22–27 October 2016 for the IGW studies. The data were averaged every 4 h over 144 h of observation. October is usually the time of receding southwest monsoons, and the vertical wind shears are less. Moreover, sea surface

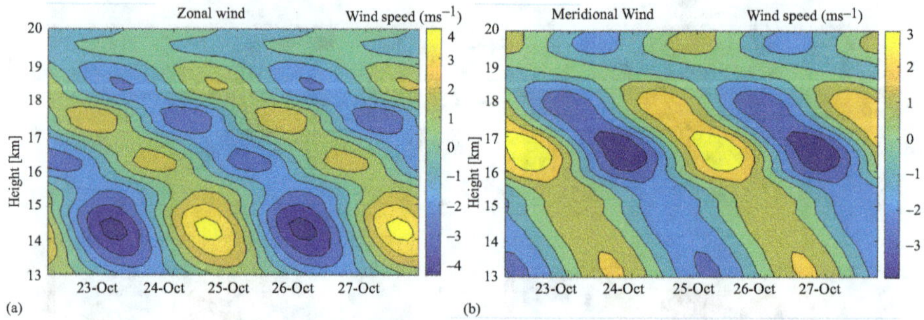

Figure 9.26 Height–time variations in zonal (a) and meridional (b) wind for a monochromatic frequency with a period of 72 h [28]

temperatures are sufficiently high over the Bay of Bengal and the Arabian Sea in October. This is an active season for the formation of tropical cyclones. A distinct low-pressure system prevailed over the east-central Bay of Bengal during the third week of October, and this depression continued to be strong and moved in an east–north–east direction. Later the system intensified into a deep depression, and the tropical cyclone developed in the Bay of Bengal on 25 October 2016.

The inertia–gravity wave parameters were derived by isolating the individual frequency components present in the wind field. The perturbations in the zonal and meridional wind corresponding to 72 h are analyzed, and the temporal evolution of the wind fluctuations is shown in Figure 9.26. The wave signatures and their time-height dependence are seen in this figure. A notable feature is the downward propagation of the phases of the zonal and meridional winds with a periodicity of around three days.

The inertia–gravity waves were detected between the altitude range of 14–20 km, and their properties were derived using the hodograph method. The intrinsic period of the wave is found to be 28.75 h. The IGWs were found to have vertical and horizontal wavelengths of 3.74 and 1,221 km, respectively, and were propagating upward at a rate of 2.5 km day^{-1}. It is found that IGW occurred when a deep depression prevailed over the Bay of Bengal and is a unique feature of IGWs occurring in the upper troposphere/lower stratosphere region.

9.5 Potential of 205 MHz ST radar for ionospheric observations

The 205 MHz ST radar at Cochin is unique; it is the first WPR operating in the far-VHF range of the 200 MHz category. It is principally designed to probe the zonal, meridional, and vertical components of winds in the troposphere and the lower stratosphere (315 m–20 km) with high resolution in time and altitude. Initially, the 205 MHz radar is alleged inept in detecting signals in the ionospheric region due to its higher frequency and low power than the conventional HF and MST radars.

However, it is noted that the 205 MHz ST radar can detect ionospheric irregularities in the E- and F-regions. Several papers were published [38,39]. These primary studies detected the ionospheric irregularities in the E- and F-regions and found submeter scale irregularities in the E-layer during summer, equinox, and winter.

9.5.1 Configuration of 205 MHz ST radar for ionospheric observations

The 205 MHz, ST radar at Cochin, is located at the geomagnetic coordinates: 1.88°N, 149.51°E, and geographic coordinates: 10.04°N, 76.33°E, declination: −1.71°, inclination: 7.49°. It constitutes 619 three-element Yagi–Uda antennas with a PAP of $1.33*10^8$ Wm^2 and can provide accurate 3D wind profiles for an altitude range of 315 m–20 km. Wind profiling radars detect the backscattered signals resulting from refractive index variations, and the corresponding Doppler shift is used to calculate air velocities. The main beam can be positioned at any angle in the range of ±30° in the off-zenith while in all 360° azimuth directions with 1° resolution. The system has a main lobe beamwidth of 3.2° and an overall gain of 35 dBi with a side lobe level of −19.7 dB.

Plasma density irregularities in the ionosphere are generally elongated along the geomagnetic field. We used the International Geomagnetic Reference Field Model (IGRF12) to compute the zenith angle at which the antenna beam is perpendicular to the geomagnetic field. We theoretically calculated 8° as the tilt required for satisfying the perpendicularity condition. Subsequently, we experimented with verifying it by tilting between 6° and 10°. Other experiment configurations and parameters used are presented in Varadarajan *et al.* [38,39].

9.5.2 Significant ionospheric observations using the 205 MHz ST radar

Figure 9.27 illustrates the E- and F-region irregularity echoes observed over Cochin on (a) 20 March 2021, F-region irregularity echoes detected during the intense storm event on (b) 23 March 2021, (c) 25 March 2021, and (d) 29 March 2021. The reflections from the E-region are mainly from 90 to 110 km, and such events are rarely observed beyond 110 km. Most observed layers have either maintained a constant altitude level or descended over time, as evidenced in the height–time–intensity (HTI) plots. Both continuous and quiet period (QP) echoes are observed during the daytime, evening, and night. The daytime lower altitude QP echoes are identified up to 102 km with periodicity varying from several seconds to 8–10 min. The night-time spread-F observed with CUSAT radar is bottom-type mainly events. All the events are within an altitude range of 250–450 km and last less than 30 min. No topside irregularities have been observed so far. The bottom-type spread-F events appeared either in a patchy or streak-like manner.

Most night-time F-region irregularities have preferentially occurred during the post-sunset hours (19:00–19:59 LT). The gradual dominance of the westward electric field reduces the effects of Rayleigh–Taylor (RT) instability and the probability of occurrence of equatorial plasma bubble (EPB) throughout the post-

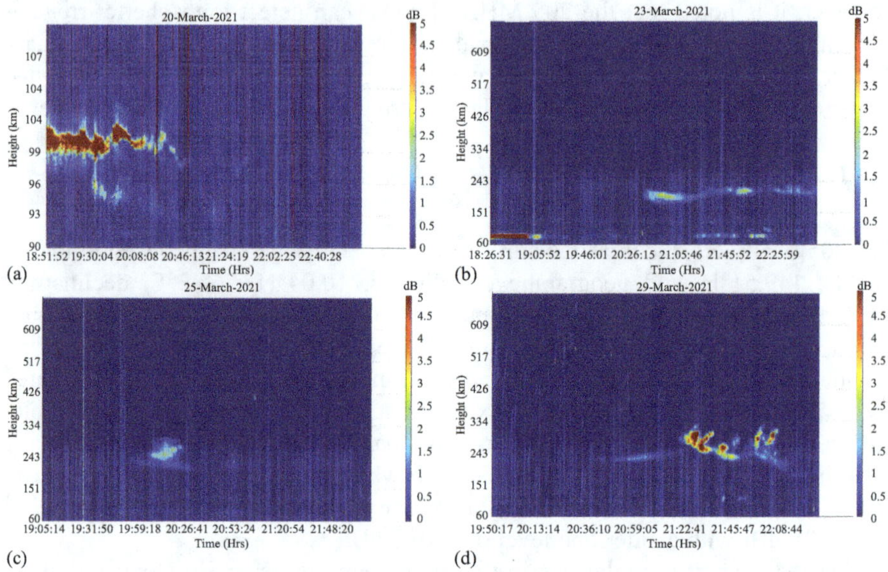

Figure 9.27 During the intense storm event 19–29 March 2021, an E-region irregularity echo was observed over Cochin on (a) 20 March 2021, and F-region irregularity echoes were observed on (b) 23 March 2021, (c) 25 March 2021, and (d) 29 March

midnight hours. It is estimated that night-time irregularities are most commonly observed around 200–250 km heights with minimum possibility for long duration (~ 200–250 min) events. In the radar observations of ESF irregularities at Cochin, short-duration events that lasted 50 min have a high predominance. Only 2.5% of the observed irregularities have a vertical width within the range of 250–300 km, while the vertical widths of 52.5% of the estimated events fall within 1–100 km.

F-region irregularity echoes over Cochin observed amid the moderate storm event during 2–30 November 2020 are displayed in Figure 9.28. ESF/EPB irregularities during midnight differ from those during post-sunset hours. In the near-equatorial sector of Cochin, the dominant westward electric field prevents the F-layer from ascending to higher heights by the upward $E \times B$ drift. In turn, upwelling plumes are mainly caused by RT instability. Apart from RT instability, a significant factor contributing to the upwelling plumes in the post-midnight sector is the convergence of equatorward meridional wind streams near local midnight that can raise the F-layer height off-equatorial region.

The 205 MHz ST radar is suitable for observing ionospheric irregularities during geomagnetic storms. Continuous and quasi-periodic E-region echoes are observed over Cochin in connection with counter electrojet (CEJ)-like conditions. The plume-like equatorial spread-F (ESF) irregularity with topside extension is also associated with the enhanced eastward electric field due to delayed disturbance dynamo electric field in the recovery phase of the intense storm. These findings substantiate the

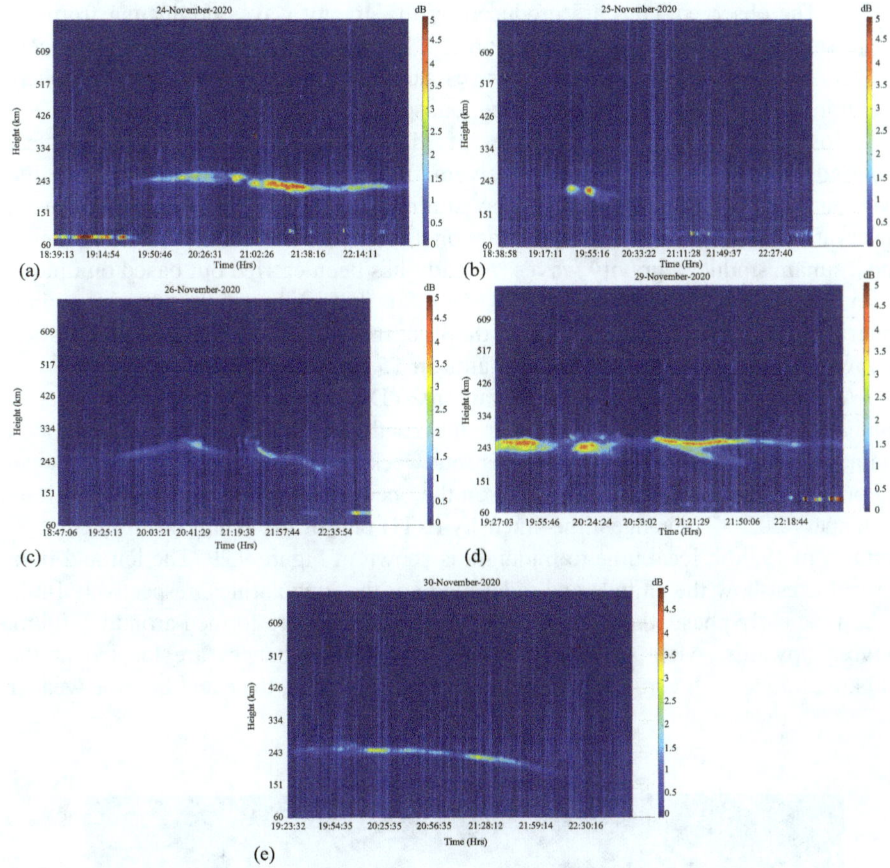

Figure 9.28 The observed ionospheric F-region irregularity echoes over Cochin on (a) 24 November 2020, (b) 25 November 2020, (c) 26 November 2020, (d) 29 November 2020, and 30 November 2020 observed amid the moderate storm event: 2–30 November 2020 [39]

potential of 205 MHz VHF radar in monitoring near-equatorial ionospheric irregularities over the Indian equatorial ionization anomaly (EIA) trough sector amid intense geomagnetic storms. Experiments are in progress to characterize the irregularity in E- and F-regions and estimate the drift velocities with the VHF radar.

9.5.3 Detection of ionospheric disturbances with volcanic eruption and Tsunami

Several studies indicate that tsunamis can generate atmospheric gravity waves that propagate up to ionospheric heights and produce perturbations in the electron densities in E and F regions. Observational and theoretical studies established that the gravity waves produced by tsunamis cause the formation of traveling ionospheric disturbances

(TIDs). The observed TIDs are produced by the gravity waves originating from the ocean surfaces where the Tsunami is formed. Even with wave heights of a few centimeters in the deep ocean, Tsunami waves can propagate upward in the atmosphere and ultimately cause perturbations in the ionosphere's total electron content (TEC).

Most of the observational studies of TIDs during the Tsunami events were reported based on ground-based receivers from the Global Navigation Satellite Systems (GNSS) and ionosonde. A few studies are available about the continuous observational evidence from VHF radars on TIDs associated with volcanic eruption and Tsunami-induced gravity waves. A study has been carried out based on unique observational evidence from the 205 MHz far-VHF radar of the traveling ionospheric disturbances associated with the occurrence of Tonga volcanic eruption followed by the incidence of the Tsunami on 15 January 2022.

A submarine volcanic eruption began on 20 December 2021 at Hunga Tonga, an uninhabited volcanic island situated in the southern Pacific Ocean. The eruption attained a tremendous peak intensity after four weeks, on 15 January 2022. In the Asian region, near Japan, a Tsunami of 1.2 m on the open ocean was reported at 23.15 h on 15 January 2022. The range–time–intensity (RTI) plot of the zenith beam tilted at 8° north from 19.00 h local time to midnight is shown in Figure 9.29. The left and right vertical axes show the altitude and radar range of the irregularities, respectively [40].

In the early phase, the sub-metre-scale irregularities are formed around 280 km, moving upwards. After an hour, the ionospheric disturbances are located at the 400 km altitude with vertical movement. In the later phase, the event became weaker

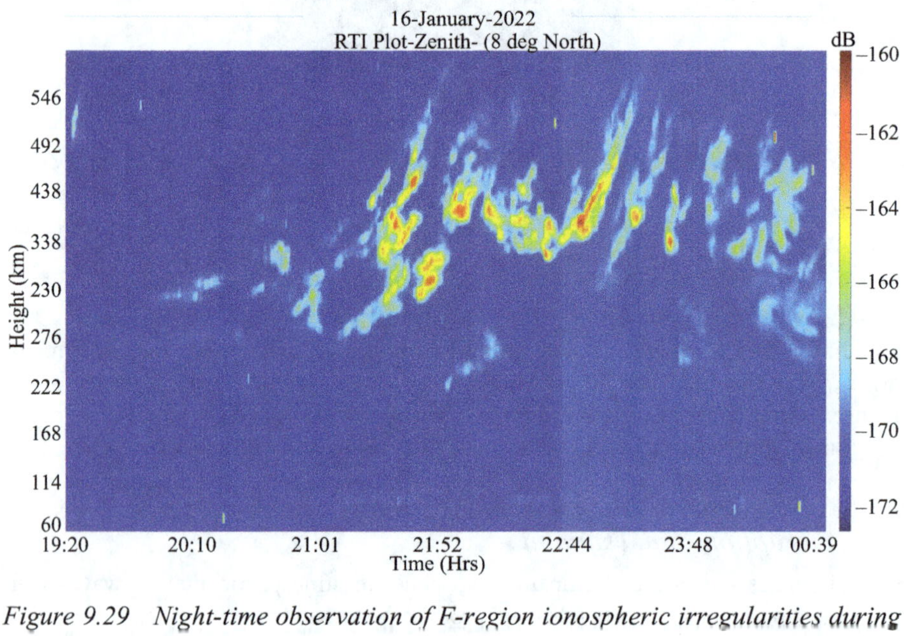

Figure 9.29 Night-time observation of F-region ionospheric irregularities during the Tonga volcanic eruption and associated Tsunami on 15 January 2022 [40]

and showed a wavy nature, almost disappearing after midnight. This observational evidence of night-time traveling ionospheric disturbances detected in the F-region from the 205 MHz VHF radar is further evidenced from GNSS observations. These sub-metre-scale ionospheric disturbances are the first-time observations from a VHF radar at 205 MHz in the near-equatorial site at Cochin, India.

9.6 Other observations by 205 MHz ST radar

In addition to the atmospheric studies, the 205 MHz ST radar installed at Cochin has extensive applications. This unique radar provides enormous information on the lower and upper atmospheric features in the tropical region. Some of the exciting observational evidence from the CUSAT ST radar is presented in the following sub-sections, which open windows for further investigation.

9.6.1 Detection of electric charges in the upper atmosphere

Thunderstorms are exciting phenomena that produce lightning from the electrically charged cloud and flash to the upper atmosphere. Flashes observed above the thunderclouds are generally called transient luminous events (TLE) [41,42]. There are many TLE categories, of which two are Sprites and Blue jets. Sprites are flashes of red light that occur in a fraction of a second above thundercloud, extending up to 100 km in altitude. Sprites are not directly connected to the thundercloud. Bluejets are flashes that begin from the thundercloud to a maximum height of 45 km. Figure 9.30 shows the pictorial representation of TLE concerning the altitude. An experiment was configured to check the possibility of observing TLEs using ST

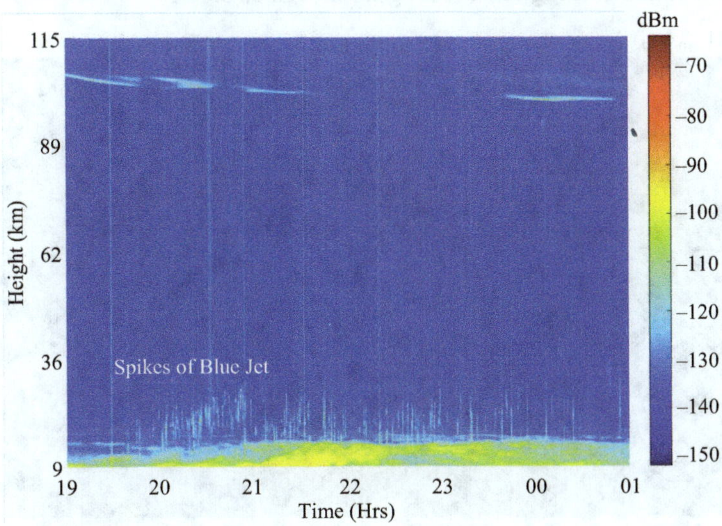

Figure 9.30 SNR plot of north beam with off-zenith angle eight on a thunderstorm event up to an altitude of 115 km

radar. The investigation was configured only for the North beam with an off-zenith angle of 8°, height coverage from 9 to 115 km, and a range resolution of 135 m. The experiment is configured for 8° north to understand whether the thunderstorm creates any ionosphere disturbances.

Radar was operated on a thunderstorm event with a time resolution of 13 s. The SNR plot shown in Figure 9.30 displays random spikes-like structures observed up to 35–40 km, and it is detected only at the time of thunderstorms. The vertical lines marked in the SNR plot are due to the system noise.

Figure 9.31 shows three different Doppler spectra of the experiment. The first Doppler spectrum is a normal condition without spikes, and the remaining Doppler spectrum shows the signature of the Blue jet. At the same time, some ionospheric disturbances were observed at an altitude of 90–100 km. Further observations have to be conducted to do a detailed analysis of the properties and features of the Blue jet.

9.6.2 Flight tracking using ST radar

Observations were made to track the Doppler signature to open up interdisciplinary research areas of tracking radars and signal processing. A strong reflected signal is observed along with the wind Doppler whenever a flight moves above the radar, and this signature is the noise-to-wind estimation. A high-speed zenith beam experiment is configured with a time resolution of 3 s to track a flight.

Figure 9.32 shows the Doppler spectrums where scattering from the flight is observed. A strong positive Doppler is observed at 14 km, i.e., a flight is approaching the radar and moving away from the radar with a negative Doppler. The exact height, speed, direction, etc., can be interpreted by developing a new digital signal processing algorithm for flight tracking. It can have applications in airports for flight tracking and locating its movements.

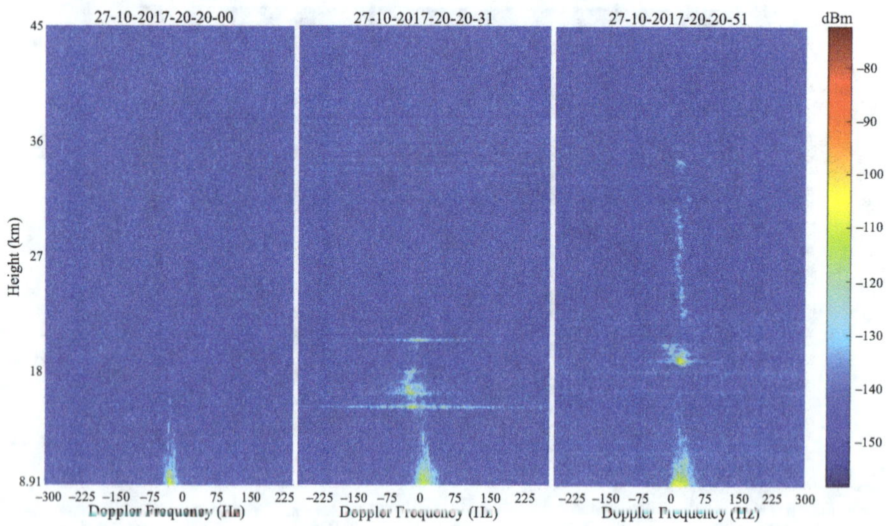

Figure 9.31 Doppler spectrum without Blue jet (1st) and with Blue jet (2nd and 3rd)

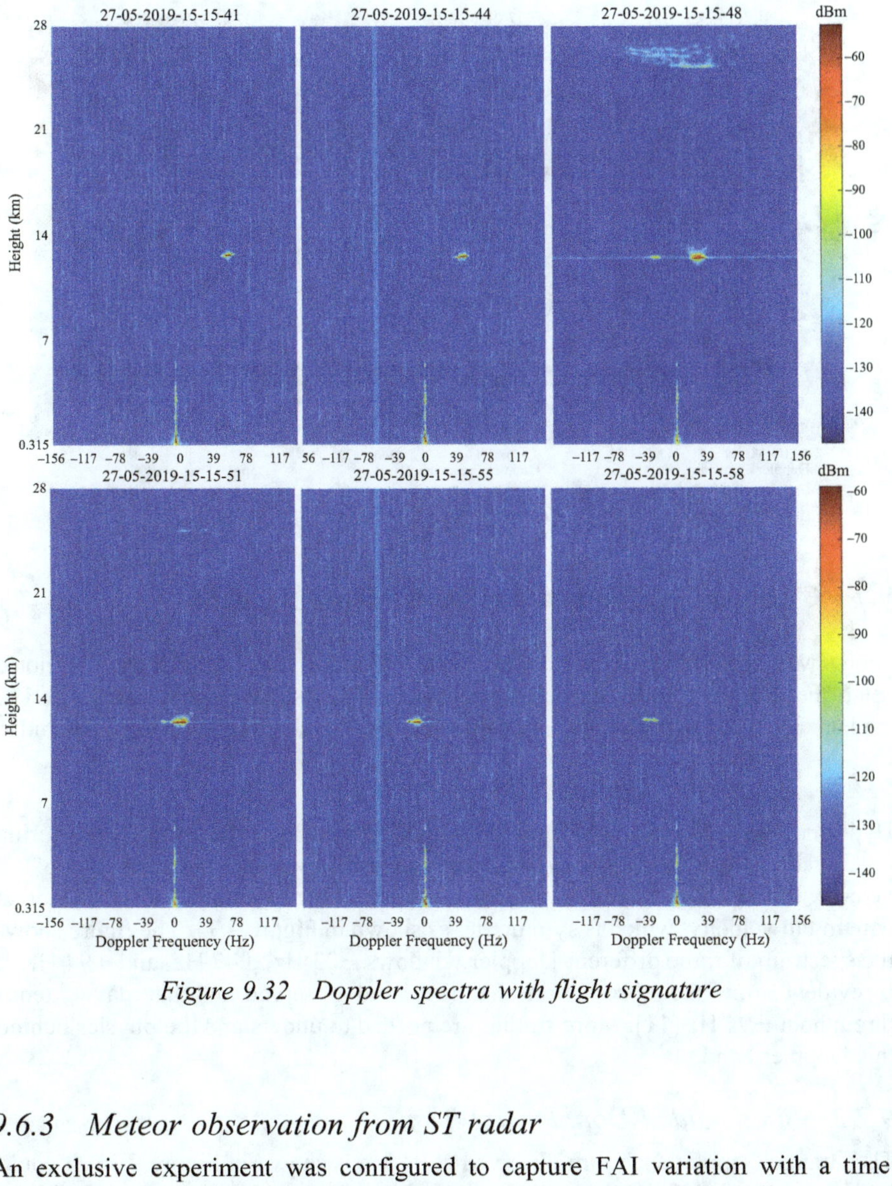

Figure 9.32 Doppler spectra with flight signature

9.6.3 Meteor observation from ST radar

An exclusive experiment was configured to capture FAI variation with a time resolution of 2 s and point toward north with off-zenith 8°. Accidentally a meteor passing was captured after sunset at 96 km. In the subsequent spectrum, the ionization trails were observed and decayed in less than 10 s. Figure 9.33 shows the observed spectrum of the meteor trail. From the first spectrum at 96 km, ionization trails are found, and a slanting line is observed, which is identified as a meteor. In the following spectrum, there is no signal after 2 s, and the meteor could disappear from the line of sight. These types of sudden echoes are seen irrespective of day and night.

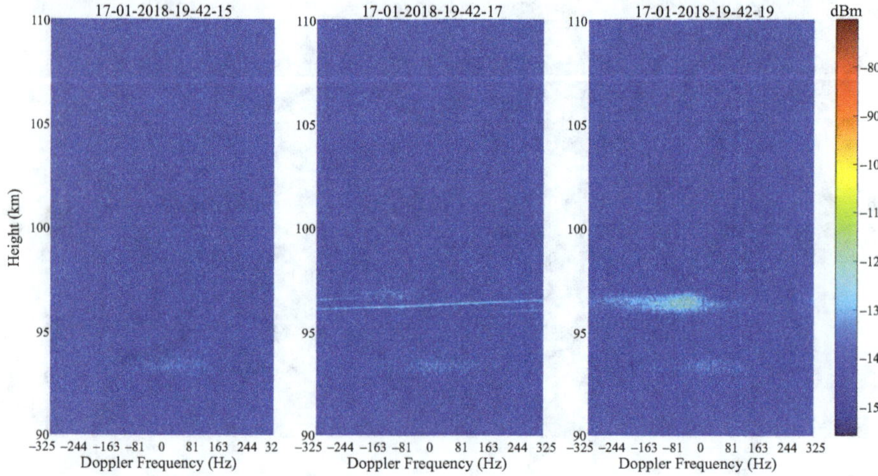

Figure 9.33 Doppler spectrum observed for meteor passing

9.7 Curious observational evidence from ST radar

Along with atmospheric observations, some intriguing observations are also noted using the 205 MHz radar in the tropics. Detailed investigations are necessitated to explain the mechanism for this unidentified observational information from this radar.

9.7.1 Rotational Doppler at melting layer

Hydrometeors fall from higher heights and melt to form water droplets during precipitation. The fall velocity of hydrometeors is small compared to the fall velocity of the water droplet. A band is observed at the exact time of water droplet gain in fall velocity, which is symmetric, as shown in Figure 9.34. The figure shows the spectrum of three different Doppler windows ±23 Hz, ±47 Hz, and ±94 Hz. It is evident from the observation that the band observed on that day extends throughout ±94 Hz [11]. More studies are needed to understand the physics behind this Doppler band.

9.7.2 Single-sided Doppler at 90 km

The strange Doppler spectrum observed at 94 km is shown in Figure 9.35. It could be a meteor trail, but the Doppler spectrum is concentrated only to the left side of the spectrum with a sharp cut-off at 0 Hz. More investigations have to be done to understand the single-sided spectrum.

9.7.3 Anomalous precipitation-like pattern observed from the tropopause level

Water vapor intrusion is observed at the thunderstorm from a lower height to a peak of 12–14 km. In this case, the presence of moisture is moving down from the

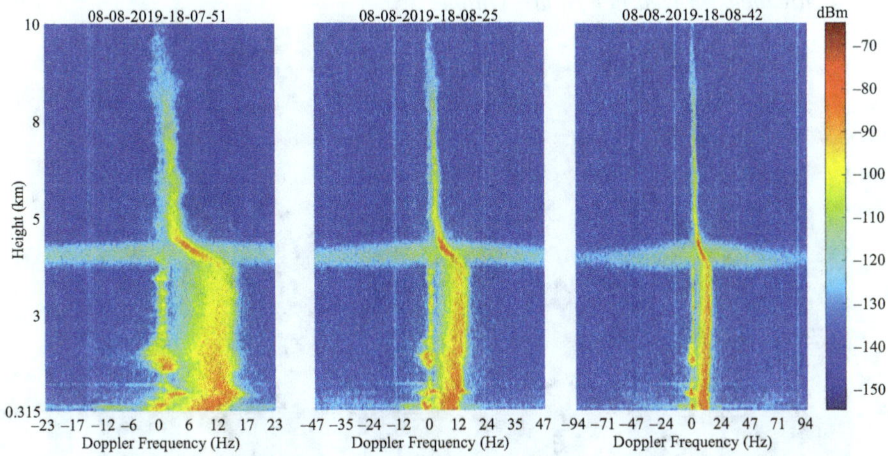

Figure 9.34 Rotational Doppler cap observed at different Doppler windows of ±23 Hz, ±47 Hz, and ±94 Hz

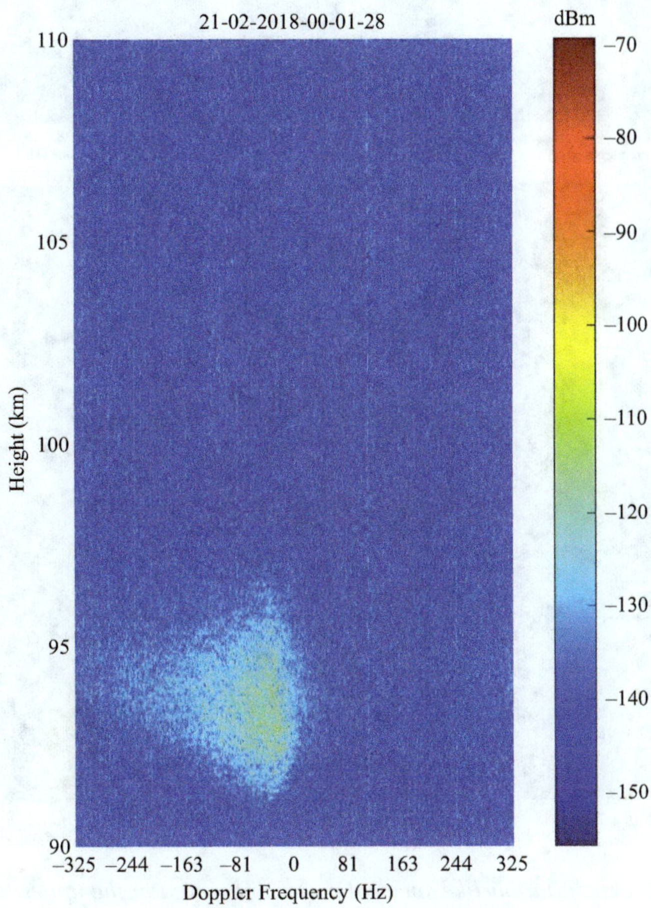

Figure 9.35 Single-sided spectrum observed at the height of 94 km

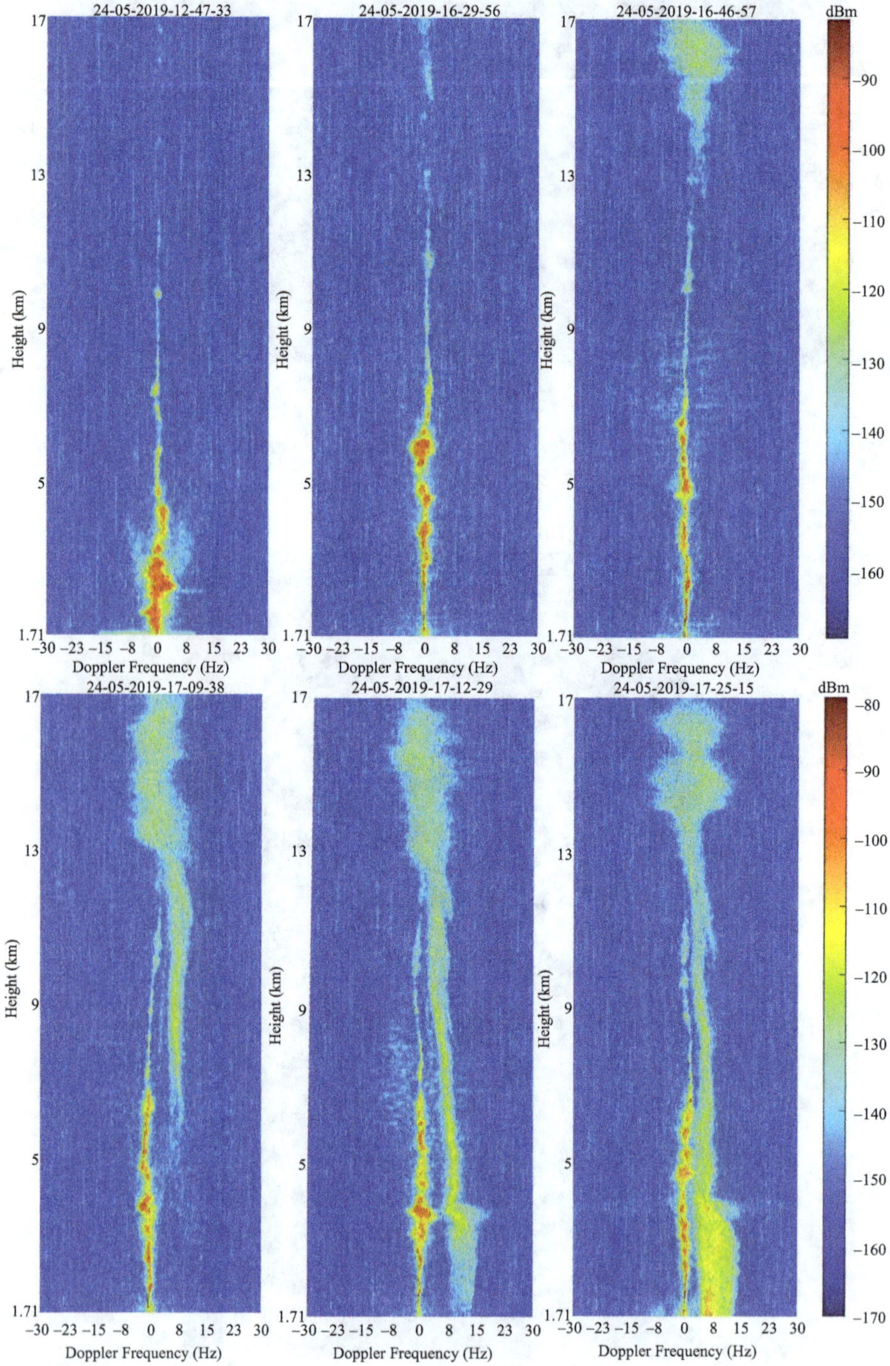

Figure 9.36 Precipitation-like spectral pattern observed in the upper troposphere

altitude of 17 km and falling to a height of 4 km like hydro-meteor signature and melt to form water droplets and precipitation happens. The evolution of the precipitation-like pattern from higher altitudes is depicted in Figure 9.36. Initially, there was no signal at the height of 17 km; as time progressed, the signal strength increased, fell, and melted to form rain [11]. A detailed study is needed to understand the mystery of how this much humidity got trapped at 17 km height.

There are more rooms in the upper atmosphere to investigate [43]. This can be easily achieved by effectively using the enormous power and potential of the 205 MHz ST radar for further studies in atmospheric science and space applications.

9.8 Application of 205 MHz radar for radio astronomical studies

The sizable effective collecting area (~400 sq. m) of the CUSAT radar underlines its significant potential for use in radio astronomical studies, with its 5 MHz bandwidth and system noise temperature of about 600 K used in their "passive" receiving mode. Such a "receive only" mode should be free from interference and transmission signals, including any switching signals relevant to its radar mode. This mode would readily enable routine observations of one or more strong radio astronomical sources, providing useful and independent calibration of the ST radar system. Preliminary tests using the ST-radar antenna with a portable receiver from the Raman Research Institute, Bengaluru, have shown encouraging results, even in some interfering switching signals.

To achieve a clean passive mode operation and extend the beam steering range to larger zenith angles, open a range of attractive opportunities for sensitive radio astronomical pursuits when used with a suitable back-end receiver system. More specifically, these include, but are not limited to: (1) study of solar bursts at 205 MHz, using dynamic spectral intensity measurements with high time and spectral resolutions; (2) stand-alone monitoring of about half a dozen strong known pulsars at this wavelength for changes of both intrinsic and extrinsic origins, apparent in their received radiation; and (3) active participation in the Indian SWAN (the sky wave, ionization trails are found tch array network, a recent strategic initiative in radio astronomy), as one of its unique stations for some of its campaign mode observations.

9.9 Future perspective

9.9.1 Imaging of the atmosphere

ST radar consists of 619 numbers of TRM and one channel DBS receiver. With this system, we can probe the atmosphere in one direction at a time. The studies using middle and upper atmosphere radar (MUR), Japan, have proved that atmospheric imaging is possible by increasing the number of receiving channels. MU radar has demonstrated 3D imaging with Capon's method [44], providing information about structure morphology. MU radar consists of 25 receiving channels. The upgraded system of Indian MST at Gadanki has got the facility of atmospheric imaging with 16 channel receivers. Using a 25-channel, angular imaging has been demonstrated.

The existing ST radar at Cochin can also be upgraded to get the atmospheric imaging by introducing a 91-channel receiver system.

Using the existing ST radar, the digital beamforming technique can be demonstrated without modification. ST radar consists of 619 TRM connected to a digital receiver using 3-stage power combiner/divider network. The first stage power combiner/divider is 1:8. Each port in the first stage power combiner/divider is connected to TRMs to form the basic sub-cluster. The eighth port is unused and terminated using a 50-Ohm terminator. Likewise, we have 91 numbers of unused ports. As per the datasheet of the 1:8 power combiner/divider used in the system, the port-to-port isolation is a minimum of 44–20 dB. The power received calculation of ST radar at different stages of power combiner/divider shows that the power received at the first stage of power combiner/divider is a minimum of −123.5 to −8.5 dBm. It is clear that at the eighth port, we will get excellent atmospheric echoes. Connecting 91 digital receivers to the unused ports will make it possible to convert the analog atmospheric echoes to the digital domain. An algorithm must be developed to adjust the time delay for implementing the digital beam forming in the digital domain.

3D imaging could detect detailed layer patterns in the troposphere and stratosphere and possible Kelvin–Helmholtz instability billow structures, obtaining valuable angular information from the interior of the radar beam. The study of the orientation of radar echo structures and their relation to wind and shear should permit a more detailed analysis of the mechanisms controlling the observed systems. 3D imaging is used helpful resolve the non-uniform echoes in the angular and the range dimension. It can show thin layers as they appear at any location in the range, isolated from the range shift of the layer across the beam width.

Acknowledgments

The authors would like to thank the research team of the Advanced Centre for Science Radar Research for their excellent work carrying out using the 205 MHz ST radar facility. We are happy to acknowledge the Science Engineering Research Board (SERB), Department of Science and Technology (DST), Government of India for the support in establishing the ST radar at CUSAT, Cochin. Thanks to M/s Data Patterns India (P) Ltd, Chennai, India, for designing, developing, installing, and maintaining the 205 MHz ST radar. Sustenance support from the Ministry of Earth Science, Government of India and Cochin University of Science and Technology is highly acknowledged. The authors thank Mr Rejoy Rebello for converting the manuscript into LaTeX format.

List of acronyms

A	ampere
ADC	analog to digital converter
A_e	effective aperture area (m^2)
ALADIN	atmospheric laser Doppler instrument

AWS automatic weather station
BPSK binary phase shift keying
$\mathbf{C}_n^2$ refractive index structure constant
CPT cold-point tropopause
CST computer simulation technology
CUSAT Cochin University of Science and Technology
DAC digital-to-analog converter
DBS Doppler beam swinging
dB decibel
dBi decibel related to isotropic gain
dBm decibel referenced to milliwatts
DC direct current
DDC direct digital converter
DST Department of Science and Technology
EIA equatorial ionization anomaly
EPB equatorial plasma bubble
ERA ECMWF re-analysis
ESF equatorial spread-F
e partial pressure of water vapor
FAI field-aligned irregularity
FFT fast Fourier transform
F/B front-to-back lobe ratio
f_d Doppler frequency
FPGA field programmable gate array
G array gain
GPS global positioning system
GRAW German-made Radiosonde
GNSS global navigation satellite system
HF high frequency
HPBW half-power beam width
HTI height–time intensity
Hz Hertz
ICWP ice cloud water path
IGRF International Geophysical Reference Field model
IGW inertia–gravity waves
IMD India Meteorological Department
IRPHA intensification of research in high priority area
ISM Indian Summer Monsoon

I/Q	in-phase/quadrature-phase
K	Kelvin
kHz	kilo Hertz
kW	kilo Watt
L_0	outer scale of turbulence
LLJ	low-level jet
LT	local time
M	potential refractive index
M0	reflected power moment
M1	mean Doppler moment
M2	Doppler width moment
M3	SNR moment
M4	noise-level moment
MERRA	Modern Era-Retrospective Analysis for Research and Applications
MHz	mega Hertz
MOK	Monsoon Onset over Kerala
MST	mesosphere–stratosphere–troposphere
MUR	middle and upper atmosphere radar
NASA	National Aeronautics and Space Administration
NCEP	National Centre for Environmental Prediction
NZC	number of zero crossing
N_c	critical plasma density
N_e	number density of electrons
n	refractive index
P	atmospheric pressure
PC	personal computer
PAP	power aperture product
PRF	pulse repetition frequency
P_r	received power
P_t	transmitted power
QP	quiet period
R	range (m)
RPC	radar-processing computer
RF	radio frequency
RT	Rayleigh–Taylor
RTI	range–time–intensity
R_r	receiver
SAM	spaced antenna mode

SERB	Science Engineering Research Board
SFDR	spurious-free dynamic range
ST	stratosphere–troposphere
STR	stratosphere–troposphere radar
SMAD	scaled median absolute deviation
SNR	signal-to-noise ratio
SSW	sudden stratospheric warming
SWAN	sky watch array network
TEC	total electron content
TEJ	tropical easterly jet
TID	traveling ionospheric disturbances
TKE	turbulent kinetic energy
TLE	transient luminous events
TRM	transmission–receiver module
TV	television
T_x	transmitter
T/R	transmit/receive
UHF	ultra-high frequency
UVW	zonal, meridional, and vertical wind components
u	zonal (east–west) wind
VHF	very high frequency
V_r	radial velocity
V_{re}	east radial velocity
V_{rw}	west radial velocity
V_{rn}	north radial velocity
V_{rs}	south radial velocity
V_{rz}	zenith radial velocity
VSWR	voltage-standing wave ratio
VTI	voltage-tunable inductor
UTLS	upper troposphere lower stratosphere
v	meridional (north–south) wind
W	Watt
WPR	wind profiler radar
WPC	wireless planning and coordination
w	vertical (upward–downward) wind
2D, 3D	two-dimensional, three dimensional
Δn	variation in refractive index
ΔR	range resolution (m)

η	volume reflectivity
η_r	volume reflectivity factor
λ	wavelength

References

[1] Fukao S and Hamazu K. *Radar for Meteorological and Atmospheric Observations*. New York, NY: Springer; 2013.

[2] Röttger J. Reflection and scattering of VHF radar signals from atmospheric refractivity structures. *Radio Science*. 1980;15(2):259–276. https://agupubs.onlinelibrary.wiley.com/doi/abs/10.1029/RS015i002p00259.

[3] Hardy KR and Gage KS. In: Atlas D (ed.), *The History of Radar Studies of the Clear Atmosphere*. Boston, MA: American Meteorological Society; 1990. https://doi.org/10.1007/978-1-935704-15-7_17.

[4] Hocking WK, Rottger J, Palmer RD, *et al. The History of Radar in Atmospheric Investigations.* Cambridge: Cambridge University Press; 2016.

[5] Balsley BB and Gage KS. On the use of radars for operational wind profiling. *Bulletin of the American Meteorological Society*. 1982;63:1009–1018.

[6] Battan LJ. Radar observation of the atmosphere. *Quarterly Journal of the Royal Meteorological Society*. 1973;99(422):793–793. https://rmets.onlinelibrary.wiley.com/doi/abs/10.1002/qj.49709942229.

[7] Chadwick RB. A feasibility study on using wind profilers to support space shuttle launches. NASA: NASA; 1984. Report 3861.

[8] Mohanakumar K, Kottayil A, Anandan VK, *et al.* Technical details of a novel wind profiler radar at 205 MHz. *Journal of Atmospheric and Oceanic Technology*. 2017;34:2659–2671.

[9] Turtle AJ and Baldwin JE. A survey of galactic radiation at 178 Mc/s. *Monthly Notices of the Royal Astronomical Society*. 1962;124(6):459–476. https://doi.org/10.1093/mnras/124.6.459.

[10] Ben-yosef N, Tirosh E, Weitz A, *et al.* Refractive-index structure constant dependence on height. *Journal of the Optical Society of America*. 1979;69:1616–1618.

[11] Titu Samson PM. *Design development calibration and validation of 205 MHz wind profiler radar [dissertation]*. Cochin University of Science and Technology, Kerala, India; 2020.

[12] Balsley BB. In: Atlas D, (ed.). *Clear-Air Radar Technology: Panel Report*. Boston, MA: American Meteorological Society; 1990. pp. 282–286. https://doi.org/10.1007/978-1-935704-15-7_24.

[13] Samson TK, Kottayil A, Manoj MG, *et al.* Technical aspects of 205 MHz VHF mini wind profiler radar for tropospheric probing. *IEEE Geoscience and Remote Sensing Letters*. 2016;13(7):1027–1031.

[14] Samson TK, Babu B, Anandan VK, *et al.* Phased array of 619-element Yagi–Uda antenna for Wind Profiler Radar at Cochin University of Science and

Technology. In: *2019 URSI Asia-Pacific Radio Science Conference (AP-RASC)*; 2019. pp. 1–4.

[15] Balanis CA. *Antenna Theory: Analysis and Design*. New York, NY: Wiley-Interscience; 2005.

[16] Frank G and Peter Gölz RH. Weather Radar Calibration and Testing Using the Moon as Reference Target. 2011. https://ams.confex.com/ams/35Radar/webprogram/Paper191872.html.

[17] Kottayil A, Mohanakumar K, Samson T, *et al.* Validation of 205 MHz wind profiler radar located at Cochin, India, using radiosonde wind measurements. *Radio Science*. 2016;51(3):106–117.

[18] Mohanakumar K, Santosh KR, Mohanan P, *et al.* A versatile 205 MHz stratosphere–troposphere radar at Cochin – scientific applications. *Current Science*. 2018;20:2459–2466.

[19] Sivan C, Rakesh V, Abhilash S, *et al.* Evaluation of global reanalysis winds and high-resolution regional model outputs with the 205 MHz stratosphere–troposphere wind profiler radar observations. *Quarterly Journal of the Royal Meteorological Society*. 2021;147:2562–2579.

[20] Dhanya R, Pradeep A, Abhiram Nirmal CS, *et al.* A novel hybrid approach for the identification of clear air and rain echoes in the 205 MHz ST radar spectrum. *IEEE Geoscience and Remote Sensing Letters*. 2022;19:1–5.

[21] Radhakrishna BP and Rao TN. Statistical characteristics of multipeak raindrop size distributions at the surface and aloft in different rain regimes. *Monthly Weather Review*. 2009;137:3501–3518.

[22] Fukao S, Hashiguchi H, Yamamoto M, *et al.* Equatorial atmosphere radar (EAR): system description and first results. *Radio Science*. 2003;38(3):1053. https://agupubs.onlinelibrary.wiley.com/doi/abs/10.1029/2002RS002767.

[23] Kottayil A, Prajwal K, Devika MV, *et al.* Assessing the quality of Aeolus wind over a tropical location (10.04 N, 76.9 E) using 205 MHz wind profiler radar. *International Journal of Remote Sensing*. 2022;43(9):3320–3335. https://doi.org/10.1080/01431161.2022.2090871.

[24] Pattanaik DR and Satyan V. Fluctuations of tropical easterly jet during contrasting monsoons over India: a GCM study. *Meteorology and Atmospheric Physics*. 2000;75:51–60.

[25] Narayanan S, Kottayil A, and Mohanakumar K. Monsoon low-level jet over the gateway of Indian summer monsoon: a comparative study for two distinct monsoon years. *Meteorology and Atmospheric Physics*. 2016 12;128.

[26] Kottayil A, Xavier P, Satheesan K, *et al.* Vertical structure and evolution of monsoon circulation as observed by 205-MHz wind profiler radar. *Meteorology and Atmospheric Physics*. 2019;132:531–545.

[27] Xavier A, Manoj MG, and Mohankumar K. On the dynamics of an extreme rainfall event in northern India in 2013. *Journal of Earth System Science*. 2018;127:1–13.

[28] Xavier A, Kottayil A, Mohanakumar K, *et al.* The role of monsoon low-level jet in modulating heavy rainfall events. *International Journal of Climatology*. 2018;38:e569–e576.

[29] Kottayil A, Xavier A, Xavier P, *et al.* Evolution of large-scale factors influencing extreme rainfall over South Western Coast of India. *International Journal of Climatology*. 2021;11.

[30] Rona MS AA and Manoj MG. Diurnal cycle of winds as observed with a 205 MHz VHF radar during the onset and northward progression phases of summer monsoon 2021. *Presented at the National Symposium on Tropical Meteorology (TROPMET-2022)*; 2022. Indian Institute of Science Education and Research (IISER), Bhopal, India during 29 November–2 December 2022.

[31] Jain AR, Rao YJ, Rao PB, *et al.* Preliminary observations using ST mode of Indian MST radar: detection of the signature of the tropopause. *Journal of Atmospheric and Solar-Terrestrial Physics*. 1994;56:1157–1162.

[32] Satheesan K and Murthy BVK. A study of tropical tropopause using MST radar. *Annales Geophysicae*. 2005;23:2441–2448.

[33] Nithya K, Kottayil A, and Mohanakumar K. Determining the tropopause height from 205 MHz stratosphere troposphere wind profiler radar and study the factors affecting its variability during monsoon. *Journal of Atmospheric and Solar-Terrestrial Physics*. 2019;182:79–84.

[34] Fujiwara M, Xie SP, Shiotani M, *et al.* Upper-tropospheric inversion and easterly jet in the tropics. *Journal of Geophysical Research: Atmospheres*. 2003;108(D24):4796. https://agupubs.onlinelibrary.wiley.com/doi/abs/10.1029/2003JD003928.

[35] Goswami BN and Xavier P. ENSO control on the south Asian monsoon through the length of the rainy season. *Geophysical Research Letters*. 2005;32:L18717:1–4.

[36] Remya R, Manoj MG, Rakesh V, *et al.* Influence of high latitude sudden stratospheric warming on tropical weather: observations from a 205 MHz stratosphere troposphere radar and surface meteorological parameters. *Earth and Space Science*. 2021;8(4):e2020EA001418. E2020EA001418 2020EA001418. https://agupubs.onlinelibrary.wiley.com/doi/abs/10.1029/2020EA001418.

[37] Kottayil A, Satheesan K, Mohankumar K, *et al.* An investigation into the characteristics of inertia gravity waves in the upper troposphere/lower stratosphere using a 205 MHz wind profiling radar. *Remote Sensing Letters*. 2018;9:284–293.

[38] Varadarajan R, Manoj MG, Mohankumar K, *et al.* Detection of submeter scale irregularities in the low-latitude ionospheric E layer using high VHF radar at 205 MHz. *Journal of Geophysical Research: Space Physics*. 2019;124(5):3752–3760. https://agupubs.onlinelibrary.wiley.com/doi/abs/10.1029/2018JA026302.

[39] Varadarajan R, Haridas S, Manoj MG, *et al.* Ionospheric nighttime F-region irregularities during geomagnetically quiet conditions as observed with 205 MHz VHF radar at an Equatorial Trough Station, Cochin. *Journal of Geophysical Research: Space Physics*. 2022a;127(6):e2021JA030129. E2021JA030129 2021JA030129. https://agupubs.onlinelibrary.wiley.com/doi/abs/10.1029/2021JA030129.

[40] Varadarajan R, Haridas S, Sivan C, *et al.* Impact of the Hunga Tonga–Hunga Ha'apai volcanic eruption on the changes observed over the Indian near-equatorial ionosphere. *Advances in Space Research*. 2022b;70(8):2480–2493. https://www.sciencedirect.com/science/article/pii/S0273117722005786.

[41] Rycroft MJ. Introduction to the physics of sprites, elves and intense lightning discharges. In: Füllekrug M, Mareev EA, and Rycroft MJ (eds.), *Sprites, Elves and Intense Lightning Discharges*. Dordrecht: Springer Netherlands; 2006. pp. 1–18.

[42] Williams E and Yair Y. The microphysical and electrical properties of sprite-producing thunderstorms. In: Füllekrug M, Mareev EA, and Rycroft MJ (eds.), *Sprites, Elves and Intense Lightning Discharges*. Dordrecht: Springer Netherlands; 2006. pp. 57–83.

[43] Seo BC, Krajewski WF, and Mishra KV. Using the new dual-polarimetric capability of WSR-88D to eliminate anomalous propagation and wind turbine effects in radar-rainfall. *Atmospheric Research*. 2015;153:296–309. https://www.sciencedirect.com/science/article/pii/S0169809514003603.

[44] Hassenpflug G, Yamamoto M, Luce H, *et al.* Description and demonstration of the new middle and upper atmosphere radar imaging system: 1-D, 2-D, and 3-D imaging of troposphere and stratosphere. *Radio Science*. 2008;43.

Chapter 10
An integrated future US weather radar architecture for aviation

Mark E. Weber[1], John Y.N. Cho[1], Henry G. Thomas[1] and James M. Kurdzo[1]

10.1 Introduction

Aviation operations are particularly sensitive to adverse weather and, as a result, radar's capability to monitor the location, movement, and intensity of storms is invaluable to the safety and efficiency of flight. Radar was introduced in the late 1940s to aid air traffic controllers in maintaining safe separation of aircraft, and the "precipitation clutter" present on their displays was quickly recognized as valuable in providing weather advisories. By the 1980s, the transition to "digital" air traffic control radars such as the airport surveillance radar, model 9 (ASR-9) and Air Route Surveillance Radar, Model 4 (ARSR-4) enabled the Federal Aviation Administration (FAA) to include parallel processing channels that provided quantitative information on the intensity of precipitation echoes for display to controllers.

The US National Weather Service or NWS (then known as the Weather Bureau) also began work to exploit radar operationally in the 1940s, using surplus radars donated by the US Navy. This led to the deployment of the Weather Surveillance Radar 1957 (WSR-57) and the subsequent WSR-74. In parallel with the FAA's ASR-9 and ARSR-4 acquisitions, the NWS fielded the WSR-88D, with advanced signal processing, automated meteorological feature detection algorithms, and Doppler capabilities. Both the FAA and the Department of Defense (DoD) were partners on the WSR-88D acquisition and had substantial input to the radar network siting, scan strategies, and the underlying technology.

A series of fatal commercial aircraft accidents in the 1970s and 1980s due to microburst[*] wind shear at airports resulted in extreme pressure on the FAA to field a warning system. The FAA determined that a Doppler radar-based system was the preferred approach at many airports and that the locations and scanning patterns of

[1]MIT Lincoln Laboratory, USA

[*]Microbursts are intense, small-scale downdrafts and outflows forced by precipitation loading and evaporation, which result in rapid, dangerous changes in lift for aircraft at low altitude.

the WSR-88D would not provide the necessary wind shear detection reliability and warning timeliness. Thus, after a fast-track system development program, FAA began deploying the Terminal Doppler Weather Radar (TDWR) to 45 large airports and subsequently modified 34 ASR-9s with a weather systems processor (WSP) to provide wind shear warning services at medium-sized airports.

By the early 1990s then, the architecture for the current US national operational meteorological radar network was established. Over the subsequent decades, this network has been fully built-out and has undergone significant enhancements such as the upgrade of the WSR-88D to provide dual-polarization observations. Its basic characteristics remain unchanged, however:

1. **National scale, high-quality meteorological radar observations** are provided by the 156 radars comprising the US WSR-88D network.† Coverage greater than 1,500 m above ground level (AGL) is nearly continuous in the eastern two-thirds of the contiguous United States (CONUS) although significant gaps are present in the mountainous western United States, Alaska, and Hawaii. Within the planetary boundary layer (PBL), which typically extends only to 500 m (AGL), coverage is much sparser due to the large spacing between radar sites (~200 km) and the effects of terrain. The volumetric update rate for the WSR-88D is low owing to the large number of beam positions scanned by its 1° pencil beam and its use of relatively long coherent processing intervals (CPI) to assure high-quality meteorological variable estimates (see, e.g., Chapter 4 in this book for details on data quality assurance). WSR-88D observations are used operationally by NWS forecasters to provide public severe weather and flood warnings and to support water resource management. Controller scopes at regional Air Route Traffic Control Centers (ARTCCs) primarily display WSR-88D observations rather than the ARSR-based precipitation reflectivity product because of their superior data quality. FAA and DoD air traffic controllers utilize the WSR-88D data tactically to provide pilot advisories on the location and movement of storms. Its slow volume scan update and processing latency have been somewhat problematic for this application since indicated positions for fast-moving storms may be inaccurate‡ [1,2]. Regional and national-scale WSR-88D displays of storm location, vertically integrated liquid (VIL) water content, and radar echo tops have proven to be very valuable in monitoring flow disruptions to commercial aviation operations and in establishing traffic management initiatives (TMIs) used to mitigate the operational impact [3].
2. **Airport wind shear advisories** are provided at 80 large- and medium-sized US airports by a dedicated TDWR or an ASR-9 retrofitted with the WSP. Both systems emphasize very short-range wind shear observations (0–10 km from the airport corresponding to the ranges at which landing or departing aircraft

†In addition, there are WSR-88Ds in Japan (Kadena) and South Korea (Kunsan and Camp Humphreys).
‡The latency issue will be mitigated when the NextGen Weather Processor replaces the Weather and Radar Processor (WARP) as input to the ARTCC En Route Automation Modernization (ERAM) display.

are within a few hundreds of meters of the ground), although the meteorological processing algorithms operate out to approximately 60 km to support broad area weather situational awareness. TDWR and WSP products are used by air traffic control tower (ATCT) controllers to advise landing or departing aircraft on near-airport wind shear and other thunderstorm hazards. Terminal radar approach control (TRACON) facility traffic management personnel utilize TDWR and WSP information on airport wind shear and the location/movement of storms affecting "gateposts" and other constrained airspace to manage operations strategically [4,5]. Finally, the NWS has implemented Supplemental Product Generator (SPG) interfaces for all TDWRs which provide data from each TDWR to the appropriate NWS forecast office. We have interviewed a number of NWS forecast offices on their use of TDWR data and uniformly, they confirm the value of TDWR data in issuing severe weather warnings [6,7]. This is due to TDWR's frequent (once per minute) low-elevation angle scans, its narrow antenna beam (0.5°), and near-airport siting that often places the radar closer to densely populated areas than the WSR-88D.

3. **Real-time precipitation reflectivity contours** on ATCT and TRACON controller's scopes are generated by "weather processing channels" on the approximately three hundred US ASR-8s, ASR-9s, and ASR-11s. These are quantized based on the six video integrator processor (VIP) reflectivity thresholds (18, 30, 38, 44, 50, and 57 dBZ). ASRs utilize a broad, cosecant-squared elevation beam pattern to detect a 1 square meter cross-section aircraft to approximately 6 km (AGL) at 100 km range, and to 11 km (AGL) at 50 km range [8]. Thus, their precipitation reflectivity measurements reflect a vertically integrated measure of hydrometeor number density and size. ASR antennae rotate in azimuth through a full 360° in 4.8 s. Their weather processors integrate data over 6 revolutions to produce a new weather reflectivity map each 30 s. TRACON controllers use this rapid-update, low-latency storm intensity information on their air traffic control (ATC) radar scopes to provide safety-essential weather advisories and "vectors" to aircraft. ASR-based weather reflectivity data are subject to significant ground-clutter contamination during super-refractive propagation conditions (anomalous propagation or AP) although processing algorithm upgrades have provided some mitigation for this problem. The NWS has experimented with ingest of ASR six-level weather data into its operational facilities as gap-fillers in locations where WSR-88D radar coverage is poor. This has not been particularly successful owing to data quality impacts associated with ASRs' broad elevation beams and coarsely quantized reflectivity outputs.

In this chapter, we discuss the opportunity for a future, highly integrated national meteorological radar network, optimized to jointly support the observing missions discussed above. The next section describes the development of the FAA's airport-based weather radar capabilities and our work to enhance their data quality and operational applicability. As discussed in Section 10.3, this led us to propose a future national operational radar architecture where surveillance radars at more than two hundred civilian and military airfields also provide high-quality meteorological observations, vastly increasing coverage in the PBL relative to today's WSR-88D network. We show that resulting benefits for thunderstorm forecasting, severe weather warnings, and quantitative precipitation estimation (QPE) can be quantified and that these would be significant. Our original concept was based on multifunction phased array radar (MPAR) technology, which would have enabled all of the radars in the network to simultaneously perform both aircraft and weather surveillance functions using a common technical architecture. In Section 10.4, we extend this work to show that, even without introduction of MPAR technology (and resolving the associated technical challenges), significant opportunities to enhance both aviation and public weather services can be realized through a holistic approach to radar siting and the observation requirements levied on the individual radars.

10.2 Airport meteorological radar development

10.2.1 Terminal Doppler Weather Radar

The TDWR was developed and fielded on a remarkably short time-scale in response to the air carrier accidents mentioned above. The University of Chicago meteorologist T. Fujita correctly determined that an intense, small-scale, and short-lived phenomena he called a "microburst" was responsible. Subsequent field programs in the Midwestern and high-plains United States[§] confirmed that microbursts were readily detected using Doppler radar and documented their characteristics. Because the locations of planned WSR-88D installations were not close enough to many airports to reliably detect shallow microburst outflows and because the radar's scan strategies did not revisit low-elevation angles frequently enough, the FAA determined that dedicated airport wind shear detection systems would be required.

A quantitative cost–benefit analysis (CBA) led by representatives from Massachusetts Institute of Technology Lincoln Laboratory (MIT LL), the National Center for Atmospheric Research (NCAR), Martin-Marietta Corporation, and the FAA evaluated life-cycle costs and operational benefits for TDWR and two alternative wind shear detection systems – ASR-9 WSP and the anemometer-based Low-Level Wind Shear Alert System (LLWAS). This analysis led to deployments of 45 TDWRs at large airports subject to frequent wind shear risk exposure. Thirty-four ASR-WSPs were fielded at intermediate-sized airports and LLWAS was used at locations where a radar-based solution was not cost effective.[¶] The preference

[§]Significant field programs focused on microbursts included the Northern Illinois Meteorological Research on Downburst (NIMROD) project [9], the Joint Airport Weather Studies (JAWS) project [10], and the Classify, Locate, and Avoid Wind Shear (CLAWS) project [11]. These projects provided the scientific and operational foundation for the development of the TDWR.

[¶]In addition, larger LLWAS anemometer networks – LLWAS Network Expansion – were deployed at nine TDWR-equipped airports, and their outputs were integrated with the TDWRs'.

for radar-based systems was motivated by the benefits provided by their capability to observe outflow boundaries ("gust fronts") approaching an airport that affect the runway configuration, and the movement of storms towards an airport or important airspace resources.

In parallel, a wind shear systems users' group – comprised of representatives from the airlines, air traffic controller, and technical communities – developed detailed concepts of operations for the systems which in turn established many of their technical requirements. Leveraging experience from the Classify, Locate, and Avoid Wind Shear (CLAWS) experiments at Stapleton Airport in Denver [11], this group determined that fully automated detections of wind shear would be presented to tower controllers as an alphanumeric warning specifying the type of wind shear (airspeed gain or loss), its intensity, and its location in relation to airport runways. The controllers would then read these advisories verbatim to pilots of affected aircraft who would decide on the appropriate action. Because of the significant hazard of microburst wind shear, airlines have advised their pilots to terminate approaches or departures through a microburst whenever feasible. This concept of operations (CONOPS), therefore, requires that the automated wind shear detections have low latency, a very high probability of detection, and a low false alarm rate.

TDWR's design was optimized for this very specific mission: rapid-update observations of the low-altitude wind field within approximately 10 km of the airport. High data quality is crucial to assure that its automated wind shear detection algorithms will not be compromised by low signal-to-noise, ground clutter, or range-ambiguous weather returns. As an operational meteorological radar, it is remarkable for its angular resolution. At its 5-cm wavelength operating frequency, TDWR's 7.6-m diameter reflector antenna provides a 3-dB beamwidth of 0.5°, half that of the WSR-88D. This allows its main beam to observe the most-intense microburst winds, typically within a few hundred meters of the surface, while minimizing the illumination of ground clutter. TDWRs initially performed sector scans, centered on the airports they protected, so as to maximize the observation update rate for near-surface winds and microburst precursor signatures aloft. The initially deployed signal- and data-processing algorithms focused on the quality of the very short-range, "on-airport" wind observations. As an example, the radar pulse repetition interval (PRI) was adaptively adjusted to move range-ambiguous weather echoes away from the airport's wind shear Areas Noted for Attention (ARENA), sometimes at the expense of data quality in other areas where severe weather might affect more strategic flight operations decisions, and/or safety of the general public.

A TDWR prototype was evaluated operationally at Denver, CO (1988), Kansas City, MO (1989), and Orlando, FL (1990–1992) to refine both the meteorological detection algorithms and the user decision support concepts. These test locations represented dramatically different wind shear environments – "dry" (i.e., low radar reflectivity) microbursts at Denver, microbursts from severe thunderstorms with strong vertical variation in the horizontal wind fields as well as tilted reflectivity and updraft structures in Kansas City, and numerous, rapidly developing microbursts from air mass thunderstorms in Orlando. Commendably, the FAA supported this ongoing test and refinement program in parallel with the TDWR acquisition so that the fielded system would best meet user needs in the diverse meteorological and operational environments in which it was to be deployed. McCarthy *et al.* [12] provide an excellent summary of the TDWR development program.

Enhancements over the years following TDWR's initial deployment have expanded the operational applications it can support. Rather than sector scans, its antenna now performs full 360° plan position indicator (PPI) tilts at each elevation angle, still returning to the surface once per minute, which is critical for low-altitude wind shear detection. Although this scan pattern change was implemented by the FAA primarily to alleviate wear on the radar's azimuth drive system, it has significantly increased the value of TDWR data for NWS operations. In fact, based on NWS forecaster experience using the TDWR, the scan patterns employed by the WSR-88D have been modified to include multiple surface scans within each full-volume scan. This enhances severe weather warning performance as discussed in Section 10.3.2.2.

On the signal processing front, improvements to the TDWR algorithms used to mitigate range- and/or Doppler-ambiguity impacts have also substantially enhanced its value for severe weather surveillance over its entire instrumented range. Innovative methods for adaptive transmission and processing of multi-PRI, phase-coded, or staggered-PRI waveforms [13] have been implemented and, along with improvements to the originally fielded ground clutter suppression algorithms, have further improved the performance of TDWR's wind shear detection algorithms. Importantly, by assuring more accurate radial velocity estimates, and by minimizing obscuration from range-ambiguous returns (see, e.g., Figure 10.1), these algorithms also enhance the value of TDWR observations for NWS forecasters. In a subsequent section, we model the benefits of TDWR data for improving public severe weather warnings. Interestingly, these approach the estimated value of the wind shear accident aversion capability for which the system was originally deployed.

10.2.2 Airport surveillance radar

ASRs operate in a preferred band (2.7–2.9 GHz) for meteorological observations and transmit a suitable high-peak power[1] (1.1 MW), high-range resolution (150 m) pulse. It is not surprising then that the robust capability of these radars to detect precipitation echoes was exploited by the FAA to provide real-time overlays of storm location and intensity on controller's scopes, and subsequently for wind shear detection.

A key difference between ASRs and meteorological radars is their elevation antenna pattern. Modern ASRs utilize two vertically displaced feedhorns which illuminate the reflector so as to produce the broad (5° half-power beamwidth) "low" and "high" elevation patterns shown in Figure 10.2. For both aircraft

[1]The ASR-11 utilizes a solid state transmitter with much lower peak-power, but provides equivalent pulse-energy (i.e., sensitivity) by transmitting a long (90 μs), frequency-modulated pulse. Pulse-compression on receive is used to achieve the required range resolution.

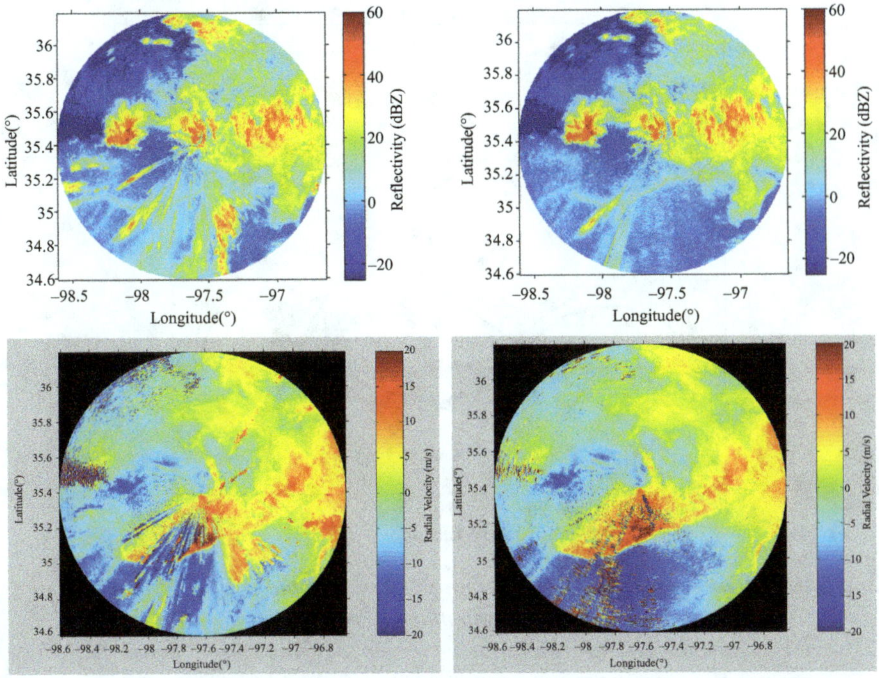

Figure 10.1 Legacy (left) and enhanced (right) TDWR processing [13]. Adaptive pulse waveforms and processing significantly reduce the impact of range-ambiguous echoes on both reflectivity (upper) and radial velocity (lower) fields.

detection and precipitation reflectivity processing, the high beam is used at short range to reduce ground clutter illumination. Beyond that range,** signals from the low beam are processed.

10.2.2.1 Six-level weather processor

Weber [14] used pencil-beam Doppler weather radar, combined with on-airport ground clutter measurements to simulate and assess the output of the then-under-development ASR-9 six-level weather processing channel. He found that its adaptively selected, high-pass Doppler filters should normally prevent ground clutter break-through on controller's displays and that the filter selection method would result in insignificant bias in the reflectivity estimates, even given the intense ground clutter environment experienced by ASRs at many US airports. The simulations showed that even though ASR-9 CPIs are very short (8 or 10 pulses at an average PRI of 0.9 ms), the weather processor's spatial and temporal smoothing algorithms would result in precipitation reflectivity contours that are statistically

**The high-to-low beam transition range for the ASR-9, for example, is typically 7 km for radars equipped with the WSP and 28 km for radars without the WSP modification.

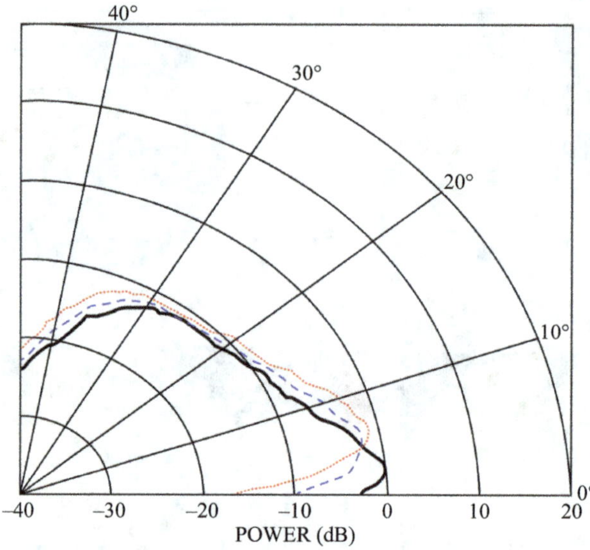

Figure 10.2 One-way elevation antenna patterns for the ASR-9 [14]. The configurable physical antenna tilt angle is assumed to be at its normal setting which places the nose of the "low" beam at 2°. The solid line is the pattern for the low beam, which is used for transmission, and for reception beyond a configurable high-low beam transition range. The dotted line is the pattern for the receive-only high beam. The dashed line is the effective high beam pattern (the square root of the product of the low and high beam patterns), given that signal transmission is on the low beam.

stable on a scan-to-scan basis, reinforcing controller confidence in the validity of the data.

The report discusses how the range-dependent thresholds used to convert received power to reported VIP-level can be adjusted – based on models for storm reflectivity structure versus height – to reflect a defined mapping of the three-dimensional (3D) reflectivity field onto the two-dimensional (2D) contours displayed on controllers' scopes. This mapping, for example, could reflect a vertical average of the reflectivity over a defined altitude interval or the maximum reflectivity at any altitude within the storm. The fidelity of these model-based mappings was shown to be quite good as validated by further, more extensive analysis [15]. The authors suggest that location-specific, seasonally adjusted storm reflectivity models could be employed by ASRs to provide an accurate, defined 2D parameterization of storm reflectivity vertical structure. More recent work at MIT LL [16] has shown that one such parameterization, VIL, is strongly correlated with the likelihood that commercial aircraft pilots will choose to deviate around storms. The use of ASR-9 weather channel data as an input to the Integrated Terminal Weather

System (ITWS) [17] reflects its value as a complement to other meteorological radars for air traffic management (ATM) decision support.

The ASR-11, fielded in the early 2000s, likewise includes a six-level weather reflectivity channel that uses digital time-series processing similar to that of the ASR-9. The older ASR-8 has been retrofitted with a "Common Terminal Digitizer" (CTD) which operates on the radar's analog outputs to provide aircraft and six-level weather reports suitable for input to the FAA's Standard Terminal Automation Replacement System (STARS). All currently operational ASRs thus provide quantitative precipitation reflectivity data, although the characteristics of these data differ owing to differences in the technical parameters of the ASR-8s, -9s, and -11s, and the implemented processing methods. Establishing a common, high-quality weather radar observation capability at airports where this is operationally justified is one significant benefit for the future weather radar architecture we propose.

10.2.2.2 Weather systems processor

Although well-suited for observation of the location and intensity of storms, the ASR's cosecant-squared elevation beams are decidedly problematic for the detection of low-altitude wind shear. Because microbursts and other thunderstorm outflows are confined to a layer within a few hundred meters of the surface, the ASR beam will typically intercept both the outflow layer and scatterers aloft moving with markedly different radial velocities. This can result in a significant bias for estimates of the near-surface wind when a conventional meteorological radial velocity estimator is used.

Approaches for mitigating this bias take advantage of the significantly different low- and high-beam responses below about 5° elevation angle (Figure 10.2). In a patent filing, Atlas [18] proposed that power spectra from the two beams be differenced in a way that emphasizes spectral components associated with scatterers below this elevation angle. Stone and Anderson [19] described a novel approach using the phase of the cross-spectrum between the high- and low-beam signals as a means to map each spectrum component to elevation angle: in essence, the two feed horns would be used as an interferometer with the signal amplitudes used to resolve ambiguities. Weber and Moser [20] developed a computationally simple, time-domain method that used high- and low-beam autocorrelation estimates to model the parameters of an assumed two-moment spectrum generated by the ASR-9's broad elevation beams. One of the parameters, the mean velocity of the "low-altitude" spectrum moment, was used as the input for the WSP's wind shear detection algorithms. This approach, successfully implemented for the fielded WSP, is planned to be replaced with a spectral processing algorithm [21] that was shown to improve the agreement between simultaneous measurements with the TDWR and WSP test systems at the FAA wind shear systems support facility in Oklahoma City, OK.

The ASR-9's transmitted pulse waveform also posed challenges for measuring radial wind velocity. The radar scans azimuthally at 75° per second, transmitting only about 18 pulses during the interval in which the antenna scans one beamwidth.

To reduce aircraft "blind speeds," this short data sequence is further broken up into two blocks consisting of eight pulses at a "low" PRF followed by ten pulses at a "high" PRF. Weber [22] proposed an "extended CPI" for the WSP (Figure 10.3) that provided sufficient samples to effectively suppress ground clutter with high-pass Doppler filters, and provide statistically stable estimates of reflectivity and radial velocity. The design of shift-variant Doppler finite impulse response filters that operate on the extended CPIs are described by Chornoboy [23] and Cho and Chornoboy [24]. Signals are filtered at four clutter suppression levels (e.g., all pass, 20, 40, and 60 dB) and compared with clutter residue maps (CRMs) at the same suppression levels collected during clear weather. The least-attenuating filter that generates a weather-signal-to-ground clutter ratio large enough for quality velocity estimates is selected for subsequent processing. This adaptive clutter suppression approach has proven to be robust, even in severe clutter environments and during the presence of super-refraction or "anomalous propagation."

The WSP's meteorological detection and processing algorithms – microburst detection, gust front detection and 20-min forecast, storm motion tracking and 20-min location extrapolation – were adapted from the corresponding algorithms developed for the TDWR. During WSP prototype operations, the legacy TDWR gust front detection algorithm proved to be inadequate due to the more poorly defined reflectivity "thin line" and velocity convergence signatures from the ASR. To address this challenge, a "machine intelligent" gust front algorithm (MIGFA)

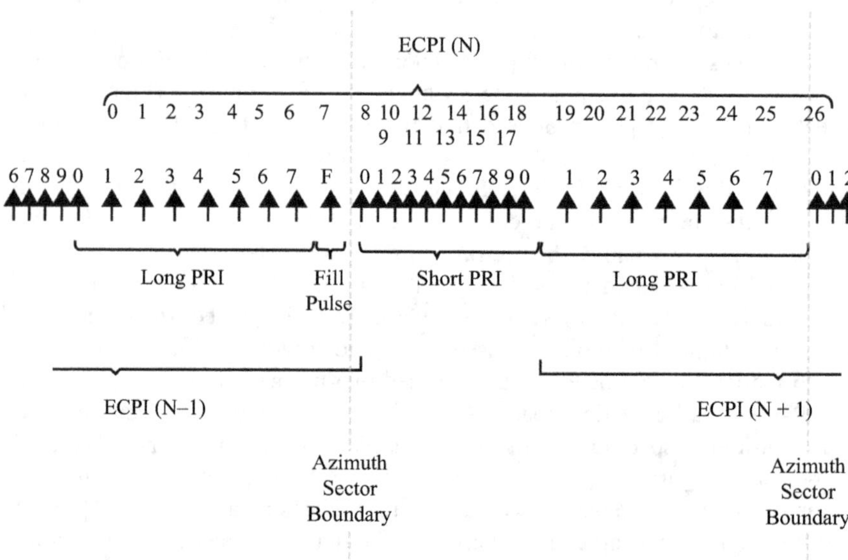

Figure 10.3 Extended CPI used by ASR-9 WSP [22]. The 8-pulse low-PRF CPI and the 10-pulse high-PRF CPIs are transmitted with a constant PRF ratio of 7:9 thus providing the same dwell time. The site-dependent average PRF is typically about 1,100 s^{-1}.

[25] was developed. MIGFA leveraged low-level machine intelligence techniques, originally developed at MIT LL to recognize military ground vehicles obscured by camouflage and foliage, to improve the probability of detection for gust fronts by more than 50% while decreasing the false alarm rate. MIGFA was subsequently retrofitted to the TDWR's algorithm suite, and core technology components such as functional template correlation and interest images [26,27] have since been successfully applied to other meteorological radar detection and classification challenges.

MIT LL's ASR-9 WSP testbed supported operational demonstrations with ATCT and TRACON controllers in Orlando, FL (1991–1992), Albuquerque, NM (1993–2004), and Austin, TX (2000–2004). Feedback from controllers at each of these facilities was very positive and, along with quantitative evaluation of the WSP's wind shear detection and false-alarm performance [28], supported the FAA's decision to field the system nationally beginning in 2002. Controllers' assertions on the broad operational value of meteorological observations from the dual-use ASR-9 and experience with both FAA and NWS utilization of TDWR informs our vision for a future where the successors to these terminal radars are fully integrated into the national meteorological radar network.

10.3 Multi-agency weather and aircraft surveillance radar concept development

10.3.1 Multifunction phased array radar

In 2006, the FAA asked MIT LL to evaluate phased array radar (PAR) technology and its potential benefits for civilian aviation. The FAA had, at that time, partnered with the National Oceanic and Atmospheric Administration (NOAA) to explore PAR as an alternative for future weather and aircraft surveillance radars. Our work led to the development of a detailed concept for meeting the surveillance radar needs of both agencies using multifunction phased array radar (MPAR) technology [29]. We described how the US Government's four distinct national radar networks (WSR, TDWR, ASR, and ARSR) could be moved to a common technology basis – active electronically scanned array panels – which could enhance weather and aircraft surveillance services with potentially lower life-cycle costs than multiple single-function radar networks. By reducing overlapping airspace coverage, MPAR could reduce the total number of radars required by approximately one-third. A key finding was that weather surveillance requirements dictate the core parameters of a multi-mission radar – airspace coverage, aperture size, radiated power, and angular resolution. Aircraft surveillance capability can, therefore, be added to a phased array weather radar at low incremental cost if its agile, electronically steered beams would allow the radar to achieve the much more rapid scan update rates needed for aircraft volume search and track missions.

We hypothesized two MPAR configurations, both consisting of four non-rotating, planar arrays deployed on a frustum-shaped housing. Replicating the angular resolution, sensitivity, and maximum range of the national scale WSR and

ARSR networks would require that each array face be approximately 8 m in diameter, therefore requiring about 20,000 transmit/receive (T/R) elements per face. Array faces of this size, which would provide a 1° pencil beam at their maximum off-broadside steering angle (45°), would also be appropriate as a replacement for the TDWR. A terminal MPAR (TMPAR) comprised of 4-m diameter array faces would provide aircraft and weather surveillance services for airports, a concept that had also been explored earlier under the FAA's Terminal Area Surveillance System (TASS) program [30]. In contrast to today's ASRs, TMPAR's elevation beamwidth would be relatively narrow (1.4°) and steerable, thereby providing vertically resolved weather variable fields.

Cho [31] performed a detailed analysis of MPAR/TMPAR siting scenarios that would provide airspace coverage and cross-range (i.e., angular) resolution equal to or greater than today's networks. Table 10.1 summarizes the most extensive replacement scenario, where services provided by WSR-88D, Common ARSR (CARSR), ARSR-4, TDWR, and ASRs at civilian and military airfields are subsumed by the MPAR/TMPAR network. The current network of 629 radars could be replaced with 411 MPAR/TMPAR systems, a 35% reduction in the number of radars. Equally important, areal coverage for high-quality weather radar observations in the PBL would be increased by a factor of two, from 15% to 30%, as shown in Figure 10.4. Much of the improved coverage would be realized in densely populated, weather-vulnerable areas: the Atlantic and Gulf of Mexico coastlines, the upper Midwest, and the Pacific coastline. Analysis described in subsequent sections indicates that the resulting benefits can be monetized and would be significant.

Weber *et al.* [29] assumed that MPAR would operate simultaneously at three frequencies, all within the 2.7–2.9 GHz range but displaced sufficiently so that they could operate independently to support its weather and aircraft surveillance missions. Subsequent technology prototyping, described below, indicated that this approach would significantly increase the complexity and costs of the array electronics thereby confounding the objective of an affordable MPAR network. A

Table 10.1 Comparison of current and MPAR/TMPAR radar numbers required to achieve equivalent airspace coverage. This assumes that all FAA, NWS, and DoD surveillance radar services are subsumed by the MPAR network [31]

	Number of faces[1]				
Type	**1**	**2**	**3**	**4**	**Total**
Current	N/A	N/A	N/A	N/A	629
MPAR	1	6	11	197	215
TMPAR	0	2	0	194	196

[1]Assumes planar phased array faces oriented at right angles to each other with one to four faces specified at each site.

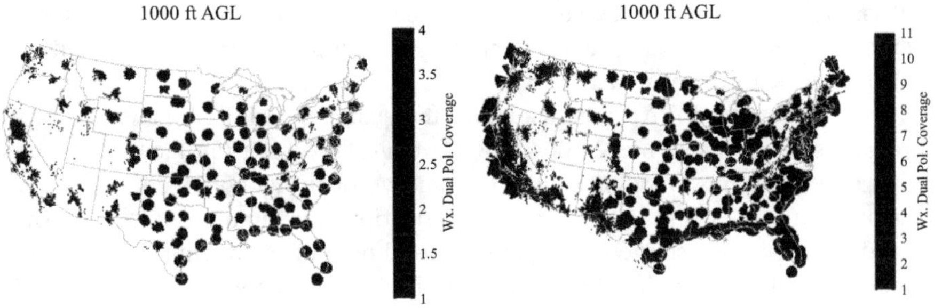

Figure 10.4 Legacy radar (left) and MPAR/TMPAR (right) coverage below 1,000 ft (305 m) (AGL) [31]

Table 10.2 Summary of notional requirements for aircraft and weather surveillance

Mission area	Coverage envelope	Sensitivity[1]	Update period (required)	Update period (goal)
Near-range non-cooperative aircraft	0.25–20 NM range 0–24 kft altitude	0.25 m^2 @ 20 NM $P_d = 0.95$, $P_{fa} = 10^{-6}$	Unspecified	Unspecified
Short-range non-cooperative aircraft	0.25–60 NM range 0–24 kft altitude	0.1 m^2 @ 60 NM $P_d = 0.90$, $P_{fa} = 10^{-6}$	3 s	1 s
Long-range non-cooperative aircraft	5–250 NM range 0-100 kft altitude	0.1 m^2 @ 92 NM $P_d = 0.80$, $P_{fa} = 10^{-6}$	12 s	12 s
ATC weather	0.25–60 NM range 0–24 kft altitude	−6 dBZ @ 6 NM 0 dB single pulse SNR	30 s	30 s
High-resolution weather	0.5–250 NM range 0–70 kft altitude	−9.5 dBZ @ 27 NM 0 dB single pulse SNR	180 s	60 s

Short- and long-range non-cooperative aircraft mission areas are those served today by the ASR and ARSR radars. The high-resolution and ATC weather missions are those served today by the WSR-88D and ASR six-level weather processors [32].
[1] P_d is the detection probability, P_{fa} is the false alarm probability.

single-frequency (per face) array is, therefore, preferred but makes the challenge of multi-mission operations more challenging since the radar's scan time would need to be partitioned amongst the missions.

Based on notional functional requirements, Weber *et al.* [32] described a detailed MPAR scanning concept. The full-sized MPAR would subsume long-range aircraft surveillance (ARSR), high-resolution weather surveillance (WSR), and – at sites where the radar is located near an air terminal – short-range aircraft surveillance (ASR). Table 10.2 is a summary of the requirements pertinent to our analysis. Note that some of the surveillance goals are more stringent than those of the current radars – for example, the capability to complete short-range aircraft and weather scans significantly faster.

An 8-m diameter MPAR antenna is assumed using the MIT LL panel described below as its basis. There are 24,064 T/R elements in each array face, each transmitting 6 W peak power in horizontal (H) and vertical (V) polarization channels. The maximum duty cycle is 5% if both polarizations are simultaneously transmitting and 10% if only a single polarization is used per pulse. Antenna beamwidth at broadside is 0.9° and, in the absence of amplitude or phase windowing across the array, its gain is 48 dB. The high power-aperture product of this antenna significantly exceeds that required to meet the short- and long-range aircraft surveillance sensitivity requirements in Table 10.2. Thus, it is possible to "spoil" the transmit beam pattern so as to spread the transmitted energy across a cluster of azimuth/elevation beam positions. Digital beamforming is used on receive – with the full array's angle resolution – to determine target position. Our assumed MPAR array is segmented into 159 overlapped subarrays whose outputs would be digitized and processed to generate the desired receive beam clusters. Figure 10.5 illustrates the angular response pattern of an example cluster that allows 33 beam positions to be simultaneously observed, reducing scan time correspondingly.

With the assumed array configuration, Weber *et al.* [32] present a notional volume search pattern for the long-range aircraft surveillance mission that would

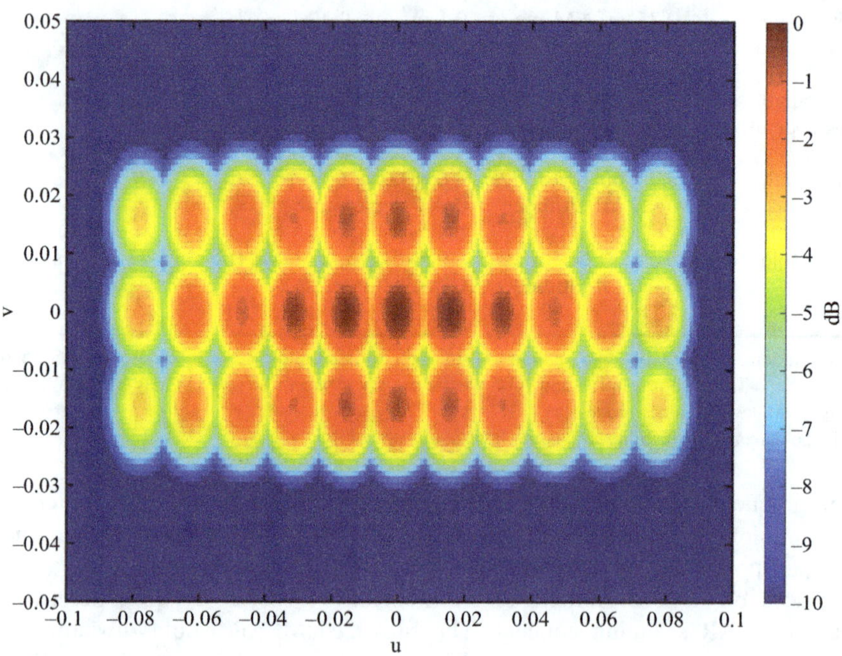

Figure 10.5 Notional MPAR receive beam cluster for long-range non-cooperative aircraft surveillance. © 2017 IEEE. Reprinted, with permission, from [32].

use 36% of the MPAR's scan time. Relatively small beam clusters (7 beams) are used at low elevation angle where surveillance to over 450 km range is required. At higher elevation angles, the size of the clusters increases to a maximum of 48 simultaneous beams. At sites where needed, we showed that the short-range aircraft surveillance mission would require an additional 25% of the radar's timeline. Thus, depending on whether the MPAR is required to provide short-, as well as long-range aircraft surveillance, 39–64% of the radar timeline is available for the high-resolution weather mission. For the frequently used WSR-88D precipitation-mode volume coverage pattern VCP-212, 5,040 beam positions are surveilled with an average dwell time of 48 ms, resulting in a volume scan time of 240 s. With the above weather mission timeline allocations, performing this scan with MPAR would require that, on average, two (64% allocation) or three (39% allocation) simultaneous beams from each of the four array faces be active for the weather scanning.

This small number of simultaneous beams is readily achieved with the assumed subarray architecture, and the 3- to 5-dB loss caused by spoiling the transmit pattern would not reduce sensitivity below the requirement in Table 10.2. Relatively high sidelobes within the angular interval defined by the spoiled transmit beam might bias weather estimates due to "smearing" of returns in volumes where reflectivity gradients are large. Figure 10.6 summarizes a method for minimizing such biases. Within the constraint that the average beam cluster size be two (or three), their size is varied – based on a fast "surveillance scan" – so as to use small clusters where reflectivity gradients in angle are large. Simulations of this method using WSR-88D images as both input and "truth" show that the resulting MPAR reflectivity and radial velocity fields differ measurably from the WSR-88D

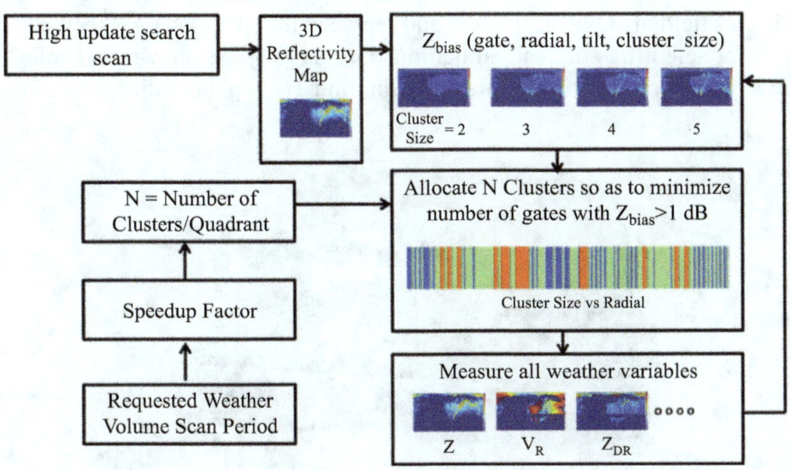

Figure 10.6 Adaptive allocation of beam cluster size to minimize variable estimate biases due to intracluster gradients in reflectivity.

in less than 2% of all resolution cells and are visually essentially identical. Overall then, our analysis shows that from a scanning timeline perspective, the MPAR concept is consistent with plausible multi-agency requirements. Specified aircraft and weather sensitivity can be achieved while satisfying the dwell time necessary for clutter suppression and target parameter estimation.

When we began developing the MPAR concept, the cost of phased array technology – typically $2,000 per T/R element – was prohibitive. To address this concern, we initiated a prototyping effort to demonstrate that array components meeting MPAR requirements could be manufactured at more than an order of magnitude reduced cost. Herd and Conway [33] discuss the cost considerations guiding the MPAR prototype design. A tiled or panel architecture, where the array is constructed of layers oriented parallel to the face of the array, was chosen to significantly reduce the number of connectors/cables and the area of the printed circuit boards (PCBs) hosting the T/R elements. Each panel is comprised of 64 (8 × 8) radiating antennae, the T/R elements, an analog subarray beam-former, cooling plenums, as well as power conditioning and control components. The T/R element monolithic microwave integrated circuits (MMICs) are packaged in low-cost commercial packages, directly soldered onto an inexpensive PCB which in turn is soldered to the back of the tile. The radio frequency (RF) and power distribution components are chosen to reduce cost and size, and support automated assembly and test. In high volume, they estimate that the MPAR T/R element cost would be ~$25. Figure 10.7 shows a fully assembled MPAR tile, which provides a low-cost, scalable building block for large array apertures.

Under FAA and NOAA sponsorship, engineers from MIT LL, General Dynamics, and the National Severe Storms Laboratory (NSSL) leveraged the MPAR panel technology to develop the Advanced Technology Demonstrator (ATD), a 4-m diameter, 10-cm wavelength, dual-polarization PAR (Figure 10.8). The ATD was fielded at NSSL in 2018 and replaced the SPY-1A PAR testbed, used until 2016 for scientific studies and demonstrations [34]. As shown in Table 10.3, its radiated power, beamwidth, antenna gain, and H–V polarization isolation are

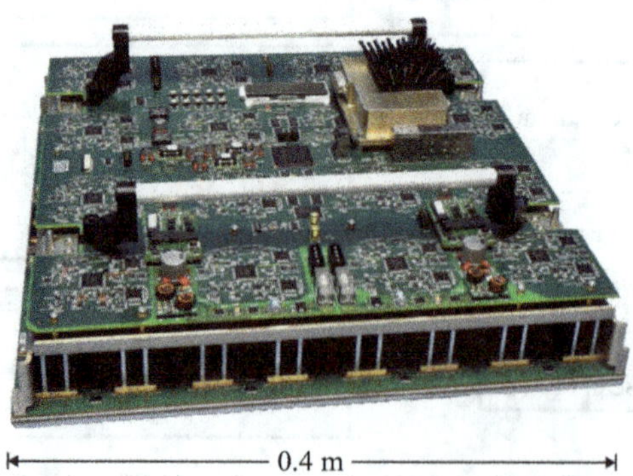

Figure 10.7 Prototype MPAR panel. © 2015 IEEE. Reprinted, with permission, from [33].

Figure 10.8 ATD inside the National Weather Radar Testbed radome in Norman, OK, USA. Courtesy James Murnan, NOAA/NSSL.

Table 10.3 Advanced technology demonstrator parameters and comparison to WSR-88D

Parameter	**ATD**	**WSR-88D**
Antenna diameter (m)	4	8.5
Gain (dB)	42 (Tx), 40 (Rx)	46
Beamwidth (deg)	1.4 (broadside)	0.9
First antenna sidelobe, two-way (dB)	−40	−60
Peak radiated power (kW)	49	500
Maximum duty cycle (%)	8	0.15
H–V polarization isolation (dB)	40 (Tx), 35 (Rx)	50

less favorable than that of the four-times-larger (in antenna area) WSR-88D. These parameters are sufficient, however, to explore key performance challenges and demonstrate operational benefits for highly digital, polarimetric PAR (see Chapter 13 for a discussion of the challenges facing polarimetric weather parameter measurements with PARs). As an example, Weber *et al.* [35] compare polarimetric observations of a thunderstorm complex by the ATD with those from the co-located WSR-88D (KCRI) which is considered as "truth" (Figure 10.9). Corrections for steering-angle-dependent biases in the ATD polarimetric variables have been applied, based on prior characterization of its antenna in a near-field anechoic chamber at MIT LL [36]. As expected, the uncorrected ATD estimates (top row) show significant biases in differential reflectivity (Z_{DR}) and differential phase

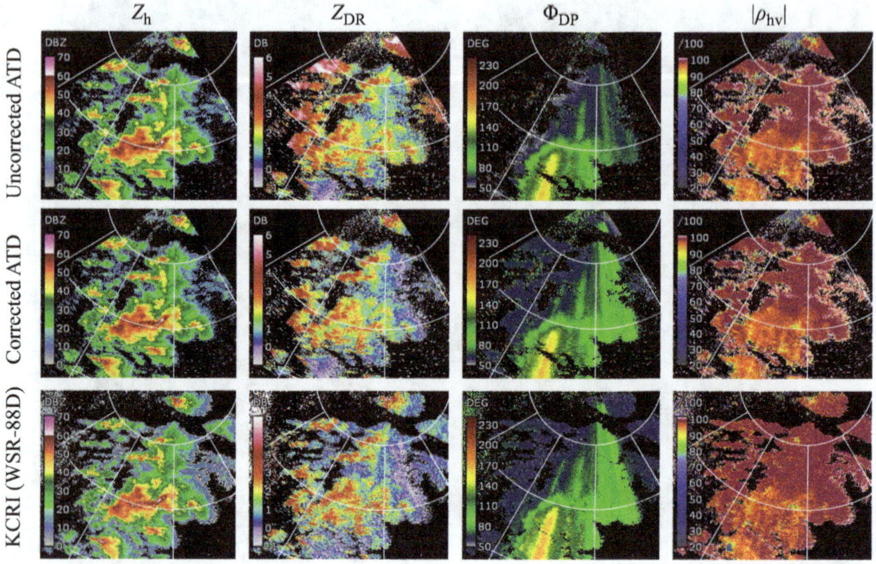

Figure 10.9 ATD reflectivity and polarimetric variable fields on 1 May 2019, (top row) without and (middle row) with bias corrections derived from near-field array characterization. The bottom row is data from the co-located WSR-88D (KCRI) operated by NOAA's Radar Operations Center (ROC) in Norman. The ATD and KCRI data were collected at the same time (19:57:43 Z) and elevation angle (0.5°). Range rings are at 50 km and 100 km [35]. © American Meteorological Society. Used with permission.

(Φ_{DP}) at electronic steering angles well away from the array normal. After the application of the chamber-determined bias corrections, the variable estimates from the ATD correspond very well with those of KCRI. Resolution cell-by-cell comparison shows the average difference between observations from the two radars to be less than NWS requirements for polarimetric variable biases, although there are a significant number of outliers where the differences are much larger than would be expected if they followed a normal distribution. This is not surprising, given the differences in beamwidth and sensitivity between the two radars, particularly in the presence of ground clutter or at the edges of storms where there may be large gradients in the weather variables.

10.3.2 Benefits for public weather warning and water management

In 2017, NOAA received supplemental funding to evaluate requirements and benefits for a future WSR that would be compatible with FAA, DoD, and Department of Homeland Security (DHS) surveillance radars. Research by investigators at NSSL, the University of Oklahoma (OU), MIT LL, and NCAR evaluated phased array technology, operational benefits of PAR rapid scanning, and benefits that would result from improved low-altitude coverage due to possible changes in the numbers and/or locations of weather radars in the future network. Weber *et al.* [35] summarize this research which included storm observations and analyses, numerical weather prediction (NWP) model data assimilation (DA) experiments with real PAR data, and observing system simulation experiments (OSSEs). Although NOAA's operational focus was public weather warnings, many aspects of the research were directly applicable to aviation. Charles Kuster *et al.* at OU used rapid-update, dual-polarization radar data to show that PAR observations of mid-level specific differential phase (K_{DP}) cores might significantly enhance the ability to predict downburst phenomena like microbursts [35]. Jenny Sun *et al.* at NCAR performed OSSEs to assess the possible benefits of PAR rapid scanning and/or "gap filler" radar deployments for 0- to 2-h convective "nowcasts" [35]. Nowcasts of thunderstorm location, vertical depth, and intensity are critical inputs to weather processing systems that support ATM decision-making [3,37]. Sun *et al.* show that DA of rapid-update PAR observations improves model analysis and 0- to 2-h forecasts of reflectivity, vector wind, rainwater mixing ratio, and other aviation-relevant variables. The improvement is larger when observations from TDWRs and gap filler radars are added to those from current WSR sites, indicating that rapid-update observations and increased low-altitude coverage work synergistically. In the experiments, PAR 1-min VCPs are especially beneficial during convective initiation when discrete storms are evolving rapidly, a scenario that can be particularly disruptive to aviation operations.

10.3.2.1 Benefits of a denser radar network

An important area of NOAA's research was our assessment of the benefits of a potentially denser future national network of weather radars, whether phased array or conventional. Our analysis focused on NOAA's severe weather warning and hydrological monitoring missions and demonstrated that integrating observations from future, more widespread airport-based weather radar capabilities could significantly enhance government weather services beyond the aviation domain. Cho and Kurdzo [6,38–40] performed regression analyses for 22 years (1998–2019) of archived storm warning (Iowa Environmental Mesonet, https://mesonet.agron.iastate.edu/request/gis/watchwarn.phtml) and storm event data (National Center for Environmental Information, https://www.ncdc.noaa.gov/stormevents/) using radar spatial coverage as the independent variable. A metric termed fraction of vertical space observed (FVO) accounts for terrain blockage, Earth's curvature effects, and the "cone of silence" above each radar. As the key radar performance metric, we used the cross-radial horizontal resolution (CHR). The analysis clearly showed that better radar coverage (FVO) and higher spatial resolution (CHR) improve severe weather detection probability and false-alarm ratio. For tornadoes and thunderstorm winds, the volume-scan update rate was included as an additional variable in the warning performance model, because the PAR Innovative Sensing Experiment (PARISE) [41,42] showed that tornado and severe thunderstorm warning performance depended on the volume update rate.

For tornadoes, flash floods, and non-tornadic severe winds, we show that the casualty rate is statistically correlated to parameters such as population density inside the tornado path, tornado or wind event intensity, fraction of population residing in mobile housing, local historical false-alarm ratio, and warning lead-time. The resulting regression model can be used to generate casualty-rate estimates on a geospatial grid, given the outputs of the severe weather warning performance models and geospatially resolved data on the population and severe weather parameters that influence casualty rates. These, in turn, can be monetized using the Department of Transportation's value of statistical life (VSL) which was $12.5M in 2020 dollars. Injuries were valued as fractions of VSL at $3.3M (hospitalized) and $0.59M (treated and released). For tornadoes, in addition to casualty cost, we also estimated the cost of time lost by people taking shelter due to false warnings which could be averted with improved radar coverage and performance.

Table 10.4 summarizes estimated societal annual costs of severe weather for scenarios where a national weather radar network does not exist (first row) and where four alternative national weather radar architectures are employed. As a baseline, casualty and unnecessary-sheltering-time costs for the WSR-88D and TDWR networks used today by NWS forecasters in issuing warnings are shown in the second row. The third row hypothesizes that the replacement of these radars with PAR technology enables rapid scanning. The fourth row hypothesizes that the current widely spaced radars are replaced with a very dense network, thereby providing "perfect" coverage in terms of FVO and achievable CHR, and the fifth row adds the benefits of rapid scanning to this very dense network. In aggregate, the current weather radar network provides $1.1B yr^{-1} benefit in terms of reduced severe weather casualties and aversion to unnecessary sheltering. The hypothesized "perfect" network with dense coverage and universal rapid scanning increases this benefit by 70%.

As shown in Figure 10.10, the "perfect" network benefit opportunities are concentrated geospatially so that the deployment of a small number of additional

Table 10.4 Annual CONUS severe storm cost estimates (millions of dollars) [6]

Scenario	TOR	TOR delta baseline	FF	FF delta baseline	SVR wind	SVR wind delta baseline	Total	Total delta baseline
No radar coverage	4192	575	1058	341	665	207	5915	1123
Current radar coverage	3617	—	717	—	458	—	4792	—
Current radar coverage, rapid scan	3259	−358	717	0	454	−4	4430	−362
Perfect coverage	3421	−196	702	−15	428	−30	4551	−241
Perfect coverage, rapid scan	2890	−727	702	−15	422	−36	4014	−778

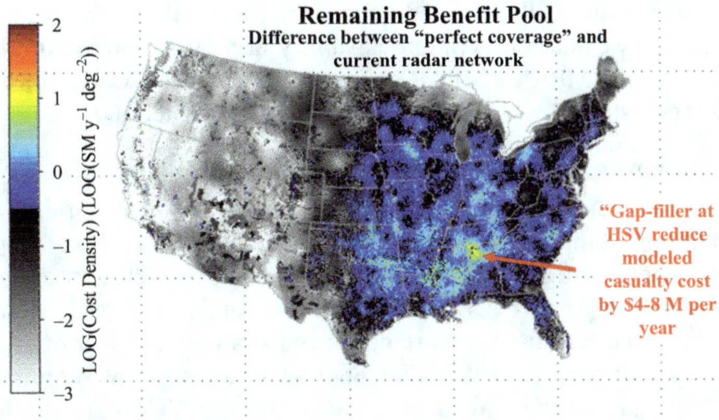

Figure 10.10 Geospatially resolved remaining benefits for a "perfect" (versus today's) weather radar network

weather radars could be significant. As an example, using the model developed for this NOAA-sponsored research, deployment of a single dual-polarization, rapid-scan weather radar in the vicinity of the Huntsville, AL airport would realize a severe weather cost aversion benefit of \$8M yr^{-1}. Other areas with significant potential benefits include northwest Pennsylvania, northeast Texas, east-central Mississippi, and along the east-central border between North and South Carolina. Our model indicates that deployment of high-quality weather radar observing capability at a large number of US airports, as envisioned in the MPAR/TMPAR concept would recover 52% of the remaining \$800M yr^{-1} benefit.

Kurdzo *et al.* [43] analyzed the potential impact of a denser weather radar network on QPE accuracy. Using Automated Surface Observing System (ASOS) gauge rain rate as truth, we analyzed the statistical accuracy of QPE estimates as a function of range from the radar and corresponding beam height, and coincident ASOS gauge rain rate. Our analysis used the specific-attenuation-based rain-rate, R (A), estimator [44], since this is the technique the NWS is adopting for the Multi-Radar Multi-Sensor (MRMS) system (see Chapter 2 for a historical perspective on radar QPE). A total of 4,750 1-h cases across three consecutive summer seasons (May–August of 2015–2017) were processed.

For the operationally significant scenario of heavy rain with flash flood potential, we found that the rain rate is on average underestimated and this bias increases with range from the radar. From this regression, we developed a geospatial analysis of maximum rain rate bias using Atlas-derived (https://hdsc.nws.noaa.gov/hdsc/pfds/index.html) 2-year/1-h rainfall return rates and current WSR-88D radar locations. Maps of significant QPE error show the largest errors in areas of the south central, southeast, and mid-Atlantic United States where peak rainfall return rates are high and distances from the nearest WSR-88D are significant. The mean QPE error across the United States for the selected rainfall return rate is an underestimate of approximately 6 mm/h. As was the case with severe weather, areas of large potential QPE error reduction are concentrated. With the addition of dual-polarization weather observation capability at a large number of airports, as envisioned under the MPAR/TMPAR concept, the observed CONUS means "peak-rate" QPE underestimate would be reduced by approximately 60%.

10.3.2.2 Benefits of more frequent low elevation scans

We have recently extended these studies of severe weather warning and QPE improvement opportunities to consider the impacts of optimized scanning, as well as radar network configuration changes. This analysis helps inform scanning requirements for the future US operational meteorological radar network and documents potential benefits for more optimized strategies. Cho *et al.* [45] exploit continuous, on-line records of the WSR-88D VCPs selected by Weather Forecast Offices (WFOs) across the United States and correlate VCP usage with warning performance. Of particular interest are VCP options implemented within the last decade to increase the update rate for at least some elevation tilts: (1) automated volume scan evaluation and termination (AVSET) [46], (2) supplemental adaptive intra-volume low-level scan (SAILS) [47], and (3) mid-volume rescan of low-level elevations (MRLE) [48]. Briefly, AVSET shortens the volume update time whenever possible by adaptively skipping high-elevation-angle scans that contain no precipitation returns. SAILS inserts 1–3 extra lowest-elevation-angle (base) scan(s) dispersed evenly in time throughout the VCP cycle. MRLE is similar to SAILS, except other low-level elevation scans (e.g., 0.9°, 1.3°, and 1.8°) are updated more frequently as well. After an initial trial period, AVSET has been on by default at all sites since 2012, although it can be manually turned off when desired. SAILS and MRLE are options that are selected in real time by WFO forecasters and have been available since 2014–2015 and 2018–2019, depending on the date that each site was updated with the corresponding software build.

Briefly, our analysis determines the scan mode that was operating on the primary WSR-88D used by the forecaster while making the decision to issue or not to issue a severe weather warning. The warning performance metrics – probability of detection (POD), mean lead time (MLT), and false alarm ratio (FAR) – were then computed as in our previous studies and parsed by VCP type. We chose mean lead time over median lead time, because confidence intervals for the mean are more straightforwardly computed and more commonly agreed upon than confidence intervals for the median, and confidence intervals are crucial in establishing statistically significant differences.

Table 10.5 summarizes the correlation between warning performance metrics and SAILS selection. The rows indicate whether SAILS is "off" or "on," and in the latter case whether one (×1), two (×2), or three (×3) supplemental base scans are inserted. Warning performance metrics for severe thunderstorms (SVRs), flash floods (FFs), and tornadoes (TORs) are tabulated in the columns: statistically significant (95% confidence) improvement over the SAILS-off case is indicated by green fill. Clearly, more-rapid base scan updates are correlated with significant increases in forecaster warning performance for each of the weather hazards considered and, with one exception, for each of the performance metrics.

Table 10.5 Severe weather warning performance statistics versus SAILS selection

SAILS status		Detection probability (%)			False alarm rate (%)			Mean lead time (min)		
		SVR	FF	TOR	SVR	FF	TOR	SVR	FF	TOR
Off		74.9	82.4	44.2	48.0	51.3	79.2	20.0	65.8	10.4
On	All	78.2	84.9	67.6	46.1	46.4	70.7	21.6	74.8	12.1
	×1	76.6	84.4	60.4	46.8	47.9	72.5	21.1	71.3	11.7
	×2	80.6	84.6	68.8	44.4	44.3	72.3	22.3	72.4	11.0
	×3	80.2	87.4	74.9	45.6	41.6	67.5	22.4	90.6	13.1

This exception is TOR mean lead times, which are not statistically different between the SAILS-on and SAILS-off scenarios. This may reflect the lower frequency of TOR warnings relative to the other categories, resulting in a smaller sample size and larger statistical confidence intervals. Overall, however, these findings are consistent with other studies [41,42,49,50] that indicate that high temporal resolution radar observations improve thunderstorm warning performance, presumably by facilitating cognitive modeling of the storm environment and making precursor signatures more readily identifiable.

Additionally, we have investigated the effects of SAILS usage on QPE accuracy [51]. The WSR-88D QPE methodology uses only the lowest elevation angle for calculations (unless clutter is present), meaning that even in the fastest VCPs, the update period is 4.2 min without SAILS. This means that a single rainfall rate is integrated over those 4.2 min, which can be quite problematic when added up during an extreme rainfall event (when QPE and associated warnings are arguably the most important). It is reasonable to expect that because SAILS provides more frequent base scans for the QPE products, the integration times would be less, leading to higher accuracy and lower overall bias.

Our analysis used a similar approach to Cho *et al.* [45], such that the operational VCPs for all of 2019 were recorded and logged whenever rainfall was observed. The data were then compared with the MRMS system, which fuses multiple radars and ground observations to create an integrated QPE product across the CONUS. MRMS gauge-corrected data, fed in part by the WSR-88Ds, were then compared with SAILS modes. With over 300,000 individual data points, SAILS was shown to statistically significantly outperform cases where SAILS was not used. Additionally, each SAILS mode (from one extra-base scan to three) showed increasing, statistically significant improvements in bias. This was especially true at convective rainfall rates, which fits the initial hypothesis. This analysis shows that even the three extra base scans have not "saturated" the improvement factor, leading to the conclusion that even faster base scans may further improve QPE estimates.

Since airport-based weather radars must provide frequent near-surface updates to reliably detect low-altitude wind shear, our SAILS analyses indicate that their incorporation into the observation system supporting public severe weather warnings would improve lead time and accuracy and would reduce QPE errors. We conclude by looking ahead to opportunities for this and other beneficial changes.

10.4 The future national weather radar system

A weather radar research and risk reduction plan [52] has been submitted to relevant US Congressional committees and the Office of Management and Budget (OMB). This describes the work needed for a decision in 2028 on the replacement architecture for the WSR-88D. Alternatives under consideration are sustaining the WSR-88D beyond its current end-of-service life (2035–2040), replacing it using a similar mechanically rotating reflector–antenna radar, or replacing it with a network of dual-polarization PARs.

The FAA is evaluating follow-on technologies for currently operating ASR-8s, ASR-9s, and ASR-11s, and the feasibility of decommissioning the radars at airports where adequate surveillance is provided by other systems. A key consideration is how to maintain weather services – six-level precipitation reflectivity provided by all ASRs, and wind shear detection provided by 34 ASR-9 WSPs.

Finally, FAA plans to sustain TDWRs through at least 2035. By then, most of these radars will have been deployed for more than 35 years.

Thus, it is likely that major sustainment or replacement of at least two of the three sub-networks within the current US operational weather radar system will take place within the next two decades. The authors believe that this offers a once-in-a-generation opportunity to examine whether the placement of the radars, and their technology basis, can be optimized to enhance operational services, support the evolving needs of current and future stakeholders, and maximize the value of the Government's investments. In the remainder of this chapter, we will identify several areas of "low hanging fruit" for this optimized weather radar system, articulate a high-level framework for its architecture, and describe the activities and decisions needed to initiate and manage the transition. We emphasize that this discussion reflects the opinions and judgments of the authors and is not the position of any US government agency.

10.4.1 Opportunities

Cho and Hallowell [53] describe an objective wind-shear detection probability model that can be applied to radar, lidar, and combinations of these sensors. The model includes effects of system sensitivity, site-specific wind-shear, ground clutter, and terrain blockage characteristics, range-aliased obscuration statistics, antenna beam-filling, attenuation, and signal processing methods. Figure 10.11 shows the 35 wind-shear-system-equipped airports where this model (since updated) estimates that the nearest WSR-88D supports a microburst probability of detection consistent with the FAA's requirement (0.9 or greater). This assumes that the WSR-88D would employ a VCP that revisits the surface once per minute. Evidence that the use of SAILS by NWS forecasters increases warning

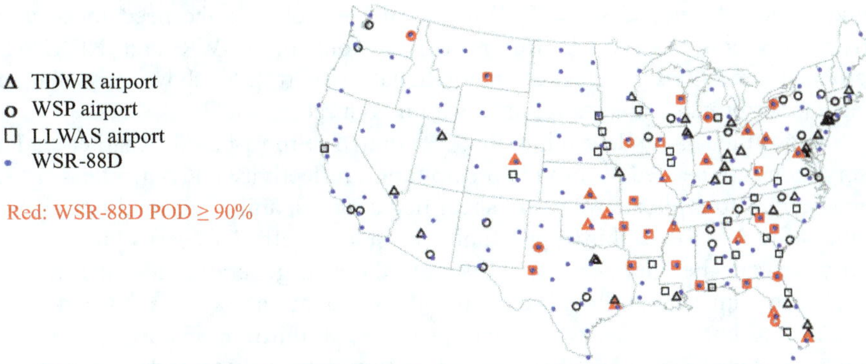

Figure 10.11 US airports with wind shear detection systems and WSR-88D locations. Red outlines on wind shear detection system icons indicate that WSR-88D siting supports microburst probability of detection equal to or greater than 0.9. The following TDWR airports are also equipped with LLWAS: Atlanta (ATL), Denver (DEN), Dallas/Fort Worth (DFW), New York-LaGuardia (LGA), Orlando (MCO), New Orleans (MSY), Chicago-O'Hare (ORD), St. Louis (STL), and Tampa (TPA).

performance indicates that it may be possible to develop VCPs that are appropriate for both airport wind shear detection and public weather warnings. If so, approximately 30% of the FAA's dedicated wind shear detection systems could be relocated or decommissioned so as to improve the overall cost-benefit ratio for these systems.

A related opportunity would be to transition the future TDWRs and WSRs to a common technology basis, presumably at 10-cm wavelength with dual-polarization capability. Eliminating the essentially co-located TDWRs and WSRs discussed in the previous paragraph would mitigate the concerns about S-band spectrum congestion that originally forced TDWR to operate in the C band. A common radar system would reduce the total acquisition and life-cycle maintenance costs for the TDWR and WSR replacement systems, and would provide flexibility in optimizing the networks to realize the operational benefits described in the preceding section. One issue that would need to be addressed is that the TDWR beamwidth (0.5°) is approximately half that of the WSR-88D (0.9°). We note that TDWR's beam is, in practice, "spoiled" to ~1° in azimuth by the use of CPIs that extend over two full physical beamwidths as the antenna rotates. This design choice was made because 0.5° azimuth resolution was not necessary for close-range wind shear detection, and achieving lower variance reflectivity and Doppler velocity estimates resulting from longer CPIs was deemed to be more useful. TDWR's narrow elevation beamwidth is, however, valuable in that it allows the radar to focus on the shallow layer of strongest outflow winds without excessive illumination of ground scatterers. We believe that improved, Doppler-spectrum-based ground clutter suppression methods [54], developed since TDWR was fielded, obviate the need for such a narrow elevation beamwidth. Recent comparisons of TDWR and NEXRAD microburst detections at Denver International Airport [55] show very similar detection performance for microbursts within both radars' fields-of-view.

Should the FAA determine that continued airport-based weather radar observations are needed to provide precipitation reflectivity on controller's radar scopes and/or to replicate the wind shear detection capability currently provided by the ASR-9 WSP, a third opportunity presents itself. This is to build high-quality weather measurement capability into the next-generation terminal radars as a system requirement. Possible technical approaches are: (1) a modified ASR antenna feed-horn configuration that produces a third, near-horizon "pencil beam" (see Appendix); (2) a PAR that could simultaneously transmit and receive beams optimized for aircraft and weather surveillance. This approach may be cost-effective if NOAA selects the PAR alternative to replace its WSR-88D network, due to economies of scale; (3) stand-alone airport weather radars in the circumstance that the future ASR replacement system does not provide necessary weather observation capabilities. For any of these technical alternatives, the future terminal radars would provide significant improvements for airport wind shear detection relative to the WSP due to increased sensitivity, radar beam focusing in the low-altitude layer where thunderstorm outflows are most intense, and greater control of the weather waveform and processing algorithms. They might also enhance the information on precipitation intensity available for terminal controller displays since the airport radars could now provide height-resolved information.

As emphasized by our research, weather-capable radars at a larger number of airports would have a significant value in improving public severe weather warning performance and hydrological monitoring by increasing coverage in the PBL. This benefit could be extended by deploying additional gap filler radars to mitigate radar coverage gaps in the Intermountain West. Observing precipitation and snowpack are critical for water management, and a modernized radar network needs to fill in these measurement gaps.

A fourth opportunity presented by our concept is the weather radar interconnectivity that would enable the FAA to integrate data from WSR and TDWR into the precipitation reflectivity overlays on terminal controllers' scopes – currently generated only with ASR input. At facilities where portions of terminal airspace are blocked by terrain near the airport, the additional observations from more favorably located WSRs or TDWRs can significantly improve coverage. Cho and Dupree [56] assess spatial and temporal coverage provided by WSRs and TDWRs in 146 TRACONs within the CONUS and Alaska. We define a weighted fraction of vertical volume observed (WFVO) where the weighting function corresponds to the site-specific distribution of precipitation reflectivity in altitude. With the exception of a handful of TRACONs, this WFVO metric for the WSR and TDWR networks averaged horizontally over the TRACON area exceeds 0.8; the median of this averaged quantity across all of the TRACONs studied is 0.94. When combined with the ASR coverage, the overall median

Table 10.6 Area-averaged WFVO for nine TRACONs where integration of WSR and TDWR precipitation reflectivity data onto controller scopes would have the largest impact [56]

TRACON	ASR	ASR + weather radars	Difference
Aspen (ASE)	0	0.836	0.836
Phoenix (P50)	0.541	0.931	0.391
Boise (BOI): Bozeman (BZN) only	0	0.344	0.344
Missoula (MSO)	0.596	0.909	0.314
Las Vegas (L30)	0.614	0.873	0.259
Boise (BOI)	0.435	0.620	0.185
Salt Lake City (S56)	0.782	0.930	0.148
N. California (NCT): Reno (RNO) only	0.825	0.927	0.103
Roanoke (ROA)	0.861	0.953	0.092

increases to 0.98 and, as shown in Table 10.6, coverage would increase by a significant amount at some sites. Although WSRs and TDWRs do not perform volume scans at the 30-s refresh rate for precipitation data on terminal radar scopes, Cho and Dupree [56] show that if observations from multiple overlapping weather radars – often available above 1,500 m (AGL) – are suitably integrated, the timeliness of their data is sufficient.

10.4.2 Architecture framework

Our vision then is a single, fully integrated national weather radar network, jointly operated by NOAA, FAA, and the DoD using a tri-agency governance model analogous to that currently in place for the WSR-88D [57]. A network of large-aperture, 10-cm wavelength, dual-polarization Doppler radars will provide observations on a national-scale supporting public weather warnings, hydrological monitoring, and air traffic management. These will also provide more specialized services such as wind shear protection and base operations support at high-density civilian airports and certain military bases. A network of smaller, primarily airport-based radars will likewise provide high-quality, dual polarization meteorological observations. These will detect wind shear and other weather hazards at the airports they support and will substantially increase coverage in the PBL with benefits as indicated by our research supporting FAA and NOAA investment analyses. Because of the relatively short operational range required of these radars, lower angular resolution is acceptable. The numbers, siting, and technology basis (e.g., reflector antenna or phased array) for both of these radar network components will be determined based on an analysis of future operational needs, associated benefits, and costs. An interesting possibility is that some low-altitude gap-filling observations might be provided to the government as a service by commercial entities operating their own radars.

Base data (i.e., full-resolution weather radar variables in polar format) from all the radars will be disseminated in real time to the Government's meteorological service providers and to designated commercial or academic stakeholders; the data will also be archived in a public-access repository. Algorithms providing data quality control, meteorological feature detection, and classification will operate within remote "automation" platforms such as the NWS's Advanced Weather Interactive Processing System (AWIPS) and MRMS, and the FAA's Nextgen Weather Processor. This more centralized data processing paradigm – contrasted, for example, with today's WSR-88D-hosted Open Radar Product Generator (ORPG) – improves the effectiveness with which information from multiple sensors can be integrated, and reduces the complexity and cost of upgrades to processing algorithms and infrastructure. The signal processing and radar control software needed to generate base data will continue to operate within the individual radars. This will be open, non-proprietary, and managed by the tri-agency team in a way that allows for efficient insertion of evolving capability improvements developed within the Government and by the broader scientific user community.

The operational and societal benefits of this architecture have been discussed throughout this chapter. By eliminating or re-siting redundant, co-located TDWRs and WSRs, and adding high-quality weather radar observation capabilities at many additional airports, the national radar network's coverage would be enhanced significantly, particularly at low altitudes. This in turn would improve both aviation weather and public safety services. Radar-based wind shear detection performance would be improved at airports currently equipped with the ASR WSP or LLWAS and would potentially be expanded to cover additional facilities. Weather radar coverage in terminal airspace would be greater, significantly so at facilities in the western United States where terrain blockage is a major factor. The addition of dual-polarization weather radar measurements at high-density airports served today by TDWR would improve the ability to predict wind shear and other severe weather events at these airports, reduce false alerts caused by birds and bats, and would facilitate nowcasting of icing conditions and winter precipitation that affect runway braking and snow removal operations.

Public weather warnings and hydro-meteorological monitoring would likewise be improved as a result of more dense radar observations in the PBL. These would improve forecaster warning performance and QPE as discussed above, and would also improve NWP model-based analyses and forecasts due to the better initial conditions provided by DA of the expanded PBL observations. If the future network uses PAR to increase the rate and efficiency of scanning, additional improvements to the accuracy of severe weather warnings, QPE, and aviation hazard advisories are expected.

Finally, life-cycle acquisition, operating, and sustainment costs on a per-radar basis will be reduced through economies of scale in deploying the network, and through consolidation of second- and facility-level maintenance responsibilities. A consolidated support facility would both reduce operating costs and facilitate the adoption of common infrastructure, software support tools, cyber security systems, data-archiving technologies, and user training.

10.4.3 Roadmap for transition to the future architecture

A decision to evolve US operational radars in these directions will require rigorous supporting analysis to quantify the operational benefits and associated costs of candidate network configurations and determine the technical characteristics of its component radars. This subsection outlines key studies, research, prototyping, and demonstration needed to support future Government acquisition decisions.

10.4.3.1 Analysis of operational benefits

Aviation and public weather hazard detection and warning

Site-specific sensor performance models analogous to those described for low-altitude wind shear detection systems [58] can be extended to cover other high-impact hazardous weather phenomena (e.g., tornadoes, damaging downbursts, outflow boundaries, and flash floods) where radar line-of-sight, sensitivity, spatial and temporal resolution are critical for detection and warning performance. In aggregate, these phenomena are responsible for the major portion of US weather-related casualties, property damage, and economic disruption that is amenable to mitigation with an improved national radar network. The economic costs these phenomena incur are well documented and geospatially segregated so that meaningful assessment of the impact of location-dependent radar performance will be possible.

Numerical weather model forecast improvement

Similarly, the operational benefits of assimilating the enhanced radar observations into NWP models must be clearly defined and monetized. "Case studies" using OSSEs or real radar observations [59,60] have shown that improved temporal resolution and/or coverage at low altitude for the assimilated radar variables improves subjective and objective agreement between the model fields and "truth" (e.g., the OSSE "nature-run" or MRMS). Quantifying the benefits of such improvements will be very challenging. A model will be needed to map radar network parameters such as coverage, sensitivity, and resolution in time and space to a meaningful forecast skill metric at various look-ahead times. This could in principle be constructed through a very large number of OSSEs with the radar parameters varied systematically over the range of plausible values for the future network.

A second set of models is then needed to map NWP forecast skill to improvements in decision support for various domains such as ATM and public severe weather warnings. These models might be developed using objective decision support tools (DSTs) that depend on NWP model input. As an example, the Traffic Flow Impact (TFI) tool [61] provides FAA traffic management specialists forecasts of achievable aircraft flow-rates through constrained airspace during thunderstorm outbreaks. The High-Resolution Rapid Refresh Ensemble (HRRRE) NWP model provides critical probabilistic input forecasts of thunderstorm VIL and radar 18-dBZ echo top (ET) height. Through post-event analysis, HRRRE objective forecast skill metrics for these variables could be correlated with TFI flow-rate forecast errors to develop the model relating the DST input-to-output performance relationship. Analogous models could be developed for decision support in other operational domains where objective DSTs are in use.

Finally, models to monetize the impact of the NWP-based operational decisions are needed. In the case of the TFI example, this is straightforward. Underestimates of achievable aircraft flow-rates result in the imposition of unnecessary flight delays or cancellations with readily monetized impacts such as airline direct operating costs and passenger value of time. Overestimates would require that aircraft be rerouted or placed into "holding patterns" with associated increases in fuel usage, arrival delay, and secondary delay impacts in airspace affected by the holding and reroutes. Analogous benefit models for other impactful operational decision domains that require NWP input (e.g., severe weather watch-to-warning products and quantitative precipitation forecasting) will be needed.

10.4.3.2 PAR risk reduction

Technology demonstration and risk-reduction activities for the future radar system must be expanded, particularly in the area of phased arrays. PAR prototypes such as the NSSL ATD will be used to address key engineering challenges such as robust dual-polarization calibration and bias correction, efficient rapid-update scanning methods, and data quality control. Processing and calibration necessary to measure the polarimetric variables with accuracy equivalent to that of the WSR-88D is particularly challenging (see Chapter 12). This must be a focus area for ongoing PAR research. Case studies of severe storms that demonstrate the benefit of rapid scanning for weather hazard warning, and for NWP forecasts, may strengthen the "business case" for PAR, especially those that demonstrate the need for rapid *volumetric* scanning rather than simply more frequent updates of the lowest elevation angle scans. Significant meteorological PAR research and demonstration activities need to spread across a wide cross-section of universities, government research laboratories, and industry using the affordable technology building blocks [e.g., highly integrated RF System-on-Chip (RFSoC)] that are emerging from the automotive radar and 5G wireless industries. We are encouraged, for example, by the development and demonstration of the Toshiba phased array weather radar in Japan [62].

10.4.3.3 Airport-based weather radar alternatives and benefits assessment

Finally, the number, siting, and system configuration for future airport-based weather radars at small and medium-density terminals must be determined. The life-cycle cost and technical performance of the alternatives described above will be determined and used to drive the parallel analysis of operational benefits. Given the opportunity to improve capability in mission areas well beyond those serviced by current ASRs, it is important that these analyses address the needs of all stakeholders in the future aviation weather radar system.

Appendix "Weather optimized" airport surveillance radar

Hensoldt AG, a multinational sensor development company headquartered in Germany, has developed a "next-generation" airport surveillance radar that utilizes three different elevation beam patterns (https://www.hensoldt.net/products/radar-iff-and-datalink/asr-ng/). In addition to high and low cosecant-squared patterns, a near-horizon pencil beam is formed by illuminating the radar's vertically extended reflector antenna at a larger incidence angle using a third feed horn. While extending the effective range for primary aircraft surveillance, a low-elevation angle pencil beam formed in this manner could also significantly improve future ASRs' capabilities to contribute high-quality near-airport weather observations to the integrated national weather radar architecture envisioned in this chapter.

To generate an example set of antenna patterns achievable with this approach, we followed the geometrical optics techniques described by Van Atta and Keary [63]. These allow us to solve a system of differential equations to arrive at the shape of the central vertical section of a reflector that can generate a desired far-field elevation pattern, accounting for the feed aim point and illumination pattern. This technique was used to generate a central section shape for an offset-feed system to generate a cosecant-squared pattern in elevation for a particular choice of feed pattern, focal distance, and angular extent subtended from the feed location. The pencil beam is incorporated into the design by extending the upper part of the central section as a paraboloid. This will perturb the cosecant-squared pattern, but the effect is relatively minor to the extent that the feed illumination on the extended paraboloid part of the reflector is very low. The pencil beam is generated by aiming a separate feed at the extended paraboloid, such that the illumination on the shaped portion of the antenna is very low. The upper cosecant-squared receive-only beam can be generated by the addition of a third feed horn that is offset from the focus used for the lower cosecant-squared beam. The radar can operate by sequencing transmit waveforms from the pencil and lower cosecant-squared beam, or by using a single transmitter and splitting power between the two feeds.

Figure 10.A1 shows the assumed reflector shape and the illumination functions corresponding to the cosecant-squared and low-elevation pencil beams. Resulting one-way elevation beam patterns are plotted in Figure 10.A2. It should be noted that for a real design, the geometry should carefully account for the fact that the three feeds cannot occupy the same space and that more care could be taken with controlling the feed pattern and spill-over onto the undesired antenna sections.

The pencil beam's half-power width is 2.7° with two-way first sidelobes of -40 dB or lower. Relative to conventional ASR's cosecant-squared patterns, this provides significantly increased vertical directivity for low-altitude weather observations. Given our concept's emphasis on short-range, "gap filling" weather observations from these weather-optimized ASRs, they would utilize a klystron or magnetron high peak-power transmitter so as to avoid the loss of sensitivity incurred from the use of a "fill pulse" for short-range surveillance, which is

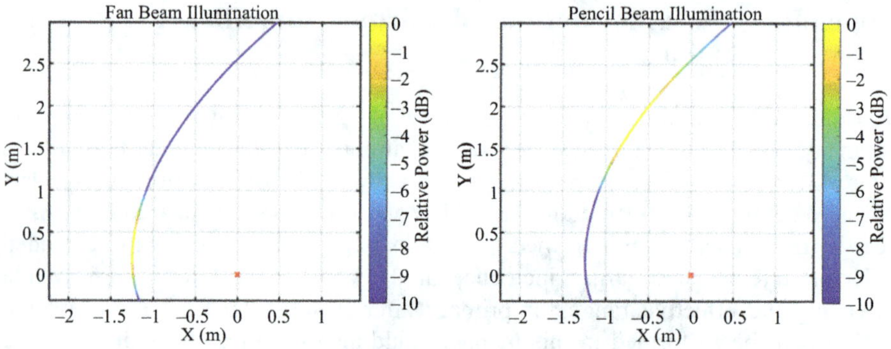

Figure 10.A1 Antenna shape and illumination patterns for (left) cosecant-squared and (right) low-elevation pencil-beam patterns

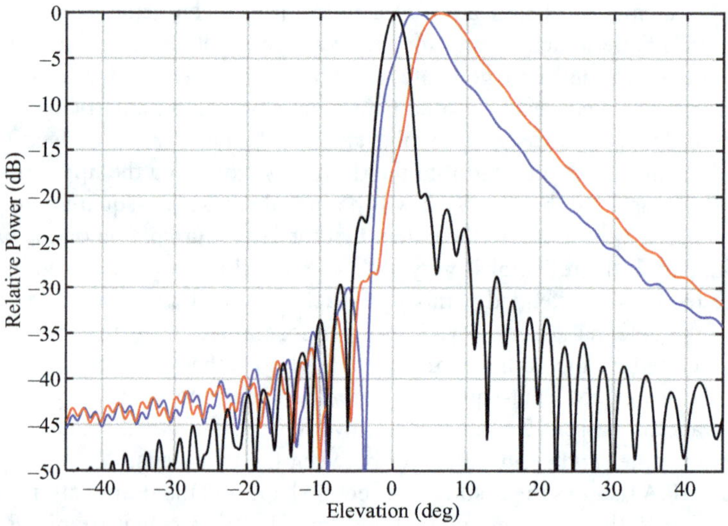

Figure 10.A2 Modeled antenna elevation patterns for "low" (blue) and "high" (red) cosecant-squared beams and near-horizon pencil beam (black)

typically used in long pulse solid-state transmitters [64]. Additional RF channels through the rotary joint would be required so that both H- and V-signals could be simultaneously transmitted and received through the cosecant-squared and pencil beams. Effective pulse transmission and receive-signal processing strategies warrant further investigation.

Acronyms and Abbreviations

2D	two-dimensional
3D	three-dimensional
AG	Aktiengesellschaft
AGL	above ground level
AP	anomalous propagation
ARENA	areas noted for attention
ARSR	air route surveillance radar
ARSR-4	air route surveillance radar, model 4
ARTCC	air route traffic control center
ASE	Aspen terminal radar approach control
ASOS	automated surface observing system
ASR	airport surveillance radar
ASR-11	airport surveillance radar, model 11
ASR-8	airport surveillance radar, model 8
ASR-9	airport surveillance radar, model 9
ATC	air traffic control
ATCT	air traffic control tower
ATD	advanced technology demonstrator
ATL	Hartsfield-Jackson Atlanta International Airport
ATM	air traffic management
AVSET	automated volume scan evaluation and termination
AWIPS	advanced weather interactive processing system
BOI	Boise terminal radar approach control
BZN	Bozeman Yellowstone International Airport
CARSR	common air route surveillance radar
CBA	cost–benefit analysis
CHR	cross-radial horizontal resolution
CLAWS	classify, locate, and avoid wind shear
CO	Colorado
CONOPS	concept of operations
CONUS	contiguous United States
CPI	coherent processing interval
CRM	clutter residue map
CTD	common terminal digitizer
DA	data assimilation
DEN	Denver International Airport

DFW	Dallas/Fort Worth International Airport
DHS	Department of Homeland Security
DoD	Department of Defense
DST	decision support tool
ERAM	En Route Automation Modernization
ET	echo top
FAA	Federal Aviation Administration
FAR	false alarm ratio
FF	flash flood
FL	Florida
FVO	fraction of vertical space observed
H	horizontal
HRRRE	high-resolution rapid refresh ensemble
ITWS	integrated terminal weather system
KCRI	Radar Operations Center testbed weather surveillance radar
$\mathbf{K_{DP}}$	specific differential phase
L30	Las Vegas terminal radar approach control
LGA	LaGuardia Airport
LL	Lincoln Laboratory
LLWAS	low-level wind shear alert system
MCO	Orlando International Airport
MIGFA	machine intelligent gust front algorithm
MIT	Massachusetts Institute of Technology
MLT	mean lead time
MMIC	monolithic microwave integrated circuit
MO	Missouri
MPAR	multifunction phased array radar
MRLE	mid-volume rescan of low-level elevations
MRMS	multi-radar multi-sensor
MSO	Missoula terminal radar approach control
MSY	Louis Armstrong New Orleans International Airport
N/A	not applicable
NCAR	National Center for Atmospheric Research
NCT	Northern California terminal radar approach control
NEXRAD	Next Generation Weather Radar
NIMROD	Northern Illinois Meteorological Research on Downburst
NM	New Mexico
NOAA	National Oceanic and Atmospheric Administration

NSSL	National Severe Storms Laboratory
NWP	numerical weather prediction
NWS	National Weather Service
OK	Oklahoma
OMB	Office of Management and Budget
ORD	Chicago O'Hare International Airport
ORPG	Open Radar Product Generator
OSSE	observing system simulation experiment
OU	University of Oklahoma
P50	Phoenix terminal radar approach control
PAR	phased array radar
PARISE	phased array radar innovative sensing experiment
PBL	planetary boundary layer
PCB	printed circuit board
P_d	detection probability
P_{fa}	false alarm probability
POD	probability of detection
PPI	plan position indicator
PRI	pulse repetition interval
QPE	quantitative precipitation estimation
R(A)	specific-attenuation-based rain-rate
RF	radio frequency
RFSoC	radio frequency system-on-chip
RNO	Reno-Tahoe International Airport
ROA	Roanoke terminal radar approach control
ROC	Radar Operations Center
Rx	receive
S56	Salt Lake City terminal radar approach control
SAILS	supplemental adaptive intra-volume low-level scan
SNR	signal-to-noise ratio
SPG	Supplemental Product Generator
STARS	Standard Terminal Automation Replacement System
STL	St. Louis Lambert International Airport
SVR	severe thunderstorm
TASS	Terminal Area Surveillance System
TDWR	Terminal Doppler Weather Radar
TFI	traffic flow impact
TMI	traffic management initiative

TMPAR	terminal multifunction phased array radar
TOR	tornado
TPA	Tampa International Airport
T/R	transmit/receive
TRACON	terminal radar approach control
TX	Texas
Tx	transmit
US	United States
USA	United States of America
V	vertical
VCP	volume coverage pattern
VIL	vertically integrated liquid
VIP	video integrator processor
VSL	value of statistical life
WARP	weather and radar processor
WFO	Weather Forecast Office
WFVO	weighted fraction of vertical volume observed
WSP	weather systems processor
WSR	weather surveillance radar
WSR-57	Weather Surveillance Radar 1957
WSR-74	Weather Surveillance Radar 1974
WSR-88D	Weather Surveillance Radar 1988 Doppler
Z_{DR}	differential reflectivity
Φ_{DP}	differential phase

References

[1] Lang J. and Ketterman J. 'Improving the WARP (Weather and Radar Processor) radar mosaic products'. In: *Air Traffic Control Association (ATCA) Annual Conference Proceedings*, Washington, DC, November 2001. Alexandria: Air Traffic Control Association; 2001. pp. 4–8.

[2] Stobie J., Moosakhanian A, Johnson J., *et al.* 'Weather and Radar Processor (WARP) products for en-route controller displays'. In: *Proceedings of the 13th Conference on Aviation, Range and Aerospace Meteorology*, New Orleans, USA, January 2008. Boston, MA: American Meteorological Society; 2008. p. 4.4.

[3] Evans J.E. and Ducot E.R. 'Corridor integrated terminal weather system'. *The Lincoln Laboratory Journal*. 2006;**16**(1):59–80.

[4] Rhoda D.A. and Weber M.E. *Assessment of the Delay Aversion Benefits of the Airport Surveillance Radar (ASR) Weather Systems Processor (WSP)*. MIT Lincoln Laboratory Project Report ATC-249, 1996.

[5] Allan S. and Evans J.E. *Operational Benefits for the Integrated Terminal Weather System (ITWS) at Atlanta*. MIT Lincoln Laboratory Project Report ATC-320, 2005.

[6] Cho J.Y.N. and Kurdzo J.M. 'Weather radar network benefit model for nontornadic thunderstorm wind casualty cost reduction'. *Weather, Climate, and Society*. 2020;**12**(4):789–804.

[7] Kurdzo J.M. and Cho J.Y.N. *National Weather Service TDWR Surveys for Tornado and Severe Thunderstorm Warning Guidance*. MIT Lincoln Laboratory Project Memorandum 43PM-Wx-0191, 2020.

[8] Taylor J.W. and Brunins G. 'Design of a new airport surveillance radar (ASR-9)'. *Proceedings of the IEEE*. 1985;**73**(2):284–289.

[9] Fujita T.T. 'Objectives, operation, and results of Project NIMROD'. In: *Proceedings of the 11th Conference on Severe Local Storms*, Kansas City, USA, October 1979. Boston, MA: American Meteorological Society; 1979. pp. 259–266.

[10] McCarthy, J., Wilson, J.W., and Fujita, T.T. 'The Joint Airport Weather Studies Project (JAWS)'. *Bulletin of The American Meteorological Society*. 1982;**63**:15–22.

[11] McCarthy J. 'The classify, locate, and avoid wind shear (CLAWS) project at Denver's Stapleton International Airport: operational testing of terminal weather hazard warnings with an emphasis on microburst wind shear'. In: *Proceedings of the 2nd International Conference on the Aviation Weather System*, Montreal, Canada, June 1985. Boston, MA: American Meteorological Society; 1985. pp. 247–256.

[12] McCarthy J., Serafin R., Wilson J., Evans J., Kessinger C., and Mahoney W. P., Jr. 'Addressing the microburst threat to aviation: research-to-operations success story'. *Bulletin of the American Meteorological Society*. 2022; **103**: E2845–E2861.

[13] Cho J.Y.N. *Signal Processing Algorithms for the Terminal Doppler Weather Radar: Build 2*. MIT Lincoln Laboratory project Report ATC-363, 2010.

[14] Weber M.E. *Assessment of ASR-9 Weather Channel Performance: Analysis and Simulation*. MIT Lincoln Laboratory Project Report ATC-138, 1986.

[15] Puzzo D.C., Troxel S.W., Meister M.A., Weber M.E., and Pieronek J.V. *ASR-9 Weather Channel Test Report*. MIT Lincoln Laboratory Project Report ATC-165, 1989.

[16] Matthews M. and DeLaura R. 'Assessment and interpretation of en route weather avoidance fields from the convective weather avoidance model'. In: *Proceedings of the 10th AIAA Aviation Technology, Integration, and Operations (ATIO) Conference*, Ft. Worth, USA, September 2010. Reston, VA: AIAA; 2010. P. 9160.

[17] Evans J.E. and Ducot E.R. 'The integrated terminal weather system (ITWS)'. *The Lincoln Laboratory Journal*. 1994;**7**(2):449–474.

[18] Atlas D. *Radar Detection of Hazardous Small Scale Weather Disturbances*. U.S. Patent US4649388A, 1985.

[19] Stone M.L. and Anderson J.R. 'Advances in primary-radar technology'. *The Lincoln Laboratory Journal*. 1989;**2**(3):363–380.

[20] Weber M.E. and Moser W.R. II. *A Preliminary Assessment of Thunderstorm Outflow Wind Measurement with Airport Surveillance Radars*. MIT Lincoln Laboratory Project Report ATC-140, 1987.

[21] Cho J.Y.N. 'Enhanced signal processing algorithms for the ASR-9 weather systems processor'. *Journal of Atmospheric and Oceanic Technology*. 2015;**32**(10):1847–1859.

[22] Weber M.E. *Ground Clutter Processing for Wind Measurements with Airport Surveillance Radars*. MIT Lincoln Laboratory Project Report ATC-143, 1987.

[23] Chornoboy E.S. 'Clutter filter design for multiple-PRT signals. In: *Proceedings of the 26th Conference on Radar Meteorology*, Norman, USA, May 1993. Boston, MA: American Meteorological Society; 1993. pp. 235–237.

[24] Cho J.Y.N. and Chornoboy E.S. 'Multi-PRI signal processing for the Terminal Doppler Weather Radar. Part I: clutter filtering'. *Journal of Atmospheric and Oceanic Technology*. 2005;**22**,(5):575–582.

[25] Delanoy R.L. and Troxel S.W. 'Machine intelligent gust front detection'. *The Lincoln Laboratory Journal*. 1993;**6**(1):187–212.

[26] Verly J.G., Delanoy R.L., and Dudgeon D.E. 'Machine intelligent technology for automatic target recognition'. *The Lincoln Laboratory Journal*. 1989;**2**(2):277–310.

[27] Wolfson M.M., Delanoy R.L., Forman B.E., Hallowell R.G., Pawlak M.L., and Smith, P.D. 'Automated microburst wind-shear prediction'. *The Lincoln Laboratory Journal*. 1994;**7**(2):399–426.

[28] Weber M.E., Cullen J.A., Troxel S.W., and Meuse C.A. *ASR-9 Weather System Processor (WSP): Wind Shear Algorithms Performance Assessment*. MIT Lincoln Laboratory Project Report ATC-247, 1996.

[29] Weber M.E., Cho J.Y.N., Herd J.S., Flavin J.M., Benner W.E., and Torok G. S. 'The next generation multi-mission U.S. surveillance radar network'. *Bulletin of the American Meteorological Society*. 2007;**88**(11):1739–1751.

[30] Rogers J.W., Tidwell C.J., and Little A.D. 'Terminal area surveillance system'. In: *Proceedings of the International Radar Conference*, Alexandria, USA, May 1995. Piscataway, NJ: IEEE; 1995. pp. 501–504.

[31] Cho J.Y.N. *Revised Multifunction Phased Array Radar (MPAR) Network Siting Analysis*. MIT Lincoln Laboratory Project Report ATC-425, 2015.

[32] Weber M.E., Cho J.Y.N., and Thomas H.G. 'Command and control for multifunction phased array radar'. *IEEE Transactions on Geoscience and Remote Sensing*. 2017;**55**,(10):5899–5912.

[33] Herd J.S. and Conway M.D. 'The evolution to modern phased array architectures'. *Proceedings of the IEEE*. 2015;**104**(3): 519–529.

[34] Forsyth D.E., Kimpel J.F., Zrnic D.S., *et al.* 'The national weather radar testbed (phased-array)'. In: *Proceedings of the 32nd Conference on Radar Meteorology*, Albuquerque, USA, October 2005. Boston, MA: American Meteorological Society; 2005. pp. 24–29.

[35] Weber M., Hondl K., Yussouf N., *et al.* 'Towards the next generation operational meteorological radar'. *Bulletin of the American Meteorological Society*. 2021;**102**(7):E1357–E1383.

[36] Conway M.D., Du Russel D., Morris A., and Parry C. 'Multifunction phased array radar advanced technology demonstrator nearfield test results'. In: *Proceedings of the 2018 IEEE Radar Conference*, Oklahoma City, USA, April 2018. Piscataway, NJ: IEEE; 2018. pp. 1412–1415.

[37] Moosakhanian A. 'FAA NextGen weather systems'. *Presented at 5th Aviation, Range, and Aerospace Meteorology Special Symposium*, New Orleans, USA, 2016.

[38] Cho J.Y.N. and Kurdzo J.M. *Monetized Weather Radar Network Benefits for Tornado Cost Reduction*. MIT Lincoln Laboratory Project Report NOAA-35, 2019.

[39] Cho J.Y.N. and Kurdzo J.M. 'Weather radar network benefit model for tornadoes'. *Journal of Applied Meteorology and Climatology*. 2019;**58**(5):971–987.

[40] Cho J.Y.N. and Kurdzo J.M. 'Weather radar network benefit model for flash flood casualty reduction'. *Journal of Applied Meteorology and Climatology*. 2020;**59**(4):589–604.

[41] Heinselman P.L, LaDue D.S., Kingfield D.M., and Hoffman R. 'Tornado warning decisions using phased-array radar data'. *Weather and Forecasting*. 2015;**30**(1):57–78.

[42] Wilson K.A., Heinselman P.L., Custer C.M., Kingfield D.M., and Kang Z. 'Forecaster performance and workload: Does radar update time matter?' *Weather and Forecasting*. 2017;**32**(1):253–274.

[43] Kurdzo J.M., Joback E.F., Kirstetter P., and Cho J.Y.N. 'Geospatial QPE accuracy dependence on weather radar network configurations'. *Journal of Applied Meteorology and Climatology*. 2020;**59**(11):1773–1792.

[44] Ryzhkov A.V., Diederich M., Zhang P., and Simmer C. 'Potential utilization of specific attenuation for rainfall estimation, mitigation of partial beam blockage, and radar networking'. *Journal of Atmospheric and Oceanic Technology*. 2014; **31**(3):599–619.

[45] Cho J.Y.N., Kurdzo J.M., Bennett B.J., *et al.* 'Impact of WSR-88D intra-volume low-level scans on severe weather warning performance'. *Weather and Forecasting*. 2022;**37**(7):1169–1189.

[46] Chrisman J.N. 'Dynamic scanning'. *NEXRAD Now*. 2013;**22**:1–3.

[47] Chrisman J.N. 'The continuing evolution of dynamic scanning'. *NEXRAD Now*. 2014;**23**:8–13.

[48] Chrisman J.N. *Mid-volume Rescan of Low-Level Elevations (MRLE): A New Approach to Enhance Sampling of Quasi-linear Convective Systems (QLCSs)* [online]. 2016. https://www.roc.noaa.gov/WSR88D/PublicDocs/NewTechnology/DQ_QLCS_MRLE_June_2016.pdf [Accessed 4 Feb 2022].

[49] Bentley E.S., Thompson R.L., Bowers B.R., Gibbs J.G., and Nelson S.E. 'An analysis of 2016–18 tornadoes and National Weather Service tornado warnings across the contiguous United States'. *Weather and Forecasting*. 2021;**36**(6):1909–1924.

[50] Kingfield D.M. and French M.M. 'The influence of WSR-88D intra-volume scanning strategies on thunderstorm observations and warnings in the dual-polarization radar era: 2011–2020'. *Weather and Forecasting*. 2022;**37**(2):283–301.

[51] Kurdzo J.M., Cho J.Y.N., Bennett B.J., and Weber M.E. 'Impact of WSR-88D SAILS usage on MRMS quantitative precipitation estimation accuracy'. *Presented at 38th Conference on Environmental Information Processing Technologies*, Houston, USA, 2022.

[52] Office of Oceanic and Atmospheric Research (USA). *Weather Radar Follow on Plan: Research and Risk Reduction to Inform Acquisition Decisions*. Washington, DC: National Oceanic and Atmospheric Administration; 2020.

[53] Cho J.Y.N. and Hallowell R.G. *Detection Probability Modeling for Airport Wind-shear Sensors*. MIT Lincoln Laboratory Project Report ATC-340, 2008.

[54] Torres S.M. and Warde D.A. 'Ground clutter mitigation for weather radars using the autocorrelation spectral density'. *Journal of Atmospheric and Oceanic Technology*. 2014;*31*(10):2049–2066.

[55] Veillette M.S., Bennett B.J., Pawlak M., and Frankel R.S. 'Microburst detection with NEXRAD AMDA'. *Extended Abstracts of the 36th Conference on Radar Meteorology*, Breckenridge, USA, September 2013. Boston, MA: American Meteorological Society; 2013. p. 15.315.

[56] Cho J.Y.N. and Dupree W.J. *Radar Coverage Analysis for the Terminal Precipitation on the Glass Program*. MIT Lincoln Laboratory Project Report ATC-450, 2022.

[57] Crum T.D. and Alberty R.L. 'The WSR-88D and the WSR-88D operational support facility'. *Bulletin of the American Meteorological Society*. 1993;**74**(9):1669–1687.

[58] Hallowell R.G., Cho J.Y.N., Huang S., Weber M.E., Paull G., and Murphy T. *Wind-Shear System Cost Benefit Analysis Update*. MIT Lincoln Laboratory Project Report ATC-341, 2009.

[59] Zhang F., Snyder C., and Sun, J. 'Impacts of initial estimate and observation availability on convective-scale data assimilation with an ensemble Kalman filter'. *Monthly Weather Review*. 2004;**132**(5):1238–1253.

[60] Stratman D.R., Yussouf N., Jung Y., *et al.* 'Optimal temporal frequency of NSSL phased-array radar observations for an experimental warn-on-forecast system'. *Weather and Forecasting*. 2020;**35**(1):193–214.

[61] Matthews M.P., DeLaura R., and Venuti J. 'Strategic forecasts of TRACON airspace capacity during convective weather impacts. In: *Proceedings of 17th AIAA Aviation Technology, Integration, and Operations Conference*, Denver, USA, June 2017. Reston, VA: AIAA; 2017. p. 3430.

[62] Ushio T., Kim D.-K., Baron P., *et al.* 'Recent progress on the phased array radar at X band'. In: *Proceedings of the 2022 IEEE Radar Conference*, New York City, USA, March 2022. Piscataway, NJ: IEEE; 2022. pp. 1–4.

[63] Van Atta L.C. and Keary T.J. 'Shaped-beam antennas'. In: Silver S. (ed.), *Microwave Antenna Theory and Design*. New York, NY: McGraw-Hill; 1949. pp. 502–509.

[64] George J., Mishra K.V., Nguyen C.M., and Chandrasekar V. 'Implementation of blind zone and range-velocity ambiguity mitigation for solid-state weather radar'. In: *Proceedings of the 2010 IEEE Radar Conference*, Arlington, USA, May 2010. Piscataway, NJ: IEEE; 2010. pp. 1434–1438.

Chapter 11

The mitigation of ground clutter

J.C. Hubbert[1], S. Ellis[1] and G. Meymaris[1]

Radar echoes that are considered as clutter depend on the data user. Biological scatterers, precipitation, buildings and other man-made structures, mountains, fields, trees, and vehicles are all considered as clutter by various radar data users. Here, we consider precipitation to be the target of interest and clutter is considered to be ground-based targets with zero mean velocity, and it is such ground-clutter echoes we wish to identify and attenuate. The contamination of weather radar echoes by ground-clutter masks the precipitation signals making it difficult or impossible to determine the characteristics of the precipitation. Radar returns from ground clutter can cause large obvious areas of high reflectivity or they can cause subtle biases when the clutter echoes contain less power than the precipitation echoes. Over the past few decades, many techniques have been developed to detect and attenuate clutter echoes so that radar meteorologists can better interpret the state of the atmosphere. Much of the improvement has been made possible by advancements in radar technology that have provided not only higher quality digital signals but also vast improvements to computer processing speeds so that real-time detection and filtering are now routine. This has led to improved hydrometer classification and quantitative precipitation estimation (QPE). The mitigation of ground-clutter echoes can be broken down into two parts: (1) identification of the presence of ground-clutter echoes in a radar resolution volume and (2) filtering clutter data from that resolution volume. These topics are addressed in this chapter. Most recently in [1], a regression-based clutter filter was analyzed, and modeled for clutter filtering and experimental examples are given herein.

11.1 Historical perspective and background

Before the advent of computers that could digitize, process, and store radar signals, ground clutter was mitigated by placing the radar at a slightly lower elevation relative to the surrounding horizon thereby blocking the sidelobes of the radar antenna pattern while allowing the main radar beam to pass over the horizon [2]. If possible, this is continued good practice. Also, ground-clutter echo effects can be reduced by using

[1]National Center for Atmospheric Research, Earth Observing Laboratory, USA

shorter wavelengths since distributed Rayleigh scattering targets, e.g. rain, are proportional to wavelength as λ^{-4} whereas larger ground-clutter targets, which manifest as specular reflectors, are less dependent on wavelength [3,4]. Such choices as radar location and wavelength are frequently not available and are only partially effective. Initial clutter mitigation efforts consisted mainly of clutter maps and manual censoring but evolved quickly as technological advances occurred.

11.1.1 Ground-clutter statistics

Temporal ground-clutter amplitude statistics have been reported to follow Ricean distributions [5,6]. Ricean distributions occur when a constant vector is added to Rayleigh fluctuations. Physically, this can be interpreted as having a relatively large dominant clutter target surrounded by smaller Rayleigh-distributed scattering centers. In [7], clutter is modeled by convolving a main-lobe Gaussian antenna pattern, in two dimensions, with clutter targets composed of one dominant scatterer surrounded by smaller Rayleigh-distributed scattering centers. Vegetation motion due to wind was accommodated by the model by adding a random phase component to the scattering centers. It was shown that clutter signals generated by this model agreed well with experimentally measured clutter signals.

Some ground-clutter gates contain two dominant targets that cause the power spectrum to be doubled-peaked with peaks close to and on each side of zero velocity. This is caused by phase interference of the scattering amplitudes of the two targets as the radar scans through the dwell angle [8]. More details on ground-clutter statistics can be found in [6].

11.1.2 Ground-clutter identification

Ground clutter consists of a myriad of stationary target types and shapes: trees, buildings, poles, power lines, bushes and other vegetation, rocks, etc., and thus, clutter is characterized by high-power variability in range and azimuth, limited vertical extent, zero mean velocity, and narrow spectrum width [9–15]. By the early 1970s, researchers realized the need for automated ground-clutter mitigation and it was known that ground-clutter echoes generally varied more slowly in time as compared to precipitation echoes. Thus, the pulse-to-pulse return power fluctuations could be used to separate regions of clutter from regions of precipitation using a power variability metric. Schaffner realized the potential of digital computer technology and described two clutter identification metrics that utilize the pulse-to-pulse power fluctuations [16,17]. One algorithm measures the difference from peak-to-mean-power differences (PM) over an integration period (referred to as a radar bin or gate) and the second measures the mean value of the power difference from pulse-to-pulse (DM). These algorithms were tested by Schaffner [17] and Geotis and Silver [18] and it was found that for scanning radar, the DM method produced superior results. However, at that time, these techniques were not yet operationally viable due to limited computer and display technology [19] but prototype systems were being developed [20]. During the 1980s, much research on clutter mitigation focused on the statistical characteristics of both normal and anomalous propagation clutter. Sirmans investigated and characterized anomalous and normal propagation clutter and filter designs for the United States (US) Next Generation Weather Radar (NEXRAD) [12,21], and Long published a book of reflectivity measurements for land and sea statistics [6] (2nd ed., 1983).

In the 1990s, most clutter recognition techniques used the integrated radar moments and their spatial texture rather than pulse-to-pulse power variability. Moszkowicz *et al.* [14] used Bayesian statistics to define both linear and quadratic discriminator functions to distinguish between anomalous propagation (AP) clutter and meteorological echoes based on both vertical and horizontal reflectivity structures. For the data tested, the total classification error was about 6% as compared to human expert classified data. Even though neural network (NN) and fuzzy logic technology had been conceived decades before, it is the work of Kosko [22] in 1992 that seems to have generated interest in these classification techniques in the radar community. A NN approach was examined by Cornelius and Gagnon [23] and Pratte *et al.* [24,25]. It used the features of spatial reflectivity texture, mean velocity, mean spectrum width, and standard deviation (Std) of velocity over a two-dimensional kernel three beams wide by 2 km in range. In Pratte *et al.* [24], three clutter recognition algorithms were tested and compared: (1) empirical, (2) NN, and (3) fuzzy logic [22]. Of the three, the NN approach yielded the best results but it was noted that it was computationally complex. Grecu and Krajewski [26] used a more sophisticated volumetric NN approach that used nine input features, however, the authors state that this is not an operational technique. Another more sophisticated NN technique has recently been reported by Lakshmanan *et al.* [27] that differentiates between non-precipitation and precipitation echoes and thus also has the capability to identify biological scatter and other non-precipitation scatter in addition to ground clutter.

The fuzzy logic algorithm in Pratte *et al.* [24], which performed nearly as well as the NN approach, was deemed a more practical approach to automated operational clutter recognition. The fuzzy logic technique was further developed by Kessinger *et al.* [28] and the algorithm, termed the radar echo classifier (REC), was deployed on the US National Weather Service (NWS) operational Weather Surveillance Radar-1988 Doppler (WSR-88D) in 2003. More recently, Cho *et al.* [29] also used a fuzzy logic approach similar to Kessinger *et al.* [28], however, they optimized the membership functions through statistical analyses. Berenguer *et al.* [30] also report on a three-dimensional fuzzy logic algorithm that used the features of echo top, vertical reflectivity gradient, spin change, reflectivity texture, and a clutter frequency map. Another approach to clutter identification is parametric time-domain estimation of spectral moments [31]. Simultaneous estimation of both the clutter and weather signal properties allowed for accurate retrievals of weather moments for high clutter-to-signal ratios (CSR). Such methods are, however, computationally expensive.

The dual-polarization variables of differential reflectivity (Z_{DR}), copolar differential phase (ϕ_{DP}), and copolar correlation coefficient (ρ_{HV}) are now commonly available for clutter identification [32,33]. Since clutter echoes exhibit large spatial variability of reflectivity, Z_{DR} also is more variable spatially in clutter than in precipitation [34–36]. The ϕ_{DP} also exhibits large spatial variability in clutter and

ρ_{HV} is typically low and spatially variable in clutter [37,38]; however, the two quantities, ρ_{HV} and the variability of ϕ_{DP}, are not independent. As ρ_{HV} decreases, the standard deviation of ϕ_{DP} also increases [32,39]. Zrnic′ and Melnikov [40] show that histograms of ρ_{HV} in clutter completely overlap histograms of ρ_{HV} in precipitation and, thus, ρ_{HV} by itself is not a robust indicator of clutter. Gourley and Tabary [41] investigated an observation-based approach to derive fuzzy logic membership functions and objective weights for the feature fields of the spatial texture of Z_{DR} and ϕ_{DP}. They also included ρ_{HV} and the pulse-to-pulse variability of reflectivity in their fuzzy logic algorithm.

There are clutter identification schemes that use various spatial dimensions of radar data: (1) three-dimensional volumetric algorithms [14,26,30,42], (2) two-dimensional, in azimuth and range, algorithms [13,28,43], (3) one-dimensional in range algorithms [44], and (4) point, a single radar gate, corresponding to a radar resolution volume, algorithms [17,18,40]. To create an operational algorithm, it is advantageous to keep the algorithm as simple as possible while still obtaining acceptable clutter identification performance.

Hubbert *et al.* [7,44] describe a one-dimensional fuzzy logic clutter identification algorithm termed clutter mitigation decision (CMD) that uses spatial texture of power (dBZ), SPIN*, Z_{DR}, ϕ_{DP}, and the single gate metric, clutter phase alignment (CPA). This algorithm was deployed on NEXRAD using only the texture of dBZ, SPIN, and CPA but subsequently in 2013 was upgraded to use the texture of Z_{DR} and ϕ_{DP} [44].

Another recent clutter mitigation algorithm that combines clutter identification and clutter filtering is clutter environment analysis using adaptive processing (CLEAN-AP) [45]. It is designed to work on a single radar gate using the lag-1 auto-spectral density function (ASD). CLEAN-AP originally had difficulties with distinguishing zero-velocity weather and clutter. Therefore, a fuzzy logic algorithm, termed weather environment thresholding (WET), based on the spatial variability of the dual-polarimetric variables, was added to help distinguish precipitation from clutter. A desirable feature of the CLEAN-AP algorithm is that it allows different time series window functions to be used on a gate-to-gate basis. As described in Section 11.2 below, window functions attenuate the time series thereby causing the Std of the radar variable estimates to increase. Therefore, lesser attenuating window functions should be used when possible.

11.1.3 Ground-clutter filtering

The first attempts at attenuating radar ground clutter were either simple censoring of clutter contaminated data or pulse cancellers; however, they were widely recognized to perform poorly [46] but little else was possible at the time due to limited computing power.

During the 1980s evaluations of the digital time domain filters for weather radar were done by Groginsky and Glover [46] and Zrnic′ *et al.* [47] who report on

*SPIN is a metric that counts the number of sign changes in the gate-to-gate slope of reflectivity [42].

infinite impulse response (IIR) and finite impulse response (FIR) filter characteristics and their initialization. Sirmans [21] investigated and characterized anomalous propagation (AP) and normal propagation clutter and filter designs for NEXRAD. They found that third-order elliptic IIR filters† had sufficiently sharp transitions from the stopband to the passband and offered 50 dB or more clutter rejection for radar systems with low-phase noise transmitters such as a Klystron and wide dynamic range receivers. Due to the feedback of IIR filters, they are prone to ringing and need to be initialized to minimize the resultant transients; thus, IIR clutter filters perform best when operating on a continuous stream of radar pulses. The filter coefficients are fixed, i.e., the filter bandwidth is fixed and, therefore, IIR filters have limited stop bandwidth adaptability. FIR filters can have a similar frequency response as a third-order IIR filter but require about 39 filter taps which makes it computationally more intensive. Thus, IIR filters were considered the superior solution [47]. The clutter filtering problem is complex since (1) ground clutter has a very large dynamic range that can vary on a gate-to-gate basis and accordingly requires various filter suppression levels and (2) AP clutter causes clutter signals to appear and disappear, depending on the vertical profile of atmospheric refractivity [49], thus rendering clutter maps ineffective for solving the general clutter problem. One solution was to have radar operators select low, medium, and high clutter rejection filters [50] depending on the needed clutter rejection.

It was already recognized by the early 1980s that the discrete Fourier transform (DFT) could be an effective tool to suppress clutter in the frequency domain [51]. The Doppler spectrum of the data from each radar gate could be examined for clutter and those spectral lines around zero velocity could then be set to zero. Subsequently, the radar variables of any possible accompanying weather signal could be calculated. Indeed by the early 2000s, when radar processors had sufficient power, the US NWS adopted such a clutter filter for their network of WSR-88Ds, NEXRAD [52,53], called Gaussian Model Adaptive Processing (GMAP) [54]. GMAP determined the bandwidth of the clutter, notched out (i.e., set to zero) the clutter signal spectral components and then also used an interpolation scheme to recover any weather signal that might have been notched out by the filter. Thus, GMAP overcame the bandwidth adaptability issue of IIR and FIR filters. Furthermore, applying IIR filters to short-length time series on a gate-to-gate basis requires filter initialization to avoid filter ringing and there are time delays and non-linear phase issues. FIR filters are also problematic since all the filter taps are not filled at the beginning and the end of the time series. Presently the NWS WSR-88Ds continue to employ GMAP spectral clutter filtering along with CMD for identifying clutter-contaminated data.

Clutter filtering of staggered pulse repetition time (PRT) signals complicates clutter attenuation due to the resulting unequally spaced time series. Using interleaved pulse trains with different PRTs provides two estimates of Doppler velocity with different Nyquist rates (i.e., folding velocities) that are used to extend the

†By the 1990s, the NWS used fifth-order IIR filters [48].

overall unambiguous velocity interval. The unambiguous range is simultaneously extended [55–57]. However, when calculating the power spectrum of staggered PRT time series with a DFT, equispaced samples are required and this is typically done by interpolating the time series with zeros. Applying a DFT to such an interpolated times series results in multiple replicas of the spectrum making clutter filtering difficult.

A clutter filtering technique first presented in [58] uses a matrix formulation to eliminate clutter in each spectral replica. The non-uniform PRT time series can be represented as a product of a higher rate uniform PRT sequence of unity members that correspond to all members of the staggered sequence and zero elsewhere. This multiplication in the time domain is represented by convolution in the frequency domain thereby creating the spectral replicas. After the clutter signal is eliminated in each replica, the technique in [58] uses magnitude deconvolution to integrate the replicas and then estimate the radar variables in the spectral domain. This technique does require the use of a window function to confine the clutter signal. More recently, the ASD technique for spectral analysis and clutter filtering was applied to staggered PRT [59,60].

In [61], a simplified staggered PRT filtering technique is described where the staggered PRT sequence is broken into two equispaced sequences that are then clutter filtered. After clutter filtering, the two sequences are recombined prior to radar variable estimation. This chapter focuses on clutter mitigation for uniform PRT data though the regression clutter filter can be applied to staggered PRT sequences [62,63].

Another ground filtering technique uses polarimetric spectral decomposition of Z_{DR}, ϕ_{DP}, and ρ_{HV} [64–66]. Doppler spectrographs (Doppler spectra versus range) of the polarimetric variable are used to identify and filter ground clutter. This technique is also applicable to the mitigation of sea clutter, birds, and chaff echoes.

11.1.4 Wind turbine clutter

Another application of Doppler spectrograph filtering is the mitigation of wind turbine clutter (WTC). Wind turbine (WT) construction has had wide proliferation in the United States, Europe, and around the world due to the increasing need for non-fossil fuels to combat climate change. When WTs are visible by weather radars, their spinning blades can distribute scattered signals across the Doppler spectrum masking weather returns and contaminating data thus presenting a problem to weather services [67]. To detect WTC, Hood *et al.* [68] used the feature fields of CPA, spectral flatness, the fourth central spectral moment, and hub-to-weather ratio (HWR) in a fuzzy logic scheme. The interpolation across censored WTC gaps is addressed in [69] with a multi-quadric interpolation scheme in two and three dimensions in the time domain. Spectral-range domain interpolations are addressed in [70]. More recently, the authors of [71,72] developed an algorithm where the weather signal is modeled as group sparse in the time–frequency domain, and the WTC signal is modeled as having a sparse time derivative. The proposed objective cost function is then optimized to separate the weather and the WTC signals.

Dual-polarimetric variables are used to identify both AP clutter and WTC in [73]. Several algorithms are developed using a decision tree based on reflectivity, ρ_{HV} and the spatial standard deviation of Z_{DR} and ϕ_{DP}. WTC is difficult to identify and remove since the spectra of the WTC are highly variable and it is particularly difficult when weather and WTC signals overlap. Research continues in this area.

11.1.5 Phased array technology

Phased array technology for weather radars has become a topic of interest due to its ability for rapid scanning and for adaptive beam forming, though calibration of dual polarimetric variables and cross-coupling continue to be challenging. In [74], a survey of phased array technology for weather radar is given and many references can be found there. If a phased array radar gathers equispaced time series samples then clutter identification and filtering are the same as presented above. If the data for a resolution volume are gathered when the antenna is stationary, clutter spectrum widths will be reduced as compared to data from a rotating dish antenna. Because of the ability of beam forming, antenna pattern sidelobes, which illuminate ground-clutter targets, can be adaptively nulled thereby reducing the received clutter signal [75]. Phased array technology is promising but calibration, cross-coupling, polarization purity at squinting angles, and other issues remain [76].

For the remainder of this chapter, we will develop a time series polynomial regression-based ground-clutter filter and the automated selection of the required polynomial order for the filter stop bandwidth. A new technique for the detection of low CSR clutter, called the RHOHV-test, is also presented that augments CMD. The new regression clutter filter directed by CMD augmented by the RHOHV-test is then compared to NCAR's spectral-based window and notch filter, directed by CMD. Only uniform PRT data will be considered.

11.2 A comparison of spectral and regression clutter filtering

Spectral clutter filters, such as GMAP, operate in the frequency domain while a regression clutter filter operates in the time domain [1,48]. Both filters operate on a single gate of data, both are adaptive and both can use Gaussian interpolation across the zero-velocity gap caused by the clutter filtering process. The two primary differences are (1) the orthogonal basis functions used to characterize the clutter signal and (2) the regression filter does not use window functions whereas the spectral filter does to contain clutter leakage.

11.2.1 Spectral clutter filter and Fourier basis functions

Spectral clutter filters use the DFT on a gate of radar data thereby expressing the time series as a summation of the Fourier series' orthogonal sine and cosine functions, in complex exponential form, from which the power spectrum is calculated. If there are N points in the time series signal, then there are N discrete complex frequency basis functions. If the radar time series has a frequency component that is not one of the N basis frequencies, then that signal must be represented by some combination of the available basis functions. This causes spectral leakage [77]. This is not particularly problematic since weather signals are not pure frequencies, do vary over the radar dwell time, and thus have a significant spectrum width. For example, see the weather time series in Figure 11.2 (b) in Section 11.2.2. Nevertheless, spectral leakage for slowly varying clutter signals over the dwell time significantly spreads the clutter power across the spectrum. The variability of the clutter signal versus time is typically much less than the lowest frequency basis function (ignoring the constant basis function), which is one cycle of a sinusoid over N points. A slowly varying clutter signal generally cannot be captured by the lowest frequency basis functions of the DFT and the result is spectral leakage.

Another way to understand clutter spectral leakage is via the time domain. When representing an N point sequence as a sum of sinusoids via the DFT, a periodic sequence, of period N, is created due to the periodicity of the basis sinusoids. The discontinuity of the time domain sequence at the periodic boundaries causes spectral leakage. The situation is more complicated since the discontinuity of the derivatives at the periodic boundary point also causes spectral leakage [77]. Therefore, to suppress clutter leakage and confine the clutter power to around zero velocity, window functions such as the Blackman, von Hann, or Hamming are used to smooth the periodic time series boundaries. The selection of the window function depends upon the clutter-to-noise ratio (CNR), CSR, and the side lobes of the spectra of the window function [45].

The drawback of the spectral filter technique is that the window function significantly attenuates the time series signal as shown in Table 11.1. In terms of just removing clutter from a clutter-only signal, the window attenuation is unimportant. However, if a weather signal is superimposed onto the clutter signal, then the weather signal also suffers this same amount of attenuation as given in Table 11.1. This in effect reduces the number of independent samples for estimating the radar variables of the weather signal due to the reduced influence of attenuated time series members. This in turn increases the measurement errors. This increase in the Std of the radar variables has been accepted as a compromise for the advantages of the spectral clutter filter but it is a major drawback [1,45,54]. Another issue is that the application of the window function, while reducing clutter leakage, also broadens the spectrum width of the clutter. For example, if the clutter signal is a constant in the time domain, or equivalently a Dirac delta function at zero frequency in the frequency domain, applying a Blackman window would increase the spectrum width of the clutter to the spectrum width of the main lobe of the spectrum of

Table 11.1 Power attenuation caused by four commonly used window functions for 64-point sequences

Window	Attenuation (dB)
Hamming	4.08
von Hann	4.19
Blackman	5.23
Blackman-Nuttall	5.89

The attenuation is greater for shorter length sequences.

the Blackman window. This is easily understood since multiplication in the time domain corresponds to convolution in the frequency domain [78]. One scanning strategy used by the NWS WSR-88Ds, at lowest elevation angles, accommodates both 0.5° azimuthal resolution (termed super-resolution) and range-velocity ambiguity mitigation via SZ phase coding of the transmit radar pulses [79,80]. This requires the use of, overlapping window functions typically the von Hann or Blackman window. This complicates the comparison of regression and spectral filtering and is not addressed in this chapter.

Because of the detrimental effects of window functions, it is of interest to identify and attenuate clutter signals without the use of window functions. This can be done by characterizing the ground-clutter signal using orthogonal polynomials as basis functions in the time domain.

11.2.2 Regression filtering and orthogonal polynomials

A conceptually different approach to ground-clutter suppression is regression filtering. Regression analysis is frequently used to estimate the trend of a time series or to analyze the residuals after the trend has been eliminated [81]. For typical longer wavelength weather radar data, the clutter signal varies slowly while the weather signal typically varies faster in time and, therefore, an appropriate polynomial order regression fit to the data can track the slowly varying trend due to the clutter targets. Regression filtering has been investigated in the biomedical field for ultrasound images to suppress stationary and slow-moving tissue signal so that the blood flow signal of interest can be measured [82,83]. For S-band weather radar, the authors of [48] examined a regression clutter filter for NEXRAD and it was found to have comparable performance to IIR filters. At that time, GMAP also became a viable option and it was adopted in lieu of regression filtering due to its adaptive filter bandwidth capability and interpolation algorithm. Additionally, GMAP was integrated into the RVP8 receiver that became the new receiver for the WSR-88D. Also at that time in [48], it was a *local regression filter* that was applied to NEXRAD experimental data. Local regression filters can be formulated as an FIR filter [84]. The frequency response of local regression filters, in terms of clutter rejection capability, is not as good as a global regression filter [1,84]. Global here means that the regression fit is executed on the entire length of the signal to be filtered instead of using a sliding window as is done with local regression filters.

The principle of the regression clutter filter is straightforward: a least squares polynomial regression fit is used to estimate the trend of a radar time series, which is assumed to be the clutter signal. To illustrate this, the I (in-phase) and Q (quadrature) parts of an experimental time series from S-Pol (S-band Polarimetric radar), operated by the National Center for Atmospheric Research (NCAR) for the National Science Foundation, are shown in Figure 11.1. S-Pol is scanning the Rocky Mountains in azimuth at a constant elevation angle at a rate of 8° s^{-1} in clear air conditions. There are 1,024 points in each plot with a PRT of 2 ms. As clutter targets enter and exit the antenna transmit beam, they influence the I and Q signals in a gradual and continuous fashion. As can be seen, the I and Q signals vary smoothly in time. The red vertical lines mark successive increments of a 1° azimuth angle, i.e., a typical dwell angle. It is easy to see that a low-order polynomial

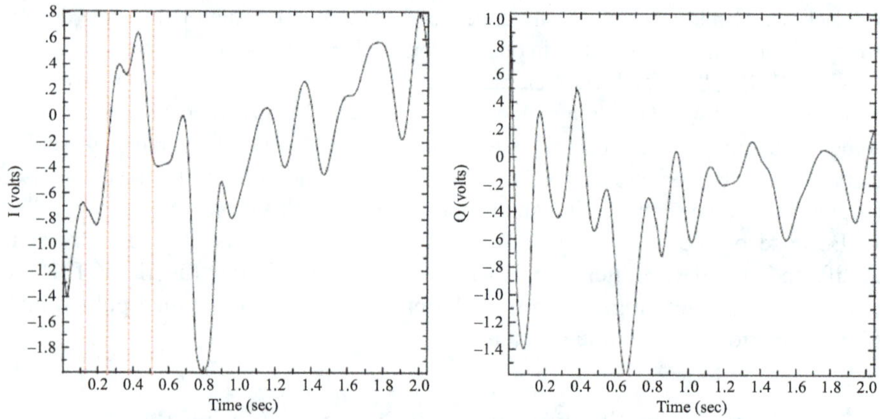

Figure 11.1 S-Pol I and Q time series of ground clutter for an 8° s^{-1} scan rate. There are 1,024 points in each plot. The vertical red lines represent increments of 1° in azimuth, the typical dwell angle for S-Pol.

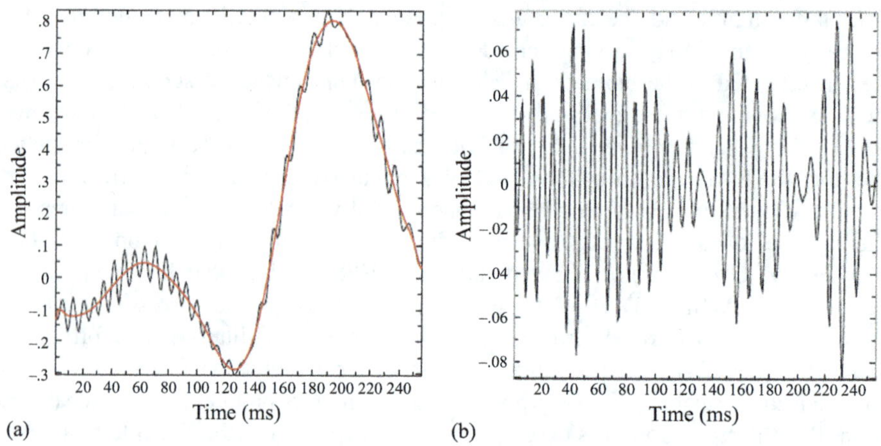

Figure 11.2 The real part of an S-Pol time series for two gates of data for illustration purposes: (a) the high-frequency weather component with a low-frequency clutter component. The red line is a polynomial regression fit. Panel (b) shows the difference between the black and red curves in panel (a), i.e., the weather signal component. From [1] © American Meteorological Society. Used with permission.

regression fit should be able to approximate the *I* and *Q* clutter curves over a 1° dwell angle. Figure 11.2(a) shows another shorter length time series that contains both weather and a clutter signal in panel (a). The red curve is a regression fit that is an estimate of the clutter signal. Figure 11.2(b) is the weather signal that results from subtracting the red curve from the black curve of Figure 11.2(a).

For low round-off error and reduced computation time, it is important to use orthonormal polynomials for the least squares fit. The Forsythe polynomial recursive fitting algorithm is used in this chapter [1,86,87]. The details of orthogonal polynomial least squares regression have been covered in many textbooks and papers.

The first seven Forsythe polynomials are plotted in Figure 11.3. Orders zero, one, and two correspond to a constant, a straight line, and a quadratic (parabolic shape), respectively. In general, these lower order Forsythe basis functions can better approximate slow varying clutter than the low-frequency Fourier basis functions. More precisely, the first few Forsythe polynomials can contain much more of the clutter power than the low-frequency Fourier sinusoids. However, if an appropriate window function is used, the clutter is then confined to zero velocity and near zero velocity Fourier basis functions [77].

11.2.3 The frequency response of the regression filter

The output of linear time-invariant filters, such as FIR and IIR, can be represented as a convolution of the filter's impulse response with the input signal. The global regression filter has no such representation, whereas the local regression filter does, such as Savitzky-Golay [84]. Nevertheless, the global regression filter is linear, time invariant. That is, if $R_F\{x_i\} = y_i$ and $R_F\{u_i\} = z_i$ where $R_F\{\cdot\}$ denotes regression filter operation then it can be shown as

$$R_F\{x_i + u_i\} = y_i + z_i \tag{11.1}$$

$$R_F\{\beta x_i\} = \beta y_i \tag{11.2}$$

where β is a constant. The regression filter operates on a block of data at a time and if the block of data is delayed in time, the regression filter will yield the same

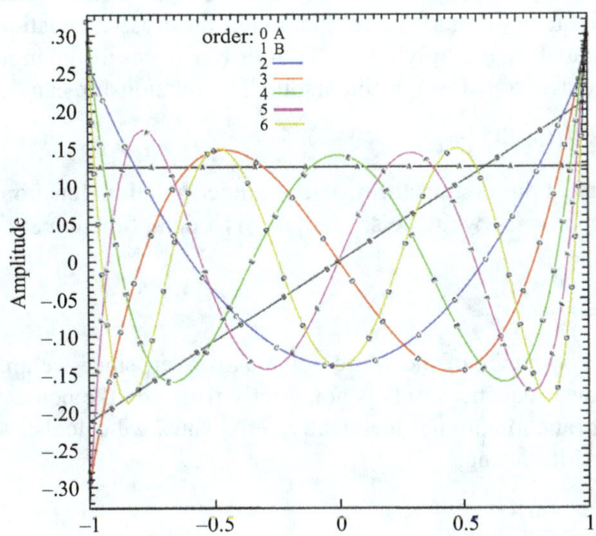

Figure 11.3 The first seven Forsythe orthogonal polynomials

output. In this sense, the regression filter is linear and time invariant. However, the global regression filter (essentially a polynomial fit) does not have an impulse response and cannot be expressed as a convolution of the impulse response and the input. Still, the frequency response of the regression filter can be defined, which has been reported before [48,82,83]. The regression clutter-filtered signal y_i can be written,

$$y_i = x_i - \sum_{j=0}^{n} \alpha_j P_j(x_i) \tag{11.3}$$

where y is the filtered sequence, x is the signal to be filtered, α are the regression coefficients, and P_j are orthonormal polynomials. It can be shown that the frequency response, $H(\omega)$, of the regression filter is

$$H(\omega) = 1 - \frac{1}{m}\sum_{i=0}^{p} |B_i(\omega)|^2 \tag{11.4}$$

where ω is the frequency and $B_i(\omega)$ are the Fourier transform of the basis polynomials P_j, p is the polynomial order and m is the filter length [48,88]. Here an alternative approach is taken to numerically generate the filter frequency response. The frequency response of the regression filter is calculated numerically by using complex white noise sequences as input into the clutter filter (11.3) [78]. Let $x(n)$ and $y(n)$ be the time domain input and output signals where n is a time index and let the cross-correlation function be denoted $x(n)\star y(n)$. The Fourier cross-correlation transform pair is,

$$\mathcal{F}\{x(n)\star y(n)\} = \mathcal{F}\{x(n)\}\mathcal{F}\{y(n)\}^* = \Phi_{xy}(\omega) \tag{11.5}$$

where * is the complex conjugation and $\mathcal{F}$ denotes the Fourier transform so that $\Phi_{xy}(\omega)$ is the frequency domain representation of the cross-correlation function and it can be calculated by multiplying the Fourier transform of the input by the conjugate of the Fourier transform of the output. The relationship can be written [78],

$$\Phi_{xy}(\omega) = H(\omega)\Phi_{xx}(\omega) \tag{11.6}$$

where Φ_{xx} is the Fourier transform of the autocorrelation function of the input signal. If the input is the white noise, then Φ_{xx} is a delta function at $\omega = 0$ and zero elsewhere so that

$$\Phi_{xy}(\omega) = H(\omega)\sigma_W^2 \tag{11.7}$$

where σ_W^2 is the variance of the white noise. However, since we are using finite-length white-noise sequences, this is not strictly true, i.e., the non-zero lags of the autocorrelation function are not identically zero. Thus, we calculate the frequency response of the filter using,

$$H(\omega) = \frac{\Phi_{xy}(\omega)}{\Phi_{xx}(\omega)}. \tag{11.8}$$

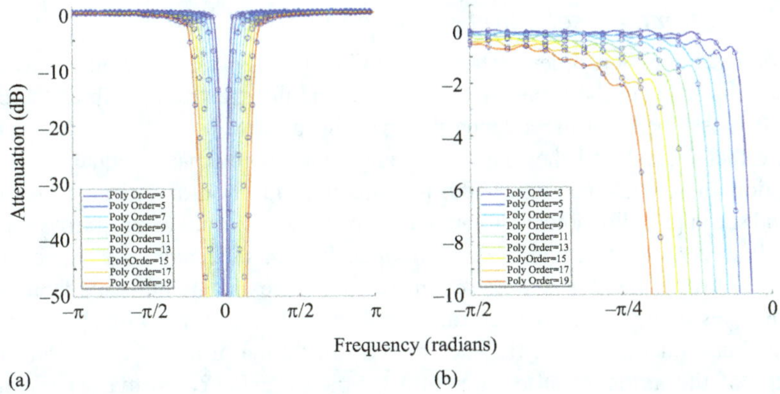

Figure 11.4 *The regression filter frequency response (power) for 64-point time series for various polynomial orders as indicated. Left: full frequency response. The small circles mark the locations of the discrete 64 points. (b) expanded version of (a) from $-\pi/2$ to 0 radians. From [1] © American Meteorological Society. Used with permission.*

Multiple simulated white noise sequences are filtered and then the corresponding $H(\omega)$ are averaged in order for the frequency response to converge to its true theoretical value.

Figure 11.4 shows the frequency response of the regression filter for length-64 time series for various polynomial orders. The small circles mark the locations of the 64 points. The curve between these points is filled in by zero padding the filtered time series before calculating the power spectrum. Figure 11.4(a) shows the full-frequency response, from $-\pi$ to π radians, while Figure 11.4(b) shows a zoomed-in version for detail in the transition region. The frequency response depends on the length of the sequence to be filtered and the polynomial order. As the polynomial order increases, the stop bandwidth of the filter increases. Longer length sequences require higher order polynomials to achieve a similar frequency response and vice versa. For very long sequences, very high polynomial orders may be required. In such cases, it is possible to divide the sequence into shorter blocks, regression filter each block separately, and then rejoin the filtered blocks [1]. The primary use of the global regression filter frequency response is to determine the needed polynomial order for sufficient clutter suppression.

11.3 Operational aspects of regression clutter filtering

Three practical aspects of implementing the regression clutter filter are (1) automated estimation of the required polynomial order, (2) estimation of the width of the zero-velocity gap, and (3) an algorithm for Gaussian interpolation across the zero-velocity gap created by the clutter filter so that reflectivity, velocity, and spectrum width biases can be reduced. The attractive aspect of GMAP is that it has an adaptable bandwidth and a Gaussian interpolation routine. Next, we show that this can also be accommodated with the regression filter.

11.3.1 Automated selection of the polynomial order

The required regression order of the clutter filter is a function of clutter power, the number of samples in the time series, the PRT, and the spectrum width of the clutter signal. It is assumed that the antenna pattern main beamwidth (about 1° for many S-band weather radars) and the radar dwell angle are approximately equal. Under this assumption, the clutter spectrum width is primarily a function of antenna rotation rate and wind speed [57]. For antenna rotation rates of 12–20° s^{-1} and wind speeds of around 10 m s^{-1}, clutter spectrum widths of approximately 0.24–0.32 m s^{-1} are expected.

To confirm these clutter width estimates, we use experimental data from S-Pol. Accurately estimating clutter spectrum widths is not straightforward. For example, the large bias and variance of the spectrum width estimator based on the zeroth and first lags of the autocorrelation function (termed the R_0/R_1 estimator) is demonstrated via simulations in [7]. In [8], experimental S-Pol data are used to estimate the spectrum width of ground clutter as a function of antenna scan rate. Since the magnitude of the autocorrelation function decreases slowly as a function of lag for narrow spectrum width clutter, a seven-lag estimator was used in [8] to decrease bias and variance of the spectrum width estimate. The result is shown in Figure 11.5 for a dwell angle of 1°. The blue line is the experimental data (blue dots are the measured data points), whereas the red line is a straight-line fit. The fit parameters are given on the plot and thus the relationship between clutter width and scan rate using NCAR's S-Pol data is:

$$w_c = 0.03 + 0.017\alpha \tag{11.9}$$

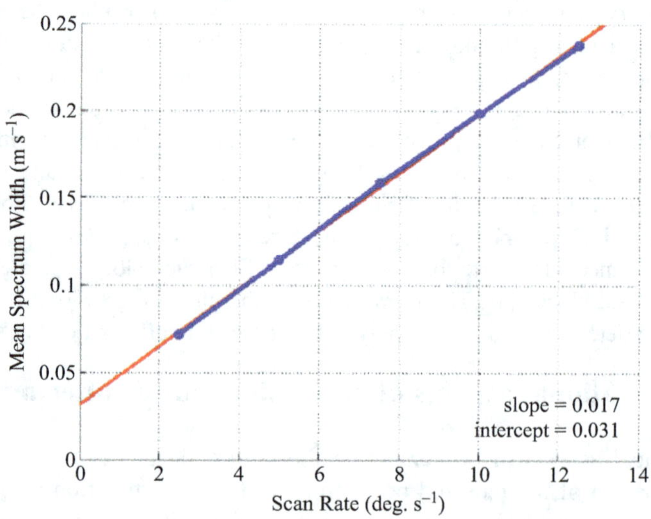

Figure 11.5 Spectrum width versus radar scan rate for experimental S-Pol clutter data. The dwell angle is 1°. The red curve is a least squares straight line fit. The blue dots are the estimated clutter spectrum widths. A Hamming window is used to limit clutter leakage.

where w_c is the clutter width in m s^{-1}. It is assumed that this relationship can be extended and used to predict clutter widths for higher scan rates to 30° s^{-1}. Other radar sites may have more vegetation that could affect the clutter widths at higher wind speeds. In this case, one could use (1.9) with an added scale factor k,

$$w_c = k(0.03 + 0.017\alpha) \tag{11.10}$$

to account for larger observed spectrum widths. For example, if it is found that the regression filter was not suppressing the clutter sufficiently, k could be increased from its nominal value of 1.

Next, simulated clutter time series for the parameters given in Table 11.2 are used to determine the polynomial order required to attenuate the clutter power to the noise floor. The used I and Q simulator was developed by [89]. The given parameters cover a range of values typically used for NEXRAD scans and for each scenario, 1,000 I and Q time series are simulated. The required polynomial order is then determined by selecting the lowest order polynomial that will, on average, attenuate the clutter signal down to the noise floor.

The results of the simulations are given in Figure 11.6. Shown is the normalized filter order (ratio of required regression order to the number of pulses) as a function of normalized clutter width (w_c/Nyq), where Nyq is the Nyquist (folding) velocity, for all of the different scenarios tested. Dots (simulation scenarios) with the same color have the same CNR, as described in the legend. The lines shown are the best quadratic least square fit lines for each CNR. It is evident that, for all scenarios tested, there are strong relationships for each CNR and w_c/Nyq. In general, as the CNR increases, higher order polynomials are required and as the normalized w_c becomes larger, higher order polynomials are required to attenuate the clutter signal. For a given scan rate, Nyquist velocity, PRT, and CNR, the vertical axis gives the normalized regression order. This value is multiplied by the number of pulses, which gives the required regression order.

Having a single-closed form relationship for predicting the polynomial order, which includes CNR, would be convenient for operational implementation. Heuristically, it was found that by dividing the normalized regression order by $CNR^{2/3}$, the curves in Figure 11.6 can be made to nearly lie on top of each other and this normalized relationship is shown in Figure 11.7. A quadratic least squares fit the data in Figure 11.7 yields

$$O_n = -1.9791w_{cn}^2 + 0.6456w_{cn} \tag{11.11}$$

Table 11.2 The parameter space for the simulation data used in Figure 11.6

Parameter	Value
Wavelength (λ)	10.68 cm
Dwell angle (θ)	1°
Scan rate (α)	5, 10, 15, 20, 25, and 30 (°) s^{-1}
PRT (T_s)	780, 1,000, 1,500, 2,000, 2,500, and 3,000 μs
Number of pulses (samples)	Based on dwell angle, PRT, and scan rate

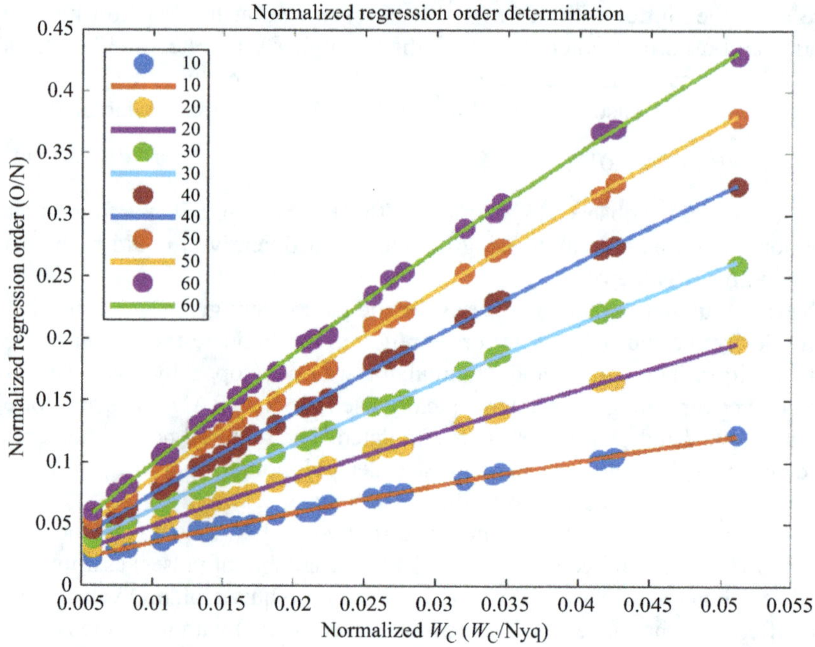

Figure 11.6 The normalized filter order (ratio of required regression order and the number of pulses) as a function of normalized clutter width (w_c/Nyq) for all the different scenarios tested. Dots with the same color have the same CNR, as described in the legend. The lines shown are the best quadratic fit lines for each CNR.

where O_n is the normalized polynomial order. Using this information, an algorithm for determining the required polynomial order can be constructed:

1. Compute the expected (unnormalized) clutter width as $w_c = k(0.03 + 0.017\alpha)$. If more clutter suppression is needed, one can increase k from one. Increasing k to 2 doubles the expected clutter spectrum width.
2. Compute the normalized expected clutter spectrum width as $w_{cn} = w_c/Nyq$.
3. Compute the normalized regression order: $O_n = -1.9791 w_{cn}^2 + 0.6456 w_{cn}$
4. Find the regression order as $O = ceil(O_n CNR^{2/3} N)$. The *ceil* function rounds numbers up to the next larger integer.

11.3.2 Interpolation across the zero-velocity gap

Clutter spectrum widths are typically narrow at about 0.3 m s^{-1}, depending on antenna rotation rate, while weather signals are typically wider at 2 m s^{-1} or greater. When applying a clutter filter to a radar signal where the weather signal is also present, portions of the weather signal can also be eliminated thus causing biases to the radar variable estimates, especially if the velocity of the weather is close to

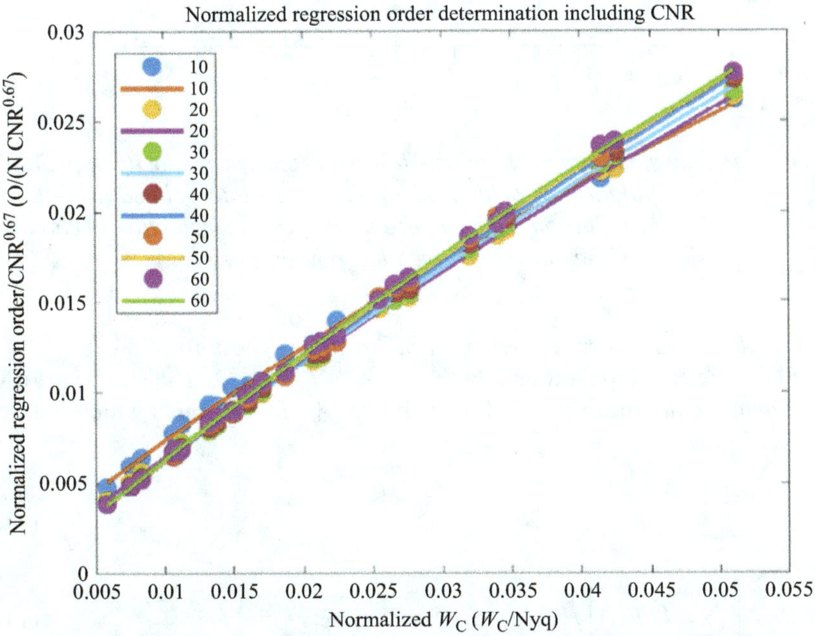

Figure 11.7 *The ratio of normalized filter order (ratio of required regression order and the number of pulses) and the CNR*$^{2/3}$ *as a function of normalized clutter width* (w_c/Nyq) *for all the different scenarios tested. Dots with the same color have the same CNR, as described in the legend. The lines shown are the best quadratic fit lines for each CNR.*

zero. To reduce bias in the estimates of reflectivity, velocity, and spectrum width, it is possible to interpolate across the zero-velocity gap created by the clutter filter. Since the power spectra of weather echoes are considered to be Gaussian in shape, a Gaussian fit routine can be employed that uses the information on either side of the zero-velocity gap.

With the regression filter, the spectral components are attenuated as indicated by the frequency response plot given in Figure 11.4 for 64-point sequences (see [1] for other frequency response lengths). The width of the gap to be interpolated is defined as the points that first fall below 1 dB of the power at the Nyquist velocity.

Next the interpolation algorithm is described. Referring to Figure 11.8 (velocity domain plot), let $F(i)$ be the power spectrum to be interpolated with i the velocity index. Let the number of points in the zero-velocity gap to be interpolated be $2L+1$ where $i=0$ is the 0-velocity component. The M points are included at each end of the $2L+1$ points to be interpolated that capture the shape of an underlying weather power spectrum (if present), which is assumed Gaussian. Based on simulations over a variety of typical scanning parameters, i.e. Table 11.2, M values of three or four perform well based on best radar variable recovery statistics. The total number of points used in the interpolation algorithm

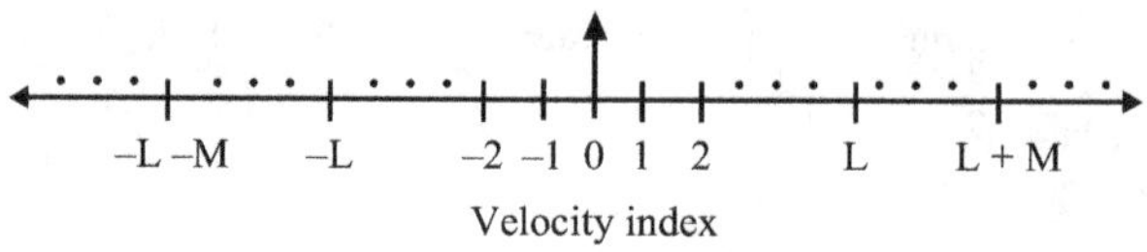

Figure 11.8 The spectrum velocity axis illustrates the zero-velocity gap Gaussian interpolation algorithm. The points to be interpolated are $-L$ to L. The M points on each end are used in the interpolation to capture the shape of the underlying weather power spectrum.

is $2M + 2L + 1$ as shown in Figure 11.8. An initial guess for the central $2L + 1$ points is a linear interpolation, $[F(-L - 1) + F(L + 1)]/2$. The Gaussian fit uses the estimate of the mean and Std over the $2M + 2L + 1$ points which are determined as follows:

$$\varsigma = \sum_{-M-L}^{M+L} F(i) \tag{11.12}$$

$$a = \left[\sum_{-M-L}^{M+L} i \cdot F(i)\right] / \varsigma \tag{11.13}$$

The Std then is,

$$s = \left\{\left[\sum_{-M-L}^{M+L} F(i)(i - a)^2\right] / \varsigma\right\}^{\sqrt{0.5}}. \tag{11.14}$$

The estimated Gaussian fit can be written,

$$\widehat{F}(i) = \frac{\varsigma}{\sqrt{2\pi}s} e^{(i-a)^2/2s^2} \tag{11.15}$$

The middle $2L + 1$ points of the power spectrum are replaced according to (11.15). The Gaussian fit process can be iterated several times to allow the Gaussian fit values to converge.

The Gaussian interpolation scheme should only be implemented when the weather velocity spectrum is close to zero velocity and, thus, the mean velocity of the weather needs to be estimated. To do this, linear interpolation across the zero-velocity gap is used after clutter filtering to reduce the bias of the initial velocity estimate. The velocity estimate, V_{est}, is then normalized by the Nyquist velocity, V_{est}/Nyq, and when the magnitude of this ratio is less than V_{th}, the Gaussian interpolation scheme is applied. Otherwise, the initial linear interpolation is used. The value of $V_{th} = 0.2$ is selected heuristically from simulations and using values from 0.15 to 0.25 yield similar results.

Reflectivity, velocity, and spectrum width are calculated from the interpolated spectrum, which can be transformed to the time domain where the autocorrelation

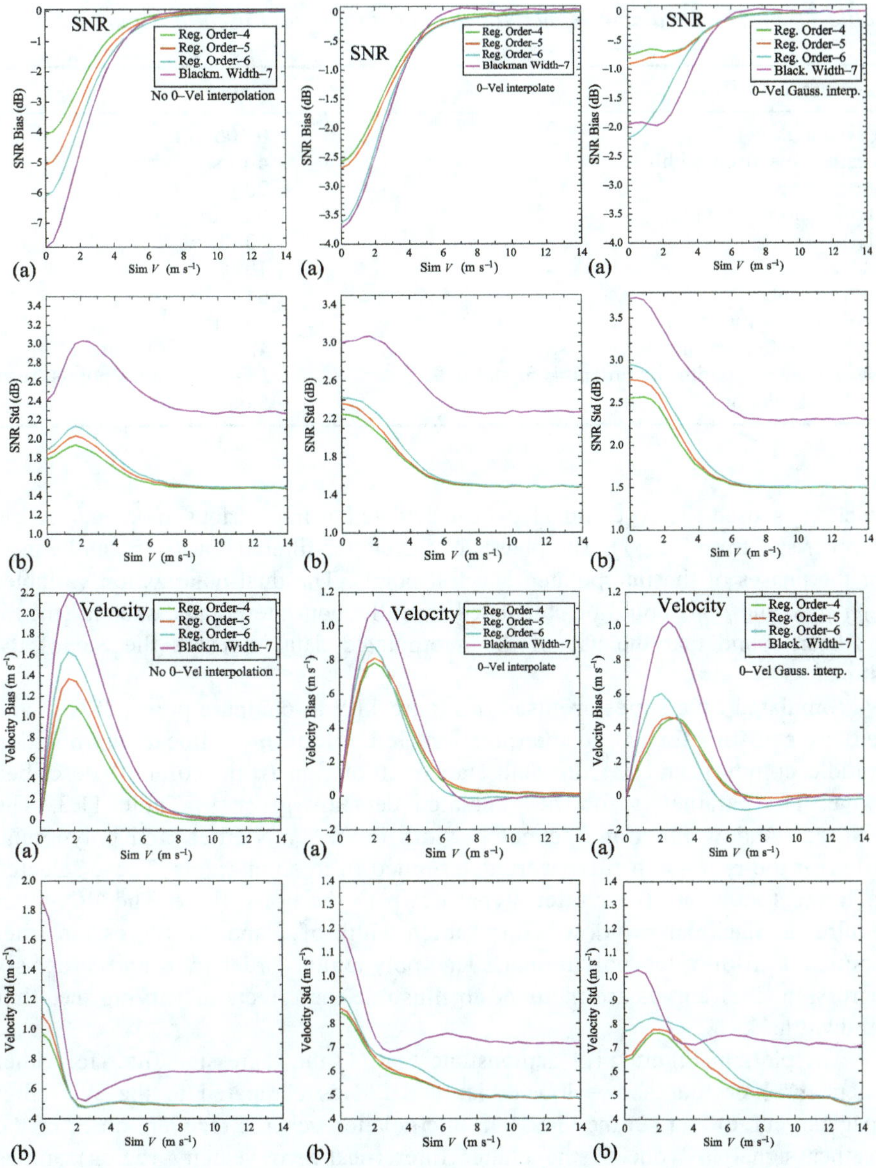

Figure 11.9 *A simulation comparison of zero-velocity gap interpolations. Left column: no interpolation; middle column: linear interpolation and right column: Gaussian interpolation. The bias (panels a), and Std (panels b) for SNR and velocity are shown. The magenta curve is for the WN filter and the other curves are for regression filter orders of 4, 5, and 6. The regression filter shows not only reduced Std as compared to the WN filter but also reduced bias for simulated velocities* < 6 *m s*$^{-1}$ *and* > 1 *m s*$^{-1}$.

Table 11.3 The parameter space for the simulation data used in Figure 11.9

Parameter	Value
Wavelength (λ)	0.1067 m
Weather spectrum width	4 m s^{-1}
Weather SNR	20 dB
Clutter spectrum width	0.28 m s^{-1}
CNR	40 dB
PRT	1 ms
No. of points	64
Nyquist velocity	26.8 m s^{-1}
Gaussian interp. points	M = 3 (see Figure 11.8)
0 V gap interp. width for orders 4, 5, and 6	L = 2, 2, and 3 points, resp.
No. of simulations	5,000

function is used. They could also be calculated using various moments of the Doppler spectrum [32,57]. The phases of the clutter-filtered spectrum can be used for the phases of the interpolated spectral points. The dual-polarization variables Z_{DR}, ϕ_{DP}, and ρ_{HV} should be calculated from the non-interpolated spectra since it has been found that the use of the interpolated data increases the Std of the estimates.

Simulated time series are used in Figure 11.9 to compare power (SNR) and velocity statistics for (1) no interpolation (left column), (2) linear interpolation (middle column), and (3) the Gaussian interpolation (right column) described above. The parameters for the simulated data are given in Table 11.3. The required window function and notch width for the WN filter, and polynomial order for the regression filter were determined from simulations. These selected parameters attenuate the clutter signal down to the noise floor. The WN filter requires a Blackman window with a notch width of 7 and the regression filter requires a fifth-order polynomial. The polynomial order-four and order-six regression filter curves are included to illustrate the effects of varying the filter bandwidth.

The plots in Figure 11.9 demonstrate that (1) the regression filter for either orders three or four shows lower bias and Std as compared to the WN filter (magenta curve) and (2) the Gaussian interpolation scheme recovers much of the weather signal lost due to the clutter filter (near-zero velocity region) and is superior to linear interpolation in the near zero-velocity region.

11.4 Identifying low CSR clutter

When the CSR is greater than zero dB and clutter dominates any underlying weather signal, clutter-contaminated gates are robustly detected with fuzzy logic algorithms such as CMD. In [44], it is estimated that about 98% of the clutter-contaminated gates are detected by CMD for CSR = 0 dB and a 50%

identification rate is obtained at about CSR = −8 dB. It is difficult to accurately estimate the true CSR from experimental radar data when the weather echo is close to zero velocity. Thus, only those gates that had velocities well away from zero velocity were used for that performance estimate in [44]. This was done to minimize the possibility that the clutter filter removed weather power thus biasing the CSR estimate. Regardless, there are numerous missed detections for these low CSR cases that affect the data quality of radar images. Clutter with low CSR can bias reflectivity, velocity, and spectrum width but it is the polarimetric variables that are the most sensitive to bias. In [90], it was shown that a significant bias of Z_{DR}, ϕ_{DP}, and ρ_{HV} occurs for CSRs down to −13 dB, −7 dB, and −18 dB, respectively. CMD and all conventional fuzzy logic clutter detection algorithms have difficulty identifying clutter robustly at these low CSRs since weather dominates the radar variables effectively masking the clutter. The most sensitive radar variable to low CSR clutter is ρ_{HV} and thus we take advantage of this fact to design a low CSR clutter detection algorithm.

It is well known that ρ_{HV} is very high in rain and in other single precipitation-type regions. Tumbling hail is an exception to this. The ρ_{HV} becomes lower in mixed phase precipitation areas such as the bright band, atop Z_{DR} columns, and at the interface of an updraft-downdraft where again mixed phase precipitation is common. When low CSR clutter and precipitation signal are mixed, ρ_{HV} is also lowered as reported in [90]. Therefore, when the clutter signal is eliminated by a clutter filter and significant weather remains, it is expected that ρ_{HV} will increase for the filtered signal. We define ρ_{HV}^{uflt} as the ρ_{HV} calculated from the unfiltered signal and ρ_{HV}^{filt} as the ρ_{HV} calculated from the clutter filtered signal. When $\rho_{HV}^{filt} > \rho_{HV}^{uflt}$, this indicates that the clutter filter removed significant clutter and thus the clutter filter should be applied to reduce bias. This is called the RHOHV-test and it is only applied on gates that were not flagged as clutter by CMD [91].

To implement the RHOHV-test, several thresholds are required. An upper threshold of $\rho_{HV}^{uflt} < 0.98$ is used since values greater than 0.98 indicate either no or insignificant clutter is present. A lower threshold of $\rho_{HV}^{filt} > 0.5$ is used since theory and experience indicate that there are no precipitation particles that yield values lower than 0.5 [36].

How much greater should ρ_{HV}^{filt} be as compared to ρ_{HV}^{uflt} to warrant the application of the clutter filter? First, an improvement factor is defined as,

$$\mathcal{F} = \frac{1 - \rho_{HV}^{uflt}}{1 - \rho_{HV}^{filt}} \tag{11.16}$$

where $\mathcal{F}$ is the amount of increase in ρ_{HV}^{filt} relative to one divided by the value of ρ_{HV}^{uflt} relative to one. Thus, larger increases in ρ_{HV}^{filt} are required for lower values of ρ_{HV}^{uflt}, as compared to higher ρ_{HV}^{uflt}, to yield the same value $\mathcal{F}$. The required amount of improvement is $\mathcal{F} > 4$ which was derived heuristically from testing with a variety of NEXRAD data sets.

11.4.1 RHOHV-test example

In this example, CMD and CMD augmented with the RHOHV-test, as used by NEXRAD, are applied to NEXRAD KDLH data from Duluth, MN. GMAP is the used clutter filter but the regression filter could equally well be used. Figures 11.10–11.13 show KDLH radar variables, as labeled, gathered on 16 March 2013 at 0110 UTC in stratiform snow. The data are from the 0.5° elevation angle. The no-clutter-filter Z field (Figure 11.10(a)) shows the stratiform snow with ground clutter near the radar and along a ridge to the south of the radar. The CMD flag (Figure 11.10(c)) shows that CMD identified a very large majority of the strong clutter, which is removed from Z by GMAP (Figure 11.10(b)). The only clutter filter bias seen is due to strong clutter overlapping the 0 m s^{-1} isodop in V_r (radial velocity) (Figure 11.10(d)), which is not recoverable using a clutter filter, indicated by the white ovals in Figure 11.10(a) and (b).

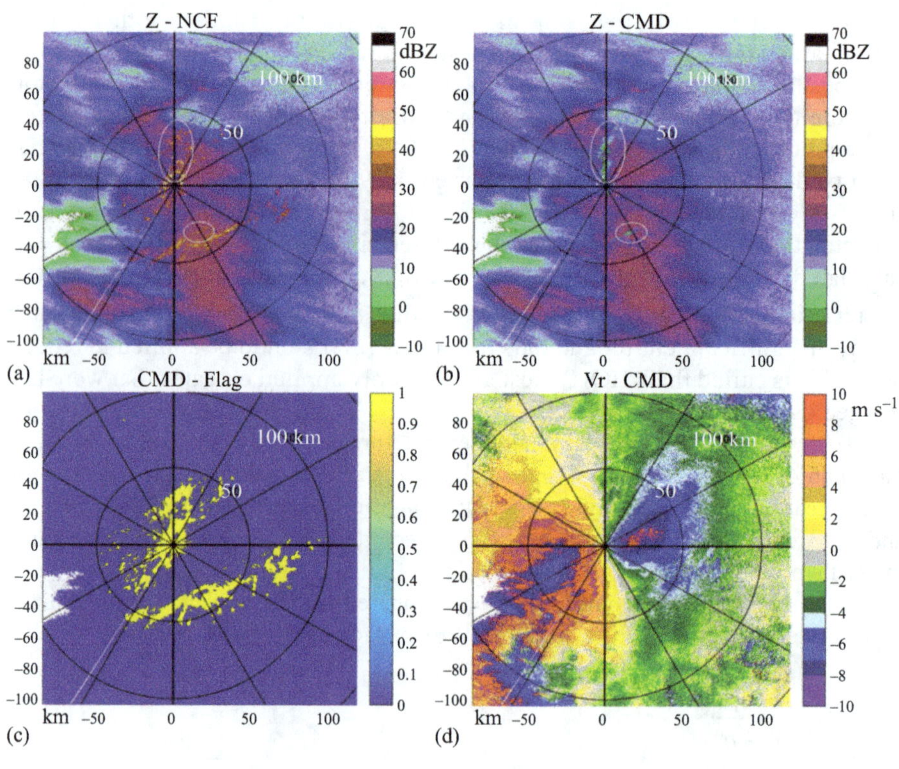

Figure 11.10 Data from KDLH taken on 16 March 2013 at 0110 UTC. The data are from the 0.5° elevation angle. Shown are Z with no clutter filtering (a), Z after CMD/GMAP processing (b), CMD flag (c), and V_r after CMD/GMAP processing (d). The ovals indicate regions of strong clutter identified by CMD within the 0 m s^{-1} isodop that have clutter filter bias.

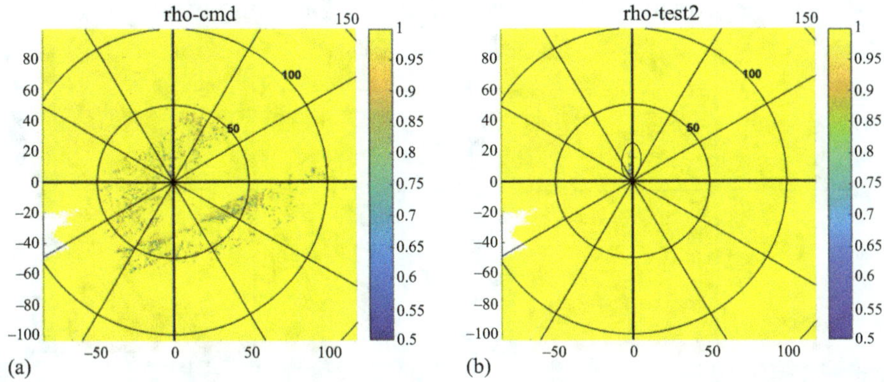

Figure 11.11 *Data corresponding to Figure 11.10, ρ_{HV} after ground-clutter filtering following CMD (a) and RHOHV-test (b). The ovals indicate regions of strong clutter identified by CMD within the 0 m s^{-1} isodop that have clutter filter bias.*

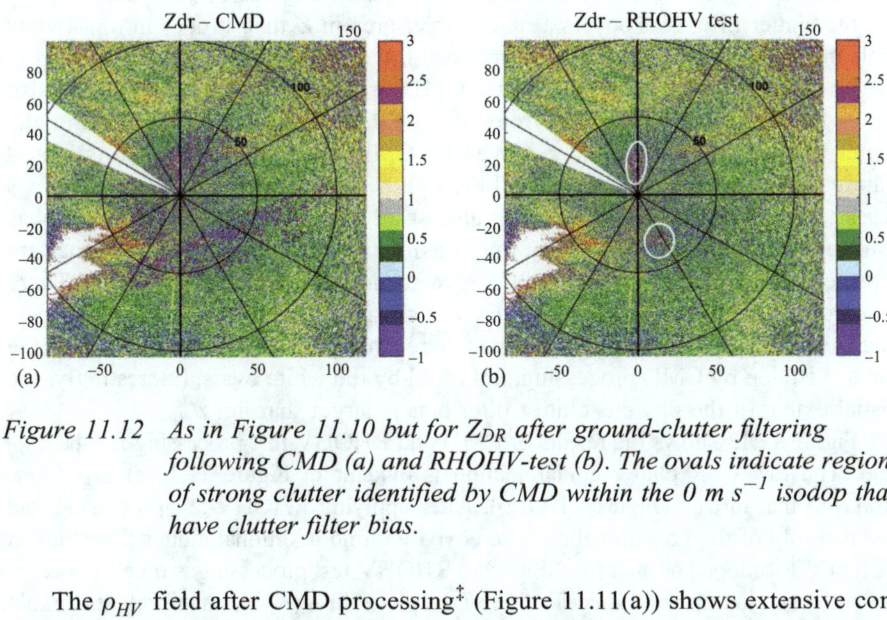

Figure 11.12 *As in Figure 11.10 but for Z_{DR} after ground-clutter filtering following CMD (a) and RHOHV-test (b). The ovals indicate regions of strong clutter identified by CMD within the 0 m s^{-1} isodop that have clutter filter bias.*

The ρ_{HV} field after CMD processing‡ (Figure 11.11(a)) shows extensive contamination from low CSR ground clutter surrounding the clutter that was identified and removed by CMD processing. After the application of RHOHV-test processing, almost all of the contamination in ρ_{HV} is removed and the values are restored to near 1.0 as expected in these snow echoes. The only remaining bias in ρ_{HV} is in the region where the strong clutter is removed by CMD processing within the 0 m s^{-1} isodop, a region that cannot be recovered using GMAP, indicated by the

‡CMD processing denotes both clutter identification and subsequent filtering by GMAP.

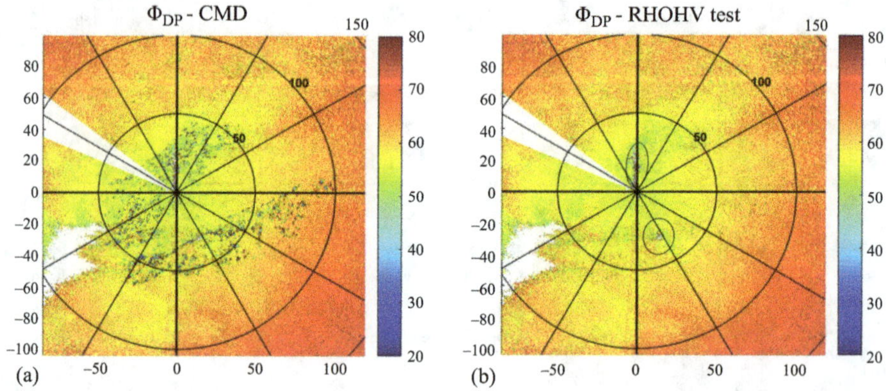

Figure 11.13 As in Figure 11.10 but for ϕ_{DP} after ground-clutter filtering following CMD (a) and RHOHV-test (b). The ovals indicate regions of strong clutter identified by CMD within the 0 m s^{-1} isodop that have clutter filter bias.

black oval in Figure 11.11(b). It is also clear comparing Figures 11.10(b) and 11.11(b) that the clutter filter bias covers a much larger area in Z than it does in ρ_{HV} (white ovals in Figure 11.10(b) and small black oval in Figure 11.11(b)).

The corresponding Z_{DR} field after CMD processing in Figure 11.12(a) also shows extensive contamination from low CSR ground clutter surrounding the clutter that was identified and removed by CMD processing. The Z_{DR} is biased mainly low by the clutter with many bins well below 0 dB. Figure 11.12(b) shows that after the RHOHV-test processing almost all of the contamination in Z_{DR} was removed, resulting in substantially improved data quality and Z_{DR} values that are expected in snow and consistent with the uncontaminated echoes in the scan. There is no additional clutter filter bias in the 0 m s^{-1} isodop after RHOHV-test processing. As before, the Z_{DR} remains biased where strong clutter was removed from the 0 m s^{-1} isodop by CMD processing, indicated by the white ovals. Interestingly, the spatial extent of the obvious clutter filter bias is larger than for ρ_{HV}.

Figure 11.13 shows the results of CMD and RHOHV-test processing for the ϕ_{DP} field. The low CSR clutter contamination is evident in Figure 11.13(a) in regions similar to that in ρ_{HV} (Figure 11.11(a)). After applying RHOHV-test processing, the vast majority of the contamination is removed with no additional clutter filter bias in the 0 m s^{-1} isodop. The ϕ_{DP} resulting from RHOHV-test processing is much smoother and will result in improved specific differential phase shift (K_{DP}), an important variable for hydrometeor classification and precipitation estimation algorithms. The regions where clutter filter bias in strong clutter within the 0 m s^{-1} isodop are about the same size as for Z_{DR} and are indicated in Figure 11.13(b) by the black ovals.

These results indicate that using RHOHV-test should substantially improve the data quality in the WSR-88D radars (and in general, other similar weather radars) and address the long-standing issue of low CSR clutter contamination of the dual-polarimetric variables. The Radar Operations Center of Norman Oklahoma has independently verified the RHOHV-test and it was implemented on the WSR-88D radars in the fall of 2019.

11.5 Regression and spectral clutter filter comparison using experimental data

Experimental S-Pol data are now used to compare the regression clutter filter using CMD plus the RHOHV-test to NCAR's WN adaptive filter, similar to GMAP, using just CMD. S-Pol was located at NCAR's Marshall Field Site close to Superior, Colorado, and was about 4 miles east from the foothills of the Rocky Mountains. PPIs (planned position indicator) of S-Pol data were gathered on 13 March 2019 at 2207 UTC at 0.5° elevation angle [92]. S-Pol was operating in fast alternating horizontal and vertical (*H* and *V*) transmit mode with a PRT of 1 ms. This means that the *H* (*V*) time series samples are separated by 2 ms. The transmit pulse is 1 μs wide so that the range resolution is 150 m. S-Pol has a 0.93° antenna beamwidth. The scan rate was 8° s^{-1} so that there are about 64 *H* and 64 *V* samples contained in a 1° azimuth angle rotation.

Figure 11.14 shows unfiltered reflectivity in dBZ. Labeled are the S-Pol location, Denver International Airport (DIA), the Denver NEXRAD KFTG, and the Rocky Mountains. Higher reflectivity areas in yellow and red color scale (Z $\gtrapprox$27

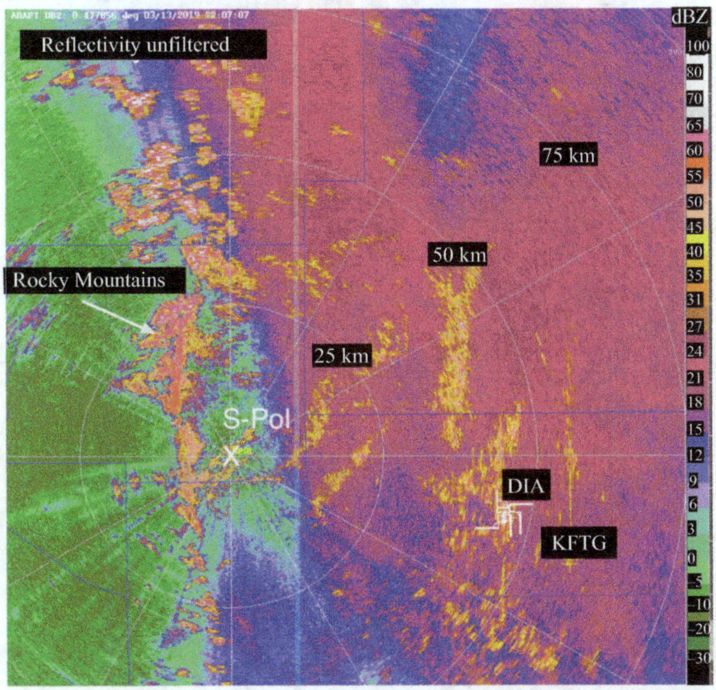

Figure 11.14 Unfiltered reflectivity PPI gathered by S-Pol on 13 March 2019 at 2207 UTC at 0.48° elevation angle. The higher reflectivities, approximately 27 dBZ and greater, are ground clutter, which are embedded in widespread precipitation. The Rocky Mountains, Denver International Airport (DIA) and KFTG, and the Denver NEXRAD are indicated.

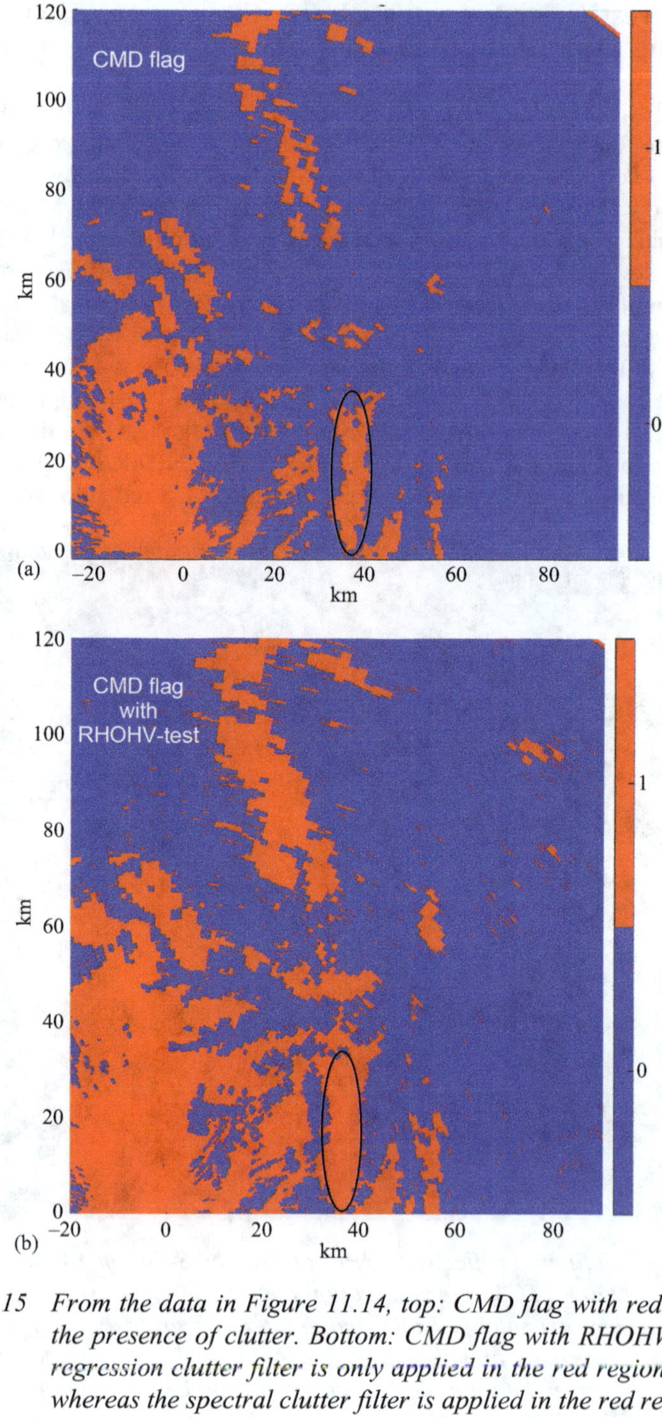

Figure 11.15 *From the data in Figure 11.14, top: CMD flag with red indicating the presence of clutter. Bottom: CMD flag with RHOHV-test. The regression clutter filter is only applied in the red region of (b) whereas the spectral clutter filter is applied in the red region of (a).*

dBZ) are predominately clutter while lower reflectivities are predominately precipitation. The data are processed by NCAR's adaptive WN filter in the red CMD flag region given in Figure 11.15(a). The regression filter is applied in the red CMD region of Figure 11.15(b), which is augmented with the RHOHV-test. The power from the central three velocity components of the Doppler spectra are used to estimate the clutter power, which is used to determine the order of the polynomial for the regression clutter filter as given in Figure 11.16. NCAR's adaptive WN filter and the regression filter are now applied to the data and interpolation is used across the zero-velocity gaps according to Section 11.3.2. The WN-filtered data use a von Hann window everywhere, which is done to illustrate the detrimental effect of the window. In regions where no clutter filter is needed, i.e., the blue area in Figure 11.15(b), the results from the WN and regression filter processed data would be identical if no window were applied. Figure 11.15 does mark an area with a black oval where both clutter filters both were applied and will serve as a comparison area in the next plots.

Shown in Figures 11.17 and 11.18 are reflectivity and velocity fields, respectively, with the WN-filtered data in panels (a) and the regression-filtered data in panels (b). Visual comparison of the WN and regression data confirms that the regression data are smoother. Examining the oval indicated area, where both filters were applied, demonstrates this also. In general, all areas where the clutter filters were applied (marked by the red color in Figure 11.15), the regression filter has superior (i.e., lower areal Stds) reflectivity and velocity recovery (if there was

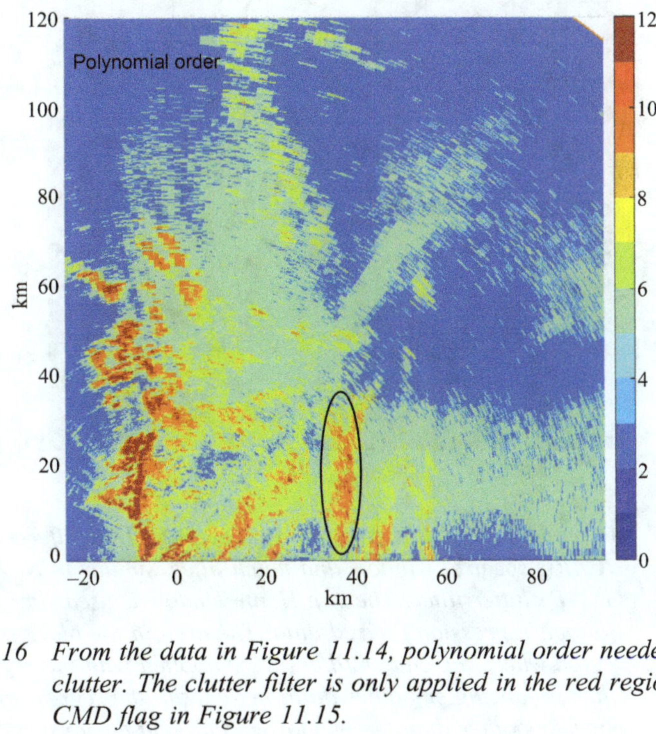

Figure 11.16 From the data in Figure 11.14, polynomial order needed to reject clutter. The clutter filter is only applied in the red region of the CMD flag in Figure 11.15.

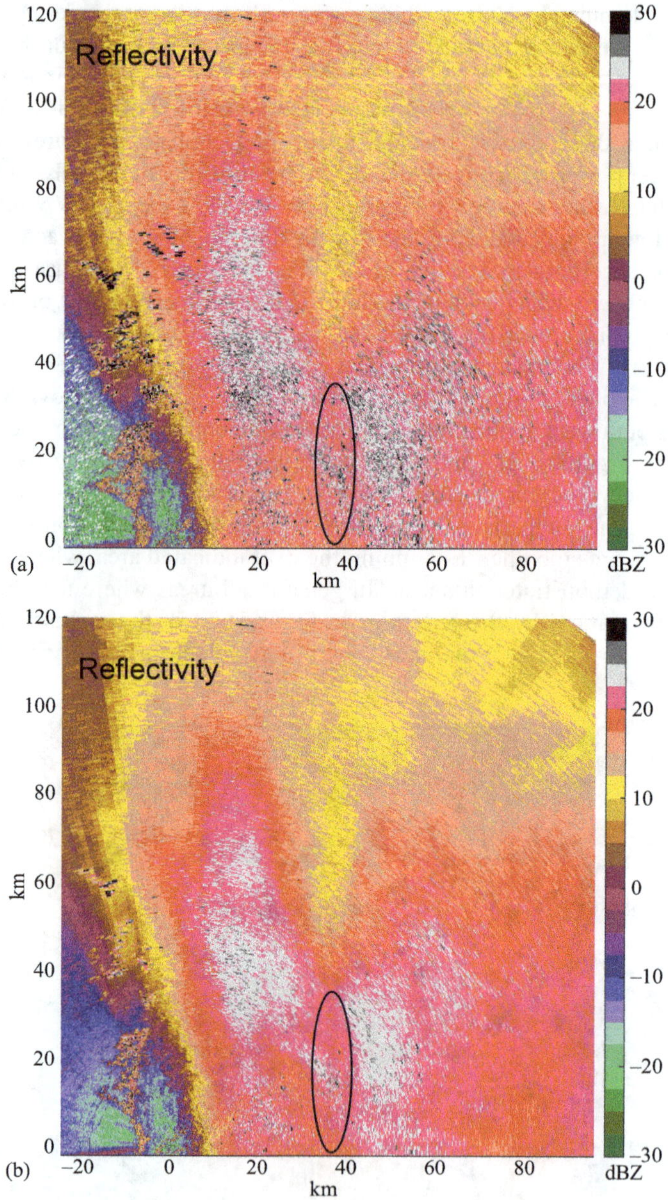

Figure 11.17 A comparison of clutter-filtered reflectivity. Top: filtered with NCAR's spectral window and notch filter, similar to NEXRAD's GMAP clutter filter. The von Hann window is used everywhere. Bottom: regression filtered data. The area in the black oval is a region where the clutter filters were applied with the regression filter producing smoother data. In general, the regression image is much smoother than the windowed data and demonstrates the deleterious effects of the window function that in effect reduces the number of independent samples.

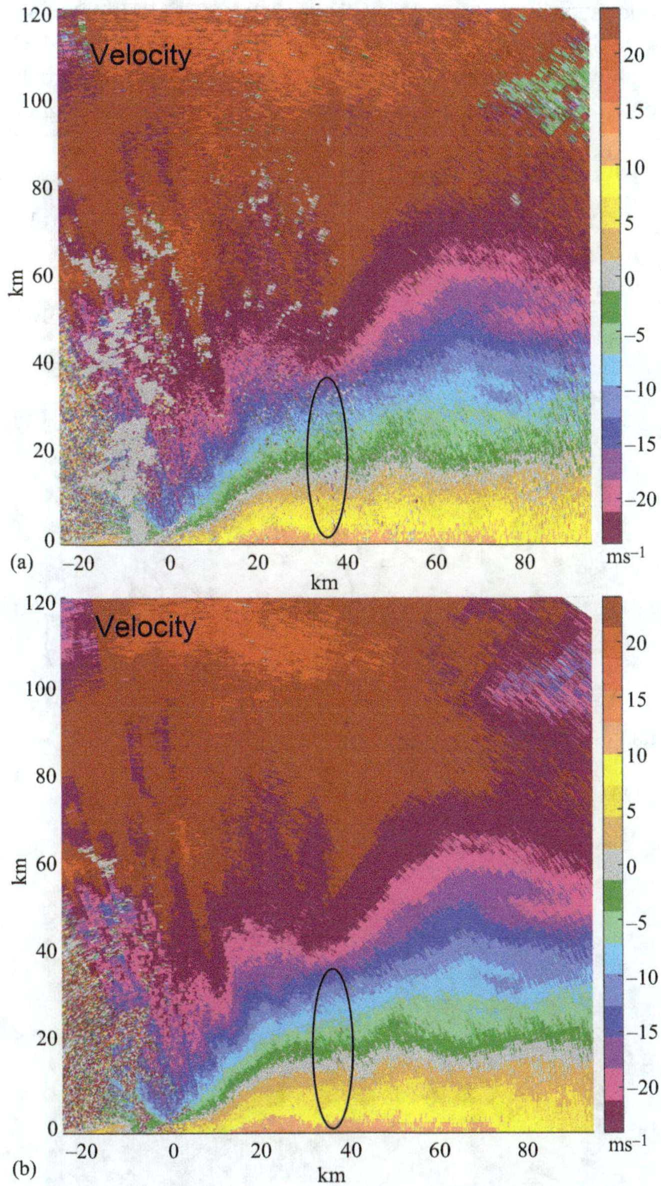

Figure 11.18 As in Figure 11.17 but for velocity. Top: WN filtered. Bottom: regression filtered.

underlying weather). Also in the velocity field comparison, the benefit of the additional clutter regions identified by the RHOHV-test is seen. The WN-filtered velocity data has small areas with gray speckles (indicating zero velocity) and other speckles around coordinates (30 km and 50 km) and in other places. In general the speckles are not seen in the regression filtered data. The RHOHV-test was able to identify low CSR gates that CMD alone was not able to detect. Finally, Figure 11.19 shows the comparison of ρ_{HV} corresponding to Figure 11.17. Again, the regression filter ρ_{HV} data are much smoother spatially and higher in value as

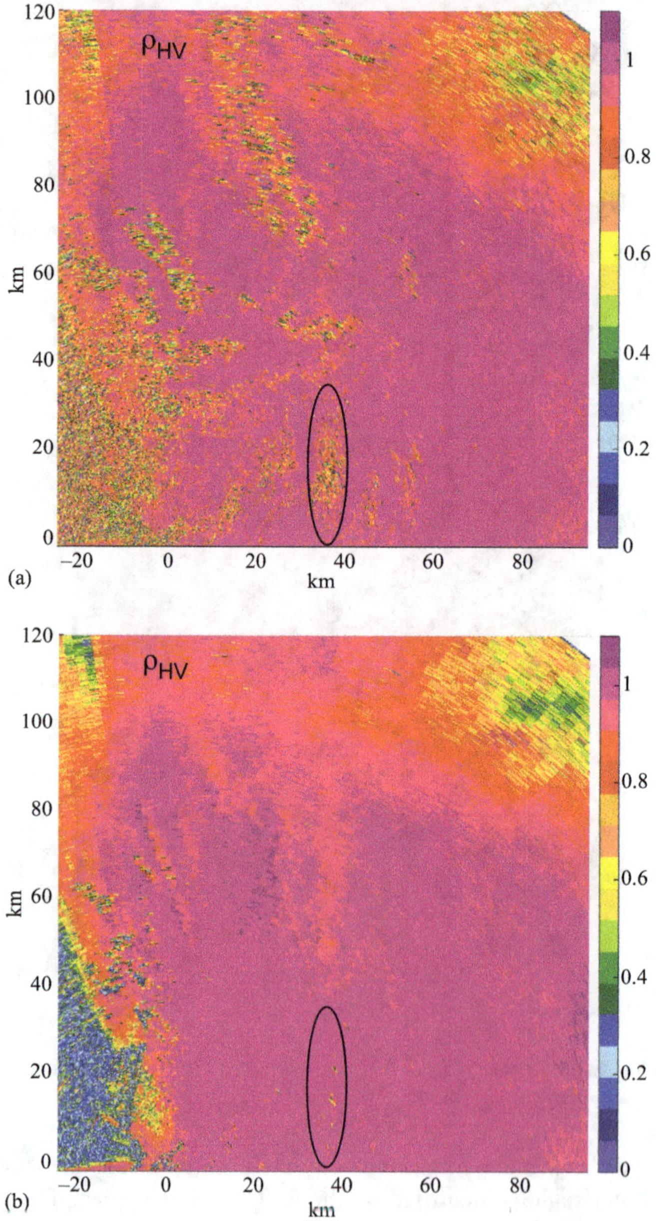

Figure 11.19 As in Figure 11.17 but for ρ_{HV}. Top: WN filtered. Bottom: regression filtered.

compared to the WN-filtered data. The ρ_{HV} is a general data quality metric with lower values indicating higher measurements errors especially for Z_{DR} and ϕ_{DP}.

11.6 Summary

There have been many different methods for mitigating ground-clutter contamination for ground-based weather radars. Modern computer capabilities make possible real-time execution of many of these techniques. Identifying ground clutter using only data from each individual gate would be ideal but this goal has been elusive and it appears that using data from several gates along a ray profile is required for robust clutter identification. Clutter echoes vary considerably from gate-to-gate, whereas precipitation echoes in general do not so that the texture of various radar variables in range are very good discriminators between precipitation and ground clutter.

When the clutter power dominates the weather power, i.e., CSR > 0 dB, fuzzy logic clutter identification algorithms utilizing texture information along the ray profile are quite robust. However, when the CSR goes below zero dB, clutter becomes more difficult to detect. Such low CSR clutter echoes can bias reflectivity, velocity, and spectrum width but the polarimetric variables are more sensitive to bias from low CSR clutter. One way to detect low CSR clutter is to use ρ_{HV}, which is the most sensitive variable to low CSR clutter contamination. To do this, ρ_{HV} is calculated from non-filtered data and from clutter-filtered data. If $\rho_{HV}^{filt} > \rho_{HV}^{unfilt} + V_{th}$, then the clutter filter should be applied. The RHOHV-test, which augments CMD, has been experimentally verified and has been deployed on NEXRAD.

Clutter power is highly variable on a gate-to-gate basis and, thus, requires a clutter filter whose stop bandwidth is adaptable. This is one reason why GMAP was adopted by the NWS and replaced IIR filters that were popular before GMAP. GMAP also incorporates a Gaussian interpolation scheme for the zero-velocity gap created by the clutter filter so that biases to reflectivity, velocity, and spectrum width estimates are reduced.

Even though the concept of regression analysis is well known, regression filtering has been largely overlooked as a viable clutter filter but has been recently reexamined in [1]. The regression filter utilizes a least squares polynomial fit to individual gates of *I* and *Q* data to estimate the trend of the data, which is considered the clutter signal. The trend is then subtracted from the signal and the residuals represent any weather signal that may remain. The regression filter is linear and can be considered time invariant in the sense that if a block of data is shifted in time and filtered, the same output results. However, the regression filter cannot be represented as the convolution of the filter input with the impulse response of the filter as can be done with traditional linear time-invariant (LTI) filters. Nevertheless, the regression filter does have a frequency response which describes the effect of the filter on the input signal. The regression filter has an adaptable bandwidth controlled by the selected order for the polynomial fit, is amenable to Gaussian interpolation across the zero-velocity gap created by the clutter filter, and rejects clutter as well as a spectral window and notch filter, such as GMAP, while using a narrower stop bandwidth. The significant advantage of the regression filter over spectral domain filters is that no window function is required. Windowing the radar time series is required by spectral domain filters to suppress clutter leakage and contain the clutter signal to zero and near-zero velocities. Windowing the time series in effect reduces the number of independent samples of the accompanying weather signal thereby increasing the standard deviation of the radar variable estimates. The increase in the standard deviations is about 35–55% [1] and is evident in the S-Pol data presented. Additionally, simulations here and in [1] show that bias is also reduced when using the regression filter. Finally, the regression ground-clutter filter is straightforward, easily implemented and runs in real time.

Acknowledgment

The automated polynomial order selection algorithm was developed by Dr Meymaris. The RHOHV-test was developed by Dr Ellis. The authors would like to thank Dr Ulrike Romatschke who made the S-Pol PPI plots and the frequency response plot, and Dr Michael Dixon who developed the NCAR window and notch filter and the accompanying S-Pol data processing software. This work was supported in part by the Radar Operations Center (ROC) of Norman, OK. This material is based on the work supported by the National Center for Atmospheric Research sponsored by the National Science Foundation under Cooperative Agreement No. 1755088. Any opinions, findings, and conclusions or recommendations expressed in this publication are those of the authors and do not necessarily reflect the views of the National Science Foundation.

List of abbreviations

ASD	auto-spectral density function
CLEAN-AP	clutter environment analysis using adaptive processing
CMD	clutter mitigation decision
CNR	clutter-to-noise ratio
CPA	clutter phase alignment
CSR	clutter-to-signal ratio
dB	decibels
dBZ	radar reflectivity factor in $mm^6\ m^{-3}$
DFT	discrete Fourier transform
FIR	finite impulse response
GMAP	Gaussian model adaptive processing
HWR	hub-to-weather ratio
I and Q	in phase and quadrature
IIR	infinite impulse response

KDLH	Duluth MN WSR-88D
KFTG	Denver CO WSR-88D
NCAR	National Center for Atmospheric Research
NN	neural network
N_{yq}	Nyquist velocity (or frequency)
NWS	National Weather Service
PRT	pulse repetition time
QPE	quantitative precipitation estimation
REC	radar echo classifier
SNR	signal-to-noise ratio
S-Pol	S-band polarimetric radar
SZ	Sachidananda Zrnić
US	United States
WSR-88D	Weather Surveillance Radar-1988 Doppler
WT	wind turbine
WTC	wind turbine clutter
WET	Weather Environment Thresholding
Z	reflectivity
Z_{DR} **or** Z_{dr}	differential reflectivity
ϕ_{DP}	copolar differential phase
ρ_{HV}	copolar correlation coefficient

References

[1] Hubbert JC, Meymaris G, Romatschke U, *et al.* Using a regression ground clutter filter to improve weather radar signal statistics: theory and simulations. *J Atmos Oceanic Technol.* 2021;38:1353–1375. DOI: 10.1175/JTECH-D-20-0026.1.

[2] Smith PL. Siting considerations for weather radars. In: Preprints, *15th Conference on Radar Meteorology*. Champaign-Urbana, IL: Amer. Meteor. Soc.; 1972. p. 99–100.

[3] Smith PL, Hardy KR, and Glover KM. Applications of radar to meteorological operations and research. *Proc IEEE*. 1974;62:724–745.

[4] Barton DK and Ward HR. *Handbook of Radar Measurements*. Dedham, MA: Artic House; 1984.

[5] Kerr DE. *Propagation of Short Radio Waves*, vol. 13. New York, NY: McGraw Hill; 1951. Massachusetts Institute of Technol., Radiation Lab.

[6] Long MW. *Radar Reflectivity of Land and Sea*. Boston, MA/London: Artech House; 2001.

[7] Hubbert JC, Dixon M, Ellis SM, *et al.* Weather radar ground clutter, Part I: identification, modeling and simulation. *J Atmos Oceanic Technol.* 2009;26:1165–1180.

[8] Hubbert JC, Dixon M, Meymaris G, *et al.* Modeling and experimental observations of weather radar ground clutter. In: *AMS 27th International Conference on Interactive Information and Processing Systems*. Seattle, WA: Amer. Meteor. Soc.; 2011. p. 7.

[9] Nathanson FE. *Radar Design Principles*. McGraw Hill, Inc.; 1969.

[10] Sekhon RS, Atlas D. Doppler radar structure of the melting zone and ground clutter effects. In: *Preprints, 15th Conference on Radar Meteorology*. Champaign-Urbana, IL: Amer. Meteor. Soc.; 1972. p. 222–227.

[11] Johnson GN, Smith PL, Nathanson FE, *et al.* An analysis of techniques for dealing with anomalous propagation clutter. In: Preprints, *16th AMS Conference on Radar Meteorology*. Houston, TX: Amer. Meteor. Soc.; 1975. p. 374.

[12] Sirmans D and Dooley JT. Ground clutter statistics of a 10-cm ground based radar. In: Preprints, *19th Conference on Radar Meteorology*. Miami Beach, FL: Amer. Meteor. Soc.; 1980. p. 184–191.

[13] Pratte JF, Gagnon R, and Cornelius R. Ground clutter characteristics and residue mapping. In: Preprints, *26th International Conference on Radar Meteorology*. Norman, OK: Amer. Meteor. Soc.; 1993. p. 50–52.

[14] Moszkowicz S, Ciach GJ, and Krajewski WF. Statistical detection of anamalous propagation in radar reflectivity patterns. *J Atmos Oceanic Technol*. 1994;11:1026–1034.

[15] Lee R, Bruna GD, and Joss J. Intensity of ground clutter and of echoes of anomalous propagation and its elimination. In: Preprints, *27th International Conference on Radar Meteorology*. Vail, CO: Amer. Meteor. Soc.; 1995. p. 651–652.

[16] Schaffner MR. On the data processing for weather radar. In: Preprints, *15th Conference on Radar Meteorology*. Champaign-Urbana, IL: Amer. Meteor. Soc.; 1972. p. 368–373.

[17] Schaffner MR. Comments on Applications of radar to meteorological operations and research. *Proc IEEE*. 1975;63:731–733.

[18] Geotis SG, Silver WM. An evaluation of techniques for automatic ground echo rejection. In: Preprints, *17th AMS Conference on Radar Meteorology*. Seattle, WA: Amer. Meteor. Soc.; 1976. p. 448–452.

[19] Smith PL, Hardy KR, and Glover KM. A reply to: applications of radar to meteorological operations and research. *Proc IEEE*. 1975;63:733.

[20] Sirmans D, Watts WL, and Horwedel JH. Weather radar signal processing and recording at the National Severe Storms Laboratory. *IEEE Trans Geosci Elect*. 1970;GE-8(2):88–94.

[21] Sirmans D. NEXRAD suppression of land clutter echo due to anamalous propagation. NSSL; 1987. Part 1. Report to NEXRAD JSPO. Revised Apr. 1988.

[22] Kosko B. *Neural Networks and Fuzzy Systems: A Dynamical Systems Approach to Machine Intelligence*. Englewood Cliffs, NJ: Prentice Hall; 1992.

[23] Cornelius R and Gagnon R. Artificial neural networks for radar data feature recognition. In: Preprints, *26th International Conference on Radar Meteorology*. Norman, OK: Amer. Meteor. Soc.; 1993. p. 340–342.

[24] Pratte F, Ecoff D, and VanAndel J. AP Ground Clutter in WSR-88D Base Data and Recommendations for Automatic AP Clutter Mitigation. NCAR, Boulder, CO; 1996. FY1996 Report to NWS/OSF.

[25] Pratte F, Ecoff D, VanAndel J, *et al.* AP clutter mitigation in the WSR-88D. In: Preprints, *28th International Conference on Radar Meteorology*. Austin, TX: Amer. Meteor. Soc.; 1997. p. 504–505.

[26] Grecu M and Krajewski WF. An efficient methodology for detection of anomalous propagation echoes in radar reflectivity data using neural networks. *J Atmos Oceanic Technol.* 2000;17(2):121–129.

[27] Lakshmanan V, Zhang J, Hondland K, *et al.* A statistical approach to mitigating persistent clutter in radar reflectivity data. *IEEE J Appl Earth Obser Rem Sen.* 2012;5:652–662.

[28] Kessinger C, Ellis S, and Andel JV. The radar echo classifier: a fuzzy logic algorithm for the WSR-88D. In: *Proceedings of 3rd Conference on Artificial Intelligence Applications to the Environmental Science.* 9–13 February. Long Beach, CA. AMS; 2003. p. 11.

[29] Cho Y, Lee G, Kim K, *et al.* Identification and removal of ground clutter echoes and anomalous propagation using the characteristics of radar echoes. *J Atmos Oceanic Technol.* 2007;23:1206–1222.

[30] Berenguer M, Sempere-Torres D, Corral C, *et al.* A fuzzy logic technique for identifying nonprecipitating echoes in radar scans. *J Atmos Oceanic Technol.* 2006;23(9):1157–1180.

[31] Nguyen CM, Moisseev DN, and Chandrasekar V. A parametric time domain method for spectral moment estimation and clutter mitigation for weather radars. *J Atmos Oceanic Technol.* 2008;25:83–92.

[32] Bringi VN and Chandrasekar V. *Polarimetric Doppler Weather Radar.* Cambridge, UK: Cambridge University Press; 2001.

[33] Hubbert JC, Weckwerth T, Ellis SE, *et al.* S-Pol's polarimetric data reveals detailed storm features (and insect behavior). *Bull Amer Meteor Soc.* 2018;99:2045–2060. Doi: 10.1175/BAMS-D-17-0317.1.

[34] Hall MPH, Goddard JWF, and Cherry SM. Identification of hydrometeors and other targets by dual-polarization radar. *Radio Sci.* 1984;19(1): 132–140.

[35] Giuli D, Gherardelli M, Freni A, *et al.* Rainfall and clutter discrimination by means of dual-linear polarization radar measurements. *J Atmos Oceanic Technol.* 1991;8:777–789.

[36] Straka JM, Zrnić DS, and Ryzhkov AV. Bulk hydrometeor classification and quantification using polarimetric radar data: synthesis of relations. *J Appl Meteor.* 2000;39(8):1341–1372.

[37] Ryzhkov AV and Zrnić DS. Polarimetric rainfall estimation in the presence of anomalous propagation. *J Atmos Oceanic Tech.* 1998;15(6): 1320–1330.

[38] Zrnić DS, Melnikov M, and Ryzhkov AV. Correlation coefficients between horizontally and vertically polarized returns from ground clutter. *J Atmos Oceanic Tech.* 2006;23(3):381–394.

[39] Hubbert JC, Bringi VN, and Brunkow D. Studies of the polarimetric covariance matrix: Part I: calibration methodology. *J Atmos Oceanic Technol.* 2003;20(5):696–706.
[40] Zrnić D and Melnikov VM. Ground clutter recognition using polarimetric spectral parameters. In: Preprints, *33rd AMS Conference on Radar Meteorology*. Cairns, Australia; 2007. p. 5.
[41] Gourley JJ, Tabary P, and du Chatelet JP. A fuzzy logic algorithm for the separation of precipitating from nonprecipitating echoes using polarimetric radar observations. *J Atmos Oceanic Technol.* 2007;24:1439–1451.
[42] Steiner M and Smith JA. Use of three-dimensional reflectivity structure for automated detection and removal of nonprecipitating echoes in radar data. *J Atmos Oceanic Technol.* 2002;19:673–686.
[43] Pratte FJ, Keeler RJ, Gagnon R, *et al.* Clutter processing during anomalous propagation conditions. In: Preprints, *27th International Conference on Radar Meteorology*. Vail, CO: Amer. Meteor. Soc.; 1995. p. 139–141.
[44] Hubbert JC, Dixon M, and Ellis SM. Weather radar ground clutter, Part II: real-time identification and filtering. *J Atmos Oceanic Technol.* 2009;26:1181–1197.
[45] Torres S and Warde DA. Ground clutter mitigation for weather radars using the autocorrelation spectral density. *J Atmos Oceanic Technol.* 2014;31 (10):2049–2066.
[46] Groginsky HL and Glover KM. Weather radar cancellor design. In: Preprints, *19th Conference on Radar Meteorology*. Miami, FL: Amer. Meteor. Soc.;1980. p.192–198.
[47] Zrnić D, Hamidi S, and Zahari A. Considerations for the Design of Ground Clutter Cancellers for Weather Radar. U.S. DOT/FAA; 1982. RD-82/68. Final Report, Systems R&D Service.
[48] Torres S and Zrnić D. Ground clutter canceling with a regression filter. *J Atmos Oceanic Technol.* 1999;16(16):1364–1372.
[49] Skolnik MI. *Introduction to Radar Systems*, 3rd ed. Elglewood Cliffs, NJ: McGraw-Hill International Editions; 2001.
[50] Ice RL, Rhoton RD, Saxion DS, *et al.* Optimizing clutter filtering in the WSR-88D. In: *Proceedings of 23rd International Conference on IIPS*. vol. 87th Annual Meeting. San Antonio, TX: Amer. Meteor. Soc.; 2007.
[51] Passarelli RE Jr, Romanik P, Geotis SG, *et al.* Ground clutter rejection in the frequency domain. In: *Proceedings of 20th Conference on Radar Meteorology*. Boston, MA: Amer. Meteor. Soc.; 1981. p. 295–300.
[52] Ice RL, Rhoton RD, Saxion DS, *et al.* ROC evaluation of the WSR-88D ORDA system signal processing. In: *Proceedings of 20th International Conference on Interactive Information Processing Systems for Meteorology, Oceanography and Hydrology*. Seattle, WA: Amer. Meteor. Soc.; 2004. p. 11–15.
[53] Ice RL, McGehee GT, Rhoton RD, *et al.* Radar operations center (ROC) evaluation of new signal processing techniques for the WSR-88D. In: *Proceedings of 21 International Conference on Interactive Information*

Processing Systems for Meteorology, Oceanography and Hydrology. San Diego, CA: Amer. Meteor. Soc.; 2005. p. 5. CD-ROM, P1.4.

[54] Siggia A and Passarelli J R. Gaussian model adaptive processing (GMAP) for improved ground clutter cancellation and moment calculation. In: *Proceedings of Third European Conference on Radar in Meteorology and Hydrology*. Visby, Gotland, Sweden: ERAD; 2004. p. 67–73.

[55] Sirmans D, Zrnić DS, and Bumgarner B. Extension of maximum unambiguous Doppler velocity by use of two sampling rates. In: Preprints, *17th Conference on Radar Meteorology*. Seattle, WA: Amer. Meteor. Soc.; 1976. p. 23–28.

[56] Zrnić DS and Mahapatra PR. Two methods of ambiguity resolution in pulse Doppler weather radars. *IEEE Trans Aerosp Electron Syst.* 1985;AES-21:470–483.

[57] Doviak RJ and Zrnić DS. *Doppler Radar and Weather Observations*, 2nd ed. New York, NY: Academic Press; 1993.

[58] Sachidananda M and Zrnić DS. Clutter filtering and spectral moment estimation for Doppler weather radars using staggered pulse repetition time (PRT). *J Atmos Oceanic Technol.* 2000;17:323–331.

[59] Torres S and Warde DA. Staggered-PRT sequences for Doppler weather radars. Part I: spectral analysis using the autocorrelation spectral density. *J Atmos Oceanic Technol.* 2017;34:51–63.

[60] Warde DA and Torres SM. Staggered-PRT sequences for Doppler weather radars. Part II: ground clutter mitigation on the NEXRAD network using the CLEAN-AP filter. *J Atmos Oceanic Technol.* 2017;34:703—716.

[61] Gray G, Hubbert J, and Meymaris G. Method and Apparatus for Clutter Filtering Staggered Pulse Repetition Time Signals; US Patent US7728765B1, 1 June 2010.

[62] Torres S and Zrnić D. *Ground Clutter Canceling with a Regression Filter*. NSSL, Norman, OK; 1998. NSSL Interim Rep.

[63] Posada D, Pellissierx JPA, Chayer B, *et al.* Staggered multiple-PRF ultrafast color Doppler. *IEEE Trans Med Imaging*. 2016;35(6):1510–1521.

[64] Moisseev DN and Chandrasekar V. Polarimetric spectral filter for adaptive clutter and noise suppression. *J Atmos Oceanic Technol.* 2009;26:215–228.

[65] Alku L, Moisseev DN, Aittomäki T, *et al.* Identification and suppression of nonmeteorological echoes using spectral polarimetric processing. *IEEE Trans Geosci Remote Sens*. 2009;53(7):3628–3638.

[66] Yin J, Unal C, and Russchenberg H. Object-orientated filter design in spectral domain for polarimetric weather radar. *IEEE Trans Geosci Remote Sens*. 2019;57(5):2725–2740.

[67] Vogt RJ, Reed J, Crum T, *et al.* Impacts of wind farms on WSR-88D operations and policy considerations. In: *Proceedings of 23th International Conference on IIPS*. vol. 87th Annual Meeting. San Antonio, TX: Amer. Meteor. Soc.; 2007.

[68] Hood K, Torres SM, and Palmer RD. Automatic detection of wind turbine clutter for weather radars. *J Atmos Oceanic Technol.* 2010;27:1868–1880.

[69] Isom BM, Palmer RD, Secrest GS, *et al.* Detailed observations of wind turbine clutter with scanning weather radars. *J Atmos Oceanic Technol.* 2009;26:894–910.

[70] Nai F, Torres SM, and Palmer RD. On the mitigation of wind turbine clutter for weather radars using range-Doppler spectral processing. *IET Radar Sonar Navig.* 2013;7:178–190.

[71] Uysal F, Selesnick I, and Isom B. Mitigation of wind turbine clutter for weather radar by signal separation. *IEEE Trans Geosci Remote Sens.* 2016;54(5):2925–2934.

[72] Hu Y, Uysal F, and Selesnick I. Wind turbine clutter mitigation via non-convex regularizers and multidimensional processing. *J Atmos Oceanic Technol.* 2019;36:1093–1104.

[73] Seo BC, Krajewski WF, and Mishra KV. Using the new dual-polarimetric capability of WSR-88D to eliminate anomalous propagation and wind turbine effects in radar-rainfall. *Atmos Res.* 2015;153:296–309.

[74] Palmer R, Bodine D, Kollias P, *et al.* A primer on phased array radar technology for the atmospheric sciences. *Bull Amer Meteor Soc.* 2022;103:2391–2416.

[75] Curtis CD, Yeary M, and Lake L. Adaptive nullforming to mitigate ground clutter on the National Weather Radar Testbed phased array radar. *IEEE Trans Geosci Elect.* 2016;54:1282–1291.

[76] Weber M, Hondl K, Yussouf N, *et al.* Towards the next generation operational meteorological radar. *Bull Amer Meteor Soc.* 2021;99:1357–1383.

[77] Harris FJ. On the use of window functions for harmonic analysis with the discrete Fourier transform. *Proc IEEE.* 1978;66:51–83.

[78] Oppenheim AV and Schafer RW. *Discrete-Time Signal Processing.* Englewood Cliffs, NJ: Prentice Hall; 1999.

[79] Torres S and Curtis C. Initial implementation of super-resolution data on the NEXRAD network. In: 87th AMS Annual Meeting; 2007.

[80] Sachidananda M and Zrnić DS. Systematic phase codes for resolving range overlaid signals in a Doppler weather radar. *J Atmos Oceanic Technol.* 1999;16:1351–1363.

[81] Darlington RB and Hayes AF. *Regression and Linear Models.* New York, NY: Guilford Press; 2017. ISBN-13: 978-1462521135.

[82] Bjaerum S, Torp H, and Kristoffersen K. Clutter filter design for ultrasound color flow imaging. *IEEE Trans Ultrason Ferroelect Freq Control.* 2002;49 (2):204–216.

[83] Kadi AP and Loupas T. On the performance of regression and step-initialized IIR clutter filters for color Doppler systems in diagnostic medical ultrasound. *IEEE Trans Ultrason Ferroelect Freq Control.* 1995;42(5):927–937.

[84] Orfanidis JR. *Introduction to Signal Processing.* Englewood Cliffs, NJ: Prentice Hall, 1996; 2010. www.ece.rutgers.edu/õrfanidi/intro2sp.

[85] Schafer RW. On the Frequency-Domain Properties of Savitzky–Golay Filters. Digital Signal Processing and Signal Processing Education Meeting (DSP/SPE). 2011; DOI: 10.1109/DSP-SPE.2011.5739186.

[86] Forsythe GE. Generation and use of orthogonal polynomials for data fitting with a digital computer. *J Soc Indust Appl Math*. 1957;5:74–88.

[87] Ruckdeschel FR. *Basic Scientific Routines, Vol. II*. New York, NY: BYTE/McGraw-Hill; 1981.

[88] Torp H. Clutter rejection filters in color flow imaging: a theoretical approach. *IEEE Trans Ultrason, Ferroelect, and Freq Control*. 1997;44: 417–424.

[89] Frehlich R and Yadlowsky MJ. Performance of mean-frequency estimators for Doppler radar and lidar. *J Atmos Oceanic Technol*. 1994;11(5):1217–1230.

[90] Friedrich K and Germann U. Influence of ground clutter contamination on polarimetric radar parameters. *J Atmos Oceanic Technol*. 2009;26(2):251–269.

[91] Ellis S, Meymaris G, Hubbert JC, *et al.* Improving the detection probability of low clutter-to-signal ratio ground clutter contamination in the WSR-88D. In: *Proceedings of the 34th Conference on EIPT*. Austin, TX: Amer. Meteor. Soc.; 2018. p. 14.

[92] NCAR/EOL Remote Sensing Facility. Marshall S-PolKa Radar Moments Data in Radial Format, S-Band. Version 1.0. UCAR/NCAR – Earth Observing Laboratory; 2020. Accessed 11 Feb. 2020. https://doi.org/10.26023/KHMH-X6EP-BW0H.

Chapter 12

Polarimetric planar phased array radar – challenges for observing weather

Dušan S. Zrnić[1,2,3], Igor I. Ivić[1,4], Dordje Mirković[1,4], Lesya Borowska[5] and Guifu Zhang[2,5]

The phased array polarimetric radar is the latest technology being considered for weather observations [1]. It has numerous advantages over radars with mechanically steered antennas. For example, fast adaptive scanning enabled by beam agility means that the phased array radar (PAR) can focus observations over volumes where severe phenomena are likely. Further, the absence of beam smearing mitigates the effects of clutter, which improves the accuracy of weather observations, and may lead to better estimation of turbulence. It has been suggested that retrieval of transverse winds may be possible [1]. Time multiplexing can be used to observe the weather, profile winds, and track objects [2]. Faster volume updates can reduce temporal errors in accumulated rainfall measurements and are better for assimilation into numerical weather prediction models. The absence of mechanically driven parts reduces maintenance and improves availability. In this chapter, we discuss issues challenging designers of polarimetric PARs and present some possibilities to overcome these.

12.1 Introduction

Despite all its advantages, the PAR has not yet reached the maturity required for operational applications. The obstacles are cost and technical issues for generating polarimetric variables comparable to the ones from radars with a parabolic dish antenna.

Conceptually it is hardest to generate good-quality polarimetric variables on the planar stationary PAR. This is because calibration is needed at every beam-pointing direction. A cylindrical PAR can scan in azimuth by commutating its

[1]NOAA National Severe Storms Laboratory, Oklahoma, USA
[2]School of Meteorology, University of Oklahoma, USA
[3]School of Electrical and Computer Engineering, University of Oklahoma, USA
[4]Cooperative Institute for Severe and High-Impact Weather Research and Operations, University of Oklahoma, USA
[5]Advanced Radar Research Center, University of Oklahoma, USA

beam and ideally would need calibration only in elevation [3], see Appendix. However, industrial experience with this geometry is limited, while weather services require routine technology for stable operation. On a radar with a planar rotating array, calibrating the polarimetric variables is the simplest of the three if electronic scanning is in the principal elevation plane. Three issues: non-orthogonality of the transmitted fields, orientation of the fields out of the array principal planes, and scanning loss are important issues specific to the PAR and must be dealt with.

We consider a PAR antenna with dual polarization. Specifically, we assume that Port 1 produces the intended horizontally (H) polarized field and Port 2 generates the intended vertically (V) polarized field in the plane of polarization at the broadside. The distribution of the cross-polar field within the beam depends on the physical structure of the radiating element and the direction of the beam. It is impossible to produce a pure horizontal or vertical field within the beam at all pointing directions with radiating elements of the same type (e.g., patch radiating elements).

If the elements are of the same type, then the cross-polar field in the principal planes would ideally be zero. If the co-polar beam is in the principal planes, then the cross-polar field at the beam center is zero. But if the beam is steered out of the principal planes, the cross-polar field will have one prominent peak at or near the beam center (i.e., the cross-polar beam's axis nearly coincides with the axis of the copolar beam). This peak is geometrically induced. It has been demonstrated [4] that it can cause significant bias in the polarimetric variables.

12.2 Radiating elements

Planar antenna arrays consist of radiating elements (radiators) arranged in a geometric lattice. The radiator type and the physical layout determine its far-field copolar and cross-polar patterns. The element pattern $f_e(\theta, \phi)$ is important because it is part of the antenna array's composite pattern

$$f(\theta, \phi) \approx f_e(\theta, \phi) f_a(\theta, \phi), \tag{12.1}$$

where the subscript a stands for the array factor (i.e., a directional two-dimensional function that weights the patterns of individual radiators adjusting relative phases and amplitudes of the microwaves emitted by the radiators in the array), and the angles θ and ϕ are the coordinates in the spherical coordinate system. If the antenna is in the y–z plane, ϕ is the azimuth angle while the elevation angle is $90° - \theta$. This relation is valid for both the copolar and cross-polar patterns. One computes the array pattern by assuming point radiating sources at the center of elements. The radar meteorology community uses Ludwig 2 definition whereby the vertically polarized field is aligned with the longitude and the horizontally polarized field is aligned with the latitude [5].

The PAR with solid-state transmitters and digital receivers can have tremendous flexibility in transmitter waveforms and in scanning strategy. In principle, the radar could also offer significant flexibility in polarization measurements. Of critical importance to dual polarization measurement is the type of radiating element and configuration of the phased array antenna. Two types of dual polarization elements are typically available. One is the electrical dipole radiator and another is the magnetic dipole. Next, we consider these and some practical radiators.

12.2.1 Electric dipole

The electric field of the dipole-type radiator is tangent to the meridian of the sphere in which the dipole is centered on the diameter connecting the poles. Two such orthogonal dipoles produce a field tangent to their respective meridians as depicted in Figure 12.1(a). It is clear that the E fields are orthogonal only in the principal planes. From visual inspection, the following ensues in the case of a vertically oriented plane of the array.

The horizontally oriented dipoles emit horizontally polarized fields in the two principal planes (horizontal and vertical). The vertically oriented dipole emits vertically oriented fields in the horizontal principal plane; in the vertical principal plane, the field is truly vertical only at the zero elevation angle. This is not detrimental to polarimetric measurements at low elevation angles where the small departure from the vertical causes insignificant bias in the polarimetric variables. The US National Service Weather Radars produce the following polarimetric variables: reflectivity Z, differential reflectivity Z_{DR}, correlation coefficient between the copolar components ρ_{hv}, and differential phase between the horizontally and vertically polarized waves, Φ_{DP}. These we consider throughout the chapter.

If the plane of the array is tilted, the field of the horizontally oriented dipole will be horizontal in the principal elevation plane as well as in the true horizontal

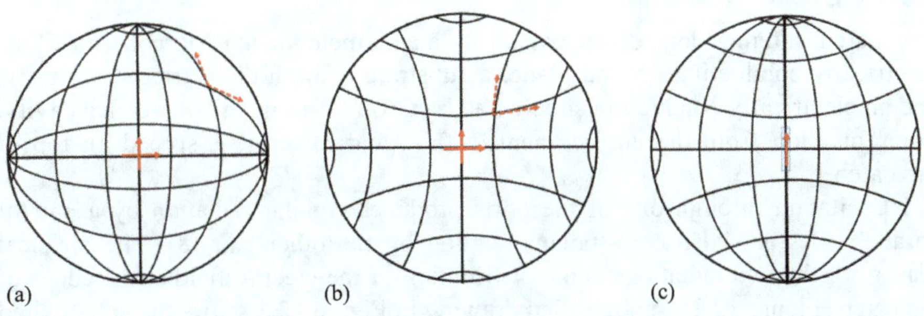

Figure 12.1 (a) Field lines of two orthogonal electric dipoles (red full arrows). The dashed arrows indicate the direction of the fields at the intersection point of the two "meridians." (b) Field lines of two orthogonal magnetic dipoles (red full arrows). The dashed arrows indicate the direction of the fields at the intersection point of the two "parallels." (c) Field lines are produced by collinear electric and magnetic dipoles. In this practical case, the magnetic dipole is a slit in a waveguide, and the electric dipole is inserted into the slit. The field lines produced by these two sources are orthogonal everywhere.

plane. The field generated by the vertically oriented dipole will be vertical only at the intersection of the vertical principal plane with the true horizontal plane.

12.2.2 Magnetic dipole

The electric field lines of a magnetic dipole are represented by the parallels (longitudinal circles) on a sphere (Figure 12.1(b)). If the plane of the array is vertically oriented, the fields of the vertically oriented magnetic dipole are truly horizontal everywhere. But, the fields of the horizontally oriented magnetic dipole are vertical only in the horizontal principal plane.

If the plane of the array is tilted, the fields of the vertically oriented dipole are horizontal only in the principal vertical plane. The fields of the horizontally oriented dipole remain vertical in the horizontal plane.

The magnetic dipole can be approximated, for example, by slots in a waveguide. Moreover, patch antennas are approximately equivalent to pairs of magnetic dipoles.

12.2.3 Collinear electric and magnetic dipoles

If the electric and magnetic dipoles are collinear, the intended H and V fields at every pointing direction are orthogonal (Figure 12.1(c)). Therefore, a vertical PAR antenna comprised of collinear magnetic and electric dipoles will produce orthogonal H and V fields in all pointing directions [6]. Nonetheless, if the array is tilted the intended H field (out of the vertical principal plane) will not be horizontal nor will the intended V field be "vertical." The physical layout of such dipoles is three-dimensional and the developments so far were exploratory.

12.2.4 Patch radiators

The patch antenna considered by [7] as an element in a PAR consists of an electrically conducting ground plane, a substrate of high dielectric permittivity, and an electrically conducting square patch on top forming an open-ended cavity. For protection from the environment, a thin coat material is spread on top of the patch.

Excitation through one of the ports produces copolar radiation by a pair of parallel sides and also cross-polar radiation by the other pair [8]. The physical reason for the two radiations is the distribution of the electric field at the edges of the patch (Figure 12.2). A simplified drawing in Figure 12.3 shows the sides exited by Ports 1 and 2.

Figure 12.4 illustrates the copolar and cross-polar patterns of a patch element positioned in the third row and the fourth column (matrix) of an 8×8 planar array. The measurements (made in an anechoic chamber) and the model capture the EM effects of the structure on the single-element patterns. A typical planar array consists of several panels and the single-element patterns embedded at the array center would be very close to the ones in Figure 12.4. Copolar components match well whereas the cross-polar components on average agree in magnitude but from about 0° to 160° are offset in phase.

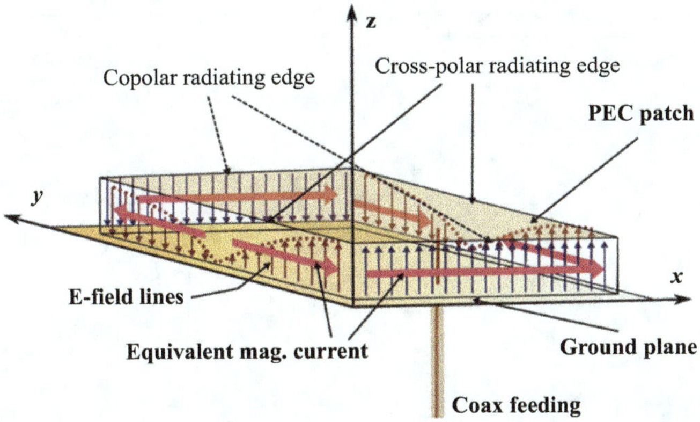

Figure 12.2 Patch antenna conceptual sketch. ©[2014] IEEE. Reprinted with permission from: S. Bhardwaj and Y. Rahmat-Samii, "Revisiting the generation of cross-polarization in rectangular patch antennas: a near-field approach," IEEE Antennas and Propagation Magazine, vol. 56, no. 1, pp. 14–38.

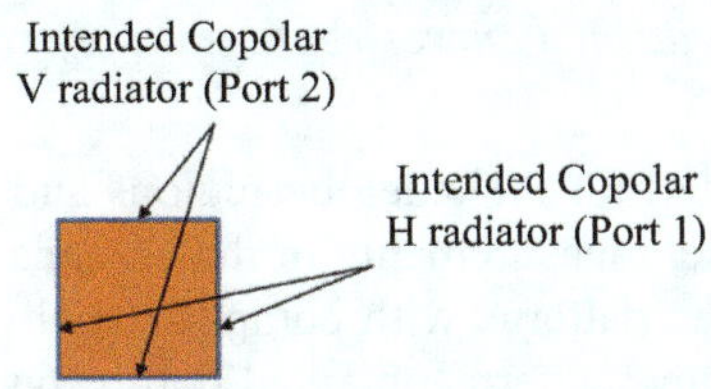

Figure 12.3 Patch schematic. The sides (radiators) of the intended polarization are indicated. The parenthesis indicates which Port is excited to enable radiation from a specific pair of sides.

The model patterns were obtained with a computational electromagnetics solver WIPL-D wherein the physical characteristics of the patch are quantified [9]. WIPL-D uses the method of moments as well as surface integral equations to simulate large EM structures. Noteworthy is the remarkable agreement between modeled and measured copolar patterns over the full 360° interval. The patterns' 3 dB widths span about 108°, quite sufficient for electronic scanning over a 90° interval. The relatively modest sidelobes are irrelevant because the sidelobes of the array pattern determine the overall levels and are much lower in large arrays. The values within the main lobe are important as they affect directly quantitative precipitation measurements.

12.2.5 Array antenna

The planar array antenna consists of hundreds of radiators spaced at about half the radar wavelength λ. Eq. (12.1) is a good approximation for the main lobe of the

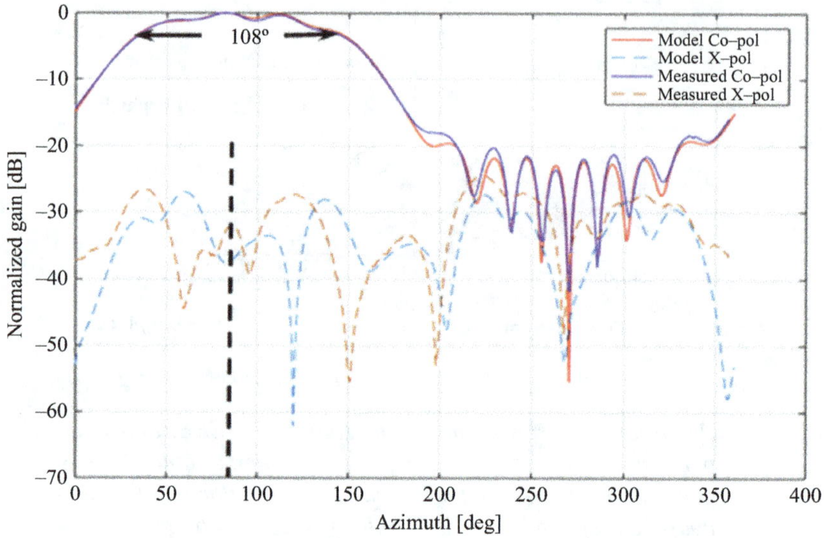

Figure 12.4 Simulated and measured patterns at H-polarization and E-plane (horizontal plane). Solid curves represent the copolar patterns, whereas the dashed curves represent the cross-polar patterns.

pattern but cannot be trusted for the sidelobe regions and the cross-polar patterns. To correctly quantify these, measurements in the antenna near field can be made, and/or one can simulate the patterns with computational electromagnetic tools. In this section, we present some patterns obtained both ways.

Currently, there are three polarimetric PARs for weather observations with fully agile beam positioning. One operates at a 3-cm wavelength [10], has multi-function capability, and has been used by researchers for observing storms. The other is a ten-panel demonstrator (TPD, Figure 12.5 [11]) meant for testing polarimetric calibration [12–14]. The third is the advanced technology demonstrator (ATD) [15,16]. Both TPD and ATD have been designed and built by LL MIT for a step-by-step learning/demonstration of this forthcoming technology. Neither of the two has been fully calibrated for precise polarimetric measurements of weather over the observation domain (90° in azimuth by 20° in elevation).

The TPD antenna is a rectangular array with dual polarization patch radiators. It consists of ten 8 × 8 panels and is mounted on a pedestal that enables mechanical antenna movement in azimuth and elevation as well as rotation around the antenna center. The latter is in addition to the electronic beam steering capability. The pedestal is mounted on a trailer for easy deployment. Researchers have developed an EM model of the antenna to compute the patterns. It uses the symmetry of the TDP to reduce the computational domain (i.e., the size of the antenna) by four [17]. Figure 12.5 shows the TPD radar (Figure 12.5(a)), modeled full antenna array (Figure 12.5(b)), a single panel of the TPD (Figure 12.5(c)), and a single element of the array with differential feeding probes for horizontal (H) and vertical (V) polarization (Figure 12.5(d)).

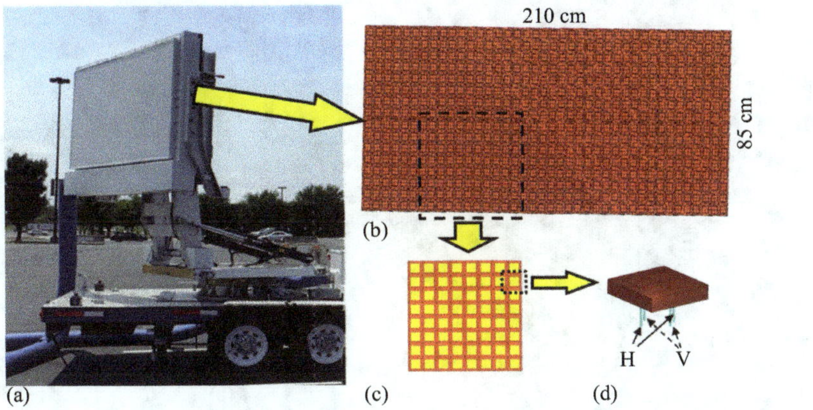

Figure 12.5 (a) TPD radar; (b) the antenna model; (c) antenna panel; (d) the antenna element with four differentially fed probes

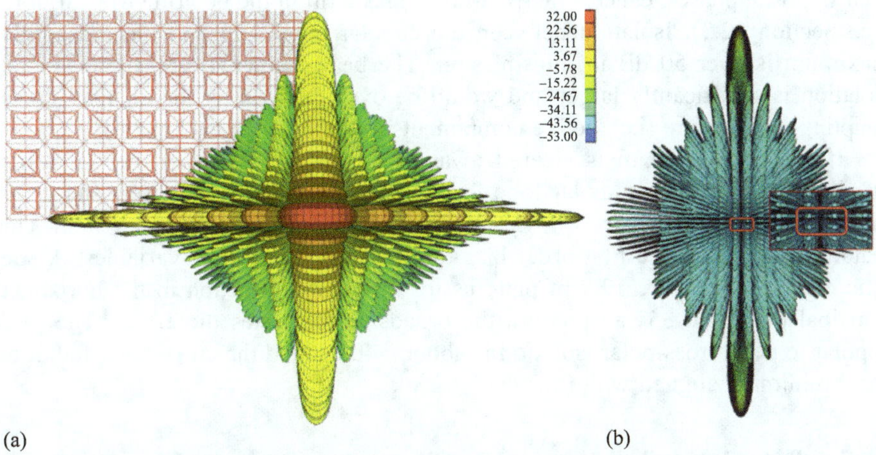

Figure 12.6 Antenna patterns of the TPD, (a) copolar $f_{hh}(\theta, \phi)$, (b) cross-polar $f_{vh}(\theta, \phi)$, and its enlarged values within the main beam (encompassed with the rectangle). The beam is pointing broadside. The cross-polar minimum is aligned with the copolar peak. The color bar indicates values in dBi.

The model yields copolar and cross-polar patterns in the Ludwig 2 definition [5]. The broadside copolar and cross-polar radiation patterns are in Figure 12.6. The beamwidth corresponding to the longer dimension (Figure 12.5(a)) is 6.7° and the one corresponding to the shorter dimension is 2.6°. The cross-polar pattern within the main copolar beam (Figure 12.6(b)) has four symmetrical lobes around the origin. This is beneficial as the contribution to the signal from these lobes, due to

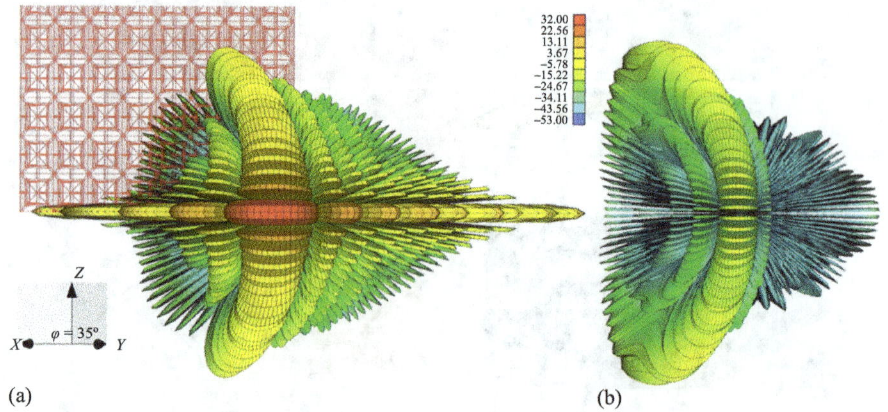

Figure 12.7 As in Figure 12.6 except the beam is pointing at 35° in azimuth and 0° in elevation

their opposite phases, cancels the first-order bias term in the polarimetric variables (see Section 12.3). Isolation between the copolar maximum and the cross-polar maximum is over 50 dB in transmission. The beam center to cross-polar center isolation is significantly larger and would be overwhelmed by the channel cross-coupling in hardware (i.e., active components in the antenna backplane).

If the copolar beam is steered away from the broadside in the horizontal principal plane, the cross-polar lobe formation changes from four opposite-phase lobes (Figure 12.6) to two. These two lobes have opposite phases as well. This feature also cancels the first-order bias term in the polarimetric variables. A specific example in Figure 12.7 of patterns for a pointing direction in the horizontal principal plane but 35° away from the broadside illustrates the effect. The peak copolar to peak cross-polar isolation is about >40 dB and the cross-polar radiation has a minimum aligned with the copolar peak.

12.3 Polarimetric modes and effects of pattern on bias in the polarimetric variables

Because beamforming is well documented in literature, we consider formed beams. We first explain two modes for polarimetric measurements because the isolation requirements between orthogonal channels are very different between the two. Acceptable biases in the polarimetric variables set the antenna requirements.

12.3.1 Operational polarimetric modes

On weather radars with parabolic reflector antennas, two modes of polarimetric measurements are common. The AHV mode refers to alternate transmission and simultaneous reception of horizontally and vertically polarized fields and it was typically used in the early days of weather observations. Isolation requirements for AHV mode are much less stringent than for the SHV (i.e., simultaneous transmission and simultaneous reception of H and V [4,18]), and compensation for coupling is relatively simple. But in AHV mode, Doppler measurements are coupled with polarimetric measurements, and dwell times are longer than for SHV mode [19,20]. PAR with electric dipole radiators could use SHV mode in directions where the coupling is minimal and AHV mode elsewhere. Moreover, the radar could transmit single polarization to acquire only Doppler and reflectivity which could be interleaved with dual polarization data acquisition. With adaptive scans [21] and pulse compression [22], very quick probing of weather over most of the observation space should be possible. But, regions with ground clutter constrain the speed of volume scans. Superior clutter filtering as on WSR-88D requires uniform PRTs and ample dwell time. A study is needed to determine if spatial filtering [23], feasible on PAR but not on WSR-88D, combined with shorter dwell time could perform as well as presently achieved with WSR-88D.

In Figure 12.8, are some processing options which might be available on PARs and an explanation follows. Suppose SHV is used. Then the Doppler variables (velocity and spectrum width) are decoupled from the polarimetric variables (Z, Z_{DR}, ρ_{hv}, and Φ_{DP}). Therefore, it is possible to estimate all four polarimetric variables from a single return by increasing the number of independent samples via pulse compression. For estimating Doppler variables, at least a pair of pulses is needed. In AHV mode, for Doppler spectrum width one pair is sufficient but for velocity, ρ_{hv}, and Φ_{DP} at least two pairs are needed. Z requires one pulse and Z_{DR} two successive pulsed one at each polarization. Mitigating range/velocity (*R*/*V*) ambiguities in AHV mode is more complicated than in SHV mode. A comprehensive comparison of the two modes [20] indicates that long PRTs (few ms) in AHV would cause unacceptable errors. Other complexities (for *R*/*V* mitigation, clutter filtering) would add to the dwell time and the increase is a good metric for comparing various designs.

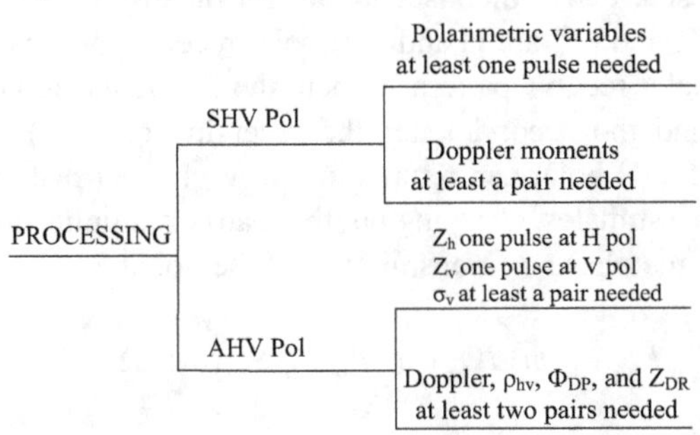

Figure 12.8 Conceptual flow chart of processing possibilities for a planar antenna PAR

12.3.2 Cross-polar pattern effects on bias in polarimetric variables

First, we briefly review the equations relating received voltages to the transmitted fields and matrix backscattering elements [24]. Start with the transmitted field by the mth pulse, $\mathbf{E}^t = [E_h^t(m) \quad E_v^t(m)]^T$ in a matrix notation as

$$\begin{bmatrix} E_h^t(m) \\ E_v^t(m) \end{bmatrix} = \begin{bmatrix} F_{hh}^t & F_{hv}^t \\ F_{vh}^t & F_{vv}^t \end{bmatrix} \begin{bmatrix} e_h(m) \\ e_v(m) \end{bmatrix} = \boldsymbol{F}^t \boldsymbol{e}\,, \tag{12.2}$$

where $\boldsymbol{F}^t$ is the transmit pattern matrix (of a formed beam), and $\boldsymbol{e}$ is the excitation vector that depends on the transmission mode. The second subscript indicates the port for intended polarization whereas the first subscript is the achieved polarization. Thus, F_{hv}^t indicates the cross-polar pattern coupled from V to H polarization and F_{vh}^t is coupled from H to V polarization.

The received voltages are obtained from integrals over the spherical space

$$\begin{bmatrix} V_h(m) \\ V_v(m) \end{bmatrix} = C\int_\Omega \left[\begin{bmatrix} F_{hh}^r & F_{vh}^r \\ F_{hv}^r & F_{vv}^r \end{bmatrix} \left\{ \begin{bmatrix} S_{hh}(m) & 0 \\ 0 & S_{vv}(m) \end{bmatrix} \right\} \mathbf{E}^t \right] d\Omega$$

$$= C\int_\Omega [\mathbf{F}^r \mathbf{S} \mathbf{E}^t] d\Omega, \tag{12.3}$$

where Ω is the solid angle (integration in the range is implied but omitted as it does not affect the results), C is a scalar factor that contains dependence on the range, attenuation, and system parameters; in this section, we set it to 1 as it has no effect on our derivations. In (12.3), $S_{pp}(m)$ (where subscript p is either "h" or "v") are sums of contributions by scatterers to the backscattering matrix element in a small incremental volume $d\Omega$. But, for practical purposes, these can be considered as densities per unit volume. The backscattering matrix $\mathbf{S}$ is assumed to be diagonal as is often the case in rain and oriented ice crystals. Nonetheless, the main point of this assumption is to enable fair comparisons of the effects under ideal conditions. Adding the off-diagonal terms complicates the analysis and obscures the essence. One more caveat is that the S_{hh} may contain the propagation differential phase and the backscatter differential phase. These are also inconsequential for the arguments that follow.

The symbols F_{hh}^r and F_{vv}^r are H and V copolar receive patterns, while F_{vh}^r and F_{hv}^r are H and V cross-polar receive patterns (where the first letter in the subscript denotes the incident field and the second letter the receiving channel). Note that F_{vh}^r/F_{hv}^r quantify coupling of V/H polarized returns into H/V channel that causes biases in the polarimetric variable estimates. Carrying out the matrix multiplications and assuming $\boldsymbol{e} = [1\ e^{j\beta}]$ where β is the phase on transmission of the voltage

$$V_h(m) = \int_\Omega F_{hh}^t F_{hh}^r S_{hh}(m) d\Omega + \int_\Omega F_{hv}^t F_{hv}^r S_{vv}(m) d\Omega$$

$$+ \left[\int_\Omega F_{hv}^t F_{hh}^r S_{hh}(m) d\Omega + \int_\Omega F_{vv}^t F_{hv}^r S_{vv}(m) d\Omega \right] e^{j\beta}. \tag{12.4}$$

A similar expression holds for $V_v(m)$. We estimate the polarimetric variables from various second-order moments of the returned voltages [25]. The example we offer is the reflectivity factor at horizontal polarization Z_h. It is proportional to the expected value $\langle |V_{hh}|^2 \rangle$ (in practice an estimate from several samples).

Assume AHV mode so that the term multiplying $e^{j\beta}$ is absent. Then the expected value becomes

$$\begin{aligned}\langle |V_h|^2 \rangle &= \left\langle \left| \int_\Omega F^t_{hh} F^r_{hh} S_{hh} d\Omega + \int_\Omega F^t_{hv} F^r_{hv} S_{vv} d\Omega \right|^2 \right\rangle = \int_\Omega \left| F^t_{hh} F^r_{hh} \right|^2 \langle |S_{hh}|^2 \rangle d\Omega \\ &\quad + \int_\Omega 2\mathrm{Re}\left(F^{t*}_{hh} F^{r*}_{hh} F^t_{hv} F^r_{hv} \langle S^*_{hh} S_{vv} \rangle \right) d\Omega \\ &\quad + \int_\Omega \left| F^t_{hv} F^r_{hv} \right|^2 \langle |S_{vv}|^2 \rangle d\Omega. \end{aligned} \tag{12.5}$$

The first term on the right comes from the equality

$$\left\langle \int_\Omega F^t_{hh} F^r_{hh} S_{hh} d\Omega \int_\Omega F^{t*}_{hh} F^{r*}_{hh} S^*_{hh} d\Omega \right\rangle = \int_\Omega \left| F^t_{hh} F^r_{hh} \right|^2 \langle |S_{hh}|^2 \rangle d\Omega \tag{12.6}$$

because the expectation $\langle S_{hh}(r_i) S^*_{hh}(r_j) \rangle$ is 0 for all $i \neq j$.

The functions (powers and cross product) of S in (12.5) are very similar hence the magnitudes of the three terms depend heavily on the product of patterns. The first term of the expended (12.5) is the desired power proportional to Z_h. Because it contains the product of copolar patterns, it is significantly larger than the other two terms. In the second term, the copolar patterns multiply the cross-polar ones. If the cross-polar transmitted and receive patterns are the same, then this term would have a cross-polar voltage pattern raised to the second power (equal to the power pattern). The third term is the product of four cross-polar patterns and is, therefore, insignificant. Clearly, the bias in Z_h is caused by the second term; it is proportional to the integral of the cross-polar pattern times the copolar pattern (i.e., the square of the voltage patterns). Although this contribution to the bias may not be negligible, it is much smaller than the contribution in SHV mode as explained next.

Let us assume the worst-case whereby β =0 in (12.4) and compute the power. The result is an expression with 10 terms. In four of these, the integrands are squares of the integrands in (12.4), and, in the six remaining terms, the integrands are products of the various integrands from (12.4). Take one of these six terms that has in the integrand a first power of the cross-polar (voltage) pattern multiplied by the copolar patterns. For example, the first term times the third term results in $2\mathrm{Re}\left(\int_\Omega F^t_{hh} F^r_{hh} F^{r*}_{hh} F^{t*}_{hv} \langle |S_{hh}|^2 \rangle d\Omega \right)$. The value of this term within the main lobe is critical. If it has a peak coincident with the main lobe of the copolar pattern, it could bias strongly Z estimate. We have seen in Section 12.2.5 that the cross-polar pattern of TPD has four lobes at the broadside and two along the principal planes. These lobes have equal magnitude and opposite signs, hence the integral of the term is zero. This important property is typical of planar array antennas.

12.3.3 Copolar pattern effects

Herein we examine the returns from scatterers illuminated by a planar array antenna pointed in an arbitrary direction. The indexes 1 and 2 correspond to the two ports on the array generating the intended H and V polarized fields as well as voltages.

In Figure 12.9, the orientation of the electric field at the beam center (θ_0, φ_0) is plotted for a tilted array and pointing out of the principal vertical plane. The geometry also applies to a vertical array pointing out of the principal planes.

To be explicit, rather than using the copolar (one-way) field pattern functions of [4]

$$F_{11}(\theta_o, \phi_o) = \sqrt{g_{11}} f_{11}(\theta_o, \phi_o); g_1(\theta, \phi) = g_{11} f_{11}^2(\theta, \phi), \text{etc.} \tag{12.7a}$$

we use

$$g_1^{1/2} = \sqrt{g_{11}} f_{11}(\theta_o, \phi_o) \tag{12.7b}$$

and

$$g_2^{1/2} = \sqrt{g_{22}} f_{22}(\theta_o, \phi_o), \tag{12.7c}$$

to scale the magnitudes of the electric fields (in the far-field region) at the beam center. For weather observations in which the radiation sphere's polar axis is aligned with the vertical direction, the gain in any direction $g_1(\theta, \phi)$ also depends on the beam direction (θ_0, ϕ_0).

We assume (with no loss in substance) that in the broadside direction, the two ports produce the same field magnitudes (these needs to be calibrated in the backend of the antenna) but have a difference in phase (on transmission) β. Moreover, we ignore the radiation of the orthogonal sides (the so-called non-radiating). This constraint holds for beams in and close to the principal planes. The difference in phase β may vary with the pointing angle. If a phase code $c(n)$ is applied to mitigate the

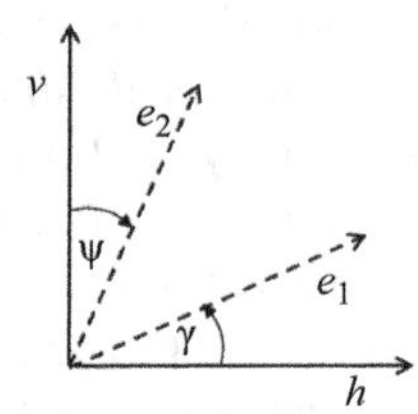

Figure 12.9 The horizontal (h) and the vertical (v) axis in the polarization plane of a propagating EM wave. The axis h is parallel to the ground; the axis v lies in a vertical plane. The axes $e_1(\theta_0, \varphi_0)$ and $e_2(\theta_0, \varphi_0)$ indicate the direction of the fields E_1, E_2 at the beam center generated by the Ports 1 and 2. On broadside, Port 1 generates horizontal polarization and Port 2 vertical, hence, ψ and γ are zero but out of the principal plane, these angles differ from zero.

effects of coupling [26], β would be a function of nT_s and can be expressed as $\beta(n) = \beta_o + c(n)$. Furthermore, assume propagation through media of oriented oblate scatterers produces differential phase Φ_{DP}. For compactness, let us use $S_h = \sum_{V_6} s_{hi} e^{-j4\pi r_i}$ and similarly $S_v = \sum_{V_6} s_{vi} e^{-j4\pi r_i}$ where the summation is over the scatters (i index in the elemental volume ΔV_6 of the resolution volume V_6) and the range to the scatterer is r_i. The s_{hi} is the element of the backscattering matrix for the horizontally polarized incident field; assume oblate spheroids (no depolarization), hence, in the notation whereby a is the symmetry axis and b the long axis, the $s_{hi} = s_{bi}$. The matrix transformation **P** relating the unit vectors e_1, e_2 to h, v (Figure 12.9) is

$$\mathbf{P} = \begin{vmatrix} \cos\gamma & \sin\psi \\ \sin\gamma & \cos\psi \end{vmatrix}, \text{ therefore } \begin{vmatrix} \mathrm{h} \\ \mathrm{v} \end{vmatrix} = \mathbf{P} \begin{vmatrix} \mathrm{e}_1 \\ \mathrm{e}_2 \end{vmatrix}. \tag{12.8}$$

Ignoring cross-polarization components induced in propagation, scattering, and antenna, the received voltage δV_1 (corresponding to the intended H field) and δV_2 (intended V field) from the elemental volume are

$$\begin{vmatrix} \delta V_1 \\ \delta V_2 \end{vmatrix} = C \begin{vmatrix} g_1^{1/2}(\theta,\phi) & 0 \\ 0 & C_R g_2^{1/2}(\theta,\phi) e^{j\xi} \end{vmatrix} \boldsymbol{P}^{\mathrm{t}} \begin{vmatrix} S_h & 0 \\ 0 & S_v \end{vmatrix} \boldsymbol{P} \begin{vmatrix} g_1^{1/2}(\theta,\phi) W_1 \\ g_2^{1/2}(\theta,\phi) C_T W_1 e^{j\beta} \end{vmatrix}$$

$$= C \begin{vmatrix} (S_h \cos^2\gamma + S_v \sin^2\gamma) g_1 W_1 + (S_h \cos\gamma \sin\psi + S_v \sin\gamma \cos\psi) g_1^{1/2} g_2^{1/2} C_T e^{j\beta} W_1 \\ (S_h \cos\gamma \sin\psi + S_v \sin\gamma \cos\psi) g_1^{1/2} g_2^{1/2} C_R e^{j\xi} W_1 + (S_h \sin^2\psi + S_v \cos^2\psi) g_2 C_T C_R e^{j(\beta+\xi)} W_1 \end{vmatrix}. \tag{12.9}$$

The following explains various assumptions in (12.9). Although S_v is complex, we set it to be real because its phase is an inconsequential reference. Therefore, $S_h = |S_h| \exp(-j\Phi_{DP})$ so that the effects of differential phase Φ_{DP} from propagation and backscatter are accounted for. C is the calibration parameter. W_1 is the voltage at Port 1 generating the field E_1 and $W_2 = C_T e^{j\beta} W_1$ is the voltage at Port 2 generating E_2 (C_T is a scaling constant relating the transmitted voltage at Port 2 to the voltage at Port 1). The phase difference in reception between the outputs from Port 2 and Port 1 is ξ and the ratio of amplitudes (Port 2 to Port 1) equals C_R. On the right side, the explicit dependences of g_1, g_2 on direction hold but are dropped to fit the equation.

Regroup the terms in (12.9) as follows:

$$CW_1 \begin{vmatrix} \left(\cos^2\gamma g_1 + \cos\gamma \sin\psi g_1^{1/2} g_2^{1/2} C_T e^{j\beta}\right) S_h + \left(\sin^2\gamma g_1 + \sin\gamma \cos\psi g_1^{1/2} g_2^{1/2} C_T e^{j\beta}\right) S_v \\ \left(\cos\gamma \sin\psi g_1^{1/2} g_2^{1/2} + \sin^2\psi g_2 C_T e^{j\beta}\right) C_R e^{j\xi} S_h + \left(\sin\gamma \cos\psi g_1^{1/2} g_2^{1/2} + \cos^2\psi g_2 C_T e^{j\beta}\right) C_R e^{j\xi} S_v \end{vmatrix}$$

$$= \begin{vmatrix} \delta V_1 \\ \delta V_2 \end{vmatrix}. \tag{12.10}$$

Eqs (12.9) and (12.10) apply to any elemental volume within the resolution volume. But if the beam is narrow as on polarimetric planar PAR (PPPAR), we can assume $g_1(\theta, \phi)$ is equal to its value at the beam center. Therefore, *for PPPAR beams steered away from the principal planes, the orientation and magnitudes of the reflected fields can be obtained from the values at the beam center* [27]. This is assumed in the sequel and the voltages from the elemental volumes are replaced with their sums V_1 and V_2.

In principle, one can invert (12.10) to express S coefficients in terms of the voltages. This requires knowledge of the orientation angles, ψ and γ, the phase differences β and ξ, and the Port 2 to Port 1 scaling factors on transmit C_T and on receive C_R, as well as the power gains g_1 and g_2 at every pointing direction. This amounts to eight numbers that need to be known plus the calibration parameter C for calculating voltages and hence reflectivity.

The voltages and S parameters change from pulse to pulse; say as a function of the sample number n of the time series data. Thus, the pulse-to-pulse inversion of (12.10) would generate two-time sequences: one for $S_h(n)$ and another for $S_v(n)$. From the average powers of these two sequences, Z_h and Z_{DR} can be computed. From the correlation of the two sequences, the differential phase Φ_{DP} and correlation coefficient, ρ_{hv} can be computed. Although promising results of inversion on a small one-dimensional array in a laboratory setup have been obtained [28], there have been no demonstrations on larger arrays yet.

A different way to estimate the second-order moments (i.e., ensemble averages$\langle|S_h|^2\rangle, \langle|S_v|^2\rangle, \langle S_h^* S_v\rangle$) is from the powers of the returned signals at the Ports 1 and 2, and the correlation of these two signals (i.e., the power estimates from the first row of (12.10) summed over M samples, similar power estimate from the second row and the estimates of the correlation between the first-row and second-row signals). The number of electric parameters that need to be known is also nine.

12.3.4 Phase coding

Phase coding can simplify the measurement/computation of the polarimetric variables. Suppose that the 0°, 180° phase code is applied to Port 2. This can be represented as $e^{j\beta(n)}$ where $\beta(n)$ changes between 0°, 180°. Fourier transform of the first row in (12.9) generates two spectra: one from the first term in row 1 is centered at the Doppler velocity and another (second term in row 1) is offset by the unambiguous velocity. Thus, one can separate these two terms as follows:

$$CW_1\left(S_h\cos^2\gamma + S_v\sin^2\gamma\right)g_1 = V_1(\bar{v}) \tag{12.11a}$$

$$CW_1(S_h \cos\gamma \sin\psi + S_v \sin\gamma \cos\psi)g_1^{1/2}g_2^{1/2}C_T e^{j\beta} = V_1(\bar{v} + v_a), \tag{12.11b}$$

where $V_1(\bar{v})$ is the sequence (measurement) obtained from the inverse Fourier transform of the spectral components corresponding to one-half the Nyquist interval containing the mean Doppler velocity $\bar{v}$. $V_1(\bar{v} + v_a)$ is the sequence corresponding to the spectrum offset by the unambiguous velocity from v. Similarly, the two terms in the second row of (12.9) can be separated so that the number of sequences is four. But, the sequences corresponding to the off-diagonal terms (i.e., cross-polar) differ by a complex multiplying factor. Therefore, there are three sequences that can be used to estimate powers and cross-correlations. Of the three, the one corresponding to the diagonal term is redundant, hence, might be useful for checking consistency or determining the initial transmitting phase.

The powers and cross-correlations of separated diagonal sequences can be used to generate three complex equations in which the ensemble averages of the unknown terms are $\langle |S_{\mathrm{h}}|^2\rangle$, $\langle |S_{\mathrm{v}}|^2\rangle$, $\langle S_{\mathrm{h}}^* S_{\mathrm{v}}\rangle$. The third term contains the differential phase. It should be expressed as one complex number. In doing so, one needs to track this number and its conjugate until the last step in the solution of the three complex equations. The angles ψ and γ, the differential phases β_{o} and ξ, the gains and amplitude calibration of the two channels on transmit and receive need to be known in addition to C.

12.3.5 Alternate transmission of H and V field (AHV)

AHV mode is similar to SHV mode with phase coding. The cross-polar signal at Port 2 (if only Port 1 is active) is the 21 term in (12.9), encompassed with brackets in the 2nd row) and if Port 2 is active, it is the 12 term in (12.9). These components are redundant but might be helpful to check the stability of the system. Computations of the second-order moments are made using the main diagonal terms in (12.9) which are estimated (measured) sequentially. Therefore, the correlation term includes the Doppler effect which needs to be eliminated [19]. From the expression (12.9), the maximum values of the angles ψ and γ for which the bias in the polarimetric variables is acceptable can be determined.

12.4 Vertically oriented array

We consider a vertically oriented array with patch radiators so that the angle $\gamma = 0$ and calibration becomes simpler. But the relation between the angle ψ and azimuth *az* and elevation *el* angles is needed. That relation is [29]

$$\cos\,\psi = \frac{\cos\,(az)}{\sqrt{1 - \cos^2(el)\sin^2(az)}}. \tag{12.12}$$

Herein, we provide more details for this geometry about the computations than are listed in Section 12.3.

12.4.1 Relations

In SHV mode, the governing relation (12.9) expressed as two equations is

$$V_1 = CW_1\left(g_1 + \sin\,\psi g_1^{1/2} g_2^{1/2} C_T e^{j\beta}\right) S_{\mathrm{h}} \tag{12.13a}$$

$$V_2 = CW_1\left[\left(\sin\,\psi g_1^{1/2} g_2^{1/2} + \sin^2\psi g_2 C_T e^{j\beta}\right) S_{\mathrm{h}} + \cos^2\psi g_2 C_T e^{j\beta} S_{\mathrm{v}}\right] C_R e^{j\xi}. \tag{12.13b}$$

Multiplying (12.13a) with $\sin\psi g_2^{1/2} C_R e^{j\xi}/g_1^{1/2}$ (from pulse to pulse) and subtracting from (12.13b) solves for the second term in (12.13b) which is

$$CW_1 \cos^2\psi g_2 C_R e^{j\xi} C_T e^{j\beta} S_\mathrm{v} = V_2 - \frac{\sin\psi g_2^{1/2} C_R e^{j\xi}}{g_1^{1/2}} V_1. \tag{12.13c}$$

Therefore, the polarimetric variables can be estimated from (12.13a) and (12.13c).

By inspection, it follows that if $\beta = 0$ and $\xi = 0$, the differential phase can be computed directly by correlating the conjugate of (12.13a) with (12.13c) and that ρ_hv equals the magnitude of the corresponding correlation coefficient. Besides the implicit dependence on the direction angle ψ of the intended V field, the polarimetric variables from (12.13) depend explicitly on this angle through the values of g_1 and g_2.

Note that Z_h and Z_v [from (12.13a) and (12.13b)] depend explicitly and implicitly (through (12.7b) and (12.7c)) on ψ. Z_DR depends on the same variables and is independent of C as it is proportional to the ratio (12.13a) to (12.13c).

An alternate way to compute the polarimetric variables is from the powers of (12.13a) and (12.13b) and the correlation between (12.13a) and (12.13b). From these, first the $\langle|S_\mathrm{h}|^2\rangle$, $\langle|S_\mathrm{v}|^2\rangle$, $\langle S_\mathrm{h}^* S_\mathrm{v}\rangle$ and Φ_DP are found and combined to generate the polarimetric variables. Thus, take the power estimates as the average of M samples:

$$\sum_{i=1}^{M} |V_{1i}|^2/M = |CW_1|^2 \left| g_1 + g_1^{1/2} g_2^{1/2} \sin\psi C_T e^{j\beta} \right|^2 \langle|S_\mathrm{h}|^2\rangle \tag{12.14a}$$

$$\sum_{i=1}^{M} |V_{2i}|^2/M = |CW_1|^2 \left\{ \begin{array}{l} \left| g_1^{1/2} g_2^{1/2} \sin\psi + g_2 \sin^2\psi C_T e^{j\beta} \right|^2 C_R^2 \langle|S_\mathrm{h}|^2\rangle + (\cos^2\psi)^2 g_2^2 C_R^2 C_T^2 \langle|S_\mathrm{v}|^2\rangle \\ +2\mathrm{Re}\left[\left(g_1^{1/2} g_2^{1/2} \sin\psi + g_2 \sin^2\psi C_T e^{-j\beta} \right) \left(g_2 \cos^2\psi e^{j\beta} \right) C_R^2 C_T \right] \langle S_\mathrm{h}^* S_\mathrm{v}\rangle \end{array} \right\} \tag{12.14b}$$

$$\sum_{i=1}^{M} V_{1i}^* V_{2i}/M = |CW_1|^2 \left[\begin{array}{l} \left(g_1 + \sin\psi g_1^{1/2} g_2^{1/2} C_T e^{-j\beta} \right) \left(\sin\psi g_1^{1/2} g_2^{1/2} + \sin^2\psi g_2 C_T e^{j\beta} \right) C_R e^{j\xi} \langle|S_\mathrm{h}|^2\rangle \\ + \left(g_1 + \sin\psi g_1^{1/2} g_2^{1/2} C_T e^{-j\beta} \right) \cos^2\psi g_2 C_R e^{j\xi} C_T e^{j\beta} \langle S_\mathrm{h}^* S_\mathrm{v}\rangle \end{array} \right]. \tag{12.14c}$$

From these equations, it is evident that all the moments depend on the pointing direction explicitly and also implicitly through (12.7b) and (12.7c). Eq. (12.14a) is not coupled to the other equations hence to compute Z_h, one only needs to integrate over the beam; its width depends on the pointing direction. Again note that adjusting β and ξ to 0 simplifies the solving process. Assuming that the calibration is acceptable (C, C_T, C_R, and (12.7b) and (12.7c) are known as well as ψ), it is in principle relatively easy to compute the polarimetric variables by solving the set (12.14). For example, (12.14b) can be properly scaled and added to (12.14c) so that the first term in (12.14c) is eliminated. Then the cross-product $\langle S_\mathrm{h}^* S_\mathrm{v}\rangle$ can be computed. Subsequent substitution in (12.14b) yields $\langle|S_\mathrm{v}|^2\rangle$. Next, we examine the number of parameters needed for calibration.

Backend (behind the antenna): The gain C, the differential gain on transmission C_T, and the differential phases β and ξ amount to 4 numbers (compare that to 2 gains for the dish antenna and the system differential phase $\beta + \xi$ which can be obtained from data). Similar holds in the receiver channels with the variable differential gain C_R and differential phase ξ; note that the overall system calibration C lumps together all the gains and losses in both receiver and transmitter chains. This amounts to 2 more numbers. We expect that these five variables would be independent of the pointing direction. Contrast this to the dish antenna where only two gains suffice for calibration.

Antenna: The gains $g_1(\theta_o, \phi_o)$ and $g_2(\theta_o, \phi_o)$, the corresponding beamwidth, and the pointing direction ψ. The two patterns should have the same elliptically shaped beam cross-sections otherwise the polarimetric variables on the weather PAR would incur unacceptable errors. This totals 4 but it may be safe to assume that the beamwidths (two needed for the elliptical shape) would be computed from the known pointing direction and the computed (calibrated via measurements) gains. That would reduce the number of "independent" variables to 3 for each pointing direction. To cover 90° in azimuth and 15 elevations with a planar array, 1,350 beam positions are needed. This translates to 4,050 calibration variables. Because of viewing symmetry (the left field of view is symmetric to the right one), the actual number to calibrate might be reduced by a factor of 2, to 2,025 (the symmetry needs verification on real systems). Some other reductions in complexity are expected in and near the principal planes (at less than 300 points).

12.4.2 Simultaneous transmission with phase coding

To condense notation, the signal with the true mean Doppler in Port 1 is written as V_{11} the one in Port 2 is V_{22} and the cross-port signals (offset by the unambiguous velocity) are V_{12} (from Port 2 coupled to Port 1) and V_{21} (Port 1 to Port 2). Furthermore, for consistency with the previous results, the relative calibration (ratio of) Port 2 to Port 1 voltage on transmission is denoted as $W_2 = C_T W_1 e^{jn\pi}$; this implies that $\beta(n) = \beta_o + e^{jn\pi}$. Here the subscript "o" on β_o is used to distinguish it from the β used in the case of alternate (AHV) mode. Upon reconstruction (separation of the components), the $e^{jn\pi}$ term is not present. Setting γ=0 in (12.11a) and (12.11b), V_{11} and V_{12} become

$$V_{11} = CW_1 g_1 S_h, \tag{12.15a}$$

$$V_{12} = CW_1 \sin\psi g_1^{1/2} g_2^{1/2} C_T e^{j\beta_o} S_h. \tag{12.15b}$$

The separation of spectra of the signal from Port 2 (12.14b) isolates the first-term spectrum from the spectra of the other two terms. These two terms are

$$V_{21} = CW_1 \sin\psi g_1^{1/2} g_2^{1/2} C_R e^{j\xi} S_h, \tag{12.15c}$$

$$V_{22} = CW_1 \left(\sin^2\psi S_h + \cos^2\psi S_v\right) g_2 C_T C_R e^{j(\beta_o+\xi)}, \tag{12.15d}$$

and it is evident that $V_{12} \neq V_{21}$.

The equations in (12.15) are shorthand expressions for these voltages at any one sample time and the four complex sequences (or four complex spectra) consist of M samples each. The powers and correlation of the first three terms are

$$\langle|V_{11}|^2\rangle = C^2|W_1|^2 g_1^2 \langle|S_h|^2\rangle, \tag{12.16a}$$

$$\langle|V_{12}|^2\rangle = C^2|W_1|^2 g_1 g_2 \sin^2\psi C_T^2 \langle|S_h|^2\rangle, \tag{12.16b}$$

$$\langle|V_{21}|^2\rangle = C^2|W_1|^2 g_1 g_2 \sin^2\psi C_R^2 \langle|S_h|^2\rangle, \tag{12.16c}$$

$$\langle V_{11}^* V_{12}\rangle = C^2|W_1|^2 g_1^{3/2} g_2^{1/2} \sin\psi C_T e^{j\beta_o} \langle|S_h|^2\rangle, \tag{12.16d}$$

$$\langle V_{11}^* V_{21}\rangle = C^2|W_1|^2 g_1^{3/2} g_2^{1/2} \sin\psi C_R e^{j\xi} \langle|S_h|^2\rangle. \tag{12.16e}$$

Inspection reveals that some relations between four parameters can be obtained easily and are independent of C, W_1, g_1, g_2, and $\sin\psi$. These are: the phases β_o and ξ from the arguments of (12.16d) and (12.16e), and thus the system differential phase $\Phi_{DPsys} = \beta_o + \xi$, and the ratio C_T/C_R by dividing (12.16d) with (12.16e), or (12.16b) with (12.16c) which is redundant and can be used for cross-checking.

Phase coding is more effective if the unambiguous velocity interval is significantly larger than the spectrum width of the weather signal so that the two offset spectral components can be easily separated. Otherwise, the original and offset spectra of weather signals would overlap precluding clean separation.

12.4.3 Polarimetric variables

The computation of the polarimetric variables in the case of a vertically oriented array and phase coding is presented in this section. Errors introduced by various approximations are evaluated and fields of view in azimuth and elevation where the errors are tolerable are plotted.

(i) *Differential reflectivity*

The differential reflectivity can be computed as follows:

$$\frac{\langle|V_{11}|^2\rangle}{\langle|V_{22}|^2\rangle} = \frac{g_1^2\langle|S_h|^2\rangle}{\langle|\sin^2\psi S_h + \cos^2\psi S_v|^2\rangle g_2^2 C_T^2 C_R^2}. \tag{12.17a}$$

Take $A_{pc} = \frac{\langle|V_{12}|^2\rangle\langle|V_{21}|^2\rangle}{|\langle V_{11}^* V_{12}\rangle||\langle V_{11}^* V_{21}\rangle|} = \frac{g_2}{g_1}\sin^2\psi C_T C_R$ or

$$\left(\frac{g_2}{g_1} C_T C_R\right)^2 = A_{pc}^2/\sin^4\psi; \tag{12.17b}$$

substitute (12.17b) in (12.17a) to obtain

$$B_{pc} = \frac{\langle|V_{11}|^2\rangle A_{pc}^2}{\langle|V_{22}|^2\rangle \sin^4\psi} = \frac{\langle|V_{11}|^2\rangle\langle|V_{12}|^2\rangle\langle|V_{21}|^2\rangle}{\langle|V_{22}|^2\rangle|\langle V_{11}^* V_{12}\rangle||\langle V_{11}^* V_{21}\rangle|\sin^4\psi} = \frac{\langle|S_h|^2\rangle}{\langle|\sin^2\psi S_h + \cos^2\psi S_v|^2\rangle}. \tag{12.17c}$$

Then take

$$\langle V_{11}^* V_{22} \rangle = C^2 |W_1|^2 g_1 g_2 C_T C_R (\sin^2\psi \langle |S_h|^2 \rangle + \cos^2\psi \langle S_h^* S_v \rangle) e^{j(\beta_o + \xi)}, \tag{12.17d}$$

and note that β_o and ξ are available in (12.16d) and (12.16e), therefore can be removed from (12.17d); also $\langle |S_h|^2 \rangle$ from (12.16a) can be substituted in (12.16d) so that the $R_{hv} = \langle S_h^* S_v \rangle$ can be computed. With this, one can solve for $\langle |S_v|^2 \rangle$ as follows:

$$\langle |S_v|^2 \rangle = \frac{\langle |S_h|^2 \rangle - B_{pc} \sin^4\psi \langle |S_h|^2 \rangle - 2R_{hv} \sin^2\psi \cos^2\psi}{B_{pc} \cos^2\psi}, \tag{12.17e}$$

and combine with (12.16a) to compute Z_{dr}. The peril in this approach is that the coefficients entering these equations (through B_{pc} and others) need to be known.

It may be tempting to use B_{pc} in (12.17c) directly to estimate Z_{dr} as

$$\widehat{Z}_{dr} = B_{pc} = \frac{Z_{dr}}{\sin^4\psi Z_{dr} + 2\rho_{hv} Z_{dr}^{1/2} \cos \Phi_{DP} \sin^2\psi \cos^2\psi + \cos^4\psi}, \tag{12.17f}$$

where the multiplicative bias is the reciprocal of the denominator. Note the $\sin^4(\psi)$ in the denominator of (12.17c). Because ψ is small, computation of (12.17c) may be problematic. If scans are in or near the principal planes, one can avoid this problem by computing the parameters g_1, g_2, C_T, and C_R and using these in (12.17a). By inspection, it is obvious that if these parameters are known, the envelope of the acceptable bias would be the same as when Z_{DR} is computed from (12.17c). At what point to transition from (12.17a) to (12.17c) will depend on the relative contribution of errors from the uncertainty in the parameters versus the contribution from the error in the uncertainty of the angle ψ.

For illustration, the field of view where the bias of Z_{DR} is acceptable is examined next. This is made under the following conditions. Z_{DR} values of 0, 1, and 2 dB; ρ_{hv} values of 0.5 and 1, and Φ_{DP} of 0 and 180° are tested. The triplet Z_{DR} = 2dB, ρ_{hv} = 1, and Φ_{DP} = 0° produced the largest field of view (Figure 12.10(a)) and the triplet Z_{DR} = 2dB, Φ_{DP} = 180°, and ρ_{hv} = 1 produced the smallest field of view (Figure 12.10(b)). In these and subsequent figures, only the first quadrant of the field of view is presented because the total field of view is symmetric with respect to the abscissa and the ordinate.

Relaxing Z_{DR} bias upper bound to 0.13 dB, and keeping the triplet (Z_{DR} = 2dB, Φ_{DP} = 0°, ρ_{hv} = 1, which produced the largest view in Figure 12.10(a)) gave a clear field of view for elevations up to 20° and all azimuths. Further relaxing the bias to 0.2 dB increases marginally the smallest field of view hence is not shown.

(ii) *Correlation coefficient*

For the vertically oriented antenna and phase coding with filtering, the four components of voltages in (12.14) are available. Then the approximate relation for computing the correlation coefficient is

$$\rho_{hv} \exp(j\Phi_{DP}) = \frac{\langle V_{11}^* V_{22} \rangle}{\left(\langle |V_{11}|^2 \rangle \langle |V_{22}|^2 \rangle \right)^{1/2}}, \tag{12.18}$$

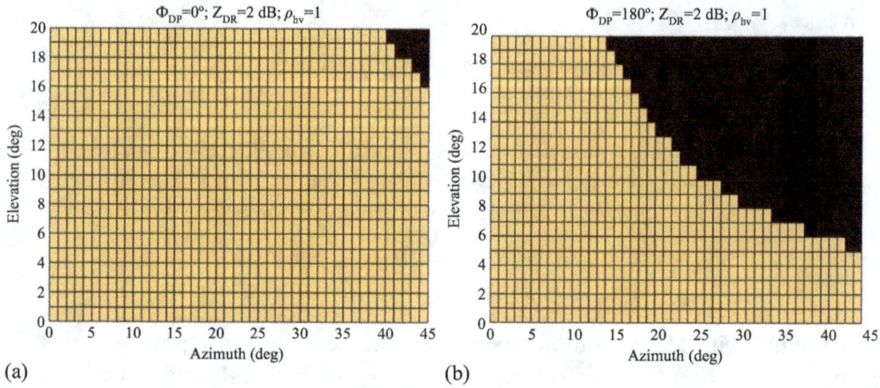

Figure 12.10 (a) The field of view in azimuth and elevation where the bias in Z_{DR} is smaller than 0.1 dB. Z_{DR} is computed from (12.16c) and the triplet Z_{DR} = 2 dB, Φ_{DP} = 0°, and ρ_{hv} = 1 was assumed in (12.16c); it yielded the largest field of view. (b) Eq. (12.16c) is used to compute Z_{DR} and it produced the smallest field of view for the following triplet: Z_{DR} = 2 dB, Φ_{DP} = 180°, and ρ_{hv} = 1. Z_{DR} bias is constrained to be less than 0.1 dB.

and it is compatible with the approximation (12.17a) for computing Z_{DR}. Obviously, this relation is independent of the various gains, but it produces a biased ρ_{hv} dependent on the polarimetric variables and pointing direction. By substituting the values for V_{11} and V_{22}, (12.15a) and (12.15d), and taking ensemble averages of the powers and correlations, the ρ_{hv} estimate becomes

$$\widehat{\rho}_{hv} = \rho_{hv} \frac{\left(1 + 2Z_{dr}^{1/2}\rho_{hv}^{-1} \cos \Phi_{DP} \tan^2\psi + Z_{dr}\rho_{hv}^{-2}\tan^4\psi\right)^{1/2}}{\left(1 + 2Z_{dr}^{1/2}\rho_{hv} \cos \Phi_{DP} + Z_{dr}\tan^4\psi\right)^{1/2}}. \tag{12.19}$$

Subtracting (12.19) from the true ρ_{hv} reveals the bias. Therefore,

$$\begin{aligned}\mathrm{bias}(\widehat{\rho}_{hv}) &= \rho_{hv} - \rho_{hv} \frac{\left(1 + 2Z_{dr}^{1/2}\rho_{hv}^{-1} \cos \Phi_{DP} \tan^2\psi + Z_{dr}\rho_{hv}^{-2}\tan^4\psi\right)^{1/2}}{\left(1 + 2Z_{dr}^{1/2}\rho_{hv} \cos \Phi_{DP} + Z_{dr}\tan^4\psi\right)^{1/2}} \\ &= \left(1 - \frac{A}{B}\right)\rho_{hv}.\end{aligned} \tag{12.20}$$

The bias (12.20) depends on several parameters and can be positive or negative. At Φ_{DP} =180°, the bias has a maximum positive value. At high values of ρ_{hv} near 1 (0.96–0.99), a change of 0.01 can signify a significant difference in types of scatterers. That is why the errors for WSR-88D are specified to be less than 0.003. Assuming this value, one can compute the field of view (Figure 12.11) within which the bias is acceptable. Maximum values of bias occur at Φ_{DP} of 0° and 180° and these maximums become larger as ρ_{hv} decreases and Z_{DR} increases. We chose ρ_{hv} of 0.96, a value at the transition between rain (or snow) and the melting layer because the gradients at this and higher values are sensitive to the hydrometeor types. Lower ρ_{hv} indicates mostly non-meteorological scatterers (with the exception of hail) and, in that range, larger errors can be tolerated. We use $Z_{DR} = 4$ dB as an upper representative value for precipitation like growing dendrites or melting graupel.

Also plotted in Figure 12.11 is the field of view where the bias in Z_{DR} is smaller than 0.2 dB. Clearly, the fields of view are similar and compatible.

(iii) *Differential phase*

Similar to and consistent with computations of Z_{DR} and ρ_{hv}, we can approximate the differential phase by the argument of the right side of (12.18). It is

$$\widehat{\Phi}_{DP} = \arg\left(\rho_{hv} Z_{dr}^{-1/2} e^{j\Phi_{DP}} \cot^2\psi + 1\right) \tag{12.21a}$$

or

$$\widehat{\Phi}_{DP} = \arctan\left(\frac{\rho_{hv} Z_{dr}^{-1/2} \sin \Phi_{DP} \cot^2\psi}{\rho_{hv} Z_{dr}^{-1/2} \cos \Phi_{DP} \cot^2\psi + 1}\right). \tag{12.21b}$$

The bias $\widehat{\Phi}_{DP} - \Phi_{DP}$ has a maximum if Φ_{DP} is 90° multiplied by an odd number. For the same parameters as in Figure 12.11 and an acceptable bias of 1° or 2°, the fields of view are plotted in Figure 12.12.

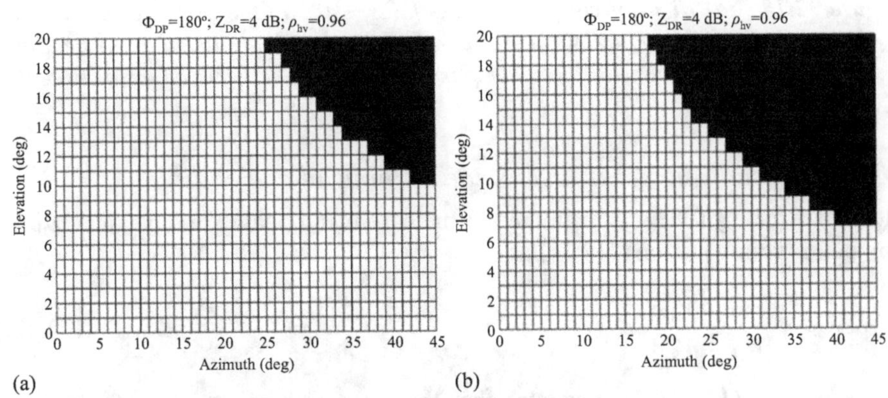

Figure 12.11 (a) Field of view where ρ_{hv} |bias| <0.003 in SHV mode with phase coding or AHV mode. The plot is valid for Φ_{DP} = 180°, Z_{DR} = 4 dB, and ρ_{hv} = 0.96. (b) The field of view where the bias in Z_{DR} is smaller than 0.2 dB and the other parameters are the same as in (a).

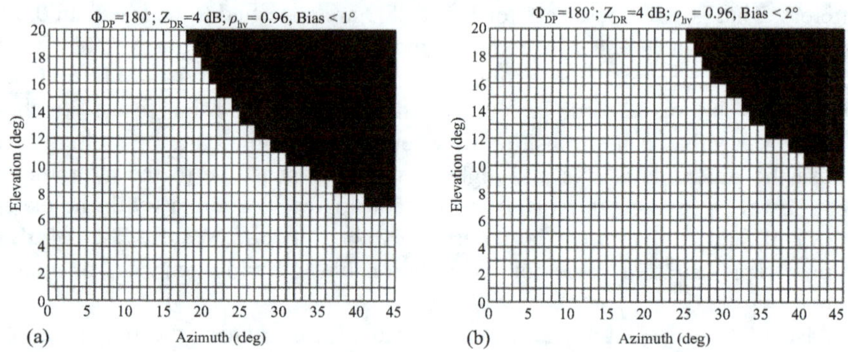

Figure 12.12 (a) Field of view where Φ_{DP} |bias| <1° in the SHV mode with phase coding or AHV mode. The plot is valid for Φ_{DP}= 90°, Z_{DR} = 4 dB, and ρ_{hv} = 0.96. (b) The same as in (a) except Φ_{DP} |bias| <2°.

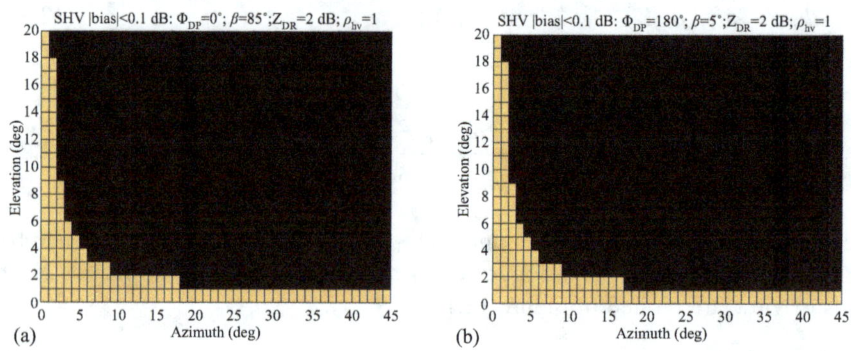

Figure 12.13 (a) Field of view where Z_{DR} |bias| <0.1 in the SHV mode, and no phase coding. Φ_{DP} = 0° and β = 85° which are listed in the heading together with the other parameters. (b) The same as in (a) except Φ_{DP} =180°, β = 5°.

12.4.4 Simultaneous transmission without phase coding

Herein the field of view is examined in the case of no phase coding. Thus, Z_{dr} is computed from equations (12.15) as

$$Z_{dr} = \frac{|V_{11} + V_{12}|^2}{|V_{21} + V_{22}|^2}. \tag{12.22}$$

Evaluation of (12.22) demonstrates that the size of the field of view depends heavily on the combination of the differential phase Φ_{DP} and the differential phase on transmission β. It has insignificant dependence on the true values of Z_{DR} (i.e., between 0 and 2 dB) and ρ_{hv} (between 0.5 and 1). For illustration, the fields of view in which the |bias| is smaller than 0.1 are plotted in Figure 12.13. In Figure 12.13(a) the $\Phi_{DP} = 0°$, $\beta = 85°$ whereas in Figure 12.13(b), $\Phi_{DP} = 180°$, $\beta = 5°$. A careful comparison of the two figures reveals that the only difference is at az = 16°, el = 2° whereby that pixel in Figure 12.13(a) has a bias smaller than 0.1 dB whereas in Figure 12.13(b), the bias is larger. This indicates that there is no fixed optimum value for the differential phase on transmission β. The optimum depends on Φ_{DP} which can have any value. Relaxing the bias to 0.2 dB produces a 45% increase in the viewing area.

12.4.5 Alternate transmission AHV

The second-order moments of voltages in the case of AHV mode are given with similar expressions in (12.16) except that there are some slight differences due to the effects of Doppler shift between successive returns [4]. It is included in pairs of pulses as follows in Figure 12.14.

Assume that the sequence starts with V_{11} so that the accumulation of pairs exemplified with $\langle V_{11}^* V_{22} \rangle$ would have a positive Doppler phase Ψ_d. Thus, we write

$$\langle V_{11}^* V_{22} \rangle = C^2 |W_1|^2 g_1 g_2 C_T C_R \left(\sin^2\psi \langle |S_h|^2 \rangle + \cos^2\psi \langle S_h^* S_v \rangle \right) \rho e^{j(\beta + \xi + \Psi_d)} \tag{12.23a}$$

and

$$\langle V_{22}^* V_{11} \rangle = C^2 |W_1|^2 g_1 g_2 C_T C_R \left(\sin^2\psi \langle |S_h|^2 \rangle + \cos^2\psi \langle S_h^* S_v \rangle \right) \rho e^{j(-\beta - \xi + \Psi_d)}, \tag{12.23b}$$

where ρ is the temporal correlation coefficient at lag T_s. The accumulation in (12.23a) starts from the zeroth pair and takes every second pair (all even pairs) and, in (12.23b), it starts with the odd pair (first) and accumulates all the odd pairs [30].

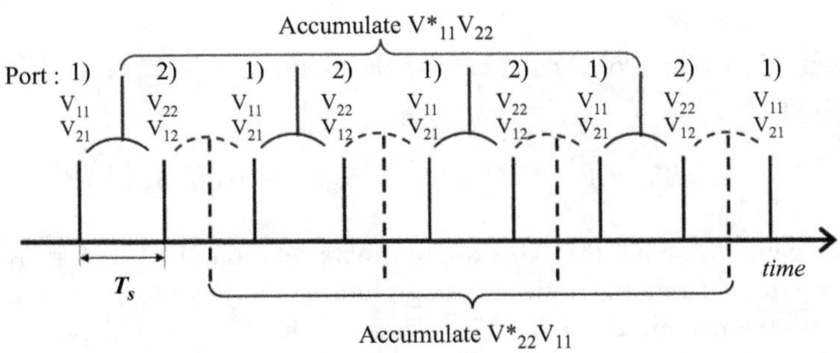

Figure 12.14. *Sequence of returns. Ports 1 and 2 are excited sequentially but the returns are received by both. That is Port 1 transmits and receives V_{11} while Port 2 receives the coupled component V_{21}. Then Port 2 transmits and so on. The products $V_{11}^* V_{22}$ are indicated with solid arcs and these are accumulated. Similarly, the products $V_{22}^* V_{11}$ are indicated with dashed arcs and are accumulated.*

The products of copolar Port 1 and offset (by lag 1) cross-polar (Port 2) returns are:

for the even pairs

$$\langle V_{11}^* V_{12}\rangle = C^2|W_1|^2 \sin\psi g_1^{3/2} g_2^{1/2} C_T\langle|S_h|^2\rangle\rho e^{j(\beta+\Psi_d)} \tag{12.24a}$$

and for the odd pairs

$$\langle V_{11} V_{12}^*\rangle = C^2|W_1|^2 \sin\psi g_1^{3/2} g_2^{1/2} C_T\langle|S_h|^2\rangle\rho e^{j(-\beta+\Psi_d)}. \tag{12.24b}$$

The product of copolar Port 1 and cross-polar returns (from Port 1 in Port 2 at the same time) is

$$\langle V_{11}^* V_{21}\rangle = C^2|W_1|^2 \sin\psi g_1^{3/2} g_2^{1/2} C_R\langle|S_h|^2\rangle e^{j\xi}. \tag{12.24c}$$

The spectrum width is retrieved from ρ (ratio of magnitudes (12.24b) and (12.24c)), ξ is the argument of (12.24c) which after substitution in (12.24a) retrieves the Doppler phase Ψ_d. Because these relations are proportional to $\sin(\psi)$, they may be inaccurate at small ψ and small SNR. Thus, if scans are in or near the principal planes, one can use (12.17a) directly. But then, the four parameters g_1, g_2, C_T, and C_R must be determined.

Similarly, the relation involving V_{22} for even pairs is

$$\langle V_{22}^* V_{21}\rangle = C^2|W_1|^2 g_1 g_2^{3/2} C_T C_R^2\left(\sin^2\psi\langle|S_h|^2\rangle + \cos^2\psi\langle S_h^* S_v\rangle\right)\rho e^{j(-\beta+\Psi_d)}, \tag{12.25a}$$

and, for odd pairs, it is

$$\langle V_{22} V_{21}^*\rangle = C^2|W_1|^2 g_1 g_2^{3/2} C_T C_R^2\left(\sin^2\psi\langle|S_h|^2\rangle + \cos^2\psi\langle S_h^* S_v\rangle\right)\rho e^{j(\beta+\Psi_d)}. \tag{12.25b}$$

The product of copolar (Port 2) and cross-polar return (from Port 2 in Port 1 at the same time) is

$$\langle V_{22} V_{21}^*\rangle = C^2|W_1|^2 g_1 g_2^{3/2} C_T^2 C_R\left(\sin^2\psi\langle|S_h|^2\rangle + \cos^2\psi\langle S_h^* S_v\rangle\right)e^{j\xi}. \tag{12.25c}$$

It is possible to obtain the same approximate relation (12.17f) for Z_{DR} as follows. Compute ρ from the ratio of magnitudes, i.e., divide the magnitude of (12.25b) with the magnitude of (12.25c). Then divide (12.24b) with ρ and form the product $\left(\langle V_{11}V_{12}^*\rangle/\rho\right)\langle V_{11}^* V_{21}\rangle$. Dividing this product with the power of (12.15a) squared produces

$$\frac{\left(\langle V_{11}V_{12}^*\rangle/\rho\right)\langle V_{11}^* V_{21}\rangle}{\langle|V_{11}|^2\rangle} = \frac{g_2}{g_1}\sin^2\Psi C_T C_R = A_{pc} \tag{12.26}$$

the same as (12.17b). Thus, one can proceed as in (12.17c) and compute Z_{DR}.

12.4.6 Principal planes

Calibration for the principal planes simplifies considerably as can be seen from (12.13a) and (12.13b) after substituting $\gamma = 0$ and $\psi = 0$. The simplified form becomes

$$V_1 = CW_1 g_1 S_h \tag{12.27a}$$

$$V_2 = CW_1 g_2 C_R e^{j\xi} C_T e^{j\beta} S_v. \tag{12.27b}$$

Because it is assumed that coupling is insignificant, only the product of terms in (12.27) needs to be calibrated (known) to obtain reflectivity and differential reflectivity. These products are functions of azimuth for the horizontal principal plane or elevation for the vertical one. Therefore, calibration becomes analogous to the one on the radar with a parabolic dish antenna except it needs to be done at each beam position because both g_1 and g_2 depend on the pointing direction. Although expressions for the gains g_{p1} and g_{p2} of patch antenna exist [31, (14–44)] in practice these do not apply exactly, hence, the gain dependence on the pointing direction would need to be measured or simulated.

12.5 Tilted antenna array

Here, it is suggested how to replicate performance of WSR-88D at the lowest elevations with the tilted phased array antenna. For that geometry, the angles ψ and γ must be known. The dependence of ψ on azimuth α, and elevation e is the same as for the vertically pointed array. The relation between ψ and pointing direction (α, e) is

$$\cos\,\psi = \frac{\cos\,\alpha}{\sqrt{1 - \cos^2(e)\sin^2(\alpha)}}. \tag{12.28}$$

A somewhat more complicated relation is between γ, pointing direction (α, e) and tilt angle θ_o with respect to the vertical z-axis. It is derived in [29].

Assume that θ_o is small so that the polarimetric variables can be computed directly from voltages at Port 1 and Port 2 (V_1 and V_2 in (12.10)) with no adjustment. For this to hold, the contributions by the "non-radiating" sides of the patch to the copolar powers and correlations must be small (similar to the ones on a parabolic dish antenna). We expect this will hold close to the principal planes. In (12.10), take the term $g_1\cos^2\gamma$ multiply S_h in the equation for V_1 and divide with $g_2\cos^2\psi C_T C_R$ multiplying S_v in the equation for V_2. This ratio squared equals $g_1^2/(C_R C_T g_2)^2$ and needs to be determined. Ideally, it should be 1 and the deviation causes constant bias to Z_{DR} that can be accounted for. Assume that is the case so with no loss of substance let $C_R = C_T = 1$ and $g_1 = g_2$. Then we can compute biases in the polarimetric variables obtained from V_1 and V_2 in (12.10) and compare these with unbiased variables. The field of view can be found by inserting the relation between the ψ, γ and the azimuth, elevation pairs. The bias depends on the intrinsic values of the polarimetric variables.

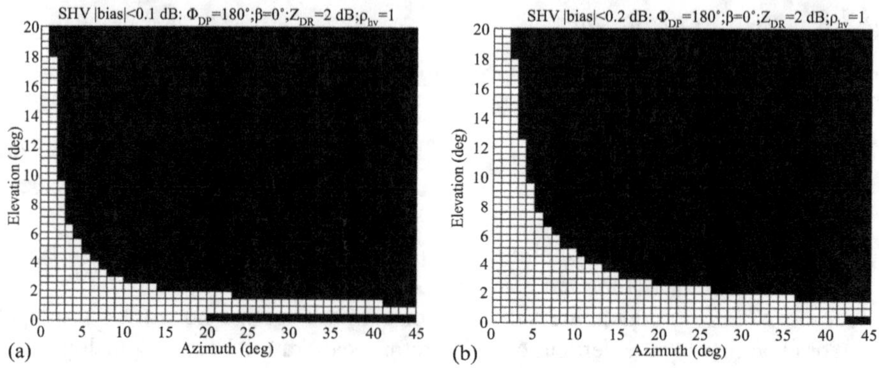

Figure 12.15 *(a) The field of view where Z_{DR} bias is less than 0.1 dB. The intrinsic polarimetric variables and the differential phase on transmission β are listed at the top. (b) The same as in (a) except Z_{DR} bias is less than 0.2 dB.*

In Figure 12.15(a) is a field of view (worst case) where Z_{DR} bias is less than 0.1 dB for a small tilt of $\theta_o = 1.5°$. The strip from 20° to 45° of larger bias is caused by the effects of non-orthogonality of the two fields and deviation of the intended horizontally polarized field from the true horizontal direction. An increase of allowed bias to 0.2 dB almost clears the bias at 0.5° elevation and expands the field of view to 2° in elevation almost everywhere.

12.6 Examples

We present some examples of data collected with TPD and compare these to data collected with the research WSR-88D (designated as KOUN). The two radars were 100 m apart and recorded almost simultaneously (within 59 s) data. In Table 12.1, are the parameters of the two radars.

Because the peak-transmitted power is weak, the radar's detectability is low; a 20 dBZ reflectivity produces an SNR of 0 dB at 25 km range. Therefore, linear frequency modulation (FM) with the frequency deviation $\Delta f = 0.33$ MHz was used. This produces the same range resolution (300 m) as the scan without pulse compression and pulse length $\tau = 2$ µs.

TPD recorded time series data (in phase I and quadrature-phase Q) while KOUN recorded spectral moments and polarimetric variables. The polarimetric variables for TPD were computed off-line.

To compute TPD reflectivity, we identified the area where TPD and KOUN beams overlap and made histograms of *Z*. Assuming that the one from KOUN is correct, we determined the radar constant *C* for TPD is 22.6 dB. The reflectivity is then $Z_h = P_h + 20\log(r) - C - 10\log[\cos(azb - az)]$, where P_h is the recorded power, *r* is the range (km), *azb* is the azimuth at the broadside, and *az* is the current azimuth. The cosine dependence accounts for the beamwidth and gain change off

Table 12.1 Scanning radar parameters: TPD versus KOUN

Parameter	TPD	KOUN
Mode	Pulse compression	Super resolution
Frequency	2.87 GHz	2.705 GHz
Peak transmitter power	3 kW	750 kW
Compression ratio	10	1.57 μs
Compressed pulse	2 μs	N/A
Linear FM Δf	330 kHz	N/A
Weighting	Uniform	N/A
PRT	1 ms	1 ms
Number of pulses	128	64
Beamwidth	Az = 6.3°; El = 2.5°	0.95°
Center azimuth angle	236°	N/A
Az sector	80°	172°
Az step	2°	0.5°
Elevation angles	1°, 2°	0.9°, 1.4°, 1.8°, 2.2°, 2.7°

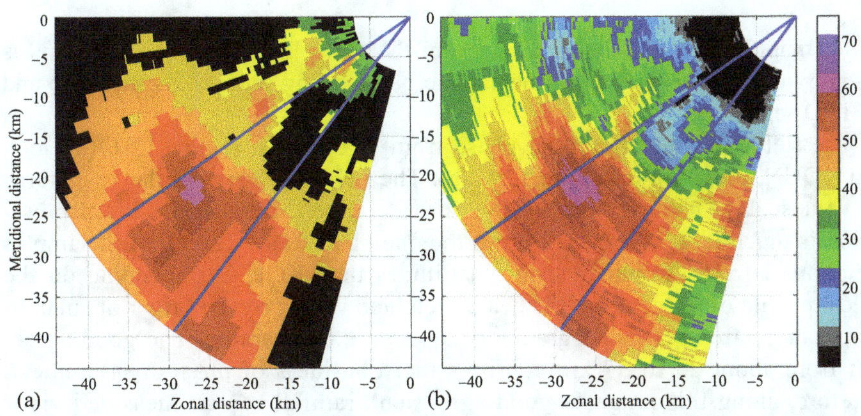

Figure 12.16 (a) Reflectivity field over a sector scanned with TPD at elevation equal 1°. (b) The same as in (a), but with KOUN at an elevation equal 1.8°. Scans made at 1759 UTC, on 15 May 2018. SNR= threshold is 2 dB.

broadside in azimuth. However, we did not include the gain change due to the element pattern because its gain is almost constant over ±25° (Figure 12.4).

Figures 12.16(a) and 12.16(b) depict the fields of Z_H measured by TPD and KOUN radars from conical PPI scan at 1° elevation (for TPD) and 1.8° elevation (for KOUN). The precipitation band is characterized by high Z_h (exceeding 55 dBZ) and Z_{DR} over 3–4 dB. Comparing Z field between two blue solid lines in Figure 12.16(a) with the one in Figure 12.16(b), we note a similar shape of the area with the high Z_h. TPD's Z_h is about 2 dB lower and overwhelmed by noise (for Z_h

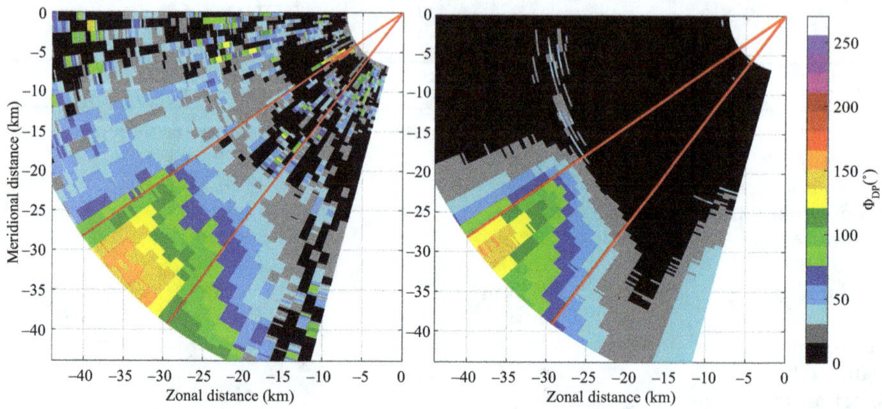

Figure 12.17 *(a) Differential phase field over a sector scanned with TPD radar at elevation 1°. (b) The same as in (a), but over a sector scanned with KOUN radar at elevation 1.8°.*

smaller than 35 dBZ) at the distance from the radar larger than 10 km. This is caused by the large beamwidth (incomplete beam filling reduces Z estimates) and low TPD sensitivity.

The fields of the differential phase from TPD are in Figure 12.17(a), and, from KOUN, these are in Figure 12.17(b). The fields are very similar in patterns and values. Φ_{DP} pattern from TDP is broader likely because its beam cross-section is much larger (about 12 times the one of KOUN). Therefore, it samples precipitation from the wider span of altitudes than KOUN. Depending on the vertical profile of reflectivity, the beam-weighted values can be larger or smaller. What matters are Φ_{DP} radial gradients. These are comparable. The good agreement from about 25 to 37 km confirms the robustness of phase measurement. Therefore, along this range of good agreement, rainfall can be measured via R (K_{DP}) relations or from specific attenuation [31]. In addition, compensation of reflectivities for attenuation is feasible. These good attributes should become even better on larger PARs.

12.7 Conclusions

This chapter exposes the challenges that PARs with planar antenna have for achieving the quality of the polarimetric variables as good as on WSR-88Ds. The principal cause of this difficulty is the dual polarization radiating elements. Standard radiating elements are either electric dipoles or patches which radiate as pairs of magnetic dipoles. We demonstrate with simple analogies to meridians and latitude circles that these radiators cannot produce orthogonal fields in all directions except principal planes.

To attain good polarimetric purity, the cross-polar pattern should be well below the copolar one. This is especially important in SHV polarimetric mode because coupling in that mode is twice stronger (in dBs) than in AHV mode. We show through a heuristic argument that if the cross-polar pattern on broadside has four lobes centered on the copolar pattern, a significant reduction of bias in the polarimetric variables ensues. A similar reduction in bias occurs if the cross-polar pattern has two lobes within the main lobe in the principal planes. In either case, the cross-polar lobes have opposite signs and thus cancel the contribution of the first-order bias terms.

Our bias analysis is for the array consisting of patch radiators. We assume that calibration can be obtained from the gains at the beam center. Furthermore, we ignore the effects of cross-coupling due to cross-polar radiation of the patches. Therefore, the main contributors to biases are (a) the non-orthogonality of intended H and V polarized fields, and (b) the non-collinearity of the intended H with the true H direction and/or non-collinearity of the intended V with the true V direction. We demonstrate that nine calibration parameters are needed to construct a set of linear equations relating second-order moments of the received signals (at Ports 1 and 2 corresponding to intended H and V polarizations) to the pertinent elements of the polarimetric covariance matrix. Depending on the orientation of the array and the beam pointing direction, the number of parameters and the solving complexity change.

To further quantify the challenge of dual-pol measurements with PPAR, we consider a vertically oriented planar array and investigate three modes of operation. We define the field of view in the azimuth elevation plane as the region where the bias in differential reflectivity is below a prescribed value (0.1 dB). The size of the field of view is defined as a metric for comparison. Similarly, we consider a slightly tilted antenna array. We show the fields of view for which it suffices to extend the simpler solutions valid in the principal planes.

We show examples of polarimetric data collected with TPD which is the first built dual polarization (10-cm wavelength) phase array weather radar. Our examination of the fields of reflectivity, differential phase, Doppler velocity, and spectrum width revealed the issues and virtues of TPD. Its fields of reflectivity agree well with the ones from KOUN and so do the velocity fields. Although the fields of spectrum width are biased, the features are preserved suggesting that bias removal is possible. Therefore, the radar has a potential for quantitative observation of weather similar to non-polarimetric weather radars. The radar can also make good measurements of differential phase. Therefore, rain measurements based on this variable would be good. Moreover, adequate measurements of rain from estimates of the attenuation is possible as well as correction of attenuated powers.

Of all the variables, Doppler velocity and differential phase are the most robust. This is expected as these measurements use the phases of the weather signals. Noises, nonlinearities, and other artifacts affect much less the signal's phases than amplitudes.

Appendix

Radar equation for PAR and beam cross-sections

The radar equation for PAR at the broadside is very similar to [25, (4.34)] except the different gains on transmission and reception must be accounted for. In the case of a rectangular vertically oriented array, the beamwidth in the H and E planes differ. Start with the equation for broadside

$$P_r(r_o) = \frac{\pi^3 P_t g_t g_r \theta_a \theta_b c\tau |K_w|^2 Z_e}{2^{10}(\ln 2)\lambda^2 r^2 l_r}, \tag{A.1}$$

where g_t, g_r are the effective gains of the transmitted beam and receiving beam, and θ_a, θ_b are the 3 dB beamwidths at the broadside corresponding to the long a and short b dimensions of the panel.

Transmitted power is P_t, τ is the pulse length, $|K_w|^2$ is a factor that depends on the refractive index and at microwave frequencies, it is between 0.91 and 0.93, and l_r is the cumulative loss (larger than unity). The beamwidths are of the composite two-way pattern computed as a product of the voltage pattern on transmission with the voltage pattern on reception. With no loss of substance, we ignore propagation losses through the atmosphere and the gain transfer g_s relating the power at the antenna to the voltages in the receiver.

Next, we adjust (A.1) to accommodate the beam pointing in directions other than the broadside. First, consider isotropic radiators (elements) so that the only dependence in returned power is due to the reduction of the antenna effective area and concomitant increase in beamwidth. The reduced effective area A_{reff} equals the projection of the effective area at broadside A_{eff} on the polarization plane, i.e.,

$$A_{reff} = A_{eff} \cos(az)\cos(el), \tag{A.2}$$

where *az* and *el* are azimuths and elevations relative to the broadside beam axis of a vertically oriented array antenna. The gains are also reduced and the beamwidth product is inversely proportional to the A_{reff}. This means that for pointing at *az, el* the product $\theta_a\theta_b$ in (A.1) should be replaced with $\theta_a\theta_b/[\cos(az)\cos(el)]$ and the product $g_t\, g_r$ with $g_t\, g_r[\cos(az)\cos(el)]^2$.

We need one more step to complete the derivation. Recall that we assumed isotropic radiators. To account for the element pattern's gain variation with pointing direction, we take the ratio squared of the gain $g_e(az, el)$ to the gain at broadside g_e and incorporate it into (A.1). The final result is the following radar equation:

$$P_r(r_o) = \frac{\pi^3 P_t g_t g_r g_e^2(az, el)\cos(az)\cos(el)\theta_a\theta_b c\tau |K_w|^2 Z_e}{2^{10}(\ln 2) g_e^2 \lambda^2 r^2 l_r}. \tag{A.3}$$

This equation is identical to the one derived by [32] who started with fundamental principles and derived radar equations for frequency agile and phased array weather radars. In practice rather than separating the element pattern gains, it may be easier to measure the gains and beamwidth at various pointing directions. In that case, the

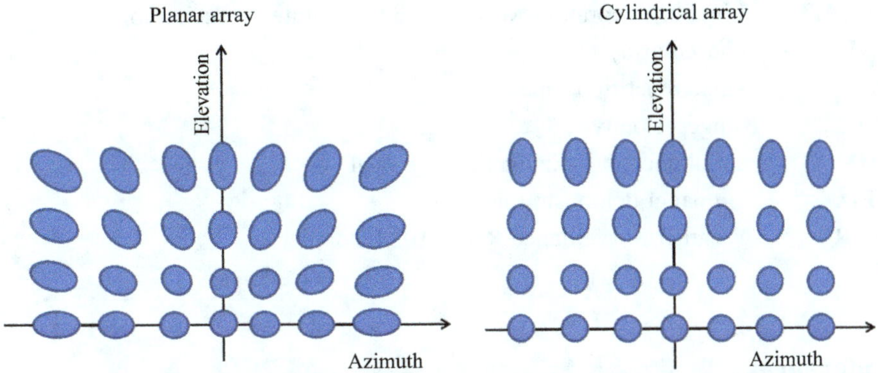

Figure A.1 *(a) Beam cross-sections as a function or azimuth and elevation for a planar vertical array. (b) The same as in (a) but for a cylindrical array.*

measured values would replace the ones in (A.3) and the angular dependence would be implicit in the measured parameters.

Planar array antennas often have a hexagonal shape to mimic beam cross-section dependence on the pointing direction of a circular aperture. The beam cross-sections of a circular array depend on the pointing direction. If the angle between the broadside direction and the beam is α, then the smallest projection of the diameter D is $D\cos(\alpha)$ while the projection of the perpendicular diameter is the largest equaling D. Consequently, the beamwidth along the projections is inversely proportional to the projections and (A.3) relates the affective area to the azimuth and elevation angles. The sketch in Figure A.1 illustrates the beam cross-sections in the azimuth elevation space for a circular planar array antenna and for a cylindrical antenna. A salient advantage of the cylindrical geometry is that with an increasing elevation, the beam cross-section increases uniformly. Therefore, the resolution decreases gradually with elevation. Assuming a fixed top of storms, it follows that as the elevation increases, the range of observable weather decreases; hence, the two effects on the resolution are opposite. The farthest distance where the weather can be observed also decreases with the range due to elevation angles and earth curvature. That means that the cylindrical antenna resolution dependence on elevation is better matched to weather observations than the resolution of the planar antenna.

List of acronyms

AHV alternate horizontal and vertical
ATD advanced technology demonstrator
FM frequency modulation
KOUN designation of the research WSR-88D (Oklahoma University Norman)

LL MIT	Lincoln Laboratory Massachusetts Institute of Technology
PAR	phased array radar
PRT	pulse repetition time
R/V	range/velocity
SHV	simultaneous horizontal and vertical
TPD	ten panel demonstrator
WSR-88D	Weather Surveillance Radar 1988 Doppler

References

[1] Zrnic, D.S., Kimpel, J.F., Forsyth, D.F., *et al.* Agile beam phased array radar for weather observations. *Bull. Amer. Meteorol. Soc.* 2007;88(11):1753–1766.

[2] Zrnic, D.S., Melnikov, V.M., Doviak, R.J., and Palmer, R. Scanning strategy for the multifunction phased-array radar to satisfy aviation and meteorological needs. *IEEE Geosci. Remote Sens. Lett.* 2015;12 (6):1204–1208.

[3] Zhang, G., Doviak, R.J., Zrnic, D.S., Palmer, R., Lei, L., and Al-Rashid, Y. Polarimetric phased array radar for weather measurement: a planar or cylindrical configuration? *J. Atmos. Oceanic Technol.* 2011;28(1):63–73.

[4] Zrnic, D.S., Doviak, R.J., Zhang G., and Ryzhkov, A.V. Bias in differential reflectivity due to cross-coupling through the radiation patterns of polarimetric weather radars. *J. Atmos. Oceanic Techn.* 2010;27(10):1624–1637.

[5] Ludwig, A. The definition of cross polarization. *IEEE Trans. Antennas Propag.* 1973;21(1):116–117.

[6] Crain, G.E. and Staiman, D. Polarization selection for phased array weather radar. *Presented at 25th International Conference on IIPS for Meteorology, Oceanography, and Hydrology*, New Orleans, USA, 2009.

[7] Lei, L., Zhang G., and Doviak, R.J. Bias correction for polarimetric phased-array radar with idealized aperture and patch antenna elements. *IEEE Trans. Geosci. Remote Sens. Lett.* 2013;51(1):473–486.

[8] Bhardwaj, S. and Rahmat-Samii, Y. Revisiting the generation of cross-polarization in rectangular patch antennas: near-field approach. *IEEE Antennas Prop. Mag.* 2014;56(1):14–38.

[9] WIPL d.o.o, WIPL-D Electromagnetic Simulation Software, 2020. http://www.wipl-d.com.

[10] Raytheon Intelligence & Space, "Skyler," 2022. https://www.raytheonintelligenceandspace.com/capabilities/products/skyler.

[11] Mirkovic, Dj. and Zrnic, D.S. Computational Electromagnetic Tools Applied to the Polarimetric Phased Array Antenna. NOAA/NSSL Report, 2019. https://www.nssl.noaa.gov/publications/par_reports/.

[12] Ivić, I.R. Effects of phase coding on Doppler spectra in PPAR weather radar. *IEEE Trans. Geosci. Rem. Sens.* 2018;56(4):2043–2065. doi:10.1109/TGRS.2017.2772962.

[13] Zhe, L., Zhang, Y., Borowska, L., *et al.* Polarimetric phased array weather radar data quality evaluation through combined analysis, simulation, and measurements. *IEEE Trans. Geosci. Rem. Sens. Lett.* 2021;18(6):1029–1033.

[14] Ivić, I.R. Quantification of polarimetric PAR effects on weather observables in the phase coded STSR mode. *IEEE Trans. Geosci. Rem. Sens.* 2022;60(1): Art. no. 5110022. doi:10.1109/TGRS.2022.3146358.

[15] Ivić, I.R., Mendoza, R., Schvartzman, D., Torres, S., and Wasielewski, D. Preliminary report on polarimetric calibration for the advanced technology demonstrator, NOAA/NSSL Report, 2020. https://www.nssl.noaa.gov/publications/par_reports/.

[16] Gregg, S., Curtis, C., and Torres, S. Advanced Technology Demonstrator System Testing Summary. NOAA/NSSL Report, 2020. https://www.nssl.noaa.gov/publications/par_reports/.

[17] Mirkovic, D. and Zrnic, D.S. Polarimetric weather radar calibration by computational electromagnetics. *Appl. Comp. Electromag. Soc. J.* 2019;34 (2):342–346.

[18] Galletti, M. and Zrnic, D.S. Bias in copolar correlation coefficient caused by the antenna radiation patterns. *IEEE Trans. Geosci. Rem. Sens.* 2011;49 (6):2274–2280.

[19] Zrnic, D.S., Zhang, G., and Doviak, R.J. Bias correction and Doppler measurement for polarimetric phased array radar. *IEEE Trans. Geosci. Rem. Sens.* 2011;49(2):843–853.

[20] Melnikov, M.V. and Zrnic, D.S. On the alternate transmission mode for polarimetric phased array weather radar. *J. Atmos Oceanic Technol.* 2015;32 (2):220–233.

[21] Torres, S., Adams, R., Curtis, C., *et al.* Adaptive-weather-surveillance and multifunction capabilities of the national radar testbed phased array radar. *Proc. IEEE* 2016;104(3):660–672.

[22] Zrnic, D.S., Melnikov, V.M., and Doviak, R.J. Issues and challenges for polarimetric measurement of weather with an agile beam phased array radar. NOAA/NSSL Report, 2012. https://www.nssl.noaa.gov/publications/par_reports/.

[23] Guerci, J. *Space–Time Adaptive Processing for Radar*. Boston, MA: Artech House; 2014. p. 283.

[24] Ivic, I.R., Conway, M., Torres, S.M., Herd, J.S., Zrnic, D.S., and Weber, M. E. Meteorological polarimetric phased array radar. In *Polarimetric Radar and SAR Signal Processing*, Stevenage: The Institution of Engineering and Technology (IET), Michael Faraday House, 2021.

[25] Doviak, R.J. and Zrnic, D.S. *Doppler Radar and Weather Observations*. Mineola: Dover Publication; 2006. p. 562

[26] Zrnic, D.S., Doviak, R.J., Melnikov, V.M., and Ivic, I.R. Signal design to suppress coupling in the polarimetric phased array radar. *J. Atmos. Oceanic Technol.* 2014;31(5):1063–1077.

[27] Lei, L., Zhang, G., Doviak, R.J., and Karimkashi, S. Comparison of theoretical biases in estimating polarimetric properties of precipitation with

weather radar using parabolic reflector, or planar and cylindrical arrays. *IEEE Trans. Geosci. Rem. Sens.* 2015;53(8):4313–4327.

[28] Fulton, C. and Chappell, W.J. Calibration of a digital phased array for polarimetric radar. In *20100 IEEE MTT-S International Symposium*, Anaheim, USA, 2010. p. 161–164.

[29] Zrnić, D.S., Doviak, R.J., Zhang, G., Zhang, Y., and Fulton, C. *Propagation and Backscattering Challenges for Planar Polarimetric Phased Array Radars,* 2017. https://www.nssl.noaa.gov/publications/par_reports/.

[30] Sachidananda, M. and Zrnic, D.S. Efficient processing of alternately polarized radar *echoes. J. Atmos. Oceanic. Tech.* 1989;6(1):173–181.

[31] Ryzhkov, A.V. and Zrnic, D.S. *Radar Polarimetry for Weather Observations*. Cham: Springer Nature Switzerland; 2019. p. 486.

[32] Knorr, J.B. Weather radar correction for frequency agile and phase array radars. *IEEE Trans. Aerospace Electron. Syst.* 2007;4(3):220–1227.

www.ingramcontent.com/pod-product-compliance
Lightning Source LLC
LaVergne TN
LVHW011547200926
848484LV00001B/293